T0176575

Molecular Modeling of Geochemical Reactions

Dedication

To see a World in a grain of sand...

—William Blake

To my wife, Doris, and son, Cody, who bring much joy to my life.

Molecular Modeling of Geochemical Reactions

An Introduction

Edited by

JAMES D. KUBICKI
University of Texas at El Paso, USA

WILEY

This edition first published 2016
© 2016 John Wiley & Sons, Ltd.

Registered Office
John Wiley & Sons, Ltd, The Atrium, Southern Gate, Chichester, West Sussex, PO19 8SQ, United Kingdom

For details of our global editorial offices, for customer services and for information about how to apply for permission to reuse the copyright material in this book please see our website at www.wiley.com.

The right of the author to be identified as the author of this work has been asserted in accordance with the Copyright, Designs and Patents Act 1988.

All rights reserved. No part of this publication may be reproduced, stored in a retrieval system, or transmitted, in any form or by any means, electronic, mechanical, photocopying, recording or otherwise, except as permitted by the UK Copyright, Designs and Patents Act 1988, without the prior permission of the publisher.

Wiley also publishes its books in a variety of electronic formats. Some content that appears in print may not be available in electronic books.

Designations used by companies to distinguish their products are often claimed as trademarks. All brand names and product names used in this book are trade names, service marks, trademarks or registered trademarks of their respective owners. The publisher is not associated with any product or vendor mentioned in this book.

Limit of Liability/Disclaimer of Warranty: While the publisher and author have used their best efforts in preparing this book, they make no representations or warranties with respect to the accuracy or completeness of the contents of this book and specifically disclaim any implied warranties of merchantability or fitness for a particular purpose. It is sold on the understanding that the publisher is not engaged in rendering professional services and neither the publisher nor the author shall be liable for damages arising herefrom. If professional advice or other expert assistance is required, the services of a competent professional should be sought.

The advice and strategies contained herein may not be suitable for every situation. In view of ongoing research, equipment modifications, changes in governmental regulations, and the constant flow of information relating to the use of experimental reagents, equipment, and devices, the reader is urged to review and evaluate the information provided in the package insert or instructions for each chemical, piece of equipment, reagent, or device for, among other things, any changes in the instructions or indication of usage and for added warnings and precautions. The fact that an organization or Website is referred to in this work as a citation and/or a potential source of further information does not mean that the author or the publisher endorses the information the organization or Website may provide or recommendations it may make. Further, readers should be aware that Internet Websites listed in this work may have changed or disappeared between when this work was written and when it is read. No warranty may be created or extended by any promotional statements for this work. Neither the publisher nor the author shall be liable for any damages arising herefrom.

Library of Congress Cataloging-in-Publication data applied for

ISBN: 9781118845080

A catalogue record for this book is available from the British Library.

Set in 10/12pt Times New Roman by SPi Global, Pondicherry, India

Printed and bound in Singapore by Markono Print Media Pte Ltd

1   2016

Contents

List of Contributors

Adelia J. A. Aquino Institute for Soil Research, University of Natural Resources and Life Sciences, Vienna, Austria and Department of Chemistry and Biochemistry, Texas Tech University, Lubbock, TX, USA

Andrey V. Brukhno Department of Chemistry, University of Bath, Bath, UK

Eric Bylaska Environmental Molecular Sciences Laboratory, Pacific Northwest National Laboratory, Richland, WA, USA

Ying Chen Chemistry and Biochemistry Department, University of California, San Diego, La Jolla, CA, USA

Marco De La Pierre Nanochemistry Research Institute, Curtin Institute for Computation, and Department of Chemistry, Curtin University, Perth, Western Australia, Australia

Raffaella Demichelis Nanochemistry Research Institute, Curtin Institute for Computation, and Department of Chemistry, Curtin University, Perth, Western Australia, Australia

Roberto Dovesi Dipartimento di Chimica, Università degli Studi di Torino and NIS Centre of Excellence "Nanostructured Interfaces and Surfaces", Torino, Italy

Alessandro Erba Dipartimento di Chimica, Università degli Studi di Torino, Torino, Italy

Marie-Pierre Gaigeot LAMBE CNRS UMR 8587, Université d'Evry val d'Essonne, Evry, France and Institut Universitaire de France, Paris, France

Martin H. Gerzabek Institute for Soil Research, University of Natural Resources and Life Sciences, Vienna, Austria

Georg Haberhauer Institute for Soil Research, University of Natural Resources and Life Sciences, Vienna, Austria

James D. Kubicki Department of Geological Sciences, University of Texas at El Paso, El Paso, TX, USA

Hans Lischka Department of Chemistry and Biochemistry, Texas Tech University, Lubbock, TX, USA and Institute for Theoretical Chemistry, University of Vienna, Vienna, Austria

Qisheng Ma Department of Computational and Molecular Simulation, GeoIsoChem Corporation, Covina, CA, USA

Marco Molinari Department of Chemistry, University of Bath, Bath, UK

Stephen C. Parker Department of Chemistry, University of Bath, Bath, UK

Kevin M. Rosso Physical Sciences Division, Pacific Northwest National Laboratory, Richland, WA, USA

James R. Rustad Corning Incorporated, Corning, NY, USA

Nita Sahai Department of Polymer Science, Department of Geology, and Integrated Bioscience Program, University of Akron, Akron, OH, USA

David M. Sherman School of Earth Sciences, University of Bristol, Bristol, UK

Dino Spagnoli School of Chemistry and Biochemistry, University of Western Australia, Crawley, Western Australia, Australia

Marialore Sulpizi Department of Physics, Johannes Gutenberg Universitat, Mainz, Germany

Yongchun Tang Geochemistry Division, Power Environmental Energy Research Institute, Covina, CA, USA

Daniel Tunega Institute for Soil Research, University of Natural Resources and Life Sciences, Vienna, Austria

John Weare Chemistry and Biochemistry Department, University of California, San Diego, La Jolla, CA, USA

Zhijun Xu Department of Polymer Science, University of Akron, Akron, OH, USA and Department of Chemical Engineering, Nanjing University, Nanjing, China

Weilong Zhao Department of Polymer Science, University of Akron, Akron, OH, USA

Preface

Humility is an underrated scientific personality characteristic. When I think of William Blake's famous lithograph of Sir Isaac Newton toiling away at the bottom of a dark ocean, I am always reminded of how much we do not know. Science is a humbling enterprise because even our most notable achievements will likely be replaced by greater understanding at some date in the future. I am allowing myself an exception in the case of publication of this volume, however. I am proud of this book because so many leaders in the field of computational geochemistry have agreed to be a part of it. We all know that the best people are so busy with projects that it is difficult to take time away from writing papers and proposals to dedicate time to a chapter. The authors who have contributed to this volume deserve a great deal of appreciation for taking the time to help explain computational geochemistry to those who are considering using these techniques in their research or trying to gain a better understanding of the field in order to apply its results to a given problem. I am proud to be associated with this group of scientists.

When my scientific career began in 1983, computational geochemistry was just getting a toehold in the effort to explain geochemical reactions at an atomic level. People such as Gerry V. Gibbs and John (Jack) A. Tossell were applying quantum chemistry to model geologic materials, and C. Austen Angell and coworkers were simulating melts with classical molecular dynamics. As an undergraduate, I had become interested in magmatic processes, especially the generation of magmas in subduction zones and the nucleation of crystals from melts. Organic chemistry exposed me to the world of reaction mechanisms which were not being studied extensively at the time in geochemistry. When the opportunity arose in graduate school to use MD simulations to model melt and glass behavior, I jumped at the chance to combine these interests in melts and mechanisms naïve to the challenges that lie ahead. Fortunately, through the guidance of people such as Russell J. Hemley, Ron E. Cohen, Anne M. Hofmeister, Greg E. Muncill, and Bjorn O. Mysen at the Geophysical Laboratory, I was able to complement the computational approach with experimental data on diffusion rates and vibrational spectra. This approach helped benchmark the simulations and provide insights into the problems at hand that were difficult to attain with computation alone. This strategy has worked throughout my career and has led to numerous fascinating collaborations.

A key step in this process occurred while I was working as a postdoc at Caltech under Geoffrey A. Blake and Edward M. Stolper. I met another postdoc, Dan G. Sykes, who also shared a passion for melt and glass structure. As I was learning how to apply quantum mechanics to geochemistry, Dan and I discussed his models for explaining the vibrational spectra of silica and aluminosilicate glasses. Dan's model differed from the prevailing interpretations of IR and Raman spectra, but his hypotheses were testable via construction of the three- and four-membered ring structures he thought gave rise to the observed trends in vibrational frequencies with composition. We argued constantly over the details of his model and came up with several tests to disprove it, but, in the end, the calculations and observed spectra agreed well enough that we were able to publish a series of papers over the objections of reviewers who were skeptical of the views of two young postdocs. Among these papers, a key study was published with the help of George R. Rossman whose patience and insight inspired more confidence in me that the path we were following would be fruitful. This simple paper comparing calculated versus observed H-bond frequencies ended up being more significant than I had known at the time because this connection is critical in model mineral–water interactions that became a theme later in my career.

When I could not find work any longer doing igneous-related research, I turned to a friend from undergraduate chemistry at Cal State Fullerton, Sabine E. Apitz, to employ me as a postdoc working on environmental chemistry. Fortunately, the techniques I had learned were transferable to studying organic–mineral interactions. This research involving mineral surfaces eventually led to contacts with Susan L. Brantley and Carlo G. Pantano who were instrumental in landing a job for me at Penn State. Numerous collaborations blossomed during my tenure in the Department of Geosciences, and all these interdisciplinary projects kept me constantly excited about learning new disciplines in science. Recently, I made the decision to move to the University of Texas at El Paso to join a team of people who are creating an interdisciplinary research environment while simultaneously providing access to excellent education and social mobility.

The rapid developments in hardware, software, and theory that have occurred since 1983 have propelled research in computational geochemistry. All of us appreciate the efforts of all those developing new architectures and algorithms that make our research possible. We offer this book as a stepping stone for those interested in learning these techniques to get started in their endeavors, and we hope the reviews of literature and future directions offered will help guide many new exciting discoveries to come.

James D. Kubicki
Department of Geological Sciences
The University of Texas at El Paso
October 3, 2015

1

Introduction to the Theory and Methods of Computational Chemistry

David M. Sherman

School of Earth Sciences, University of Bristol, Bristol, UK

1.1 Introduction

The goal of geochemistry is to understand how the Earth formed and how it has chemically differentiated among the different reservoirs (e.g., core, mantle, crust, hydrosphere, atmosphere, and biosphere) that make up our planet. In the early years of geochemistry, the primary concern was the chemical analysis of geological materials to assess the overall composition of the Earth and to identify processes that control the Earth's chemical differentiation. The theoretical underpinning of geochemistry was very primitive: elements were classified as chalcophile, lithophile, and siderophile (Goldschmidt, 1937), and the chemistry of the lithophile elements was explained in terms of simple models of ionic bonding (Pauling, 1929). It was not possible to develop a predictive quantitative theory of how elements partition among different phases.

In the 1950s, experimental studies began to measure how elements are partitioned between coexisting phases (e.g., solid, melt, and fluid) as a function of pressure and temperature. This motivated the use of thermodynamics so that experimental results could be extrapolated from one system to another. Equations of state were developed that were based on simple atomistic (hard-sphere) or continuum models (Born model) of liquids (e.g., Helgeson and Kirkham, 1974). This work continued on into the 1980s. By this time, computers had become sufficiently fast that atomistic simulations of geologically interesting materials were possible. However, the computational atomistic simulations were based on classical or ionic models of interatomic interactions. Minerals were modeled as being composed of ions that interact via empirical or *ab initio*-derived interatomic potential functions (e.g., Catlow et al., 1982; Bukowinski, 1985). Aqueous solutions were composed of ions solvated by (usually) rigid water molecules modeled as point charges (Berendsen et al., 1987). Many of these simulations have been very successful and classical models of minerals and aqueous

Molecular Modeling of Geochemical Reactions: An Introduction, First Edition. Edited by James D. Kubicki.
© 2016 John Wiley & Sons, Ltd. Published 2016 by John Wiley & Sons, Ltd.

solutions are still in use today. However, ultimately, these models will be limited in application insofar as they are not based on the real physics of the problem.

The physics underlying geochemistry is quantum mechanics. As early as the 1970s, approximate quantum mechanical calculations were starting to be used to investigate bonding and electronic structure in minerals (e.g., Tossell et al., 1973; Tossell and Gibbs, 1977). This continued into the 1980s with an emphasis on understanding how chemical bonds dictate mineral structures (e.g., Gibbs, 1982) and how the pressures of the deep earth might change chemical bonding and electronic structure (Sherman, 1991). Early work also applied quantum chemistry to understand geochemical reaction mechanisms by predicting the structures and energetics of reactive intermediates (Lasaga and Gibbs, 1990). By the 1990s, it became possible to predict the equations of state of simple minerals and the structures and vibrational spectra of gas-phase metal complexes (Sherman, 2001). As computers have become faster, it now possible to simulate liquids, such as silicate melts or aqueous solutions, using *ab initio* molecular dynamics.

We are now at the point where computational quantum chemistry can be used to provide a great deal on insight on the mechanisms and thermodynamics of chemical reactions of interest in geochemistry. We can predict the structures and stabilities of metal complexes on mineral surfaces (Sherman and Randall, 2003; Kwon et al., 2009) that control the fate of pollutants and micronutrients in the environment. We can predict the complexation of metals in hydrothermal fluids that determine the solubility and transport of metals leading to hydrothermal ore deposits (Sherman, 2007; Mei et al., 2013, 2015). We can predict the phase transitions of minerals that may occur in the Earth's deep interior (Oganov and Ono, 2004; Oganov and Price, 2005). Computational quantum chemistry is now becoming a mainstream activity among geochemists, and investigations using computational quantum chemistry are now a significant contribution to work presented at major conferences on geochemistry.

Many geochemists want to use these tools, but may have come from a traditional Earth science background. The goal of this chapter is to give the reader an outline of the essential concepts that must be understood before using computational quantum chemistry codes to solve problems in geochemistry. Geochemical systems are usually very complex and many of the high-level methods (e.g., configuration interaction) that might be applied to small molecules are not practical. In this chapter, I will focus on those methods that can be usefully applied to earth materials. I will avoid being too formal and will emphasize what equations are being solved rather than how they are solved. (This has largely been done for us!) It is crucial, however, that those who use this technology be aware of the approximations and limitations. To this end, there are some deep fundamental concepts that must be faced, and it is worth starting at fundamental ideas of quantum mechanics.

1.2   Essentials of Quantum Mechanics

By the late nineteenth and early twentieth centuries, it was established that matter comprised atoms which, in turn, were made up of protons, neutrons, and electrons. The differences among chemical elements and their isotopes were beginning to be understood and systematized. Why different chemical elements combined together to form compounds, however, was still a mystery. Theories of the role of electrons in chemical bonding were put forth (e.g., Lewis, 1923), but these models had no obvious physical basis. At the same time, physicists were discovering that classical physics of Newton and Maxwell failed to explain the interaction of light and electrons with matter. The energy of thermal radiation emitted from black bodies could only be explained in terms of the frequency of light and not its intensity (Planck, 1900). Moreover, light (viewed as a wave since Young's experiment in 1801) was found to have the properties of particles with discrete energies and momenta (Einstein, 1905). This suggests that light was both a particle and a wave. Whereas a classical particle could have any value for its kinetic and potential energies, the electrons bound to atoms were found to only have discrete (quantized) energies (Bohr, 1913). It was then hypothesized that particles such as electrons could also be viewed as waves (de Broglie, 1925); this was experimentally verified by the discovery of electron diffraction (Davisson and Germer, 1927). Readers can find an accessible account of the early experiments and ideas that led to quantum mechanics in Feynman et al. (2011).

The experimentally observed wave–particle duality and quantization of energy were explained by the quantum mechanics formalism developed by Heisenberg (1925), Dirac (1925), and Schrodinger (1926). The implication of quantum mechanics for understanding chemical bonding was almost immediately demonstrated when Heitler and London (1927) developed a quantum mechanical model of bonding in the H_2 molecule. However, the real beginning of computational quantum chemistry occurred at the University of Bristol in 1929 when Lennard-Jones presented a molecular orbital theory of bonding in diatomic molecules (Lennard-Jones, 1929).

The mathematical structure of quantum mechanics is based on set of postulates:

Postulate 1:

A system (e.g., an atom, molecule or, really, anything) is described by a *wavefunction* $\Psi(\mathbf{r}_1, \mathbf{r}_2, \ldots, \mathbf{r}_N, t)$ over the coordinates $\{\mathbf{r}_N\}$, the N-particles of the system, and time t. The physical meaning of this wavefunction is that the probability of finding the system at a set of values for the coordinates $\mathbf{r}_1, \mathbf{r}_2, \ldots, \mathbf{r}_N$ at a time t is $|\Psi(\mathbf{r}_1, \mathbf{r}_2, \ldots, \mathbf{r}_N, t)|^2$.

Postulate 2:

For every observable (measurable) property λ of the system, there corresponds a mathematical operator $(\hat{L})$ that acts on the wavefunction.

Mathematically, this is expressed as follows:

$$\hat{L}\Psi = \lambda\Psi \tag{1.1}$$

Ψ is an eigenfunction of the operator $\hat{L}$ with eigenvalue λ. An eigenfunction is a function associated with an operator such that if the function is operated on by the operator, the function is unchanged except for being multiplied by a scalar quantity λ. This is very abstract, but it leads to the idea of the states of a system (the eigenfunctions) that have defined observable properties (the eigenvalues). Observable properties are quantities such as energy, momentum, or position. For example, the operator for the momentum of a particle moving in the x-direction is

$$\hat{p} = i\hbar\frac{\partial}{\partial x}\vec{\mathbf{i}} \tag{1.2}$$

where i is $\sqrt{-1}$, $\hbar$ is Planck's constant divided by 2π, and $\vec{\mathbf{i}}$ is the unit vector in the x-direction. Since the kinetic energy of a particle with mass m and momentum p is

$$T = \frac{p^2}{2m},$$

the operator for the kinetic energy of a particle of mass m that is free to move in three directions (x, y, z) is

$$\hat{T} = \frac{-\hbar^2}{2m}\left(\frac{\partial^2}{\partial x^2} + \frac{\partial^2}{\partial y^2} + \frac{\partial^2}{\partial z^2}\right) = \frac{-\hbar^2}{2m}\nabla^2 \tag{1.3}$$

In general, the operator for the potential energy $(\hat{V})$ of a system is a scalar operator such that $\hat{V} = V$. That is, we multiply the wavefunction by the function that defines the potential energy. The operator $\hat{E}$ for the total energy E of a system is

$$\hat{E} = i\hbar \frac{\partial}{\partial t} \qquad (1.4)$$

It is important to recognize whether or not a quantity is a "quantum mechanical observable." Chemists (and geochemists) often invoke quantities such as "ionicity," "bond valence," "ionic radius," etc., that are not observables. These quantities are not real; they exist only as theoretical constructs. They cannot be measured.

1.2.1   The Schrödinger Equation

In classical mechanics, we express the concept of conservation of energy in terms of the Hamiltonian *H* of the system:

$$H = E = T + V \qquad (1.5)$$

In quantum mechanics, we express the Hamiltonian in terms of the operators corresponding to *E*, *T*, and *V*:

$$\hat{H}\Psi = \left(\hat{T} + \hat{V}\right)\Psi = \hat{E}\Psi \qquad (1.6)$$

or

$$\hat{H}\Psi = \left(\hat{T} + \hat{V}\right)\Psi = i\hbar \frac{\partial \Psi}{\partial t} \qquad (1.7)$$

This is the time-dependent Schrödinger equation. If the kinetic *T* and potential *V* energies of the system are not varying with time, then we can write:

$$\Psi(r_1, r_2, \ldots, r_N, t) = \Psi(r_1, r_2, \ldots, r_N)e^{-iEt/\hbar} \qquad (1.8)$$

Substituting this into the Hamiltonian gives:

$$\hat{H}\Psi = \left(\hat{T} + \hat{V}\right)\Psi = E\Psi \qquad (1.9)$$

This is the time-independent Schrödinger equation, and it is what we usually seek to solve in order to obtain a quantum mechanical description of the system in terms of the wavefunction and energy of each state.

1.2.2   Fundamental Examples

At this point, it is worthwhile to briefly explore several fundamental examples that illustrate the key aspects of quantum mechanics.

1.2.2.1   *Particle in a Box*

This is, perhaps the simplest problem yet it illustrates some of the fundamental features of quantum reality. Consider a particle of mass *m* inside a one-dimensional box of length *L* (Figure 1.1). The potential energy *V* of the system is 0 inside the box but infinite outside the box. Therefore, inside the box, the Schrödinger equation is

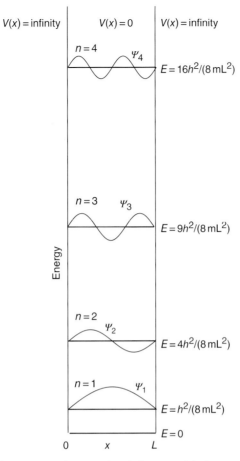

Figure 1.1 *Wavefunctions and energy levels for a particle in a one-dimensional box.*

$$\frac{-\hbar^2}{2m}\left(\frac{d^2\Psi(x)}{dx^2}\right) = E\Psi(x) \tag{1.10}$$

The solution to this differential equation is of the form:

$$\Psi(x) = A\sin(kx) + B\cos(kx) \tag{1.11}$$

Since the potential energy is infinite outside the box, the particle cannot be at $x = 0$ or at $x = L$. That is, we have $\Psi(0) = \Psi(L) = 0$. Hence,

$$\Psi(0) = A\sin(0) + B\cos(0) = 0 \tag{1.12}$$

which implies that $B = 0$. However, since

$$\Psi(L) = A\sin(kL) = 0$$

we find that $kL = n\pi$, where $n = 1, 2, 3, \ldots$

If we substitute $\Psi(x)$ back into the Schrödinger equation, we find that

$$E = \frac{\hbar^2 k^2}{2m} = \frac{\hbar^2}{2m}\left(\frac{n^2\pi^2}{L^2}\right) \tag{1.13}$$

That is, the energy is quantized to have only specific allowed values because n can only take on integer values. The quantization results from putting the particle in a potential energy well (the box). However, the quantization is only significant if the dimensions of the box and the mass of the particle are on the order of Planck's constant ($h = 6.6262 \times 10^{-34}$ J/s, i.e., if the box is angstroms to nanometers in size). The formalism of quantum mechanics certainly applies to our macroscopic world, but the quantum spacing of a 1 g object in a box of, say, 10 cm in length is too infinitesimal to measure.

1.2.2.2   The Hydrogen Atom

Now, let's consider the hydrogen atom consisting of one electron and one proton as solved by Schrödinger (1926). We will consider only the motion of the electron relative to the position of the proton and not consider the motion of the hydrogen atom as a whole. Hence, our wavefunction for the system is $\Psi(\mathbf{r})$ where $\mathbf{r}$ is the position of the electron (in three dimensions) relative to the proton (located at the origin). The Schrödinger equation for this problem is then

$$\frac{-\hbar^2}{2m}\nabla^2\Psi(\mathbf{r}) - \frac{e^2}{4\pi\varepsilon_0 r}\Psi(\mathbf{r}) = E\Psi(\mathbf{r}) \tag{1.14}$$

where e is the charge of the electron and ε_0 is the permittivity of free space. To avoid having to write all the physical constants, it is convenient to adopt "atomic units," where m, $\hbar^2$, and $e^2/4\pi\varepsilon_0$ are all set to unity. The unit of energy is now the Hartree (1 H = 27.21 eV = 4.360×10^{-18} J) and the unit of distance is the Bohr (10^{-10} m = 1 Å = 0.529177 Bohr). Computer programs in quantum chemistry often express their results in hartrees and bohrs; it is important that the user be aware of these units and know how to convert to more conventional units such as kJ/mol and angstrom (Å).

In atomic units, the hydrogen Schrödinger equation becomes

$$-\frac{1}{2}\nabla^2\Psi(\mathbf{r}) - \frac{1}{r}\Psi(\mathbf{r}) = E\Psi(\mathbf{r}) \tag{1.15}$$

Since the problem has spherical symmetry, it is more convenient to use spherical coordinates $\Psi(\mathbf{r}) = \Psi(r, \theta, \phi)$ rather than Cartesian coordinates (Figure 1.2). Since the coordinates are independent of each other, we can write

$$\Psi(r,\theta,\phi) = R_{nl}(r)\,Y_{lm}(\theta,\phi) \tag{1.16}$$

where $Y_{lm}(\theta, \phi)$ are the spherical harmonic functions; they give the angular shape of the wavefunctions. The radial wavefunction is

$$R_{nl}(r) = \sqrt{\left(\frac{2}{n}\right)^3 \frac{(n-l-1)!}{2n[(n+l)!]}}\, e^{-r\ln}\left(\frac{2r}{n}\right)^l L_{n-l-1}^{2l+1}\left(\frac{2r}{n}\right) \tag{1.17}$$

where L_{n-l-1}^{2l+1} are special functions called Laguerre polynomials. The important result that emerges from this solution is that the wavefunction can be specified by three quantum numbers n, l, and m with values

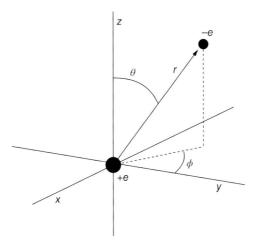

Figure 1.2   *Spherical coordinates used for solution of the hydrogen atom.*

$$n = 1, 2, 3, \ldots$$
$$l = 0, 1, 2, \ldots, n-1$$
$$m = -l, -l+1, \ldots, 0, \ldots, l-1, l.$$

The energy of the electron is quantized with values

$$E_n = -\left(\frac{1}{2}\right)\frac{1}{n^2} \tag{1.18}$$

This predicted quantization of the hydrogen electron energies triumphantly explains the empirical model proposed by Bohr for the energies of the lines observed hydrogen atom spectrum. It is standard practice to denote orbitals with $l = 0$ as "s" (not to be confused with the spin-quantum number described later), orbitals with $l = 1$ as "p," orbitals with $l = 2$ as "d," and orbitals with $l = 3$ as "f." The three m quantum numbers for the p-orbitals are denoted as p_x, p_y, and p_z. For the d-orbitals, we take linear combinations of the orbitals corresponding to the different m quantum numbers to recast them as d_{xy}, d_{xz}, d_{yz}, $d_{3z^2-r^2}$, and $d_{x^2-y^2}$. The schematic energy-level diagram and the shapes of the hydrogenic orbitals are shown in Figure 1.3.

Unfortunately, we cannot find an exact analytical solution for the Schrödinger equation for an atom with more than one electron. However, the hydrogenic orbitals and their quantum numbers enable us to rationalize the electronic structures of the multielectronic elements and the structure of the periodic table. As will be shown later, we will use the hydrogenic orbitals as building blocks to approximate the wavefunctions for multielectronic atoms.

1.3   Multielectronic Atoms

1.3.1   The Hartree and Hartree–Fock Approximations

Consider the helium atom with two electrons and a nucleus of charge +2. The coordinate of electron 1 is $\mathbf{r}_1$ and the coordinate of electron 2 is $\mathbf{r}_2$. We will assume that the nucleus is fixed at the origin. The Schrödinger equation for the system is then

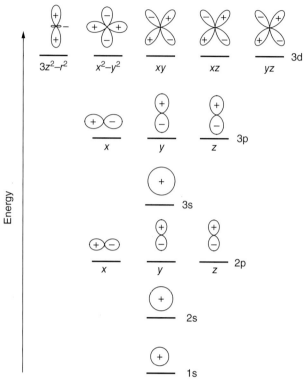

Figure 1.3 *Schematic energy levels and orbital shapes for the hydrogen atom.*

$$\left(-\frac{1}{2}\nabla_1^2 - \frac{1}{2}\nabla_2^2 - \frac{2}{r_1} - \frac{2}{r_2} + \frac{1}{|\mathbf{r}_1 - \mathbf{r}_2|}\right)\Psi(\mathbf{r}_1, \mathbf{r}_2) = E\Psi(\mathbf{r}_1, \mathbf{r}_2) \tag{1.19}$$

A reasonable approach to solving this might be to assume that

$$\Psi(\mathbf{r}_1, \mathbf{r}_2) = \psi_a(\mathbf{r}_1)\psi_b(\mathbf{r}_2) \tag{1.20}$$

This is known as the Hartree approximation; it provides a very important conceptual reference point because it introduces the idea of expressing our many-body problem in terms of single-particle functions ("one-electron orbitals"). However, because of the interelectronic repulsion, described by the term

$$\frac{1}{|\mathbf{r}_1 - \mathbf{r}_2|} \tag{1.21}$$

the Hartree approximation is too crude to be quantitatively useful; we cannot really separate out the motions of the electrons. We say that the electrons are *correlated*. In spite of this shortcoming, we will still express the wavefunction for a multielectronic system in terms of single-particle wavefunctions. However, we must go beyond the pure Hartree approximation because, in a multielectronic system, the electrons must be indistinguishable from each other. This is more fundamental than stating that the electrons have identical mass, charge, etc. It means that observing one electron in a system is the same as observing any other electron. Hence, even if we ignore the interelectronic interaction, the wavefunction for a two-electron system must have the following form:

$$\Psi(\mathbf{r}_1, \mathbf{r}_2) = \psi_a(\mathbf{r}_1)\psi_b(\mathbf{r}_2) \pm \psi_a(\mathbf{r}_2)\psi_b(\mathbf{r}_1) \tag{1.22}$$

Here, we must digress: All fundamental particles have a property called "spin"; this is an intrinsic angular momentum with magnitude $S = \hbar\sqrt{s(s+1)}$. For an electron, $s = 1/2$. The classical analogy is that an electron is a like a little sphere spinning on its axis; however, this is not what is really happening. Spin is a purely quantum mechanical phenomenon. Nevertheless, since spin is a type of angular momentum, it has a z-axis component, s_z. However, s_z can take on only quantized values of $m_s\hbar$ where $m_s = \pm 1/2$. If $m_s = +1/2$, we say the spin is "up" or α-spin; if $m_s = -1/2$, we say the spin is "down" or β-spin. Fundamental particles in the universe are either Fermions (with half-integer values of $s = 1/2, 3/2, \ldots$) or Bosons (with integer values of $s = 0, 1, 2, \ldots$). Fermions must have "antisymmetric" wavefunctions:

$$\Psi(\mathbf{r}_1, \mathbf{r}_2) = \psi_a(\mathbf{r}_1)\psi_b(\mathbf{r}_2) - \psi_a(\mathbf{r}_2)\psi_b(\mathbf{r}_1) \tag{1.23}$$

Electrons are Fermions and the electronic wavefunctions must have this antisymmetry. This is a formal and abstract way of stating the *Pauli exclusion principle*. The antisymmetry requirement means that no two electrons can be in the same quantum state or have the same single-particle orbital. An antisymmetric wavefunction obeys the Pauli exclusion principle since, if $\mathbf{r}_1 = \mathbf{r}_2$, then $\Psi(\mathbf{r}_1, \mathbf{r}_2) = 0$ unless $\psi_a(\mathbf{r}_1) \neq \psi_b(\mathbf{r}_1)$. If we build a multielectronic atom in terms of single-particle hydrogenic orbitals, the Pauli exclusion principle means that no two electrons can have the same four quantum numbers (n, l, m_l, m_s). From now on, instead of using m_s, we will designate the spin of an electron by α or β. The spin coordinate of an electron will be accounted for by using separate wavefunctions for α- and β-spin electrons designated as ψ_1^α, ψ_1^β. Using separate wavefunctions for α- and β-spin electrons is referred to as a *spin-unrestricted formalism*. As we will see later, the two wavefunctions ψ_1^α, ψ_1^β will be numerically different if the number of α-spin electrons differs from the number of β-spin electrons because the interelectronic repulsion and electron experiences will depend on its spin.

The construction of antisymmetric wavefunctions from the single-particle (hydrogenic) orbitals is much easier if we use the algebraic trick of expressing the wavefunction as the determinant of a matrix of the one-electron orbitals:

$$\Psi(\mathbf{r}_1, \mathbf{r}_2) = \begin{vmatrix} \psi_1(\mathbf{r}_1) & \psi_2(\mathbf{r}_1) \\ \psi_1(\mathbf{r}_2) & \psi_2(\mathbf{r}_2) \end{vmatrix} \equiv |\psi_1\psi_2| \tag{1.24}$$

(on the right-hand side is a shorthand notation for the determinant). Or, for an N-electron atom,

$$\Psi(\mathbf{r}_1, \mathbf{r}_2, \ldots, \mathbf{r}_N) = \begin{vmatrix} \psi_1(\mathbf{r}_1) & \psi_1(\mathbf{r}_2) & \cdots & \psi_1(\mathbf{r}_N) \\ \psi_2(\mathbf{r}_1) & \psi_2(\mathbf{r}_2) & \cdots & \psi_2(\mathbf{r}_N) \\ \vdots & \vdots & \vdots & \vdots \\ \psi_N(\mathbf{r}_1) & \psi_N(\mathbf{r}_2) & \cdots & \psi_N(\mathbf{r}_N) \end{vmatrix} \equiv |\psi_1\psi_2\ldots\psi_N| \tag{1.25}$$

These are called *Slater determinants*. If any two columns of a matrix are identical, the determinant of the matrix is zero. Hence, if any two electrons occupy the same orbital, we will have two columns to be the same and the determinant (and, hence, the wavefunction) will be zero.

The *Hartree–Fock approximation* is that we can express our multielectronic wavefunction using a single-Slater determinant. This is an important starting point as it gives us a conceptual framework to understand electronic structure and the powerful concept of electron configuration. We use the Hartree–Fock approximation and construct a wavefunction for the multielectronic atom using the hydrogenic orbitals (1s, 2s, 2p, etc.) and populate those orbitals according to the Pauli exclusion principle. Hence, using the shorthand notation for a determinant, the Hartree–Fock wavefunction for Mg is

$$\Psi = \left| \phi_{1s}^{\alpha} \phi_{1s}^{\beta} \phi_{2s}^{\alpha} \phi_{2s}^{\beta} \phi_{2p_x}^{\alpha} \phi_{2p_x}^{\beta} \phi_{2p_y}^{\alpha} \phi_{2p_y}^{\beta} \phi_{2p_z}^{\alpha} \phi_{2p_z}^{\beta} \phi_{3s}^{\alpha} \phi_{3s}^{\beta} \right| \tag{1.26}$$

which, for simplicity, is written as the electron configuration:

$$(1s)^2 (2s)^2 (2p)^6 (3s)^2.$$

Although this is not a quantitative solution, the concept of electronic configuration is an immensely powerful tool in predicting the chemical behavior of the elements. The Hartree–Fock approximation also gives us a starting point in calculating the energies and wavefunctions of a multi-electronic system.

1.3.1.1   The Variational Principle and the Hartree–Fock Equations

Suppose that our system is described by a Hamiltonian $\hat{H}$ and a wavefunction Ψ. The expectation value (the mean value of an observable quantity) of the total energy is given by

$$\langle E \rangle = \frac{\int \Psi^* \hat{H} \Psi^* \, d\mathbf{r}}{\int \Psi^* \Psi^* \, d\mathbf{r}}$$

where the asterisk ($*$) means the complex conjugate (i.e., replace i by $-i$). Suppose that we did not know Ψ for a given $\hat{H}$ but had a trial guess for it (of course, this is true for nearly all of our problems). The variational principle states that the expectation value of the total energy we obtain from our trial wavefunction will always be greater than the true total energy. This is extremely useful because the problem becomes one of minimizing the total energy with respect to the trial wavefunction, and we will obtain the best approximation we can. Formally, we require that

$$\frac{\delta \langle E \rangle}{\delta \Psi} = 0$$

The δ symbol refers to the functional derivative; a functional is a function of a function. The functional derivative comes from the calculus of variations; we will not go into the details here, but a discussion is given in Parr and Yang (1989).

Now, suppose our unknown wavefunction Ψ is that of a multielectronic atom with N electrons and nuclear charge Z. If we use the Hartree–Fock approximation and express the wavefunction as a single-Slater determinant over N single-particle orbitals, then the expectation value of the total energy will be

$$\begin{aligned}
\langle E \rangle = &\sum_{j}^{N} \int \Psi_j^*(\mathbf{r}) \left(-\frac{1}{2}\nabla^2 - \frac{Z}{r} \right) \psi_j(\mathbf{r}) \, d\mathbf{r} \\
&+ \sum_{j<k}^{N} \int \frac{|\psi_j(\mathbf{r}_1)|^2 |\psi_k(\mathbf{r}_2)|}{|\mathbf{r}_1 - \mathbf{r}_2|} \, d\mathbf{r}_1 d\mathbf{r}_2 \\
&- \sum_{j<k}^{N} \delta(\sigma_j, \sigma_k) \int \frac{\psi_j^*(\mathbf{r}_1)\psi_k^*(\mathbf{r}_2)\psi_j(\mathbf{r}_1)\psi_k(\mathbf{r}_2)}{|\mathbf{r}_1 - \mathbf{r}_2|} \, d\mathbf{r}_1 d\mathbf{r}_2
\end{aligned} \tag{1.27}$$

Here, $\delta(\sigma_j, \sigma_k) = 1$ if the spins of electrons j and k are the same but is 0 if the spins are different. We can now use the variational principle: we minimize the total energy with respect to the single-particle orbitals subject to the constraint that the total number of electrons is held constant. We end up with a set of simultaneous equations for the individual orbitals $\psi_j(\mathbf{r}_j)$ and their energies ε_j:

$$
\frac{-1}{2}\nabla^2 \psi_j(\mathbf{r}_1) - \frac{Z}{r_1}\psi_j(\mathbf{r}_1) + \left(\sum_k^N \frac{|\psi_k(\mathbf{r}_2)|^2}{|\mathbf{r}_1 - \mathbf{r}_2|} d\mathbf{r}_2 \right)\psi_j(\mathbf{r}_1)
$$

$$
- \left(\sum_k^N \delta(\sigma_j, \sigma_k) \right) \int \frac{\psi_k^*(\mathbf{r}_2)\psi_j(\mathbf{r}_2)}{|\mathbf{r}_1 - \mathbf{r}_2|} d\mathbf{r}_2 \right) \psi_j(\mathbf{r}_1) = \varepsilon_j \psi_j(\mathbf{r}_1)
$$

(1.28)

These are the Hartree–Fock equations. The first summation term (the *Coulomb potential*) describes the repulsive potential experienced by an electron in orbital j at $\mathbf{r}_1$ due to the presence of all the other electrons in orbitals k at $\mathbf{r}_2$. Note, however, that this summation also contains a term where the electron interacts with itself (when $j = k$). This self-interaction must be compensated for. The second summation (the *exchange potential*) modifies the Coulomb potential to remove the interactions between electrons with the same spin that are in the same orbital. It is important to note that the exchange potential also removes the self-interaction of each electron since it cancels the Coulomb potential when $j = k$.

Before we can solve the Hartree–Fock equations for the orbitals $\psi_j(\mathbf{r}_j)$, we need to evaluate the Coulomb and exchange potentials. However, we cannot evaluate the Coulomb and exchange potentials until we know the orbitals $\psi_j(\mathbf{r}_j)$! We get around this problem by starting with an initial guess for $\psi_j(\mathbf{r}_j)$, evaluating the Coulomb and exchange potentials using that guess, and then solving for a better set of $\psi_j(\mathbf{r}_j)$. We then take the new set of wavefunctions and evaluate a new Coulomb and exchange potential. After so many iterations, we should converge to a *self-consistent field* (SCF) solution. All electronic structure methods need to iteratively arrive at a self-consistent solution. To confuse matters, however, the quantum chemistry literature and many textbooks often will equate "SCF" with the Hartree–Fock approximation.

The Hartree–Fock approximation is a major improvement over the Hartree approximation since it accounts for the interelectronic repulsion between electrons with the same spin. However, it completely neglects the correlation between electrons with opposite spin. The consequence of this illustrated in the simple molecule H_2 (Figure 1.4a). As we will discuss later, we can express the one-electron orbitals of a molecule (the molecular orbitals) as linear combination of atomic orbitals centered on the different atoms. If we label the two H atoms in H_2 as A and B, with atomic orbitals ϕ_A and ϕ_B, then the molecular orbitals can be constructed as follows:

$$
\psi_+ = \frac{1}{\sqrt{2}}(\phi_A + \phi_B)
$$

$$
\psi_- = \frac{1}{\sqrt{2}}(\phi_A - \phi_B)
$$

(1.29)

These correspond to the bonding and antibonding orbitals (Figure 1.4b). These are not the actual states of the system; those are given by the multielectronic wavefunctions expressed as Slater determinants. The lowest energy state (bonding orbital) is the Slater determinant:

$$
\Psi(1,2) = \begin{vmatrix} \psi_+^\alpha(1) & \psi_+^\alpha(2) \\ \psi_+^\beta(1) & \psi_+^\beta(2) \end{vmatrix}
$$

(1.30)

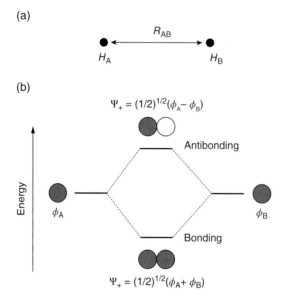

Figure 1.4 *(a) The H_2 molecule and (b) schematic energy-level diagram for the first two one-electron molecular orbitals showing the bonding and antibonding combinations of the atomic orbitals.*

If we expand the Slater determinant in terms of the atomic orbitals, we find that the wavefunction is

$$\Psi(1,2) = \frac{1}{2}\left(\phi_A^\alpha(1) + \phi_B^\alpha(1)\right)\left(\phi_A^\beta(2) + \phi_B^\beta(2)\right) - \frac{1}{2}\left(\phi_A^\beta(1) + \phi_B^\beta(1)\right)\left(\phi_A^\alpha(2) + \phi_B^\alpha(2)\right)$$

$$= \frac{1}{2}\left(\phi_A^\alpha(1)\phi_A^\beta(2) + \phi_A^\alpha(1)\phi_B^\beta(2) + \phi_B^\alpha(1)\phi_A^\beta(2) + \phi_B^\alpha(1)\phi_B^\beta(2)\right) \tag{1.31}$$

$$- \frac{1}{2}\left(\phi_A^\beta(1)\phi_A^\alpha(2) + \phi_A^\beta(1)\phi_B^\alpha(2) + \phi_B^\beta(1)\phi_A^\alpha(2) + \phi_B^\beta(1)\phi_B^\alpha(2)\right)$$

The two summations simply differ by flipping the spins of electrons 1 and 2 in order to have an antisymmetric wavefunction. Now, consider the physical meaning of each term in the two summations: $\phi_A^\alpha(1)\phi_A^\beta(2)$ corresponds to both electrons being localized to atom A, the term $\phi_A^\alpha(1)\phi_B^\beta(2) + \phi_B^\alpha(1)\phi_A^\beta(2)$ corresponds to the two electrons being delocalized over the two atoms, and the term $\phi_B^\alpha(1)\phi_B^\beta(2)$ corresponds to the two electrons localized to atom B.

Suppose that we dissociate the H_2 molecule. It should dissociate into two neutral H atoms, and the wavefunction should be as follows:

$$\Psi_+ = \frac{1}{2}\left(\phi_A^\alpha(1)\phi_B^\beta(2) + \phi_B^\alpha(1)\phi_A^\beta(2)\right) - \frac{1}{2}\left(\phi_A^\beta(1)\phi_B^\alpha(2) + \phi_B^\beta(1)\phi_A^\alpha(2)\right) \tag{1.32}$$

However, this is not what is predicted with the single-determinantal wavefunction. Instead, H_2 can only dissociate into a state where the probability of having H + H and H^+ + H^- are the same. This means that the dissociation energy of H_2 will be greatly underestimated since the energy of the ion pair is about 12.85 eV higher than that of the two neutral atoms. The only way to correct this is to mix in other configurations to cancel out the terms where both electrons are localized on the same atom. This could be done using Valence Bond theory or by expressing the wavefunction in terms of more than one Slater determinant. Neither approach is practical for any system of geochemical interest.

The Hartree–Fock total energy provides a reasonable first approximation from which we can calculate physical properties of a mineral such as compressibility, vibrational frequencies, etc. However it is not usually accurate enough to reliably address the energetics of chemical reactions. The Hartree–Fock approximation fails to completely describe the repulsion of electrons with opposite spin since the single-Slater determinant wavefunction does not go to zero when two electrons with opposite spin occupy the same position in the same orbital. This excess repulsion energy means that the Hartree–Fock energy will always be greater than the true energy.

The difference between the exact energy and the Hartree–Fock energy is known (by convention) as the *correlation energy*. However, the correlation of the motions between electrons with like spin (the "exchange correlation") is accounted for by the Hartree–Fock formalism; the "correlation energy" refers to the correlation motion between electrons with opposite spin (Coulomb correlation). Although the exchange correlation is much larger, the Coulomb correlation is still significant and, because it is neglected, Hartree–Fock energies cannot be used to reliably predict chemical reactions and the energetics of atoms, molecules, and crystals. For example, van der Waals interactions cannot be described at the Hartree–Fock level because the induced dipole is an effect of electron correlation between atoms.

A more accurate approximate wavefunction for a multielectronic system can be obtained by a finite expansion (linear combination) of Slater determinants resulting from the different possible electronic configurations over the one-electron orbitals. This approach is known as *configurational interaction*. It is the gold standard for quantum chemical calculations and can be applied to small molecules to predict properties to high accuracy (Sherrill and Schaefer, 1999). However, configuration–interaction calculations are impractical for the complex systems of interest in geochemistry. Instead, a completely different approach is needed.

1.3.2 Density Functional Theory

Density functional theory or DFT (Parr and Yang, 1989) is an approach to bonding and electronic structure that was developed in the physics community. Until the 1990s, it was not believed to be accurate enough to deal with chemical systems; however, subsequent developments have made DFT the major tool of quantum chemistry. Nearly all quantum mechanical calculations applied to geochemical systems are based on DFT. Note, however, that many traditional quantum chemists will not use the phrase "ab initio" for calculations based on DFT. This is because applied DFT must always use a fundamental approximation in how it describes the interelectronic interactions.

The basis of density functional theory is a theorem (Hohenberg and Kohn, 1964) that the ground-state total energy E of a system of particles (e.g., electrons) subject to any kind of external potential V_{ext} can be expressed (exactly) in terms of functionals of the particle density $\rho(\mathbf{r})$:

$$E[\rho(\mathbf{r})] = T[\rho(\mathbf{r})] + U[\rho(\mathbf{r})] + \int V_{\text{ext}}(\mathbf{r})\rho(\mathbf{r})d\mathbf{r} \qquad (1.33)$$

The $T[\rho(\mathbf{r})]$ functional describes the kinetic energy of the system, while the $U[\rho(\mathbf{r})]$ functional describes the potential energy due to interelectronic repulsion. Here, the external potential V_{ext} would include the electron–nuclear interactions in an atom along with the nuclear–nuclear interactions in a molecule. The problem with using the Hohenberg–Kohn theorem, however, is that, for a system of interacting electrons, we do not know what the $T[\rho(\mathbf{r})]$ and $U[\rho(\mathbf{r})]$ functionals are.

Nevertheless, we do know some aspects of these functionals. Our strategy is to first separate out the parts of the functionals that we know. To do this, we first note that part of the interelectronic repulsion is a classical Hartree Coulombic term:

$$U[\rho(\mathbf{r})] = G[\rho(\mathbf{r})] + \frac{1}{2} \int \int \frac{\rho(\mathbf{r})\rho(\mathbf{r}')}{|\mathbf{r}-\mathbf{r}'|} d\mathbf{r} d\mathbf{r}' \tag{1.34}$$

Here, we have temporarily defined a new functional $G[\rho(\mathbf{r})]$ that includes that part of $U[\rho(\mathbf{r})]$ that we do not know. We also know that if the particles did not interact with each other, than the wavefunction of the system would be a single-Slater determinant over single-particle orbitals $\psi_j(\mathbf{r})$ and we could express the charge density in terms of these single-particle orbitals:

$$\rho(\mathbf{r}) = \sum_{j}^{\text{occup}} |\psi_j(\mathbf{r})|^2 \tag{1.35}$$

The kinetic energy functional would then be simply

$$T_0 = -\frac{1}{2} \sum_{j}^{\text{occ}} \int \psi_j^*(\mathbf{r}) \nabla^2 \psi_j(\mathbf{r}) d\mathbf{r} \tag{1.36}$$

We can then write out total energy functional as

$$
\begin{aligned}
E[\rho(\mathbf{r})] = &-\frac{1}{2} \sum_{j}^{\text{occ}} \int \psi_j^*(\mathbf{r}) \nabla^2 \psi_j(\mathbf{r}) d\mathbf{r} \\
&+ \frac{1}{2} \int \int \frac{\rho(\mathbf{r})\rho(\mathbf{r}')}{|\mathbf{r}-\mathbf{r}'|} d\mathbf{r} d\mathbf{r}' \\
&+ \int V_{\text{ext}}(\mathbf{r})\rho(\mathbf{r})d\mathbf{r} + E_{\text{xc}}[\rho(\mathbf{r})]
\end{aligned}
\tag{1.37}
$$

where we have defined a new functional $E_{\text{xc}}[\rho(\mathbf{r})]$ that describes the correction to the kinetic energy relative to that used for noninteracting electrons and that part of the interelectronic repulsion that we do not know. The $E_{\text{xc}}[\rho(\mathbf{r})]$ term must, therefore, describe the Coulomb correlation (repulsion between electrons with opposite spin) and the exchange energy (repulsion between electrons with like spin). However, even though we are defining the charge density in terms of noninteracting particles, our treatment is still exact since the Hohenberg–Kohn theorem is valid for any potential. This means that we can express our system of electrons in terms of noninteracting quasiparticles. These quasiparticles will have single-particle orbitals $\psi_j(\mathbf{r})$. The single-particle orbitals are solutions to the Kohn–Sham equations (Kohn and Sham, 1965) that take the form of one-electron Schrödinger equations:

$$\left(-\frac{1}{2}\nabla^2 + v_{\text{xc}} + V_{\text{ext}}(\mathbf{r}) \right)\psi_j(\mathbf{r}) = \varepsilon_j \psi_j(\mathbf{r}) \tag{1.38}$$

with

$$v_{\text{xc}} = \int \frac{\rho(\mathbf{r})}{|\mathbf{r}-\mathbf{r}'|} d\mathbf{r}' + \frac{\delta E_{\text{xc}}[\rho(\mathbf{r})]}{\delta\rho(\mathbf{r})} \tag{1.39}$$

1.3.2.1 Meaning of Kohn–Sham Eigenvalues

Before discussing the exchange–correlation functional and potential, we should think about the meaning of the Kohn–Sham orbitals and their eigenvalues (energies). Strictly speaking, the Kohn–Sham single-particle orbitals have no actual physical meaning; they are wavefunctions for fictitious particles that give a density from which we can determine the total ground-state energy of a system. However, the Kohn–Sham orbitals will have the same symmetry as the one-electron orbitals obtained in a Hartree–Fock (single determinant) picture. It is common practice to relate Kohn–Sham orbitals and their eigenvalues (energies) to the actual electronic states in a system (e.g., Stowasser and Hoffmann, 1999; Cramer and Turhlar, 2009; Sherman, 2009). However, there is nothing in DFT that formally says this is so. The wavefunction for the collection of Kohn–Sham noninteracting quasiparticles for any system is a single-Slater determinant. However, the actual electronic states of many systems (e.g., FeO, the complex $Fe(H_2O)_6^{2+}$) cannot be described by a single-Slater determinant. This has been a subject of much discussion (e.g., Stowasser and Hoffmann, 1999) but goes beyond the scope of this chapter. We would like, however, to be able to relate the electronic structure of a system to its oxidation-reduction potential (see Chapter 7). In this regard, there is a physical meaning for the Kohn–Sham orbital energies, however, since these relate to the total energy as

$$\varepsilon_i = \left(\frac{\partial E}{\partial n_i} \right)_{j \neq i} \tag{1.40}$$

where E is the total energy of the system and n_i is the occupancy of orbital i; that is, the Kohn–Sham eigenvalues correspond to chemical potentials of the electrons. This is to be compared to the Hartree–Fock orbital energies, which obey Koopman's theorem:

$$\varepsilon_i = E(N+1) - E(N) \tag{1.41}$$

where $E(N)$ is the total energy of the system with N electrons. This means that the energy of the highest occupied orbital will be equal to the first ionization energy of the atom or molecule.

1.3.2.2 Local Density and Generalized Gradient Approximations

Although the Kohn–Sham equations are exact, we cannot solve them for a real multielectronic system because we do not know the exchange–correlation functional $E_{xc}[\rho(\mathbf{r})]$.

The first approximation to $E_{xc}[\rho(\mathbf{r})]$ is to consider the case of a system where the electron density is uniform. For such a system, we can separate the exchange and correlation energy:

$$E_{xc}[\rho(\mathbf{r})] = E_x[\rho(\mathbf{r})] + E_c[\rho(\mathbf{r})] \tag{1.42}$$

The exchange energy for a uniform electron gas is

$$E_x[\rho(\mathbf{r})] = -\frac{3}{4}(3/\pi)^{1/3} \int \rho(\mathbf{r})^{4/3} d\mathbf{r} \tag{1.43}$$

There is no analytical expression for the correlation energy $E_c[\rho(\mathbf{r})]$ of a uniform electron gas. However, $E_c[\rho(\mathbf{r})]$ can be estimated using stochastic simulations in the limits of high and low electron density; the resulting correlation energy as a function of electron density have been parameterized in several schemes. A recommended version is that provided by Vosko et al. (1980).

Table 1.1   *Geometry and vibrational frequencies of gas-phase water calculated using a 6-311G* basis set with different exchange–correlation functionals.*

XC functional	Hartree–Fock	LDA	PW91	PBE96	B3LYP	Experimental[a]
R(O–H) Å	0.9393	0.9714	0.9701	0.9710	0.9626	0.9572
α(HOH)°	107.48	106.11	105.13	105.02	105.87	104.52
ν_3	1840	1622	1671	1671	1705	1595
ν_2	4141	3672	3658	3656	3766	3657
ν_1	4248	3800	3779	3776	3881	3756

[a] Benedict et al. (1956).

The $E_{xc}[\rho(\mathbf{r})]$ for an electron in uniform electron gas only depends on the charge density at the electron's position. For this reason, we call the uniform electron gas the *local density approximation (LDA)*. The uniform electron gas is a drastic approximation to most chemical systems, but it often works reasonably well. In a real system, $E_{xc}[\rho(\mathbf{r})]$ will depend on some complex way on the electron density at distances away from the electron's position. One way to account for nonuniform charge distributions is to come up with an exchange–correlation functional that depends not only on $\rho(\mathbf{r})$ but also on the gradient of the charge density $\nabla\rho(\mathbf{r})$. This would still be a local functional, but at least it contains some influence from the neighboring density. Functionals of the form $E_{xc}[\rho(\mathbf{r}), \nabla\rho(\mathbf{r})]$ are known as the *generalized gradient approximation* (GGA). There are a variety of GGA functionals currently in use (Perdew et al., 1996), and they offer a substantial improvement over the LDA at little extra computational cost. In the past 20 years, there have been many attempts to develop improved approximations for $E_{xc}[\rho(\mathbf{r})]$, and now the user has a choice of many possibilities (Cramer and Turhlar, 2009). It is important that the user be aware of the strengths and weaknesses of the different versions of the $E_{xc}[\rho(\mathbf{r})]$ functionals. To this end, it is prudent to explore how the results of a simulation are dependent upon the choice of exchange–correlation functionals. Table 1.1 shows the calculated geometries and vibrational frequencies of gas-phase water molecule obtained using Hartree–Fock and different exchange–correlation functionals. Hartree–Fock exaggerates the bonding interactions making the OH-bond length too short and the vibrational frequencies too high. The LDA (which includes some correlation energy) is a significant improvement, but the generalized gradient schemes are more accurate. The hybrid scheme of B3LYP (discussed later) is closer to experiment.

1.3.2.3   Self-Interaction Error: Hybrid Functionals

Both the LDA and all versions of the GGA suffer from a major error. Recall that in the Hartree–Fock equation (Eq. 1.28), one of the terms in the exchange summation also serves to correct the Coulomb summation for the term that inadvertently describes an electron interacting with itself. Because both the LDA and GGA give the exact Coulombic term but only an approximate exchange term, the self-interaction error is not fully corrected for in these exchange–correlation functionals. The physical consequence of the self-interaction error is that electrons will repel themselves and over-delocalize. This is not a problem for systems that are metallic in the first place, but for systems with localized electrons, it gives rise to erroneous results. For example, the molecule H_2^+ with one electron should dissociate to give $H^+ + H$. However, all LDA and GGA functionals will cause H_2^+ to dissociate to give $H^{+0.5} + H^{+0.5}$ even if the H atoms are infinitely separated. In mixed-valence systems such as magnetite ${\{Fe^{+3}\}}_A{\{Fe^{+2}Fe^{+3}\}}_B O_4$, the self-interaction error will cause the system to have a ground-state configuration ${\{Fe^{+3}\}}_A{\{Fe^{+2.5}Fe^{+2.5}\}}_B O_4$. More generally, band gaps in solids and the HOMO–LUMO gaps (the gap between the highest occupied and lowest unoccupied molecular orbitals) in molecules are underestimated (Perdew, 1986; Cramer and Truhlar, 2009).

Within DFT, several methods have been developed to approximately correct for the self-interaction error (e.g., Perdew and Zunger, 1981). Recall, however, that the Hartree–Fock equations correctly accounts for the self-interaction correction. One approach is to use "hybrid functionals" that mix in some degree of Hartree–Fock exchange with a DFT exchange–correlation functional. The most popular in this regard is the B3LYP functional

$$E_{xc}[\rho(r)] = E_X^{LDA} + a_0\left(E_X^{HF} - E_X^{LDA}\right) + a_X\left(E_X^{GGA} - E_X^{LDA}\right) + E_C^{LDA} + a_C\left(E_C^{GGA} - E_C^{LDA}\right) \quad (1.44)$$

with

$$a_0 = 0.2, \ a_X = 0.72, \ a_C = 0.81$$

where E_X^{HF} is the Hartree–Fock exchange, E_X^{LDA} is the LDA exchange, E_X^{GGA} is the GGA exchange (Becke, 1993), E_C^{LDA} is the LDA correlation energy of Vosko et al. (1980), and E_C^{GGA} is the GGA correlation of (Lee et al., 1988). The parameters are variable and one can choose how much Hartree–Fock exchange to mix in. As can be seen in Table 1.1, the B3LYP calculations are an improvement over those done with the GGA functional.

1.4   Bonding in Molecules and Solids

1.4.1   The Born–Oppenheimer Approximation

Consider the hydrogen molecule H_2 (Figure 1.4). This molecule consists of two nuclei A and B located at $\mathbf{R}_A$ and $\mathbf{R}_B$ and two electrons located at $\mathbf{r}_1$ and $\mathbf{r}_2$. The Hamiltonian for this system is

$$\hat{H} = -\frac{1}{2M}\left(\nabla_A^2 + \nabla_B^2\right) - \frac{1}{2}\left(\nabla_1^2 + \nabla_2^2\right)$$
$$+ \left(\frac{1}{|\mathbf{R}_A - \mathbf{R}_B|} - \frac{1}{|\mathbf{r}_1 - \mathbf{R}_A|} - \frac{1}{|\mathbf{r}_1 - \mathbf{R}_B|} - \frac{1}{|\mathbf{r}_2 - \mathbf{R}_A|} - \frac{1}{|\mathbf{r}_2 - \mathbf{R}_B|} + \frac{1}{|\mathbf{r}_1 - \mathbf{r}_2|}\right) \quad (1.45)$$

The first two terms are the kinetic energies of the nuclei and the electrons. However, the mass of each nucleus is 1848 times greater than the mass of each electron. Accordingly, the electron–nuclear interactions cannot change the kinetic energies of the nuclei. The *Born–Oppenheimer approximation* is to neglect the kinetic energy of the nuclei and write a Hamiltonian that is a function of the nuclear coordinates. The Schrödinger equation for the electrons is then:

$$\left(-\frac{1}{2}\nabla_1^2 - \frac{1}{2}\nabla_1^2 + V(\mathbf{R}_A, \mathbf{R}_B)\right)\Psi(\mathbf{r}_1, \mathbf{r}_2, \mathbf{R}_A, \mathbf{R}_B) = E_n(\mathbf{R}_A, \mathbf{R}_B)\Psi(\mathbf{r}_1, \mathbf{r}_2, \mathbf{R}_A, \mathbf{R}_B) \quad (1.46)$$

What we have done is to remove the nuclear motion and defined electronic states with energies $E_n(\mathbf{R}_A, \mathbf{R}_B)$ that are a function of the nuclear positions.

For practical problems, rather than to deal with the multielectronic Schrödinger equation, we will use DFT and solve the Kohn–Sham equations at a fixed set of nuclear coordinates $\{\mathbf{R}\}$:

$$\left(-\frac{1}{2}\nabla^2 + v_{xc} + V_{ext}(\mathbf{r}, \{\mathbf{R}\})\right)\psi_j(\mathbf{r}) = \varepsilon_j(\{\mathbf{R}\})\psi_j(\mathbf{r}, \{\mathbf{R}\}) \quad (1.47)$$

The total electronic energy $E(\{\mathbf{R}\})$ is evaluated in DFT as follows:

$$E(\{\mathbf{R}\}) = \sum_i^N \varepsilon_i(\{\mathbf{R}\}) - V_C(\{\mathbf{R}\}) + E_{xc} \qquad (1.48)$$

From such calculations, we can minimize the total energy $E(\{\mathbf{R}\})$ with respect to $\{\mathbf{R}\}$ and predict equilibrium structures of molecules and crystals. When we neglect the energy associated with the nuclear motion, the energy we calculate is referred to as the *static* energy of the system. So, it is common to use codes to predict the static energies of molecules and crystals; however, the static energies give energy differences that neglect the zero-point and thermal energies of the system. We can correct for this once (at least in the harmonic approximation) if we calculate the vibrational modes of the system.

1.4.2   Basis Sets and the Linear Combination of Atomic Orbital Approximation

In principle, we could solve the Kohn–Sham equation for a molecule or crystal numerically. However, this would be unwieldy. In nearly all codes used in computational chemistry and solid-state physics, the electronic wavefunctions $\psi_j(\mathbf{r},\{\mathbf{R}\})$ are constructed from a *basis set* of simple functions $\phi_j(\mathbf{r},\{\mathbf{R}\})$:

$$\psi_j(\mathbf{r},\{\mathbf{R}\}) = \sum_k c_k \phi_k(\mathbf{r},\{\mathbf{R}\}) \qquad (1.49)$$

An obvious choice of a basis set would be functions that approximate the atomic orbitals of the atoms that make up our molecule. This is the way we think about chemical bonding anyway, so it also offers a very compelling conceptual framework. In this scheme, the one-electron molecular orbitals $\psi_j(\mathbf{r},\{\mathbf{R}\})$ are expressed as a linear combination of atomic orbitals. By atomic orbitals, we mean functions that are centered on the different atoms and have the same n, l, m_l quantum numbers that a hydrogenic wavefunction would have. If we express our molecular wavefunctions this way, then our solution to the Kohn–Sham equation is obtained by variationally minimizing the energy of the system with respect to the coefficients c_k in the expansion, subject to the constraint that the wavefunction is normalized:

$$\int_0^\infty \psi_j^*(\mathbf{r},\{\mathbf{R}\})\psi_j(\mathbf{r},\{\mathbf{R}\})d\mathbf{r} = 1 \qquad (1.50)$$

(The normalization requirement simply means that the probability of finding the electron somewhere must be 1.)

The choice of basis functions is crucial, and it is imperative that the user of any quantum chemistry code be aware of the adequacy of the basis set for the problem at hand. According to the variational principle, the more functions we use, the more accurate our total energy will be since we will have more coefficients c_k that we can minimize our total energy with respect to. However, with the more coefficients we have, the more computationally intensive our problem will be. The basis set, therefore, must be something that allows a good approximation with as few parameters as possible. One starting point for a basis set is to use atomic orbitals that are approximately expressed using Slater functions:

$$\phi_{nlm}(\mathbf{r}) = r^{n-1} e^{-\varsigma r} Y_l^m(\mathbf{r}) \qquad (1.51)$$

where $Y_l^m(\mathbf{r})$ is the spherical harmonic function that describes the angular shape of the wavefunction and ς is a constant that describes the "effective charge" of the nucleus (i.e., the nuclear charge that the electron sees after it has been screened by the other electrons). Note that the Slater function has a similar form to that of the actual hydrogen orbitals described earlier. However, the adjustable screening parameter ς is needed to approximately correct for the presence of the other electrons.

Slater functions, however, are technically awkward to use when solving for the various multicenter integrals that occur in the Hartree–Fock or Kohn–Sham equations. Few computational quantum chemistry codes actually use Slater Functions as basis sets, therefore. One notable exception is the Amsterdam density functional (ADF) code (te Velde et al., 2001). Most other codes, such as GAUSSIAN (Frisch et al., 2009) and NWCHEM (Valiev et al., 2010), use basis functions that express the atomic orbital wavefunctions in terms of Gaussian functions. The reason for this is that the product of two Gaussian functions is itself a Gaussian function. Hence, the evaluation of the Coulomb and exchange terms in the Hartree–Fock or Kohn–Sham equations becomes much simpler.

Early work approximated the Slater-type orbitals in terms of several Gaussian functions. This lead to basis sets referred to as "STO-NG" where STO means Slater-type orbital and NG means that it was approximated by N Gaussian functions. These basis sets were minimal in that they expressed each atomic orbital in terms of a single function. This is not a bad approximation for the core electrons (e.g., the 1s, 2s, and 2p orbitals in second row elements and transition metals). However, the valence electrons need a much more flexible basis set with more variational degrees of freedom if they are to be used in a range of bonding environments and to describe atoms in different oxidation states. To this end, split-valence basis sets were developed. Such a basis set will express a valence atomic orbital in terms of several independent functions (Gaussians or Slater functions). This provides more variational degrees of freedom.

For most routine calculations on systems of geochemical significance, a split-valence basis set such as 6-31G* is reasonably adequate. For oxygen, with an electron configurations $1s^2 2s^2 2p^4$, this basis set would represent the 1s levels with six Gaussian functions, but the 2s and 2p levels would be represented by one function made up of three Gaussians and one function made up of a single Gaussian. Tables 1.2 and 1.3 show the geometries and vibrational modes of gas-phase water molecule calculated using different basis sets and using the B3LYP exchange–correlation functional. A minimal basis set such as STO-3G or SZ gives poor results. The split-valence basis sets give significant improvement albeit with diminishing returns as more functions are added and with significant increases in computational cost.

Table 1.2 *Geometry and vibrational frequencies of gas-phase water calculated using a B3LYP functional with different Gaussian basis sets.*

Basis set	STO-3G	6-31G	6-31G*	6-311G*	cc-PVTZ	Expt.
R(O–H) Å	1.0309	0.9759	0.9687	0.9626	0.9611	0.9572
α(HOH)°	96.73	108.30	103.65	105.87	104.56	104.52
ν_3	1977	1619	1713	1705	1640	1595
ν_2	3571	3616	3727	3766	3803	3657
ν_1	3793	3781	3849	3881	3905	3756
Time	3.9	6.9	5.9	7.2	17.3	

Table 1.3 *Geometry and vibrational frequencies of gas-phase water calculated using a B3LYP functional with different Slater orbital basis sets.*

Basis set	SZ	DZ	DZP	TZP	TZ2P	Expt.
R(O–H) Å	1.025	0.983	0.968	0.965	0.963	0.9572
α(HOH)°	97.08	109.70	104.59	104.45	104.89	104.52
ν_3	2006	1566	1623	1623	1606	1595
ν_2	3551	3533	3727	3780	3800	3657
ν_1	3776	3700	3826	3874	3891	3756

The basis sets we have discussed so far (Slater orbitals or Gaussians) are localized basis sets. That is, each function is centered about some atom. This is an obvious choice for systems where the electrons are more-or-less localized to specific atoms. In metallic systems, however, the electrons are delocalized throughout the crystal. It might be more efficient to start with a basis set that has delocalized functions. A completely delocalized function is a plane wave

$$\phi_g = e^{i\mathbf{g}\mathbf{r}} \tag{1.52}$$

However, we can describe any function in terms of these plane wave (e.g., as in a Fourier series expansion). So, a linear combination of plane waves could also be used to describe fairly localized states:

$$\psi(\mathbf{r}) = \sum_{g=0}^{g=\text{cutoff}} c_g e^{i\mathbf{g}\mathbf{r}} \tag{1.53}$$

Moreover, wavefunctions expressed as plane-wave expansions are computationally convenient. The problem with plane waves, however, is that if we need to describe a highly localized orbital (e.g., a 1s, 2s core electrons of an element), then we need a very large number of plane waves. We get around this by ignoring the core-electrons in a system and using a *pseudopotential*. A pseudopotential is something we add to the Kohn–Sham equation to trick the valence electrons into thinking that the core electrons are present! There are a number of different types of pseudopotentials that are available. The Vanderbilt ultrasoft pseudopotentials (Vanderbilt, 1990) are especially useful for oxygen-based systems (i.e., most geochemical systems) as they enable accurate calculations with small values of the plane-wave cutoff for *g*.

1.4.3 Periodic Boundary Conditions

Most systems of interest in geochemistry and geophysics are crystalline phases or bulk liquids. For a crystalline phase, the potential due to the atomic nuclei has a periodic (translational) symmetry. That is, the potential in one unit cell is repeated in all of the unit cells. If a system is periodic with a translational repeat **G**, then the potential due to the nuclei must also be periodic

$$V(\mathbf{R}) = V(\mathbf{R} + \mathbf{G}) \tag{1.54}$$

The wavefunction must have the same symmetry as the potential, so

$$\psi(\mathbf{r}) = \psi(\mathbf{r} + \mathbf{G}) \tag{1.55}$$

Bloch's theorem states that any wavefunction that results from a periodic potential must be of the form

$$\psi_k(\mathbf{r}) = u(\mathbf{r})e^{i\mathbf{kr}} \tag{1.56}$$

where $\mathbf{k}$ is a wave vector that can take on a continuous range of values (it is not quantized if the crystal is infinite). Do not confuse the Bloch wavefunction with the plane-wave expansion described earlier. Usually both are implemented together so we have both the $\mathbf{k}$-vectors to describe the wavefunction and the $\mathbf{g}$-vectors to describe the elements of the plane-wave basis set. However, the range of values that $\mathbf{k}$ can have is limited. Suppose, for the sake of simplicity, that we have a one-dimensional system that is periodic; $\mathbf{G}$ is the translation vector. Then

$$e^{i\mathbf{k}(\mathbf{r}+\mathbf{G})} = e^{i\mathbf{kr}} \tag{1.57}$$

or

$$e^{i\mathbf{kG}} = 1 \tag{1.58}$$

Hence, $\mathbf{kG} = 2\pi$ or $|\mathbf{k}| = 2\pi/|\mathbf{G}|$. We say that $\mathbf{k}$ is a vector in *reciprocal space*. Now, all the $\mathbf{k}$ values from 0 to $2\pi/|\mathbf{G}|$ give unique wavefunctions, but for $\mathbf{k}$-values greater than $2\pi/|\mathbf{G}|$, the wavefunctions simply repeat. We call the region of reciprocal space where $0 < k < 2\pi/|\mathbf{G}|$ (or equivalently $-\pi/|\mathbf{G}| < k < \pi/|\mathbf{G}|$) the *Brillouin zone*. In a periodic system, all we need to do is to solve for the wavefunctions in the Brillouin zone, and we have all of the information we need to describe the periodic system.

Periodic boundary conditions are not only used for crystalline solids; they can also be used to describe liquids when we do not want to inadvertently have artificial phase boundaries (e.g., if we tried to simulate a bulk liquid with a finite cluster of atoms, we would have an unwanted liquid–vacuum interface at the edge of our cluster). In a liquid simulation, we can define a unit cell (as large as our computer resources allow) and have that cell repeat in three dimensions. If we are doing a dynamical simulation (discussed later), then whenever one atom leaves the unit cell, another atom enters the unit cell on the opposite side.

1.4.4 Nuclear Motions and Vibrational Modes

For most problems in geochemistry and geophysics, we are only interested in the lowest energy electronic state or ground state (E_0) since excited electronic states are several eV (hundreds of kJ/mol) higher in energy and are not thermally accessible.

(Exceptions might include spin pairing in the lower mantle and photochemical processes in the environment.) Even if we restrict ourselves to the ground electronic state, however, the quantum states associated with the nuclear motions are thermally accessible and are responsible for most of the thermal properties of earth materials (see Chapter 3).

Moreover, in many problems in geochemistry and mineralogy, we are interested in knowing the vibrational modes of minerals and metal–ligand complexes in aqueous solutions so that we can interpret infrared and Raman spectra (see Chapter 10).

We can obtain a Schrödinger equation for the nuclear motion of the atoms in a molecule or crystal. Recall that our wavefunction $\Psi(\{\mathbf{r}\},\{\mathbf{R}\})$ is both a function of the electronic coordinates $\{\mathbf{r}\}$ and the nuclear coordinates $\{\mathbf{R}\}$. We used the Born–Oppenheimer approximation to write

$$\hat{H}_{elec}\Psi(\{\mathbf{r}\},\{\mathbf{R}\}) = E(\{\mathbf{R}\})\Psi(\{\mathbf{r}\},\{\mathbf{R}\}) \tag{1.59}$$

The assumption here is that the electronic state of the system does not change as the nuclei move.

$$\Psi(\{\mathbf{r}\},\{\mathbf{R}\}) = \chi(\{\mathbf{R}\})\Psi(\{\mathbf{r}\}) \tag{1.60}$$

This means that the electrons and nuclei do not exchange kinetic energy. From this, we get a Schrödinger equation for the nuclear motion:

$$\left(-\sum_k^N \frac{1}{2M_k}\nabla_k^2 + E(\{\mathbf{R}\})\right)\chi(\{\mathbf{R}\}) = W\chi(\{\mathbf{R}\}) \tag{1.61}$$

The potential energy of the nuclei as a function of the nuclear coordinates $\{\mathbf{R}\}$ is given by the total electronic energy $E(\{\mathbf{R}\})$. In a bound system (e.g., a molecule or crystal), this means that for a given electronic state $E(\{\mathbf{R}\})$, the motion of the nuclei will yield a set of quantized vibrational (or rotational) modes with quantized energies W_j. If we know these energies, we could calculate thermodynamic quantities of the system using statistical mechanics (discussed later and also in Chapter 3).

In practice, we do not solve the nuclear Schrödinger equation explicitly as it is too complex. Instead, we assume that, in the vicinity of the equilibrium positions of the atoms $\{\mathbf{R}_0\}$, the nuclear motions are approximately those of a harmonic oscillator with respect to each coordinate $\mathbf{R}$. So, along one coordinate $R = |\mathbf{R}|$,

$$E(R - R_0) = \frac{1}{2}k(R - R_0)^2 \tag{1.62}$$

where k is the force constant along that coordinate. The Schrödinger equation for a harmonic oscillator is solvable and yields quantized energies:

$$W_j = \left(j + \frac{1}{2}\right)\hbar\omega \tag{1.63}$$

where $j = \{0, 1, 2, 3, \ldots\}$ and ω is the angular frequency

$$\omega = \sqrt{\frac{k}{\mu}} \tag{1.64}$$

In a molecule or crystalline solid, many vibrational modes are equivalent to each other because of symmetry. The unique vibrational modes can be obtained by finding the eigenvalues and eigenfunctions of the matrix of force constants (the dynamical matrix). All of the standard quantum chemistry codes do this automatically.

1.5 From Quantum Chemistry to Thermodynamics

Using computational quantum chemistry, we can calculate the total energies of a system as a function of the nuclear coordinates. We can also calculate the energies of the vibrational modes in a molecule or crystal. Ultimately, however, we would like to relate these quantities to macroscopic thermodynamic properties. The macroscopic thermodynamic properties of a system, however, reflect the entropy that a system has. Suppose that a system (e.g., a gas, an aqueous solution, or a mineral) is at a particular thermodynamic state. This state could be defined by any three of the variables N, V, T, P, E and μ, where N is the number of particles, V is the volume, T is the temperature, P is the pressure, E is the total energy, and μ is the chemical potential. We call this a *macro-state*. Associated with that macrostate are a nearly countless number of *microstates*. A microstate of a system is one of the particular arrangements of atomic positions and momenta that are possible within that macrostate of a system. The set of microstates that correspond to a particular macrostate of a system is referred to as a *thermodynamic ensemble*. The probability of a system being in a particular microstate i that has an energy E_i is

$$P_i = \frac{1}{Z} e^{-\beta E_i} \tag{1.65}$$

where $\beta = k_B T$ and Z is the *partition function*

$$Z = \sum_i e^{-\beta E_i} \tag{1.66}$$

From this, we can calculate the average value (or *expectation value*) $\langle O \rangle$ of any property O of the system

$$\langle O \rangle = \frac{1}{Z} \sum_i O_i e^{-\beta E_i} \tag{1.67}$$

That is, we simply take a weighted average of the values that O has at each microstate. We can now derive expressions for many thermodynamic quantities. The internal energy U is simply the average energy of all the microstate and works out to be

$$U = \langle E \rangle = \frac{\partial}{\partial \beta} \ln(Z) \tag{1.68}$$

The pressure is

$$P = \beta \left(\frac{\partial \ln(Z)}{\partial V} \right)_{N,T} \tag{1.69}$$

The Helmholtz free energy is

$$A = \beta \ln(Z) \tag{1.70}$$

Hence, we could use quantum mechanics to evaluate the energy E_i of each microstate and then calculate the partition function for the system. Then we could calculate thermodynamic quantities. The problem we are up against, however, is that the number of microstates in a thermodynamic ensemble is incomprehensibly large. The entropy (S) of a system in a particular macrostate is a measure of the number of microstates Ω consistent with that macrostate:

$$S = k_B \ln \Omega \tag{1.71}$$

Consider ice at 273 K; the absolute entropy is 41.3 J/(mol K). Since Boltzmann's constant is 1.4×10^{-23} J/K, we find that the number Ω of microstates (per mole of ice) is $\ln \Omega = 3 \times 10^{22}$. It follows that evaluating the partition function for ice is impossible. However, in crystalline ice, the periodic symmetry means that the way that the vibrational modes are coupled to each other is constrained by the periodic symmetry of the crystal (every vibrational mode must have the same periodicity of the crystal). Just as for the electronic states, the allowed vibrational modes in a crystal are enumerated using the concept of the *Brillouin zone*. In practice, we can evaluate all of the vibrational modes of a crystalline solid by simply calculating the vibrational modes at selected points in the Brillouin zone and then interpolating between them. The partition function for the crystal would then be evaluated by integrating over the vibrational modes. This methodology is known as *lattice dynamics* and is implemented at the *ab initio* level using codes such as CRYSTAL (Dovesi et al., 2014), VASP (Hafner, 2008), CASTEP (Clark et al., 2005), and SIESTA (Soler et al., 2002).

In the case of an ideal gas, we also have a tractable problem. Each molecule in the ideal gas has only a small number of vibrational/rotational modes, and these are not coupled with the vibrational/rotational modes of any of the other molecules in the gas. Hence, if we knew the vibrational/rotational modes of a single molecule, we can calculate the partition function for a single molecule and, since the molecules are uncoupled, we can calculate the partition function for one mole of an ideal gas phase by simply multiplying by Avogadro's number.

In the case of a liquid (e.g., a silicate melt or an aqueous solution), however, directly evaluating the partition function is impossible. All of the vibrational/rotational and translational modes are coupled to each other, and there is no symmetry to guide us. However, even though there are an uncountable number of microstates, some of them are far more probable than others. (For example, the microstates where one atom has all of the kinetic energy, but all the other atoms have none, are of no significant probability of occurring.) If we only sampled the most probable microstates, we might be able to estimate thermodynamic properties. This can be done using Monte Carlo simulations or by molecular dynamics (Allen and Tildesley, 1987; Haile, 1992).

1.5.1   Molecular Dynamics

Consider a box in which we have placed N atoms, each with mass $\mathbf{m}_j$ and coordinate $\mathbf{R}_j$. Each atom has a kinetic energy. Each atom also has a potential energy resulting from the attractive and repulsive interactions with the other atoms in the box. If we ignore the quantized nature of the motions of the atomic nuclei, then the equations of motion for the atoms would be given by Newton's second law:

$$m_j \frac{d^2\mathbf{R}_j}{dt^2} = \mathbf{F}_j = -\nabla_j E(\mathbf{R}_1, \mathbf{R}_2, \dots, \mathbf{R}_N) = m_j \mathbf{a}_j \qquad (1.72)$$

where $\mathbf{a}_j$ is the acceleration of atom j and $E(\mathbf{R}_1, \mathbf{R}_2, \dots, \mathbf{R}_N)$ is the potential energy of the system as a function of the atomic coordinates of the atoms. If we knew the interatomic potential function $E(\mathbf{R}_1, \mathbf{R}_2, \dots, \mathbf{R}_N)$, we could integrate the equation of motion by finite differences. To do this, we discretize the problem with respect to time t:

$$t_n = t_0 + n\Delta t$$
$$\left(\mathbf{R}_j\right)_n = \mathbf{R}_j(t_n) \qquad (1.73)$$
$$\left(\mathbf{v}_j\right)_n = \mathbf{v}_j(t_n)$$

where $n = 0, 1, 2, 3, \dots$ and Δt is the time interval between steps n and $n + 1$. We need to choose a small time step (e.g., $\Delta t = 0.0025$ fs) to ensure that we accurately integrate the equation of motion. A commonly used method for the integration is the Verlet algorithm (Allen and Tildesley, 1987):

$$(\mathbf{v}_j)_{n+1} = (\mathbf{v}_j)_n + \frac{1}{2}\Big((a_j)_n + (a_j)_{n+1}\Big)\Delta t$$

$$(\mathbf{R}_j)_{n+1} = (\mathbf{R}_j)_n + (\mathbf{v}_j)_n \Delta t + \frac{1}{2}(a_j)_n \Delta t^2 \qquad (1.74)$$

So, we start with a set of atoms with initial positions and velocities. We calculate the forces on those atoms from all the other interatomic interactions; we then move the atoms under those forces (applied for a fraction of a femtosecond) to obtain the new positions and velocities. We repeat this for many time steps (e.g., millions) and watch how the system will evolve. Suppose, for example, our box consisted of 100 water molecules along with 10 Na and Cl atoms. The ground electronic state of this system would yield a system where the Na atoms become Na^+ ions and the Cl atoms become Cl^- ions. As our simulation evolves, we would find that the atoms are rapidly moving in a chaotic motion but, on average, the Na^+ ions are surrounded by six water molecules as are the Cl^- ions. Water molecules may be moving in and out of the solvation shell surrounding the Na^+ and Cl^-, but usually there are six waters surrounding each ion. The water molecules themselves form a structure that reflects the hydrogen bonding interactions. However, the hydrogen bonding structure is constantly breaking down and reforming. Sometimes the Na^+ and Cl^- are so attracted to each other that they each discard one of their solvation waters and form an inner-sphere Na–Cl ion pair. This pair may last for a few picoseconds and then dissociate. Simulations like this can provide us a great deal of molecular-scale insight on the nature of geochemically important systems such as hydrothermal fluids and silicate melts (e.g., Sherman, 2007; Mei et al., 2013, 2015). These simulations can also give us thermodynamic information using techniques such as thermodynamic integration or constrained molecular dynamics (Sprik and Ciccotti, 1998).

Molecular dynamics provides a way to estimate thermodynamic properties of liquids without needing to directly calculate the partition function. The trick is that we can estimate the ensemble average $\langle O \rangle$ of a quantity by taking the time average of that quantity during a simulation. This is known as the *ergodic hypothesis*:

$$\langle O \rangle = \frac{1}{Z}\int_j O(r,p)e^{-E(r,p)/kT}\,drdp = \frac{1}{\tau}\int_0^\tau O(t)dt \qquad (1.75)$$

In the course of our simulation, we are sampling the most probable microstates. As we keep sampling, our time average quantity will get closer and closer to the ensemble average.

There are several types of thermodynamic ensembles: if we constrain N, V and E, we have a *microcanonical ensemble*; this is the ensemble that would result if we solved the equations of motion as described earlier. However, we could also constrain N, V, and T to yield a *canonical ensemble*. To do this, we need to impose a thermostat on the equations of motion to maintain the temperature (Nose, 1984; Hoover, 1985). We could also constrain N, P, and T to yield an *isobaric–isothermal ensemble*; for this, we need include a barostat (Andersen, 1980). Such simulations are more challenging and it is more common to use the constant-volume NVT ensemble. Because we want to constrain the simulations, it is more convenient to recast the problem using Lagrangian mechanics (this will also enable us to discuss *ab initio* MD using the Car–Parrinello method given later). The Lagrangian of a system is the difference between the kinetic and potential energies:

$$L = T - V \qquad (1.76)$$

Or, for a collection of atoms, we can express the Lagrangian as follows:

$$L = \sum_j \frac{1}{2}M_j \left|\frac{d\mathbf{R}_j}{dt}\right|^2 - E\big[\{\mathbf{R}_j\}\big] + \text{constraints} \qquad (1.77)$$

In Lagrangian Mechanics, the equation of motion is

$$\frac{d}{dt}\left(\frac{\partial L}{\partial R_j'}\right) - \left(\frac{\partial L}{\partial R_j}\right) = 0 \tag{1.78}$$

which we integrate using finite differences as in the Verlet algorithm.

The fundamental problem with setting up a molecular dynamics simulation is knowing the interatomic potential function $E(\{\mathbf{R}\})$. Until recently, nearly all molecular dynamics simulations were based on using empirically derived classical functions that describe the attractive and repulsive potentials between atoms. Usually, these functions considered only pair-wise interactions so that

$$E(\mathbf{R}_1, \mathbf{R}_2, \ldots, \mathbf{R}_N) = \sum_{ij} E_{ij}\left(|\mathbf{R}_j - \mathbf{R}_i|\right) \tag{1.79}$$

This is a very drastic approximation. In our NaCl solution, for example, it means that the interaction between a Na atom and a water molecule is independent of what that water molecule is bonded to. Nevertheless, simulations using these types of pair-wise potentials often give excellent results (Smith and Dang, 1994)! In particular, they seem to work well for highly ionic systems. However, for many systems (e.g., complexation of transition metals in aqueous solutions) this approach usually fails. More elaborate three- and four-body potentials can be developed to cope with conformational changes in covalent molecules. Such potentials play an important role in modeling conformational structures in proteins (Ponder and Case, 2003).

Ultimately, however, the interactions between atoms must be described using quantum mechanics and must be evaluated in a self-consistent manner. This means that for every positional configuration at each time step in our simulation, we must solve the Schrödinger equation (or, more likely, the Kohn–Sham equation). This approach is known as *ab initio molecular dynamics*. If we do this, we can evaluate the force on each atom using the Hellmann–Feynman theorem:

$$F_{R_j} = -\frac{\partial E}{\partial \mathbf{R}_j} = \int \Psi^* \frac{d\hat{H}}{d\mathbf{R}_j} \Psi d\mathbf{r} \tag{1.80}$$

Here, $d\mathbf{r}$ means that we are taking the integral over all the volume. Note that, even if we do this, we can still take those quantum mechanically derived forces and still carry on the simulation using classical (Newtonian) mechanics for the nuclear motion. This is usually a safe approximation because the spacings between the quantum levels of the nuclear motions in a liquid are very small. Intramolecular vibrations, in particular O–H bonds, however, will be in error as the large zero-point energy will be neglected.

Even if we neglect the quantized nature of the nuclear motion, solving the Kohn–Sham equation for the electronic motion to yield the correct interatomic forces at each time step is a formidable task. However, for small systems with modern supercomputer facilities, we are now at a point where geochemically interesting simulations can be done.

There are two approaches to *ab initio* molecular dynamics: the first approach is to simply solve the Kohn–Sham equation at each time step and evaluate the forces using the Hellman–Feynman theorem. At each time step, therefore, we bring the system to the ground electronic state. This method is known as Born–Oppenheimer MD and is used, for example, in the code CASTEP (Clark et al., 2005). The advantage of this approach is that it is very stable; however, it is computationally demanding.

The second approach to *ab initio* molecular dynamics is that developed by Car and Parrinello (1985). Consider what is happening in an *ab initio* MD simulation: as the nuclei move, the wavefunction for the system (as expressed in terms of a set of one-electron orbitals) will change in response to the new atomic positions. To calculate how the wavefunction changes, the Car–Parrinello method incorporates the one-electron orbitals into the dynamics of the system. To do this, we give the one-electron orbitals a fictitious mass (μ) and a fictitious kinetic energy. The Lagrangian therefore becomes

$$L = \frac{1}{2}\left(\sum_{j}^{\text{nuclei}} M_j \left(\frac{d\mathbf{R}_j}{dt}\right)^2 + \mu \sum_{i}^{\text{orbitals}} \int \left|\frac{d\psi_i}{dt}\right| dr\right) - E\left[\{\psi_i\},\{\mathbf{R}_j\}\right] \tag{1.81}$$

Note that, if the fictitious mass $\mu \rightarrow 0$, we approach the Born–Oppenheimer molecular dynamics. We impose the constraint that the one-electron orbitals are orthogonal (orthonormal):

$$\int \psi_i^*(\mathbf{r},t) \int \psi_k(\mathbf{r},t)d\mathbf{r} = \delta_{ik} \tag{1.82}$$

We then have equations of motion for the nuclei and the wavefunctions:

$$M_j \frac{d^2\mathbf{R}_j}{dt^2} = -\nabla_j E\left[\{\psi_i\},\{\mathbf{R}_j\}\right] + \sum_{ik} \Lambda_{ik} \frac{d}{d\mathbf{R}_j} \int \psi_i^* \psi_k d\mathbf{r} \tag{1.83}$$

$$\mu \frac{d^2\psi_i}{dt^2} = -\frac{\delta E}{\delta\psi_i^*} + \sum_{k}^{\text{orbitals}} \Lambda_{ik}\psi_k$$

where Λ_{ik} are the Lagrange multipliers for the orthonormalization constraint. The fictitious electron mass is chosen to be between 400 and 800 a.u. In typical simulations, a time step of 5 a.u. (0.12 fs) is reasonable. The Car–Parrinello method is implemented in the CPMD code (Marx and Hutter, 2009).

1.6   Available Quantum Chemistry Codes and Their Applications

Table 1.4 lists some of the more commonly used quantum chemistry codes and describes the level of theory (Hartree–Fock vs. DFT and Hybrid schemes) and the types of basis sets they employ. Many of these codes are freely available and most are well-documented. All of the codes can do static total energy calculations (e.g., energy as a function of geometry), geometry optimizations and the calculations of vibrational modes. Calculations on finite molecules (e.g., using GAUSSIAN or ADF) can be used to predict isotopic fractionation reactions (e.g., Jarzecki et al., 2004; Tossell, 2005; Rustad et al., 2010a, 2010b; Sherman, 2013) in gas-phase clusters. Usually, the calculation of thermodynamic quantities from vibrational modes is done automatically. Calculations on periodic systems can be used, for example, to predict phase transitions between solid minerals. Some of the periodic codes (e.g., VASP and CPMD) are primarily used for *ab initio* molecular dynamics simulations on liquids. Such calculations can be used to predict the properties of silicate melts in the Earth's interior (e.g., Karki and Stixrude, 2010) and metal–ligand complexation reactions in hydrothermal fluids (e.g., Mei et al., 2013).

Table 1.4 *Examples of current quantum chemistry software.*

Code	Basis set	Systems	Theory	Capabilities	Access and source URL
GAUSSIAN	Localized Gaussian	Finite	HF, DFT, and advanced	Molecular structures, energetics, and properties	Commercial www .gaussian. com
ADF	Slater orbitals	Finite	HF, DFT, and hybrid	Molecular structures, energetics, and properties	Commercial www.scm .com
GAMMES	Localized Gaussian	Finite	HF, DFT, and advanced	Molecular structures, energetics, and properties	Free upon registration www.msg .ameslab .gov
NWCHEM	Gaussian and plane waves	Finite and periodic	HF, DFT, hybrid, and advanced	Structures, energetics, and properties of molecules and periodic systems. BO molecular dynamics	Open source www .nwchem-sw .org
CRYSTAL	Localized Gaussian	Periodic	HF, DFT, and hybrid	Structures, energetics, and properties of periodic systems (crystalline solids and two-dimensional slabs)	Commercial (free to UK) www.crystal .unito.it
CPMD	Plane waves	Periodic	DFT	CP and BO molecular dynamics	Free upon registering cpmd.org
VASP	Plane waves	Periodic	HF, DFT, hybrids	Energetics and properties of periodic systems. BO molecular dynamics	Commercial www .vasp.at
Quantum Espresso	Plane waves	Periodic	DFT		Open source www .quantum-espresso.org
Abinit	Plane waves	Periodic	DFT LSD + U	Molecular and periodic systems. BO molecular dynamics	Open source www.abinit .org
CP2K (Quickstep)	Gaussian + plane waves	Periodic	HF, DFT, and hybrid	Molecular and periodic systems. BO molecular dynamics	Open source www.cp2k .org
CASTEP	Plane waves	Periodic	DFT	Energetics and properties of periodic systems. BO molecular dynamics	Commercial (free to UK) www.castep .org

References

Allen M. P. and Tildesley D. J. (1987) *Computer Simulation of Liquids*, Oxford University Press, New York.
Andersen H. C. (1980) Molecular dynamics simulations at constant pressure and/or temperature. *J. Chem. Phys.* **72**, 2384–2393.

Becke A. D. (1993) Density-functional thermochemistry. III. The role of exact exchange. *J. Chem. Phys.* **98**, 5648–5652.

Benedict W. S., Gailar N., and Plyler E. K. (1956) Rotational-vibration spectra of deuterated water vapor. *J. Chem. Phys.* **24**, 1139.

Berendsen H. J. C., Grigera J. R., and Straatsma T. P. (1987) The missing term in effective pair potentials. *J. Phys. Chem.* **91**, 6269–6271.

Bohr N. (1913) On the constitution of atoms and molecules. *Philos. Mag.* **26**, 1–25.

Bukowinski M. S. T. (1985) First principles equations of state of MgO and CaO. *Geophys. Res. Lett.* **12**, 536–539.

Car R. and Parrinello M. (1985) Unified approach for molecular dynamics and density functional theory. *Phys. Rev. Lett.* **55**, 2471–2747.

Catlow C. R. A., Thomas J. M., Parker S. C., and Jefferson D. A. (1982) Simulating silicate structures and the structural chemistry of pyroxenoids. *Nature* **295**, 668–662.

Clark S. J., Segall M. D., Pickard C. J., Hasnip P. J., Probert M. J., Refson K., and Payne M. C. (2005) First principles methods using CASTEP. *Zeitschrift fuer Krist.* **220**, 567–570.

Cramer C. J. and Truhlar D. G. (2009) Density functional theory for transition metals and transition metal chemistry. *Phys. Chem. Chem. Phys.* **11**, 10757–10816.

Davisson C. and Germer L. H. (1927) Diffraction of electrons by a crystal of nickel. *Phys. Rev.* **30**, 705–741.

De Broglie L. (1925) Rescherches sur la theorie des quanta. *Ann. Phys. (Paris).* **3**, 22–128.

Dirac P. A. M. (1925) The fundamental equations of quantum mechanics. *Proc. R. Soc. Lond. Ser. A Math. Phys. Eng. Sci.* **109**, 642–653.

Dovesi R., Orlando R., Erba A., Zicovich-Wilson C. M., Civalleri B., Casassa S., Maschio L., Ferrabone M., Pierre M. D., La D'Arco P., Noel Y., Causa M., Rerat M., and Kirtman, B. (2014) CRYSTAL14: A program for the ab initio investigation of crystalline solids. *Int. J. Quantum Chem.* **114**, 1287–1317.

Einstein A. (1905) Über einen die Erzeugung und Verwandlung des Lichtes betreffenden heuristischen Gesichtspunkt. *Ann. Phys.* **17**, 132–148.

Feynman R. P., Leighton R. B., and Sands M. (2011) *The Feynman Lectures on Physics, Vol. III: The New Millennium Edition: Quantum Mechanics (Volume 3)*, Basic Books, New York.

Frisch M. J., Trucks G. W., Schlegel H. B., Scuseria G. E., Robb M. A., Cheeseman J. R., Scalmani G., Barone V., Mennucci B., Petersson G. A., Nakatsuji H., Caricato M., Li X., Hratchian H. P., Izmaylov A. F., Bloino J., Zheng G., Sonnenberg J. L., Hada M., Ehara M., Toyota K., Fukuda R., Hasegawa J., Ishida M., Nakajima T., Honda Y., Kitao O., Nakai H., Vreven T., Montgomery J., Peralta J. E., Ogliaro F., Bearpark M., Heyd J. J., Brothers E., Kudin K. N., Staroverov V. N., Kobayashi R., Normand J., Raghavachari K., Rendell A., Burant J. C., Iyengar S. S., Tomasi J., Cossi M., Rega N., Millam J. M., Klene M., Knox J. E., Cross J. B., Bakken V., Adamo C., Jaramillo J., Gomperts R., Stratmann R. E., Yazyev O., Austin A. J., Cammi R., Pomelli C., Ochterski J. W., Martin R. L., Morokuma K., Zakrzewski V. G., Voth G. A., Salvador P., Dannenberg J. J., Dapprich S., Daniels A. D., Farkas Ö., Foresman J. B., Ortiz J. V., Cioslowski J., and Fox D. J. (2009) Gaussian 09, Revision D.01.

Gibbs G. V. (1982) Molecules as models for bonding in silicates. *Am. Mineral.* **67**, 421–450.

Goldschmidt V. (1937) The principles of distribution of chemical elements in minerals and rocks. The seventh Hugo Müller Lecture, delivered before the Chemical Society. *J. Chem. Soc.*, 655–673.

Hafner J. (2008) Ab-initio simulations of materials using VASP: Density-functional theory and beyond. *J. Comput. Chem.* **29**, 2044–2078.

Haile J. M. (1992) *Molecular Dynamics Simulation*, John Wiley & Sons, Inc, New York.

Heisenberg W. (1925) Über quantentheoretische Umdeutung kinematischer und mechanischer Beziehungen. *Zeitschrift für Phys.* **33**, 879–893.

Heitler W. and London F. (1927) Wechselwirkung neutraler Atome und homopolare Bindung nach der Quantenmechanik. *Zeitschrift für Phys.* **44**, 455–472.

Helgeson H. C. and Kirkham D. H. (1974) Theoretical prediction of the thermodynamic behavior of aqueous electrolytes at high pressures and temperatures: II Debye-Huckel parameters for activity coefficients and relative partial molal properties. *Am. J. Sci.* **274**, 1199–1261.

Hohenberg P. and Kohn W. (1964) Inhomogeneous electron gas. *Phys. Rev.* **136**, B864–B871.

Hoover W. G. (1985) Canonical dynamics: Equilibrium phase-space distributions. *Phys. Rev. A* **31**, 1695–1697.

Jarzecki A. A., Anbar A. D., and Spiro T. G. (2004) DFT analysis of $Fe(H_2O)_6^{3+}$ and $Fe(H2O)_6^{2+}$ structure and vibrations; implications for isotope fractionation. *J. Phys. Chem. A* **108**, 2726–2732.

Karki B. B. and Stixrude L. P. (2010) Viscosity of $MgSiO_3$ liquid at Earth's mantle conditions: Implications for an early magma ocean. *Science* **328**, 740–742.

Kohn W. and Sham I. J. (1965) Self-consistent equations including exchange and correlation effects. *Phys. Rev.* **140**, A1133–A1138.

Kwon K. D., Refson K., and Sposito G. (2009) Zinc surface complexes on birnessite: A density functional theory study. *Geochim. Cosmochim. Acta* **73**, 1273–1284.

Lasaga A. C. and Gibbs G. B. (1990) Ab initio quantum mechanical calculations of water-rock interactions: Adsorption and hydrolysis reactions. *Am. J. Sci.* **290**, 263–295.

Lee C. T., Yang W. T., and Parr R. G. (1988) Development of the Colle-Salvetti correlation-energy formula into a functional of the electron-density. *Phys. Rev. B* **37**, 785–789.

Lennard-Jones J. E. (1929) The electronic structure of some diatomic molecules. *Trans. Faraday Soc.* **25**, 665–668.

Lewis G. N. (1923) *Valence and the Structure of Atoms and Molecules*, The Chemical Catalog Company, Inc., New York.

Marx D. and Hutter J. (2009) *Ab Initio Molecular Dynamics: Basic Theory and Advanced Methods*, Cambridge University Press, Cambridge.

Mei Y., Sherman D. M., Liu W., and Brugger J. (2013) Ab initio molecular dynamics simulation and free energy exploration of copper(I) complexation by chloride and bisulfide in hydrothermal fluids. *Geochim. Cosmochim. Acta* **102**, 45–64.

Mei Y., Sherman D. M., Liu W., Etschmann B., Testemale D., and Brugger J. (2015) Zinc complexation in chloride-rich hydrothermal fluids (25–600°C): A thermodynamic model derived from ab initio molecular dynamics. *Geochim. Cosmochim. Acta* **150**, 265–284.

Nose S. (1984) A unified formulation of the constant temperature molecular-dynamics methods. *J. Chem. Phys.* **91**, 511–519.

Oganov A. R. and Ono S. (2004) Theoretical and experimental evidence for a post-perovskite phase of $MgSiO_3$ in Earth's D'' layer. *Nature* **430**, 445–448.

Oganov A. R. and Price G. D. (2005) Ab initio thermodynamics of $MgSiO_3$ perovskite at high pressures and temperatures. *J. Chem. Phys.* **122**, 124501.

Parr R. G. and Yang W. (1989) *Density-Functional Theory of Atoms and Molecules*, Oxford University Press, New York.

Pauling L. (1929) The principles determining the structure of complex ionic crystals. *J. Am. Chem. Soc.* **51**, 1010–1026.

Perdew J. P. (1986) Density functional theory and the band gap problem. *Int. J. Quantum Chem.* **19**, 497–523.

Perdew J. P. and Zunger A. (1981) Self-interaction correction to density-functional approximations for many-electron systems. *Phys. Rev. B* **23**, 5048–5079.

Perdew J. P., Burke K., and Ernzerhof M. (1996) Generalized gradient approximation made simple. *Phys. Rev. Lett.* **77**, 3865–3868.

Planck M. (1900) Über eine Verbesserung der Wienschen Spektralgleichung. *Verhandlungen der Dtsch. Phys. Gesellschaft* **2**, 202–204.

Ponder J. W. and Case D. A. (2003) Force fields for protein simulations. *Adv. Protein Chem.* **66**, 27–85.

Rustad J. R., Bylaska E. J., Jackson V. E., and Dixon D. A. (2010a) Calculation of boron-isotope fractionation between $B(OH)_3$(aq) and B(OH)4 -(aq). *Geochim. Cosmochim. Acta* **74**, 2843–2850.

Rustad J. R., Casey W. H., Yin Q. Z., Bylaska E. J., Felmy A. R., Bogatko S. A., Jackson V. E., and Dixon D. A. (2010b) Isotopic fractionation of Mg^{2+} (aq), Ca^{2+} (aq), and Fe^{2+} (aq) with carbonate minerals. *Geochim. Cosmochim. Acta* **74**, 6301–6323.

Schrödinger E. (1926) Quantisierung als Eigenwertproblem. *Ann. Phys.* **79**, 361–376.

Sherman D. M. (1991) The high-pressure electronic structure of magnesiowustite (Mg, Fe) O: Applications to the physics and chemistry of the lower mantle. *J. Geophys. Res. Solid Earth* **96**, 14299–14312.

Sherman D. M. (2001) Quantum chemistry and classical simulations of metal complexes in aqueous solutions. *Mol. Model. Theory Appl. Geosci.* **42**, 273–317.

Sherman D. M. (2007) Complexation of Cu+ in Hydrothermal NaCl Brines: Ab initio molecular dynamics and energetics. *Geochim. Cosmochim. Acta* **71**, 714–722.

Sherman D. M. (2009) Electronic structures of siderite (FeCO3) and rhodochrosite (MnCO3): Oxygen K-edge spectroscopy and hybrid density functional theory. *Am. Mineral.* **94**, 166–171.

Sherman D. M. (2013) Equilibrium isotopic fractionation of copper during oxidation/reduction, aqueous complexation and ore-forming processes: Predictions from hybrid density functional theory. *Geochim. Cosmochim. Acta* **118**, 85–97.

Sherman D. M. and Randall S. R. (2003) Surface complexation of arsenic(V) to iron(III) (hydr)oxides: Structural mechanism from ab initio molecular geometries and EXAFS spectroscopy. *Geochim. Cosmochim. Acta* **67**, 4223–4230.

Sherrill D. C. and Schaefer H. F. (1999) The configuration interaction method: Advances in highly correlated approaches. In *Advances in Quantum Chemistry*, Academic Press, Cambridge, MA, pp. 143–269.

Smith D. E. and Dang L. X. (1994) Computer simulations of NaCl association in polarizable water. *J. Chem. Phys.* **101**, 3757–3766.

Soler J. M., Artach E., Gale J. D., Garcia A., Junquera J., Ordejon P., and Sanchez-Portal D. (2002) The SIESTA method for ab initio order-N materials simulation. *J. Phys. Condens. Matter* **14**, 2745–2779.

Sprik M. and Ciccotti G. (1998) Free energy from constrained molecular dynamics. *J. Chem. Phys.* **109**, 7737–7744.

Stowasser R. and Hoffmann R. (1999) What do the Kohn-Sham orbitals and eigenvalues mean? *J. Am. Chem. Soc.* **121**, 3414–3420.

Te Velde G., Bickelhaupt F. M., Baerends E. J., Fonseca Guerra C., van Gisbergen S. J. A., Snijders J. G., and Ziegler T. (2001) Chemistry with ADF. *J. Comput. Chem.* **22**, 931–967.

Tossell J. (2005) Calculating the partitioning of the isotopes of Mo between oxidic and sulfidic species in aqueous solution. *Geochim. Cosmochim. Acta* **69**, 2981–2993.

Tossell J. A. and Gibbs G. V. (1977) Molecular-orbital studies of geometries and spectra of minerals and inorganic compounds. *Phys. Chem. Miner.* **2**, 21–57.

Tossell J. A., Vaughan D. J., and Johnson K. H. (1973) Electronic structure of ferric iron octahedrally coordinated to oxygen. *Nature* **244**, 42–45.

Valiev M., Bylaska E. J., Govind N., Kowalski K., Straatsma T. P., Van Dam H. J. J., Wang D., Nieplocha N., Apra E., Windus T. L., and de Jong W. A. (2010) NWChem: A comprehensive and scalable open-source solution for large scale molecular simulations. *Comput. Phys. Commun.* **181**, 1477.

Vanderbilt D. (1990) Soft self-consistent pseudopotentials in a generalized eigenvalue formalism. *Phys. Rev. B* **41**, 7892.

Vosko S. H., Wilk K., and Nusair M. (1980) Accurate spin-dependent electron liquid correlation energy for local spin density calculations: A critical analysis. *Can. J. Phys.* **58**, 1200–1205.

2

Force Field Application and Development

Marco Molinari,[1] Andrey V. Brukhno,[1] Stephen C. Parker,[1] and Dino Spagnoli[2]

[1]*Department of Chemistry, University of Bath, Bath, UK*
[2]*School of Chemistry and Biochemistry, University of Western Australia, Crawley, Western Australia, Australia*

2.1 Introduction

The need to simulate molecular systems composed of a large number of species in many different configurations is often essential for correctly interpreting experimental data and for predicting properties of complex systems. The quality of atomistic simulations depends on the accuracy to which the interatomic forces reproduce those of the specific elements. In this chapter, we describe ways of modelling the interatomic interaction and show how they can be applied successfully for a wide range of minerals, including silicates, carbonates and clay minerals.

Molecular modelling provides the molecular specificity for performing calculations on materials where the use of quantum mechanics (QM) would be computationally prohibitive. In addition, molecular modelling accounts for all force field (FF) methods, including all-atom (AA) and coarse grain (CG) methods, where all the interactions within a system are parameterised either analytically or numerically. An FF is defined as a collection of functional forms and appropriate parameters needed to provide a description of the potential energy landscape of a system of interacting particles. The physical meaning of these parameters is typically straightforward in an FF, as each functional form represents a type of interaction, for example, bonds, angles, torsions and inversions. Of course, an FF must realistically describe the system of interest. Generally, the success of an FF is related to the ease and spread of its usage, how well it can be combined with other FFs to represent all the interactions within a system and its transferability to different problems. There are different types of FFs; some limit the set of simulated atomic species in favour of high accuracy (specific FF), others lower the accuracy in favour of a broader set of atomic species and transferability (universal FF (UFF)), while still others go further and associate the species with groups of atoms (CG FF).

Molecular Modeling of Geochemical Reactions: An Introduction, First Edition. Edited by James D. Kubicki.
© 2016 John Wiley & Sons, Ltd. Published 2016 by John Wiley & Sons, Ltd.

Terminology changes with the community; 'FF' is a term developed in the organic/biological community while in the material/mineral physics community is called a 'potential model'. It is only in the last decades with the advent of hybrid systems that the term 'FF' has been more generally applied to minerals. The development of FFs started in 1950 for describing the underlying interactions in gases and liquids, where the Lennard–Jones potential was found to be both adequate and computationally feasible (Verlet, 1967, 1968; Wood and Parker, 1957) in both molecular dynamics (Allen and Tildesley, 1989; Frenkel and Smith, 1996) and Monte Carlo methodologies (Metropolis et al., 1953; Rosenbluth and Rosenbluth, 1954). Nowadays, a large variety of FFs are available for a wide range of systems such as inorganic, organic and biological systems, and their success is judged by their accuracy in representing the physical and chemical forces.

The observed experimental properties are a result of such forces, and even with the advent of improved experimental techniques and elaborate methodologies, modelling can still be beneficial. As functional structures became more and more important and complex for a variety of applications from biomedical, to catalysis, to energy conversion, FFs have been developed to integrate inorganic, organic and biological parts. Lennard–Jones (LJ)-based FFs have been well developed in the bio-organic community pushed by the strength of the pharmaceutical community, which funded many of these developments. Historically, the mineralogical community has received less funding for large-scale projects and hence has tended to develop potential models for specific systems or tasks rather than towards large libraries of potential parameters that could be used for a large number of minerals. In the materials community, the Buckingham potential form was the most used due to the preference (and better physical significance) for exponential-like decay of the repulsion between electron charge clouds rather than the r^{-12} term of the LJ potential. However, with more research focussing on organic–inorganic composites and biomedical systems, the community has moved towards FFs that treat a large number of minerals and materials using a straightforward way of mixing the potential parameters with the organic counterpart.

A number of general FFs (e.g. UFF (Rappe et al., 1992) and ESFF (Barlow et al., 1996)) have been derived to cover the entire periodic table, but they have not been entirely successful because the models used have not captured the complexity and diversity of the elements. More specific and well-established FFs are available for organic and biological molecules such as MM2 (Allinger, 1977), MM3 (Allinger et al., 1989), AMBER (Pearlman et al., 1995; Weiner et al., 1986), CHARMM (Brooks et al., 1983; MacKerell et al., 1998), MMFF (Halgren, 1996), OPLS (Jorgensen and Swenson, 1985; Jorgensen et al., 1996), VFF (Lifson et al., 1979), CVFF (Dauberosguthorpe et al., 1988), DREIDING (Mayo et al., 1990) and GROMOS (Hermans et al., 1984), which are capable, to different degrees of accuracy, of modelling all important species involved in molecular systems. In the last 10 years, research has successfully managed to build FFs for minerals that can also be used with organic/biological FFs (e.g. INTERFACE (Heinz et al., 2013) and CLAYFF (Cygan et al., 2004)). Such FFs are still in their infancy due to the complexity of the bonding in minerals. Indeed, potential parameters tend to be material dependent with limited transferability and poor representation outside their field. This is clearly the main drawback that still results in requirement to develop specific potential models for different minerals and even for particular properties of a given mineral.

FF-based molecular mechanics (MM) has the advantage of being able to deal with large systems and larger time scales compared to the QM counterpart and can be used for a wider range of systems, from inorganic, to organic molecules, to bio-macromolecules. However, it is not without limitations. Electronic transitions and electron transport-related phenomena are still beyond the capability of most of the FFs. Furthermore, to obtain a high-quality output, it is extremely important that the quality of the input is high, as an FF is a collection of empirically derived parameters. Therefore, it is important that an FF is reliable and accurate, transferable and not over-parameterised. These conditions are essential for the success of an FF.

In this chapter, we describe a number of the most successful FFs used for geological systems, including the types of data that can be generated. For more general discussions, see the books by Leach (2001) and Finnis (2003). Before giving specific examples, we describe the common potential forms currently used for AA and CG models, the issues of combining potential parameters and the techniques for deriving and testing both AA and CG FFs.

2.2   Potential Forms

Commonly, FFs reflect atom types, which are assigned based on atomic properties (charge and weight), bonding and local environment. Clearly, atoms interact via non-bonded interactions, while if part of a molecule or polyatomic ion, they may also interact with neighbours through bonded interactions. Non-bonded interactions represent the electrostatic, short-range repulsive and attractive van der Waals interactions.

2.2.1   The Non-bonded Interactions

2.2.1.1   *Coulombic Interactions*

The long-range interaction normally accounts for the majority of the interaction energy in a mineral system and can be written as in Equation 2.1, considering the net charges q_i and q_j and the interatomic distance between them r_{ij}:

$$U_C = \frac{e^2}{4\pi\varepsilon_0} \sum_{ij} \frac{q_i q_j}{r_{ij}} \tag{2.1}$$

The first term includes the electron charge and permittivity of free space ε_0. The Coulombic term is conditionally convergent – it might never converge or in fact diverge – with $1/r_{ij}$. Different methods can be applied to speed up convergence and were derived for 3D systems by Ewald (1921) and 2D systems by Parry (1975, 1976). The interaction between two charge distributions however not only is defined by the charge–charge interaction but also accounts for the interactions between the charges and the multipoles and the interactions between the multipoles themselves.

2.2.1.2   *Short-Range Two-Body Non-bonded Interactions*

The short-range two-body terms represent partially the repulsion between electron charge clouds and attractive van der Waals interactions. The Lennard–Jones (LJ) functional form is the simplest and most widely used model to describe the interaction energy between two particles, i and j at distance r_{ij} (Jones, 1924a, 1924b). The LJ potential can be written in three equivalent forms (Eqs. 2.2–2.4) with simple rules that allow for switching between the different forms (Eqs. 2.5–2.8). For each interaction between i and j, ε is the depth of the potential energy well, σ is the distance at which the potential between the particles is zero, and r_0 is the equilibrium distance at which the potential reaches the minimum. The r^{-12} and r^{-6} terms represent the Pauli repulsion at short distance and the van der Waals dispersion forces at longer separations, respectively (Figure 2.1):

$$U_{(r_{ij})\text{LJ}} = 4\varepsilon_{ij} \left[\left(\frac{\sigma_{ij}}{r_{ij}} \right)^{12} - \left(\frac{\sigma_{ij}}{r_{ij}} \right)^6 \right] \tag{2.2}$$

$$U_{(r_{ij})\mathrm{LJ}} = \varepsilon_{ij}\left[\left(\frac{r_0}{r_{ij}}\right)^{12} - 2\left(\frac{r_0}{r_{ij}}\right)^{6}\right] \tag{2.3}$$

$$U(r_{ij})_{\mathrm{LJ}} = \frac{A_{ij}}{r_{ij}^{12}} - \frac{B_{ij}}{r_{ij}^{6}} \tag{2.4}$$

$$r_0 = \sigma\sqrt[6]{2} \tag{2.5}$$

$$A = 4\varepsilon\sigma^{12}, \quad B = 4\varepsilon\sigma^{6} \tag{2.6}$$

$$A = \varepsilon r_0^{12}, \quad B = 2\varepsilon r_0^{6} \tag{2.7}$$

$$\varepsilon = \frac{B^2}{4A}, \quad r_0 = \sqrt[6]{\frac{2A}{B}} \tag{2.8}$$

The Buckingham functional form is a widely used alternative and is similar to the LJ form, but the repulsive term decays exponentially (Eq. 2.9) with A_{ij} and ρ_{ij} related to ion size and hardness. The Buckingham potential can be rewritten using Equation 2.10 (Gilbert, 1968):

$$U(r_{ij}) = A_{ij}\exp^{(-r_{ij}/\rho_{ij})} - \frac{c_{ij}}{r_{ij}^{6}} \tag{2.9}$$

$$U(r_{ij}) = f(b_i + b_j)\exp^{(a_i + a_j - r_{ij}/b_i + b_j)} - \frac{c_i c_j}{r_{ij}^{6}} \tag{2.10}$$

The disadvantage of this functional form is that at very small, normally unphysical, distances (i.e. <1 Å), the energy tends to negative infinity if the c_{ij} or c_i and c_j are greater than 0. Thus, other adaptations include either to make c_{ij} distance dependent and converging to zero when $r = 0$ or to replace the r^{-6} term. One example is the Morse potential, which can also be used for bonded inter-actions (Eq. 2.11 or 2.12), with ε_{ij} being the depth of the potential well and a_{ij} related to the curvature of the potential energy well (Eq. 2.13), μ the reduced mass and ω the frequency of the (bond) vibra-tion (Pedone et al., 2006):

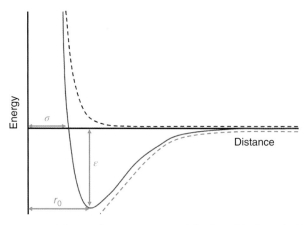

Figure 2.1   *Lennard–Jones potential curve between two particles in red is decupled in the two contributions: the attraction in green and the repulsion in blue.*

$$U\left(r_{ij}\right) = \varepsilon_{ij}\left[1 - \exp^{-a_{ij}\left(r_{ij}-r_0\right)}\right]^2 - \varepsilon_{ij} \tag{2.11}$$

$$U\left(r_{ij}\right) = 4\varepsilon_{ij}\left(\exp^{\frac{2a_{ij}}{r_0}\left(1-\frac{r_{ij}}{r_0}\right)} - \exp^{\frac{a_{ij}}{r_0}\left(1-\frac{r_{ij}}{r_0}\right)}\right) \tag{2.12}$$

$$a_{ij} = \omega\left(\mu/2\varepsilon_{ij}\right)^{0.5} \tag{2.13}$$

2.2.2 The Bonded Interactions

FFs for minerals also include functional forms representing covalently bonded interactions. An alternative two-body term to the Morse potential is the harmonic potential, which links two atoms involved in a covalent bond through a force constant k_{ij} and applies an energy penalty when the bond length deviates from the equilibrium distance r_0 (Eq. 2.14):

$$U\left(r_{ij}\right) = \frac{1}{2}k_{ij}\left(r_{ij}-r_0\right)^2 \tag{2.14}$$

Equation 2.14 requires only two parameters, compared to the three for the Morse potential, but unlike the Morse potential, it cannot account for bond breaking (Figure 2.2). Another important component of covalent bonding is the angle-dependent three-body terms, which are included to describe the angle bend of covalently bonded molecules (Figure 2.2). In a similar way to the two-body functional, they are often represented by a harmonic functional form with a force constant k_{ijk} and any deviation from the equilibrium angle θ_0 adding an energy penalty (Eq. 2.15):

$$U\left(\theta_{ijk}\right) = \frac{1}{2}k_{ijk}\left(\theta_{ijk}-\theta_0\right)^2 \tag{2.15}$$

Many other functional forms are needed to simulate complex organic and biological molecules including dihedral and inversion.

2.2.3 Polarisation Effects

Another important many-body term that some FFs seek to represent is electronic polarisation. The polarisation energy results from the change in electrostatic interaction between atoms due to the field of their neighbours and/or the distortion of the electron charge cloud from spherical symmetry. This polarisation effects can be captured with different approaches, which mimic the effect of the electronic relaxation (Dykstra, 1993). One is via charge equilibration where the charge on each atom changes and it is incorporated in reactive FFs. Another is via the induced multipoles (Wilson et al., 1996a), which are normally truncated at the level of dipolar interactions (Dick and Overhauser, 1958).

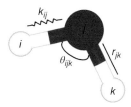

Figure 2.2 *Schematic representation of the harmonic bond, k_{ij}, distance, r_{ij}, and angle, θ_{ijk}, depicted on the water molecule (oxygen in red and hydrogen in white).*

However, the simplest type of FF, namely, the rigid ion model (RIM) (or point-charge electrostatic model), neglects the polarisability. RIM has the disadvantage of generating a crystal with a unitary high-frequency dielectric constant matrix and often has the consequence of overestimating the high-frequency vibrational modes and defect energies (Woods et al., 1960). Arguably, a way of implicitly including the polarisability into RIM is to reduce the charges on the atoms and therefore use scaled formal charges (Spagnoli and Gale, 2012). This is equivalent to having full valence charges screened by a dielectric constant.

The simplest model of introducing polarisation effect is the point-polarisable ion model (PPIM) in which a dipole is induced by a fixed polarisability. Instabilities can arise when species relax to close distances, which in extreme cases leads to a polarisation catastrophe (Faux, 1971). A way of overcoming the instability at short distances and accounting for the induced dipoles is the shell model (Dick and Overhauser, 1958), as shown in Figure 2.3. The approach assumes that the atomic species comprise a core representing the nucleus and inner electrons, which carries the entire atomic mass and a massless shell that represents the valence electrons. The core and shell are coupled via a force constant. The resulting polarisability is therefore dependent on the environment unlike in the PPIM. The major disadvantage, however, is the computational cost in deriving and using such models, which is not the case for the point ion dipolar model (Applequist, 1985). Hence, the shell models are not widely used in mineral system at present.

There are series of models containing an increasing degree of sophistication when accounting for multipoles and ion-shape changes (Aguado et al., 2003; Wilson et al., 1996b, 2004). The *quadrupole-polarisable aspherical ion model* (QUAIM) includes the ion-shape deformation up to the quadrupole order and the polarisation effects. Four parts need to be evaluated, which include charge–charge (Coulombic interaction), dispersion (dipole–dipole and dipole–quadrupole terms), overlap repulsion (short-range repulsion, e.g. Born–Mayer exponential) and polarisation effects (dipolar and quadrupolar contributions). A less computationally demanding model is the *dipole-polarisable ion model* (DIPPIM), which includes dipole polarisation effects only. There are also more complicated models, which can account for only ion compression effects (CIM) or a mixture of them and dipolar and quadrupolar interaction (QUAD-CIM) and ion-shape deformation (QUAIM). Wilson et al. (1996b) made a critical assessment of the different models on stoichiometric ZrO_2, whose ground state is monoclinic. They found that simple RIM predicts the ground state of ZrO_2 to be a rutile orthorhombic symmetry. However, on including the compression effects (CIM), the lower coordination (rutile) is destabilised over the higher coordination of Zr atoms (fluorite and monoclinic) and the ground-state results of cubic symmetry. It was only with the addition of both dipoles and quadrupoles (QUAD-CIM) that the monoclinic ground state was obtained due to the more

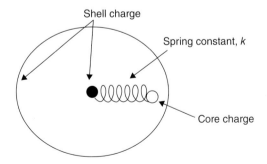

Figure 2.3   *Schematic representation of the shell model.*

asymmetric environment of the anions in the monoclinic structure compared to the rutile and fluorite structures.

2.2.4 Reactivity

Classical non-reactive FFs (i.e. where covalent bonds cannot be broken or formed) are commonly used to describe systems that are near equilibrium and have fixed charges and no reactivity. However, when pH effects or non-equilibrium processes are under study, reactive FFs provide a method to include the forming/breaking of bonds and sampling transition states of chemical reactions, therefore allowing the system to move away from equilibrium. As geological and geochemical events can profoundly be affected by surrounding aqueous environment, it is important to understand the structure, composition and reactivity at these interfaces. The effect is particularly marked when a significant number of covalent bonds are formed and broken during the course of the reactions, which leads to the formation of a significant number of more or less complex transition states. Approaches to tackle reactivity in QM are available (Santiso and Gubbins, 2004; Taylor and Neurock, 2005), but their widespread usage is still limited because of the computational expense.

There are two types of approaches in dealing with reactivity: the first is to assume fixed charges allowing molecules to dissociate and modify the bonding of the species using the coordination. Examples of this type include the work of Rustad (Rustad et al., 1996) and Garofalini (Lockwood and Garofalini, 2009). Alternatively, the models can include mixing of either state or charge redistribution and coordination dependence, such as the empirical valence bond (EVB) (Villà and Warshel, 2001; Warshel and Weiss, 1980) and the ReaxFF methodology (Mueller et al., 2010; van Duin et al., 2001). A full review of the methodologies and other reactive potentials is beyond the scope of this chapter, but Liang et al. provide an excellent review on this subject (Liang et al., 2013).

The EVB approach (Villà and Warshel, 2001; Warshel and Weiss, 1980) was originally developed for reactions in solution and in enzymes. Most implementations of EVB are based on the assumption that the environmental effect behind the reaction, and therefore the resulting potential energy surface (PES) of the reaction, is the consequence of the interaction between the PES of the reactants and the products, thus providing a way of constructing the PES of a reacting system. There is no conceptual limitation to its usage in reactions at the mineral–water interfaces, where, for example, proton exchange controls mineral dissolution and precipitation and sorption of species including organic matter and water. Here the challenge in constructing the PES for the proton transfer reactions is a combination of multiple issues including the small mass of the proton, quantum effects and small energy differences between the different states. The composition of surfaces due to pH effects leads to a local environment that is rich in protonated or charged species, which diffuse via a Grotthuss mechanism rather than a simple ion diffusion, and at high pH, greater than the pK_a of the surface species, one would expect the local environment to have high concentration of hydroxyl groups. A series of EVB model has been developed during the years for both the hydronium and the hydroxyl species (Brancato and Tuckerman, 2005; Lobaugh and Voth, 1996; Schmitt and Voth, 1999; Ufimtsev et al., 2009; Vuilleumier and Borgis, 1999; Wu et al., 2008). The EVB approach has been established since 1980, and while there is no debate on the validity of the overall method, the parameters used to reproduce the correct *ab initio* potential surface must be chosen carefully to avoid unjustified results (Kamerlin et al., 2009; Valero et al., 2009).

ReaxFF, which was originally developed by van Duin and co-workers (Mueller et al., 2010; van Duin et al., 2001), allows for a description of a variety of chemical environments; for example, a simulation using this method could involve a fully covalent hydrocarbon system (Chenoweth et al., 2008) or a completely ionic metal oxide (Raymand et al., 2008). ReaxFF uses bond order combined with bond distance relationships to retain important bond orders at transition-state geometries (Liang et al., 2013). These many-body empirical potential terms describe the total energy of the system as in Equation 2.16. E^{self} is the difference in energy between the various charge states of an atom. E^{Coul} and E^{vdW} describe the Coulombic and non-bonded interactions. E^{bond} contains the bond-order terms (to describe the reaction barriers accurately, the bond-order terms are corrected for over coordination). $E^{\text{angle}}, E^{\text{torsion}} \ E^{\text{conjugation}}, E^{\text{H-bond}}, E^{\text{lone-pair}} + E^{\text{over}} + E^{\text{under}}$ and E^{others} are terms related to the valence and torsion angle distortion, three and four-body conjugation, weak hydrogen bonding term, number of valence electrons around an atom and several other including corrections for allene-type, terminal triple bonds and C_2 species, respectively:

$$E_{\text{system}} = E^{\text{self}} + E^{\text{Coul}} + E^{\text{vdW}} + E^{\text{bond}} + E^{\text{angle}} + E^{\text{torsion}} + E^{\text{conjugation}}$$
$$+ E^{\text{H-bond}} + E^{\text{lone-pair}} + E^{\text{over}} + E^{\text{under}} + E^{\text{others}}$$

$$(2.16)$$

The energy terms are all bond-order dependent apart from the van der Waals and Coulombic terms and upon bond dissociation, the terms that rely on bond order disappear. The ReaxFF methodology has been associated with the fields of organic chemistry and catalysis but is starting to be used more and more in the field of geochemistry because, like the EVB approach, it can be used for larger systems and longer time frames than *ab initio* molecular dynamics (AIMD). In Section 2.5.1, we discuss the scope of the ReaxFF by describing its application to calcium carbonate.

The main disadvantage of the reactive FFs is the difficulty in deriving parameters, checking their transferability and to a lesser extent the computational cost. There are many problems where the emergent behaviour is more dependent on size and time. The challenge is then to approximate as much as possible retaining the essential selectivity, which is the basis of CG.

2.2.5 Fundamentals of Coarse Graining

CG modelling addresses one of the main unresolved issues in molecular simulations: the time and length scale difference between computational and experimental methods. AA simulations are currently limited to system sizes of nanometres and time scale of nanoseconds for complex FFs and to micrometres and milliseconds for the simplest models (Reith et al., 2003; Ruehle et al., 2009), the latter suggesting a route for accessing even larger times and sizes by simplifying further while maintaining the essential molecular specificity. The approach, coarse graining, reduces the number of DOF with respect to an AA model, which effectively increases the simulated time and length scale orders of magnitude (Figure 2.4). The aim of the CG approach is, therefore, to develop a less expensive (per unit length and time compared to AA modelling) and faster (in terms of accessible times) models while maintaining the correct physical behaviour of the system. Typically, this is done by simulating, in the first place, a relatively small but representative AA system (Lyubartsev, 2005; Mirzoev and Lyubartsev, 2014; Reith et al., 2003; Ruehle et al., 2009; Wang and Deserno, 2010), which provides the set of effective interactions for a CG model, which is then applied to significantly larger systems that would not be feasible for the AA treatment. Clearly, if the physics of the system is highly sensitive to the small-scale phenomena, the coarse graining may not be straightforward. To tackle such intricate cases, systematic CG methodologies have been developed, as opposed to conventional, generic CG models exploiting only hard and soft spheres; in the new generation of CG models, one can introduce reactive or mutating beads or resort to hybrid AA–CG simulations.

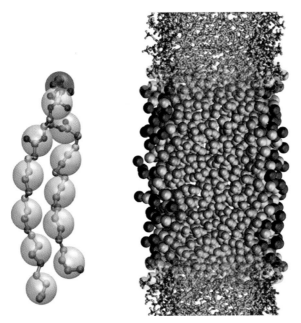

Figure 2.4　*Illustration of coarse grain (CG) modelling in the case of DOPC lipids: CG models of a single molecule and a self-assembled bilayer (membrane) are schematically depicted. The lamping of atomic groups into spherical CG units and the resulting CG representation are shown.*

There are three main steps in defining a CG model. The first step is the definition of CG particles, also referred to as beads or superatoms, which represent groups of atomic species with similar chemical structure. The choice of the atoms to be united into specific CG beads is essentially arbitrary. The obvious and relatively straightforward choice is to map common chemical structures into same CG types, for instance, polar, non-polar and charged beads (see examples of phospholipid models (Lyubartsev, 2005; Mirzoev and Lyubartsev, 2014; Marrink et al., 2004, 2007; Wang and Deserno, 2010; Wu et al., 2011) and other organic compounds (Fischer et al., 2008; Huang et al., 2010; Lee et al., 2009; Markutsya et al., 2013; Milani et al., 2011; Prasitnok and Wilson, 2013; Ruehle et al., 2009; Wang et al., 2011)). Superatoms can be as small as a few atoms or as big as hundreds species (Carof et al., 2014; Suter et al., 2009). Thus, monomers within an organic polymer molecule can be defined as single CH_3 and CH_2 units, whereas a higher level CG model could include several structural units within the same CG bead (Carof et al., 2014; Ramirez-Hernandez et al., 2013; Suter et al., 2009). Inadequate AA-to-CG mapping can lead to artefacts in CG simulations (Lyubartsev, 2005; Markutsya et al., 2013; Reith et al., 2003).

The second step is to define the potential functions, or Hamiltonian, for the CG model, which specifies the interactions between the superatoms and has to be chosen such that the properties of the CG system are representative of the original AA system or the corresponding experiment. A few numerical schemes are available for deriving effective pair interactions between the CG beads (see Section 2.3.5).

The third step is the definition of the dynamical equations suitable to evaluating the dynamical properties of the system. Currently, there are only limited methods to address this matter rigorously, which often employ stochastic thermostats and Brownian or dissipative particle dynamics (DPD) (Flekkoy et al., 2000; Frenkel and Smith, 1996; Lyubartsev et al., 2003).

The advantage of CG is that the behaviour of an evolving system is captured on a large time and length scales and, if coupled with AA models, it could provide an acceptable microscopic description of the system with atomistic detail where needed. However, the methodology is still subject to the reliability and accuracy of the descriptors of interaction between the beads – the descriptors, which are normally derived by using atomistic level techniques. Furthermore, the reduction of the DOF of the system can inadvertently lead to inaccurate or otherwise misleading results, as the choice of the 'essential' DOF remains driven in large part by common sense and depends on physical intuition. It is not surprising therefore that multi-scale modelling via hybrid AA–CG simulation methods has been attempted (Abrams and Kremer, 2003; Rzepiela et al., 2011) and is gaining increasingly more attention and research effort (di Pasquale et al., 2012; Srivastava and Voth, 2013). For an overview of modern structure-fitting CG methods and models, see the recent reviews in (Brini et al., 2013; Noid, 2013).

2.3 Fitting Procedure

We have seen that an FF is composed of parameterised analytical equations. These parameters can, of course, be modified to gain enhanced performance of the FF. Furthermore, as these equations often correspond to DOF of the system such as bond lengths, angles and torsion, obviously, visualising the corresponding macroscopic phenomena associated with the functional form helps with the optimisation of the parameters, the so-called fitting. We can consider two different scenarios: the first, where the FFs for different parts of the system are known and therefore combination rules can in principle be applied, and the second, where there is no FF available to describe the system.

2.3.1 Combining Rules Between Unlike Species

A great deal of effort has been put to deriving mixing rules for FF parameters by applying simple rules and thereby reducing the number of independent parameters. Mixing rules allow the combination of potential parameters for single atomic species to obtain all possible atom–atom interactions. It should be emphasised that it is essential to provide some evidence that the mixing rule is applicable to the system under consideration. The most common form used for mixing is the non-bonded LJ parameters assuming that the charges are constant. Conventional rules include the Lorentz arithmetic mean rule (Lorentz, 1881) (Eq. 2.17) and the Berthelot geometric mean rule (Eq. 2.18) (Berthelot, 1889; Delhommelle and Millie, 2001):

$$r_{ij} = \frac{1}{2}\left(r_{ii} + r_{jj}\right) \tag{2.17}$$

$$\varepsilon_{ij} = \left(\varepsilon_{ii}\varepsilon_{jj}\right)^{0.5} \tag{2.18}$$

Arithmetic rules for r (or σ) are used in MM2 (Allinger, 1977), MM3 (Allinger et al., 1989), MM4 (Allinger et al., 1996), AMBER (Pearlman et al., 1995; Weiner et al., 1986), CHARMM (Brooks et al., 1983; MacKerell et al., 1998), INTERFACE (Heinz et al., 2013), MMFF (Halgren, 1996) and CLAYFF (Cygan et al., 2004), and geometric rules for both r (or σ) and ε are used in UFF (Rappe et al., 1992), OPLS (Jorgensen and Swenson, 1985; Jorgensen et al., 1996), VFF (Lifson et al., 1979), CVFF (Dauberosguthorpe et al., 1988), DREIDING (Mayo et al., 1990) and GROMOS (Hermans et al., 1984). More elaborate rules (e.g. sixth power rules) are used, for example, in COMPASS (Sun, 1998) and PCFF (Sun et al., 1994). However, a variety of rules are available and can be used when the simple one fails (Delhommelle and Millie, 2001; Halgren, 1992; Kong, 1973; Waldman and Hagler, 1993).

Mixing rules are also available for the Buckingham functional. Different formulations have been proposed and applied to different systems (Eq. 2.19 (Gilbert, 1968) applies to Eq. 2.10, Eq. 2.20 (Catlow et al., 1982) applies to Eq. 2.9, Eq. 2.21 (Srivastava, 1958) applies to Eqs. 2.9 and 2.10). Arithmetic rule for ρ and both arithmetic and geometric rules for A have been suggested (Mason, 1955):

$$A_{ij} = f\left(b_i + b_j\right) \exp^{\left(a_i + a_j/b_i + b_j\right)}, \quad \rho_{ij} = b_i + b_j \tag{2.19}$$

$$A_{ij} = \frac{1}{2}\left(A_i + A_j\right), \quad \frac{1}{\rho_{ij}} = \frac{1}{2}\left(\frac{1}{\rho_i} + \frac{1}{\rho_j}\right) \tag{2.20}$$

$$C_{ij} = \left(C_{ii}C_{jj}\right)^{0.5} \tag{2.21}$$

Simple mixing rules are also available for the Morse potential form (Konowalow, 1969). For ε the simple geometric mean rule can be used, while for r, the arithmetic mean rule holds. a_{ij}/r_{ij} can be calculated by means of Equation 2.22:

$$\frac{r_{ij}}{a_{ij}} = \frac{1}{2}\left(\frac{r_{ii}}{a_{ii}} + \frac{r_{jj}}{a_{jj}}\right) \tag{2.22}$$

When introducing new elements, it is not possible to use mixing rules and hence the parameters have to be derived by fitting.

2.3.2 Optimisation Procedures for All-Atom Force Fields

The fitting procedure works by selecting the functional form, which can best describe the system, and then adjusting the parameters to be derived so that the system's properties are reproduced to a specified accuracy. The sum of squares of the differences is the statistical measure of assessing the accuracy of the model. This relies on evaluating the weighted difference between observables and calculated quantities (Eq. 2.23). The weighting factor w, for each quantity i, results in an infinite number of solutions to the equation. As with many statistical regression models, it is often advised to perform the fitting using weighted factors inversely proportional to the uncertainty of the observable and to the magnitude of the observable squared. A number of standard methods are available for solving the sum of the squares, which are borrowed from the optimisation process, for example, BFGS and Newton–Raphson algorithms as implemented efficiently in the GULP code (Gale and Rohl, 2003):

$$F = \sum_{i=1}^{N} w_i \left(f_i^{\text{known}} - f_i^{\text{calc}}\right)^2 \tag{2.23}$$

2.3.2.1 *Empirical, Simultaneous and Relax Fittings*

Historically, the most common fitting procedure has been empirical fitting where the parameters are adjusted to reproduce experimental data. However, the procedure is constrained by the limited knowledge of usually experimentally derived data (e.g. crystal structure, elastic, piezoelectric, dielectric constants, phonon frequencies, and cohesive energies), which can be absent or limited to a restricted set of thermodynamic states and therefore not representative of materials behaviour at high pressure or high temperature. With the advent of *ab initio* calculations and the increase of computer power, fitting has also been performed on potential energy surfaces obtained from quantum methods. In this case, the observables may include not just the equilibrium structure but also energies and derivatives far from equilibrium (Ercolessi and Adams, 1994; Madden et al., 2006). However, here additional care is needed to ensure that the electronic structure data is itself accurate. For example, it is advisable to include long-range dispersion, which are absent in Hartree–Fock or standard DFT if deriving an FF between two molecules.

The approach is to optimise the potential parameters by using a force-fitting procedure, in which forces, multipole moments and stress tensor components of the whole cell generated by the model for condensed phase reference configurations are matched to those calculated using *ab initio* techniques, normally AIMD in order to extract information at a range of temperatures. This approach is called force matching (FM). The key target is an accurate and transferable set of potential parameters, and therefore it is important that the number of configurations (samples) is large enough so that each species can feel a wide range of coordination environments. The procedure uses three Cartesian force components, and if three components for the dipole and six components for the quadrupole of each ion are included, then a total of 12N variables per atom are available during the fitting, which is considerably more than the usual number of variables used in fitting parameters from experimental data. In this case, the objective functions that need to be minimised act on the whole set of potential parameters (Eqs. 2.24 and 2.25) (Wilson et al., 2004). i is the number of species; α represents the x, y and z coordinates; A is the number of sample configurations; $\zeta_{i,\alpha}^{A,\text{DFT}}(\{\chi\})$ and $F_{i,\alpha}^{A,\text{DFT}}(\{\chi\})$ are the multipole (point dipole and point quadrupole) components and forces obtained from DFT calculation; and $\zeta_{i,\alpha}^{A}(\{\chi\})$ and $F_{i,\alpha}^{A}(\{\chi\})$ are the same quantities obtained from MD calculations:

$$A_P(\{\chi\}) = \frac{1}{2} \frac{\Sigma_{i,\alpha,A} \left| \zeta_{i,\alpha}^{A}(\{\chi\}) - \zeta_{i,\alpha}^{A,\text{DFT}} \right|^2}{\Sigma_{i,A} \left| \zeta_{i,\alpha}^{A,\text{DFT}} \right|^2} \tag{2.24}$$

$$A_F(\{\chi\}) = \frac{1}{2} \frac{\Sigma_{i,\alpha,A} \left| F_{i,\alpha}^{A}(\{\chi\}) - F_{i,\alpha}^{A,\text{DFT}} \right|^2}{\Sigma_{i,A} \left| F_{i,\alpha}^{A,\text{DFT}} \right|^2} \tag{2.25}$$

When using models that do not use point dipoles, such as in the shell model, particular caution must be taken in relaxing the position of the shell since its position is not defined. The method of simultaneous relaxation of shells allows the shell to relax during the fitting (Gale, 1996). The shell can be minimised at each step of the fitting, or in a simpler approach, its position is treated as a fitting variable while the shell forces become observables as implemented in the GULP code (Gale and Rohl, 2003). The two approaches become equivalent when the only observable in the fitting is the crystal structure. The simultaneous fitting is not free from flaws. In theory, the description of the curvature close to the minimum can deteriorate even though the description of the forces improves. This can be overcome by the inclusion of the curvature in the fitting, but in the case of non-zero forces, this will lead to errors as well. The relax fitting (Gale, 1996) overcomes these flaws by minimising the structure and evaluating the properties at every step of the fitting instead of the gradient. As the variables need to be minimised at a valid optimisation, it is therefore imperative that the initial potential parameters are reasonable.

2.3.2.2   *Genetic Algorithm and Monte Carlo Search*

One alternative to the least-squares technique is to perform fitting using genetic algorithms (GA), which have proven useful in cases where a complex system is being fitted, when there is no reasonable starting approximation to the FF parameters available, or where there may be multiple local minima in the parameter space and therefore the minimisation algorithm will fail. GA applies a procedure in which starting from a trial set of random configurations, the algorithm allows to evolve according to principles (tournament, crossover and mutation) and discarding those configurations that do not fit the selection criteria.

Fitting parameters has also been achieved via a combination of Monte Carlo and energy minimisation algorithms (Freeman et al., 2011). The MC procedure can select one parameter at a time or a selection of potential parameters, and by adjusting the values by random amount of the order of ±0.1 to 10% of the current value, the suitability factor can be evaluated by performing a comparison between the properties of the system at the iteration $i + 1$ and i. Clearly, when the suitability parameter is lower, the new set of parameters will be accepted, while if it is higher, the change will be accepted with an acceptance ratio. This procedure might lead to an incredibly high number of sets of parameters, which have to be tested. One way to overcome this issue is to include more and more properties in the 'on the fly' testing of the parameters. However, the problem then becomes ensuring there are sufficient properties to test the FF.

2.3.2.3   *Derivation of the Charges*

A comparison of charges attributed to the atomic types from different FFs shows a widespread of values. As the atomic charges are not experimental observables, if the charges are not simply assigned as valence charges, they can be fitted or even derived by means of QM, but in the latter case, there is no unambiguous way of calculating them from the electronic distribution of the system. Most models assume the charge is located on the nuclear centre. Such models include the population analysis and the molecular electrostatic potential. The former is perhaps the simplest arbitrary method to assign the charge; there is a simple partition of the electron distribution between adjacent species, for example, by Mulliken (1955), or by the curvature of the electron density, for example, by Bader (1985), so that each species has a number of electrons associated with it. Cox and Williams (1981) first suggested a least-squares fitting procedure to derive a set of atomic charges that best describe the electrostatic potential. Ever since, a variety of electrostatic potential methods have been derived; for example, Singh and Kollman used two of those to derive the charges of AMBER (Bayly et al., 1993; Singh and Kollman, 1984). More complicated models include distributed multipole models in which not only the charges but also the multipoles are distributed throughout the system (Stone, 1981; Stone and Alderton, 1985). More information on the derivation of the charges for FF development can be found in appropriate books such as Leach (2001).

2.3.3   **Deriving CG Force Fields**

The CG approach removes those DOF that are not crucial for understanding the driving forces in the phenomena of interest. However, the CG parameterisation needs to be sufficiently accurate to capture the essential physical aspects of the reference AA system. Typically, the parameterisation is undertaken on a small but representative atomic system that has to be modelled in AA simulation. The optimised CG model is then applied to a much larger system that could not be feasibly treated by AA modelling. CG potentials are generally transferable across chain lengths, and, therefore, parameterisation for short chains is usually acceptable for longer chains (Wang et al., 2012a). However, this does not apply to more complex systems where many-body interactions become increasingly more involved; for example, simulations of hydrated PCHD showed that the CG potentials were not transferable when the water content varies (Wang et al., 2012b).

The first step in coarse graining is to define the mapping between the CG beads and the corresponding group of atomic species. This is normally achieved by placing a bead at the centre of mass (COM) of the chosen group of atoms. The second step is to generate the effective interaction potentials between the CG beads. The methods available for this purpose are briefly described. The potential of mean force (PMF) approach is based on thermal averaging over the DOF that are omitted in the CG system. Therefore, a CG potential can be defined by the corresponding $U_{PMF}(r; \rho, T)$, in which PMF is state dependent and specific for a given concentration, ρ, and temperature, T, and it is calculated for all r distances between the CG centres (centres of mass of the respective atomic groups) within Γ– the restricted phase space corresponding to a given value of r (Eq. 2.26) (Harmandaris et al., 2006). One has to be aware that PMFs obtained at finite concentrations are likely to exaggerate the structure in CG simulation compared to the reference AA system, due to essentially double counting the many-body interactions (Fu et al., 2012; Shell, 2012):

$$\beta U_{PMF}(r; \rho, T) = -\ln \langle \exp[-\beta U(r; \Gamma)] \rangle_r \qquad (2.26)$$

FM is an alternative method to obtain CG FFs. It was originally used as a way to fit classical potential parameters to *ab initio* calculations (Ercolessi and Adams, 1994) (see Section 2.3.2.1) and later applied to CG modelling (Izvekov and Voth, 2005; Noid et al., 2008). The idea is to construct a CG FF that would optimally match the many-body potential of the mean force. The internal optimisation procedure is similar to the sum of squares as in Equation 2.23.

Another way to proceed is the structure-fitting CG (SCG), which includes the iterative Boltzmann inversion (IBI) (Reith et al., 2003; Ruehle et al., 2009; Wang and Deserno, 2010) and the inverse Monte Carlo (IMC) (Lyubartsev, 2005; Lyubartsev and Laaksonen, 1995; Lyubartsev et al., 2003; Mirzoev and Lyubartsev, 2014). The schemes are inherently numerical in that they use grid-tabulated potentials. Conceptually, the SCG relies on the Henderson theorem establishing a strict uniqueness relationship between any pair correlation function (PCF), $p_2(q)$, for example, radial distribution function (RDF), $g_{\alpha\beta}(r)$, and the corresponding pair potential, $U_{\alpha\beta}(q)$, generating it (Henderson, 1974). Thus the SCG attempts to fit the reference $p_{\alpha\beta}(q)$, known from either experiment or AA simulation, by iteratively adjusting the potentials with the aid of one of the following iterative schemes. Provided an initial guess, $U_{\alpha\beta,0}(q)$, the IBI approach iteratively updates the potential $U_{\alpha\beta,i}(q)$ as in Equation 2.27:

$$U_{\alpha\beta,i+1}(q) = U_{\alpha\beta,i}(q) + k_B T \ln \frac{p_{\alpha\beta,i}(q)}{p_{\alpha\beta}(q)} \qquad (2.27)$$

Each iteration implies performing a separate simulation for the updated CG model, and at each step, the pair correlation function is evaluated and compared to the reference $p_{\alpha\beta}(q)$. The optimisation is achieved when the updated pair correlation function converges to the reference. The IMC works in a similar manner but uses a different update procedure, which is based on the expansion of the variation, $\Delta p_{\alpha\beta}(q; U_{\alpha\beta})$, and strict statistical–mechanical formulae to compute $\Delta U_{\alpha\beta}(q)$ from a set of linear equations. A detailed description is given by Lyubartsev (Lyubartsev, 2005; Lyubartsev and Laaksonen, 1995; Mirzoev and Lyubartsev, 2014). In both IBI and IMC, more than one potential can be updated at each iteration step, but practical attempts have shown that it is more efficient to vary only a small subset at a time. It is advisable to combine IBI and IMC, by starting the iteration with a few dozen of IBI steps and fine-tuning the CG force field with the use of IMC, which appears to be much more sensitive to the variations in $U_{\alpha\beta}(q)$ (Ruehle et al., 2009). Apart from the initial AA simulation to derive the pair correlation functions (when experimental data are not available), each iteration implies performing a separate simulation of the updated CG system, which can result in a time-consuming fitting process if the initial guess of the CG potentials is unphysical.

Finally, a general and rigorous approach has been introduced by Shell (2012), which shows that measurement and minimisation of the so-called relative entropy (RE), S_{rel} in Equation 2.28, can provide a statistical–mechanically justified means for not only optimising CG models but also quantifying the effectiveness of such optimisations in terms of the excess entropy differences between the CG and AA systems, ΔS_{ex}:

$$S_{rel} = S_{map}(\boldsymbol{M}(q)) + k_{B} \int p_{ref}(q) \frac{p_{ref}(q)}{p_{CG}(\boldsymbol{M}(q))} = \ln \omega(\boldsymbol{M}(q)) - \ln \left\langle e^{\beta(\Delta U - \langle \Delta U \rangle_{AA})} \right\rangle_{AA} \qquad (2.28)$$

In Equation 2.28, $\omega(\boldsymbol{M}(q))$ is the number of AA microstates, $\{q\}$, that are reduced into the same CG microstate by the mapping operator $\boldsymbol{M}(q)$, where q denotes the original AA DOF. The RE method is based on the fact that coarse graining cannot result in a decrease of entropy, that is, $S_{rel} \geq 0$, whence the minimisation of RE is necessitated. The RE technique provides the most general statistical–mechanical basis for any type of coarse graining and thereby allows for informed optimisation of the other CG methodologies, regardless of the actual potential representation in analytical or tabulated form. Formally, the RE optimisation of a CG force field with respect to a given parameter λ is achieved when $\partial S_{rel}(\lambda)/\partial \lambda = 0$, which translates into a number of useful thermodynamic relationships, for example, Equation 2.29 and other equations directly linking the AA and CG measurable quantities: pressure, compressibility and diffusion coefficients. In practice, in each particular case, the relevant expressions are to be (re)formulated in terms of the parameters of a given CG scheme:

$$\left\langle \frac{\partial U_{CG}(\lambda)}{\partial \lambda} \right\rangle_{CG} = \left\langle \frac{\partial U_{AA}(\lambda)}{\partial \lambda} \right\rangle_{AA} \qquad (2.29)$$

The third CG step is to use dynamical equations for describing the time evolution of the CG system. The CG model explores the free energy surface of the AA system (PMF), but the resulting molecular dynamics of the CG particle cannot reflect the actual dynamical behaviour of the classical AA model. Additional terms have to be included such as generalised Langevin-like friction and random force terms, which help with mapping the CG dynamics back onto the AA counterpart.

2.3.4 Accuracy and Limitations of the Fitting

A successful fitting of a set of data does not necessarily mean that the potential model will be able to simulate all properties of the system in all conditions. An important point is that the users must establish and be able to demonstrate that the parameters are suitable for their problem. The number of possible solutions for the F function appears to be infinite (Section 2.3.2). However, there are a few checks, which should be considered. Firstly, the potential parameter should represent the observables used during the fitting to a reported degree of accuracy. Secondly, the newly derived potential parameters should be tested against properties that have not been included in the fitting: in the case of minerals, the full elastic constant matrix, phonon frequencies or structures that were not used during the fitting. Finally, potential parameters are fitted to physically significant values. For example, the dispersion terms in the Lennard–Jones and Buckingham forms can become significantly higher during the fitting, leading to a model that is generally not transferable across a wide range of structures.

2.3.5 Transferability

The ability of an FF to simulate different systems, in other words to be transferred to different problems, is crucial. Transferability could be improved by deriving suitable parameters for a wide range of systems, but this has proved a great challenge for the development of FFs. A variety of FFs are available but are often based on different functional forms, which adds a major hurdle for transferability, for example, when simulating complex interfaces between organic/biological and inorganic/mineral systems. This has been recently addressed by returning to the LJ potential for minerals, which as shown have simple combining rules that have not only reliably simulated systems containing different minerals but also combined them with organic and biological systems. More complicated functional forms are available such as spline functions (tabulate values vs. distance), the embedded atom and the bond-order potentials, but they are even more complicated by the nature of the functional form, which limits the transferability of the FF. Reactive and polarisable FFs have perhaps the greatest potential, but they are highly parameterised and cannot be easily mixed with simple harmonic functional forms, and hence, transferability has to be built into the fitting procedure. The recognition of the importance of transferability has led to the establishment of FFs that are increasingly being extended and developed to include metal ions and minerals.

2.4 Force Field Libraries

There are many examples in the literature of bespoke FFs for individual systems, and some of these will be discussed in Section 2.5, when we consider some examples from the main mineral classes. However, one of the major challenges is the derivation of FFs applicable to a wide range of minerals and compositions. In this section, we review some of the FF libraries available for geochemical applications. FFs for minerals can be highly parameterised in terms of bonded interactions while quite simple in terms of non-bonded interaction (in the sense that usually there is only a charge and a LJ potential to define each atom species). The reason behind the limited number of examples of FFs that can reliably model a significant number of minerals is whether their simplicity can capture the significant number of coordination environments (oxygen atoms are threefold coordinated in rutile, fourfold in corundum and sixfold in periclase), the deviation of highly symmetric geometries (Jahn–Teller effect) and loosely bound electrons (delocalised bonds), without the need of extra parameters.

2.4.1 General Force Fields

The idea behind developing a general FF is that ideally there would be one FF for all – or most of them – atoms of the periodic table that could be used in any situation, from minerals to organic matter. Amongst the first examples of such FFs include the DREIDING FF (Mayo et al., 1990), the UFF (Rappe et al., 1992) and the extensible systematic FF (ESFF) (Barlow et al., 1996; Shi et al., 2003). These FFs evaluate the potential energy of the system as a sum of contributing energy terms including bond, angle, torsion, inversion, van der Waals and electrostatic energies.

The DREIDING FF has the advantage of being fast to evaluate by adopting CG features such as treating hydrogen atoms implicitly coupled with terms explicitly acknowledging the hydrogen species capable of forming hydrogen bonds. A polymeric chain can be, for example, represented by atom types including CH_2 groups, thus reducing the number of species treated explicitly and therefore the computational effort. The UFF comprises a set of parameters based on the elements (atomic numbers), their oxidation states, their hybridisation and connectivity including a set of hybridisation-dependent ionic radii and angles, van der Waals parameters, torsional and inversion barriers and effective nuclear charges. One feature of UFF is the implementation of cosine Fourier series for the treatment of angle bending, which can model appropriately the distortion of angles approaching 180° compared to the harmonic term usually implemented in specific FFs. The ESFF was originally derived to model organometallic compounds containing transition metals, and it is capable of modelling organic, inorganic, and organometallic systems. It uses semi-empirical rules to derive the potential parameters from atom-based parameters, which are then used to generate a complete system-specific FF.

The usage of such FFs is still limited to organometallic compounds and rarely used to simulate crystal structures of minerals, and hence, their transferability to all minerals is not fully established. A modified version of the DREIDING was successfully used by Aicken et al. (1997) and Newman et al. (1998) to simulate layered double hydroxides and Zeng et al. that have studied the structure of nanoconfined alkylammonium in the montmorillonite (MTM) clay mineral (Zeng et al., 2003). UFF was used successfully to determine the surface characteristics of complex aluminosilicate minerals including spodumene, jadeite, feldspar and muscovite (Rai et al., 2011) and their interaction with oleate and dodecylammonium chloride used in the froth flotation of spodumene for lithium extraction. Remarkable agreement with the experimental measured contact angles and flotation results was achieved. Bains et al. have investigated the swelling properties of sodium and potassium MTMs and organo-smectites (Bains et al., 2001; Boek et al., 1995). Nanocomposites are of emerging research interest, and the combination of experiments and simulations is recognised as a valuable combination. Two examples have been studied by Tokarsky et al. (2012) who characterised the kaolinite–anatase system using SEM and molecular modelling finding larger adhesion energies of TiO_2 nanoparticles on the gibbsite terminated compared to the siloxane terminated (001) surface of kaolinite and Holesova et al. (2014) who studied antibacterial kaolinite–urea–chlorhexidine nanocomposites as mineral clays are also commonly used in pharmaceutical productions using UFF.

2.4.2   Force Field Libraries for Organics: Biomolecules with Minerals

The MM FFs developed by Allinger and co-workers MM2 (Allinger, 1977), MM3 (Allinger et al., 1989), and MM4 (Allinger et al., 1996) are widely used in calculations of organic molecules and have been parameterise to unable to distinguish between different hybridisation of an atomic species, for example, carbon atom in sp^3, sp^2, sp, carbonyl, radical, etc. Similar approaches are implemented in the AMBER FF of Kollman and co-workers (Pearlman et al., 1995; Weiner et al., 1986) and the CHARMM FF of Karplus and co-workers (Brooks et al., 1983; MacKerell et al., 1998) developed for amino and nucleic acids, but in this case, the atomic species depends also on the protonation state. These organic FFs have a number of differences (e.g. different treatments of inversion as improper torsion), and their performances usually are compared and debated in the literature (Roterman et al., 1989).

Historically, the field of computational geochemistry has evolved by using FFs specifically derived for individual minerals or small groups of minerals, contrary to the FF libraries for organic and biological molecules. In the last two decades, the idea of a library of potentials that can be used to simulate a wide range of minerals that can be combined with the organic and biological FFs became appealing. The driving force has clearly been the greater desire of simulating organic–mineral–water interfaces. CLAYFF and the INTERFACE FFs are perhaps the most widely used FFs in the study of minerals in contact with the aqueous solutions, and both use the LJ functional form to describe the non-bonded interactions. Examples of these applications are given in Section 2.5.

Another FF developed specifically for minerals, the MS-Q FF, uses the Morse function to describe the non-electrostatic terms and the charge equilibration procedure of Rappé to adjust the atomic charges as a function of the instantaneous geometry (Rappe and Goddard, 1991) and hence has features required for a reactive FF. MS-Q FF was originally developed for silicate and aluminophosphate (Demiralp et al., 1999) but later to model clay minerals including kaolinite and pyrophyllite (Hwang et al., 2001). This is indeed the most popular way of developing FFs where terms are derived and added to the main set of potential parameters of the original development. Hwang et al. (2001) have optimised the parameters for hydrogen to reproduce the structural parameters of kaolinite and pyrophyllite and their interactions with organic molecules. As for many other studies, the organic adsorbates were modelled using the DREIDING FF (Mayo et al., 1990) and the cross terms generated by using mixing rules.

Researchers have also started re-parameterising specific FFs to be able to use them within a more general procedure. Bhowmik et al. (2007) studied the polymer–hydroxyapatite composites as potential bone replacement materials by re-parameterising the Hauptmann potential energy function of apatites (Hauptmann et al., 2003) composed of interatomic terms in the Born–Mayer–Huggins functional form into CVFF LJ form that could be used in combination with the LJ form used for the description of the polymer.

Another important requirement for modelling such systems is the need to include reactivity in potential-based techniques. Examples are the EVB (Villà and Warshel, 2001; Warshel and Weiss, 1980), the ReaxFF methodology (Mueller et al., 2010; van Duin et al., 2001), the charge optimised many-body (COMB) potential (Shan et al., 2010a, 2010b), the reactive empirical bond order (REBO) (Brenner, 1990; Brenner et al., 2002), the charge equilibration method (QEq) (Rappe and Goddard, 1991) and some others (Rustad et al., 1996, 2003).

2.4.3   Potentials for the Aqueous Environment

Water plays a vital role in many geochemical processes, and hence, it is worth noting some of the models in use. A major application for modelling is the interface between water and mineral surface as it affects mineral growth, aggregation and sorption of species on mineral surfaces. There are many water potential models with different level of complexity. Simple water models treat the molecule as a rigid entity; flexible models allow for conformational changes, complex models improve the description by including polarisation and/or many-body effects, while reactive models allow the treatment of proton transfer. The choice of the best model for a particular task can be based upon previous calculations (where the model is used to simulate similar systems) or computational effort (simple models run faster with more species) or even the ease of mixing the water model with the FF in usage for the system under consideration. Again, in each case, a critical assessment of the performance of the model and possibly comparison with available experimental or *ab initio* data must be carried out. Comparison between the performances of different water models is subject of numerous studies (Bickmore et al., 2009; Lopes et al., 2009; Sirk et al., 2013; Vega and Abascal, 2011; Vega et al., 2009). However, for a more specific evaluation of the performance of the water model with the chosen mineral FF, one must carefully analyse the properties of water at the mineral interface. For example, an effective experimental technique that probes important structural features is X-ray reflectivity (Fenter et al., 2013; Skelton et al., 2011).

SPC (Berendsen et al., 1987) is a simple rigid model that treats the water molecule as a three-site entity of two positively charged hydrogen atoms compensated by a negatively charged oxygen atoms (Figure 2.5a). The interactions between different molecules are computed by using an LJ functional form only acting between the oxygen species. TIP3P (Jorgensen et al., 1983) works like the SPC while TIP4P (Abascal et al., 2005; Jorgensen et al., 1983) is a four-site model, which introduces a point along the bisector of the HOH angle with its own charge (Figure 2.5b). Five-site models are also available, and in this case, the molecules have two sites centred on the lone-pair sites of the oxygen atom (Mahoney and Jorgensen, 2000). Flexibility is usually introduced in the simple

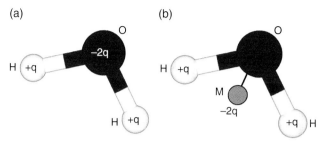

Figure 2.5 *Schematic representation of the SPC and TIP3P water models (a) left and TIP4P (b). Oxygen in red and hydrogen in white.*

models by adding bond stretching and angle bending terms (Ferguson, 1995; Wu et al., 2006). In this case, the vibrational properties can be evaluated.

Usually simple and flexible water potential models are sufficient, but when large electric stimuli are involved (e.g. charged species and highly polarisable interfaces), one might move towards more sophisticated polarisable models, which explicitly treats the polarisation and the many-body effects. There is a selection of ways to account for polarisability into water models (e.g. induced molecular point dipoles, Drude oscillators, fluctuating charges, induced atomic dipoles (Lopes et al., 2009; Tröster et al., 2013)). Examples of polarisable FFs are the PIPF-CHARMM (Xie et al., 2007) and the AMOEBA FF (Ponder et al., 2010), while dissociative water potential models have been developed for mineral–water interfaces by Rustad et al. (1996) and Garofalini (Lockwood and Garofalini, 2009; Mahadevan and Garofalini, 2007). Reactive models for water are still limited but allow the bond breaking/formation on a classical level (Bresme, 2001; Halley et al., 1993; Hofmann et al., 2007; Mahadevan and Garofalini, 2007; van Duin et al., 2001). It should be noted that this is by no way a complete list of water potentials, and there are some review articles in the literature that provide a more extensive description (Guillot, 2002; Jorgensen et al., 1983).

2.4.4 Current CGFF Potentials

As the time and length scale of many geochemical problems require techniques that go beyond atom-level simulations, coarse graining can be a valuable development for extending the scales. Transferable CG FFs optimised for general use are still rare as only recently systematic CG methodologies have been developed and exploited. The organic and biological fields have the most developed CG potentials, and it is worth mentioning two recent models, the MARTINI (Marrink et al., 2007) and the virtual atom molecular mechanics (VAMM) (Korkut and Hendrickson, 2009), which are developed for macromolecules and represent viable tools for colloids.

Historically, most of simplified CG models are based on three types of potentials: stepwise or hard models (hard disks, spheres, walls, rigid rods, freely jointed polymer chains, recent 'tube' model for proteins), soft Lennard–Jones models (atoms, ions, colloidal particles and conglomerates thereof) and, finally, very recent numerically tabulated CG models (in all the cases, the electrostatic contributions would be added). A wealth of knowledge about the fundamental physical and thermodynamic driving forces in condensed matter systems has been gained with the aid of the first two model types, which in many cases can be regarded as generic CG models. While the crystal structure in minerals does not leave much of alternative but using quantum-mechanically derived and guided AA FFs, the widespread colloidal systems, such as clays, soils, soil and sea waters, remain mostly the realm of generic CG modelling. That is, many aspects of interactions in these materials have strong overlaps with typical intermolecular and surface forces that are at play in colloidal solutions and can be measured by surface-force apparatus, atomic force microscopy and other experiments (Israelachvili, 2011). Therefore, in the succeeding text, we summarise the results that can be obtained with generic CG models and then turn to the possible extensions due to the most recent systematic CG developments.

One particularly important case is the electrostatically driven aggregation in colloids and clays, which is mediated by multivalent ions (including polyelectrolytes) (Khan et al., 1984a, 1984; Labbez et al., 2006; Marra, 1986; Plassard et al., 2004, 2005). Despite the success of the well-known DLVO theory (Derjaguin and Landau, 1993; Verwey et al., 1948), which predicts a strong electric double layer (EDL) repulsion between any like-charged surfaces immersed in an aqueous solution, there exist many situations where the forces appear to be attractive at the nanoscale, giving rise to cohesion and ultimately leading to phase separation, segregation and formation of solid deposits. These phenomena have been extensively studied in simulation with the use of very simplified, seemingly 'crude' models: hard core ionic and polyionic species constrained in between two parallel flat surfaces bearing some charge density (Attard, 2007; Guldbrand et al., 1984). Similar observations hold true in the case of charged spherical aggregates modelled by soft core (LJ-type) potentials, where competition between mono- and multivalent ions has been studied (Brukhno et al., 2009; Hynninen and Panagiotopoulos, 2009). A very recent study (Thuresson et al., 2013) of the charged platelet aggregation in clays (so-called tactoids) has also used the hard core potential for both ions and the particles of the platelet material, giving instructive simulation data on the free energies associated with adding extra platelets to an existing tactoid column. Similar to other cases, the thermodynamically favoured aggregation during tactoid formation is attributed to the presence of divalent ions, as opposed to swelling due to the EDL repulsion with monovalent ions.

Amongst other surface forces found to be important contributors to the stability, enhanced aggregation or swelling of colloids are the strong depletion and/or bridging forces in the presence of neutral or adsorbing polymers (Broukhno et al., 2000), reduced EDL repulsion mediated by polyampholytes (Broukhno et al., 2002) and stratification phenomena with confined polyelectrolytes (Jönsson et al., 2003). However, a quantitatively accurate evaluation of the EDL forces requires CG models accounting for subtle effects that have origins on the atomistic level, such as information on hydration shells (barriers) around charged species, the dielectric permittivity dependence on temperature and solute content and, perhaps even more importantly, its strong reduction (so-called dielectric saturation) in the close proximity of an ion (Hess et al., 2006; Lenart et al., 2007; Mirzoev and Lyubartsev, 2011; Zelko et al., 2010). Hydrogen bonding in aqueous solutions, within and/or between organic species, is another atomistic level phenomenon requiring an appropriate representation in CG FFs.

A simulation method, where large-scale CG models are commonly used, is DPD. This technique relies on integrating out intramolecular DOF (atomic positions, bond vibrations and rotation fluctuations) to such an extent that the soft-sphere CG 'particles' it employs do not possess any rigid core, which allows for these soft spheres to overlap and penetrate through each other, enabling in turn much larger time steps than used in standard MD (Frenkel and Smith, 1996). Importantly, it has been shown that the large-scale dynamics and diffusion can be rigorously linked with those on the atomistic scale (Flekkoy et al., 2000; Lyubartsev et al., 2003). The use of DPD in geochemistry is justified when actual mesoscale simulations are necessary, for example, the confinement of ions in large-scale clay sheets, which are not feasible for AA MD (Carof et al., 2014; Suter et al., 2009). DPD is also suited and widely used for mesoscale simulation of transport in porous media (Meakin and Xu, 2009; Zhao and Wang, 2014) including blood cells (Moreno et al., 2013; Tosenberger et al., 2011). Therefore, it may be of use in studying water flow in soils.

Models have also been developed using systematic CG methods for a range of compounds: polyisoprene methanol, propane and hexane (Brini et al., 2011), polystyrene (Brini et al., 2012), phospholipid, PEO/PEG, polysaccharides and other compounds. The developments in systematic CG and smoothed DPD methods open up a number of new opportunities to approach geochemical systems where inorganic materials occur in contact with organic compounds. In particular, one of the remaining biggest challenges is to develop realistic models for the interaction of mineral surfaces with amphiphilic molecules and biologically active liquids, including solutions containing, for example, surfactants, lipids, peptides and proteins. The increased reliability of CG methodologies inspired by biochemistry studies is such that it is merely a matter of time and extra effort before it is applied to geochemistry. Perhaps, the main difficulty on this route would be to appropriately combine the AA representation of an inorganic–organic interface (at least, specific mineral surfaces) with a pure CG model for organic substances in solution away from the interface (surface). This research could build up on the recently reported attempts to develop hybrid AA–CG models where only a small part of the system is simulated with atomistic detail while the dominant remainder is treated on the CG level (Brini et al., 2013; Noid, 2013; Srivastava and Voth, 2013).

2.4.5 Multi-scale Methodologies

The multi-scale approach couples computational methodologies specific for different length and time scales. Fixed-resolution multi-scale QM/MM (Downing et al., 2014; Scanlon et al., 2013) and AA–CG (Neri et al., 2005) methodologies have been developed and applied in many fields, but while the former are used both in reactivity of materials and biological macromolecules, the latter are normally more exploited in organic and biological investigations. The fixed-resolution multi-scale approach has the main drawback of having fixed regions treated with different resolutions and therefore preventing the exchange of particles between the regions (Figure 2.6). Hence, the recent interest in adaptive multi-scale simulations in a range of different fields.

In these methodologies, there is normally a region where the two combined techniques are mixed and therefore the number of degree of freedom changes 'on the fly' during the simulations. They can simulate much longer length and time scales but retain the high resolution required. However, strict conditions are necessary such that thermodynamic equilibrium between the different regions must be ensured, the representation of the species in the different regions must retain the same physical

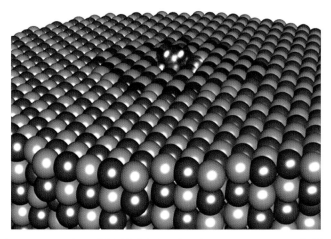

Figure 2.6 *Schematic representation of QM/MM showing three-region approach. The transparent inner region shows the adsorbed species, the second region (coloured green and red) is allowed to relax in response to the defect and the third region is held fixed.*

meaning and the exchange of species between the regions must happen in thermodynamic and statistical equilibrium.

Adaptive QM/MM multi-scale simulations are still rare as the fluctuations in the number of electrons have to be treated in a consistent way when changing the degree of freedom (DOF) on the fly. One possible route is via treating the electron in a statistical way within a macrocanonical ensemble where the number of electron is allowed to fluctuate or via mapping the quantum subsystem onto a classical one in a path integral quantum mechanical fashion (Bulo et al., 2009; Nielsen et al., 2010). More advanced are the AA–CG dynamic multi-scale methodologies, for instance, the Hamiltonian replica exchange methods (REM) and the adaptive resolution schemes (AdReSs). In the former, several independent copies of the systems, that is, replicas, are propagated via MD or MC at different temperature and potential energy functions with some finite probability to exchange their configurations (Liu and Voth, 2007; Lyman et al., 2006). The adaptive resolution scheme couples the different regions by changing the number of DOF 'on the fly' or by preventing the free exchange of species between the regions. The description of the system is achieved by changing the DOF during the simulation; the AA species loses its vibrational and rotational DOFs when passing to the CG region, passing through a stage of hybrid AA–CG representation where the species is represented by a linear combination of the AA species with the additional center of mass of the CG bead and finally reducing its representation to a sphere whose DOFs are only the translational ones of the centre of mass. Examples are limited to the water solvent (Delgado-Buscalioni et al., 2008; Matysiak et al., 2008) and hydration of molecules (Praprotnik et al., 2007, 2008).

2.5 Evolution of Force Fields for Selected Classes of Minerals

We will next focus our attention on a few representative minerals present in the terrestrial crust and subsurface, including calcite, clay minerals, hydroxides, silica and silicates and iron minerals to illustrate the capability of the current FFs and to indicate how they are developing.

2.5.1 Calcium Carbonate

Calcium carbonate ($CaCO_3$) forms a variety of different naturally occurring mineral phases; there are three crystalline polymorphs (calcite, aragonite and vaterite), the amorphous form (ACC) and two hydrates (monohydrocalcite and ikaite) (Carlson, 1983). In the process of biomineralisation, organisms can selectively control the synthesis of one of these polymorphs and generate a large variety of structures and shapes. $CaCO_3$ is also important commercially for carbon sequestration, and hence, the crystal growth mechanism has been extensively investigated by experimental (Gebauer et al., 2008) and theoretical studies (Demichelis et al., 2011; Wallace et al., 2013).

There have been numerous FFs derived to describe the bulk properties (Cygan et al., 2002; Pavese et al., 1996) and thermodynamic properties at the $CaCO_3$–water interface (Kerisit and Parker, 2004; Raiteri et al., 2010). While each FF may serve its purpose for a particular system of interest, one must also be aware of the complexity in the $CaCO_3$ system, which can limit the applicability of an FF. For instance, many of the derived FFs for $CaCO_3$ either do not give calcite as the stable phase at room temperature contrary to experiment or over-stabilise calcite by up to an order of magnitude and fail to describe the free energies of solvation of the ions accurately leading to a solubility product that is in error by tens of orders of magnitude. This has repercussions for the description of solvation effects (Cooke and Elliott, 2007; Kerisit et al., 2005a; Quigley et al., 2011).

Recently, a rigid and flexible FF that correctly represents both the thermodynamic behaviour of the $CaCO_3$–water interface and the free energy difference between calcite and aragonite has been

(a)   (b)

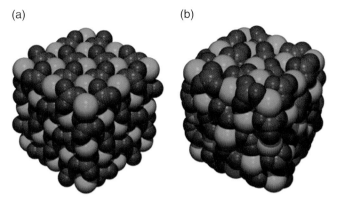

Figure 2.7   *Crystalline nanoparticles of calcite comprising of {10.4} surfaces, before (a) and after minimisation (b).*

developed (Raiteri and Gale, 2010; Raiteri et al., 2010). One of the challenges is to include the solvation and thermodynamics data in the fitting protocol. Capturing these effects is particularly important in the study of the growth of $CaCO_3$ in the aqueous environment. Recent experiments have demonstrated that growth does not follow the classical nucleation theory but rather follows stable pre-nucleation clusters prior to the formation of amorphous calcium carbonate (ACC) (Gebauer et al., 2008; Pouget et al., 2009).

The study of such mechanisms is a suitable challenge for potential-based techniques as the time and length scales of aggregation events are currently computationally prohibitive for *ab initio* techniques. Simulations of very small nanoparticles of $CaCO_3$ (less than 50 units) tend to lose the long-range order and become amorphous (Kerisit et al., 2005a). An example of nanoparticles of crystalline calcite is given in Figure 2.7 showing some loss of crystallinity (Martin et al., 2006), which has been seen both experimentally and computationally (Raiteri and Gale, 2010). However this is not the case when the nanoparticles are immersed in an aqueous environment as the solvation effects on the surface species increase the coordination of the surface species and their stabilisation (Cooke and Elliott, 2007; Kerisit et al., 2005a). Whether $CaCO_3$ nanoparticles of different sizes are preferentially amorphous or crystalline is still debated, but it has been seen that by changing the volume available to the nanoparticle and the solvent, it is possible to manipulate the stable phase, which might indicate a plausible route for nature to selectively control the synthesised polymorph (Quigley et al., 2011). Further development of the $CaCO_3$ FFs has recently concluded that amorphous nanoparticles containing a number of water molecules as suggested experimentally are lower in free energy than calcite nanoparticles up to a size of 4 nm, beyond which they become metastable (Raiteri and Gale, 2010; Raiteri et al., 2010). Furthermore, other calculations suggested that the aggregation of crystalline nanoparticles was affected by the surrounding hydration sphere while amorphous nanoparticles were not. This also supports the non-classical nature of the nucleation and growth of amorphous $CaCO_3$ (Gebauer et al., 2008). Indeed, there are proposals that suggest that pre-nucleation clusters form in one simulation study (Nielsen et al., 2014). However, observed pre-nucleation clusters are yet to be proven experimentally.

As forming and breaking of bonds at the mineral–water interface control nucleation and growth of minerals, possibly through the formation of transition states, FFs have naturally evolved to account for these effects. The current FFs all treat the carbonate anion as a molecular entity. This means that the bonding within this molecule is predefined at the start of the simulation and must remain the same throughout the simulation. While this type of assumption can be considered a reasonable approximation for bulk phases of $CaCO_3$, problems arise when considering growth mechanisms in the aqueous environment that involve significant re-bonding. This was addressed by Gale et al. (2011) applying ReaxFF. They assumed a formal ionic description to account for binding between the calcium ion and water molecules. The calcium ion was treated with a screened Coulombic interaction, which is used within ReaxFF. As the screening parameter was determined via combination rules, Gale et al. were able to introduce a single fitting parameter for calcium, prior to considering the pairwise interactions. To overcome the fact that the $Ca-H$ and $Ca-C$ short-range interactions would be not well defined, FF parameters where fitted to calcium hydride (CaH_2) and calcium carbide (CaC_2) in which hydrogen and carbon are in anionic state. The $Ca-O$ interactions were defined by fitting to the experimental structures of calcite, portlandite ($Ca(OH)_2$), calcium oxide (CaO) and *ab initio* structure and binding energy for hexa-aquo calcium $\left(Ca(H_2O)_6^{2+}\right)$. The fitted parameters for the $Ca-H$, $Ca-O$ and $Ca-C$ were found to give a good representation of the bulk lattice parameters for a number of solid phases that contain calcium. All calculated lattice parameters were within 10% error of experimental lattice parameters. The reactive FF developed was able to give the free energy difference of 0.84 kJ/mol at 298K, between calcite and aragonite in favour of calcite. This is in agreement with experimental data, but the model is not yet totally complete as the aragonite structure differs from the experimental determined structure (Gale et al., 2011). This approach shows great promise giving a good representation of the equilibria between carbonate, bicarbonate and carbonic acid. The ReaxFF results show also good agreement with first principle simulations performed using the same starting configuration and temperature. This is a useful result because when considering the speciation of carbonate species in water, validation of the FF can come from first-principles calculations.

2.5.2 Clay Minerals

Clay minerals are layered crystalline structures where each individual layer is composed of octahedrally coordinated aluminium oxide hydroxide sheets (O) and tetrahedrally coordinated hydrous silica sheets (T) (Deer, 1992). They can be divided by structural type as either 1 : 1 (TO) or 2 : 1 (TOT), where a 1 : 1 clay layer has one tetrahedral sheet and one octahedral sheet and a 2 : 1 consists of an octahedral sheet sandwiched between two tetrahedral sheets. The minerals can also be divided into groups such as kaolinites, illites, smectites and vermiculites. Substitution of Al and Si on either the structural sheets is common (including alkali and transition metals and alkaline earth) and leads to the presence of cations within the interlayer to maintain charge neutrality. The chemical composition controls the capacity to become expandable and the affinity of the minerals to water.

Clay minerals can be divided into water repellent, dispersible and hygroscopic. In the absence of cations, clay minerals are hydrophobic. In the presence of cations, the enthalpy of hydration and therefore the water affinity increases for small-sized cations $\left(Li^+ > Na^+ > K^+ > Rb^+ > Cs^+\right)$; divalent cations usually decrease the swelling properties of the clay reducing the affinity to water. Most clay minerals have swelling properties and are able to uptake water and inorganic cations and organic molecules; hence they are used in applications where cation exchange is of particular importance. Clay minerals are widely used in subsurface disposal systems for spent nuclear fuel (Arcos et al., 2008; Dixon et al., 1985), drilling fluids (Anderson et al., 2010), cosmetics, detergents and, more recently, in nanotechnology and biotechnology (Aguzzi et al., 2007; Lin et al., 2002; Paul and Robeson, 2008; Ray and Bousmina, 2005). In terms of pollutant remediation, clays are able to remove both heavy metal ions (Celis et al., 2000) and, importantly, organic pollutants. It is clear that the success of such applications will depend on the morphologies and properties of the clay minerals at the nanometre scale. As the experimental description of local interfaces is still limited, simulations become useful for gaining insights in describing local environment and computing dynamical properties at the nanoscale. Molecular computer simulations have been widely exploited to provide the atomistic descriptions of the structure and behaviour of clay minerals. Most of these have focussed on the swelling behaviour of smectite clays using either Monte Carlo or classical molecular dynamics. More rigorous QM methods have been applied to clay minerals, but they have largely been limited to simple clay structures (Alexandrov and Rosso, 2013; Austen et al., 2008; Zhang et al., 2012). Alternative strategies include fully ionic shell model potentials derived for the component oxides that have been applied to mica (Collins and Catlow, 1992) and pyrophyllite (Austen et al., 2008).

The first models derived specifically for clay minerals were proposed by Boek et al. (1995) and Skipper et al. (1995) who applied them to understand the swelling of clay minerals using MC techniques. Later, Teppen et al. (1997) successfully derived a new model to reproduce a wider range of clay minerals including pyrophyllite, kaolinite and beidellite. The advantage of this model was the improved description of aluminium by the introduction of a bending O–Al–O for both the tetrahedral and octahedral coordinated aluminium atoms; however, the bonds between all the species must be identified and specified prior to the simulation. Highly parameterised models are generally available for phyllosilicate including the FF of Sainz-Diaz et al. for highly charged structures (Sainz-Diaz et al., 2001) and of Heinz et al. (2005). The latter is called phyllosilicate FF (PFF) and was originally developed for mica, montmorillonite and pyrophyllite. Although all the bonds between the atoms need to be specified, it has the great advantage to be fully compatible with organic PCFF, CVFF, CHARMM and GROMACS. The development of this FF was carried out since the first release, and improvements have led to the development of a large number of potential parameters into the INTERFACE FF (Heinz et al., 2013) for many inorganic compounds including clay minerals, silicates, aluminates, metals, sulfates and apatites.

Cygan et al. took another approach to FF development; they developed CLAYFF. The FF is based on an ionic non-bonded description of the metal–oxygen interactions associated with hydrated phases, which gives full flexibility and transferability across different structures including layered hydroxides boehmite, portlandite and brucite and clay kaolinite, pyrophyllite, montmorillonite and hydrocalcite. CLAYFF has been successfully used to tackle different intriguing topics, such as the water transport properties in interparticle porosity of MTM, which depends on the pore water composition and the surface charge density (Churakov et al., 2014); the rotational disorder of distorted MTM, which leads to energetically favourable configurations at temperature and pressure relevant to geological carbon storage (Myshakin et al., 2014); and the behaviour of supercritical CO_2 and aqueous fluids on both the hydrophilic and hydrophobic basal surfaces of kaolinite showing that the CO_2 droplets do not interact directly with the gibbsite-like surface but do through a mixture of adsorbed CO_2 and H_2O molecules on the siliceous surface (Cygan et al., 2012; Tenney and Cygan, 2014).

The list of simulation studies on the remediative properties of clays towards pollutants and heavy metals increases every year, and it drives the development of the FF, by adding compatible parameters or by modifying the current ones. For example, Greathouse et al. studied the adsorption of uranyl onto MTM, beidellite and pyrophyllite (Greathouse and Cygan, 2005, 2006; Greathouse et al., 2005; Zaidan et al., 2003), as clay minerals are used as sealant in nuclear waste repositories, using CLAYFF and a mix of potential parameters for the uranyl (Guilbaud and Wipff, 1996) and the carbonate (Greathouse et al., 2002).

Adsorption of molecular species is not limited to inorganic complexes but can also concern organic nanoparticles and hazardous molecules. Zhu et al. studied the adsorption of C_{60}, a carbon nanoparticle, onto pyrophyllite (Zhu et al., 2013) (Figure 2.8) and phenol molecules into

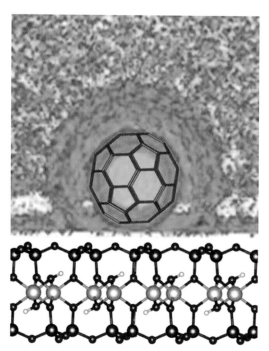

Figure 2.8 *Time-averaged image of water (blue) and C_{60} (yellow) densities above the pyrophyllite (001) surface (O in red, Si in blue, Al in light blue, H in white and C in gray). Adapted from Zhu et al. (2013). Reproduced with permission of American Chemical Society.*

the nano-sized aggregates formed by CTMA alkyl chains in an organo-MTM (Zhu et al., 2011), while Shapley et al. studied the sorption of dioxins at clay–water interfaces (Shapley et al., 2013) (Figure 2.9) showing that the adsorption was due to the hydrophobic environment of the surface created by the organic cations. All these studies used a combination of potential-based and quantum methods; this has the great advantage of validating the potential-based results with higher accuracy simulations but more importantly of validating the performance and reliability of the FF. However, as DFT calculations are still limited to much smaller systems compared to the MM counterpart, the DFT is applied to a small subset of calculations while the broader picture of the dynamics is given by potential-based MD.

Potential-based simulations are indeed utilised for studies on the structure and dynamics of inter-calated molecules in clay minerals, which are important phenomena when dispersions in nanocom-posites and soil properties are considered. As the systems can include higher level of structural and compositional complexity compared to the *ab initio* counterpart, potential-based simulations can be used to explore more complex phenomena. Heinz et al. studied alkylammonium-modified MTMs with different cation exchange capacity (CEC) finding that low CEC leads to stepwise increases of the basal plane spacing contrary to high CEC that leads to a continuous increase in basal plane spacing with increasing chain length (Heinz et al., 2006). Fu et al. studied the cleavage properties of 50 surfac-tant-modified clay minerals with different CECs (Fu and Heinz, 2010). The knowledge of mechanical properties of clays is of crucial importance for designing next-generation materials such as clay–polymer filler nanocomposites. Suter et al. estimated the bending modulus corresponding to the in-plane Young's modulus of MTM (Suter et al., 2007), while Carrier et al. computed the elastic properties of hydrated MTM finding that the major factors controlling the stiffness of the

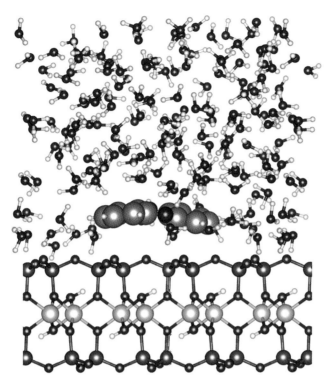

Figure 2.9 *Adsorption of dioxin at the water–(001) surface of pyrophyllite interface (O in red, Si in blue, Al in light blue and H in white). Adapted from Shapley et al. (2013). Reproduced with permission of American Chemical Society.*

material were the water content and the temperature (Carrier et al., 2014). More complex hybrid systems have been considered in the design of materials with tailored strength. Duque-Redondo et al. using CLAYFF and CHARMM found that the intercalation of organic dyes into the interlayer of laponite reduces the mechanical strength compared to the simple clay–water system and related this effect to the disruption of the hydrogen bonding network present in the interlayer space (Duque-Redondo et al., 2014).

One of the areas that is ripe for exploitation in the future is the interface with biological molecules. There have already been some progress in the area of bio-composites (see Chapter 11). Bio-composites composed of the polysaccharide xyloglucan (XG) and MTM clay, which have potential as a 'green' replacement of conventional petroleum-derived polymers in the packaging industry, have been studied using CLAYFF and GLYCAM06. Wang et al. studied the molecular interaction between XG and MTM, which are responsible for the tensile properties (Wang et al., 2014). Bio-composites are also an important area of study in the 'origin of life' where simulations are performed to understand the possible chemical pathways to the formation of biomolecules pertaining to the relative adsorption of bio-macromolecules on mineral surfaces (see Chapter 12). Folding of RNA on MTM in the presence of charge balancing cations was studied by Swadling et al. using the CLAYFF and the AMBER FFs (Swadling et al., 2010). The RNA sequences fold to characteristic secondary structural motifs on the mineral surface that were not seen in the corresponding bulk water simulations. It is clear that all these examples have been made feasibly by the ease of combining different FFs.

2.5.3 Hydroxides and Hydrates

Hydroxide minerals are another important class found in many soils, but they are also relevant in technological applications such as the cement industry. The interest in the latter has increased the need for developing transferable potential parameters for hydroxides and hydrates, particularly for Ca-silicates, aluminates and sulphates and, because of the water, a complex combination of hydrated phases (e.g. ettringite, tobermorites, calcium hydroxide and C-S-H). These mineral phases are not just important for the cement industry, but they represent a large group of naturally occurring minerals geochemically relevant. The C-S-H system is a poorly understood phase. Much of the work is based on modelling the mineral–water interfaces of minerals related to cement phase.

Kalinichev et al. applied first CLAYFF (Kalinichev et al., 2007) to tobermorite, showing that on the (001) surface water shows strong structuring above the surface and in the channels between the drietkette silicate chains due to the development of an integrated H-bond network involving the water and the surface sites. The calculated diffusion coefficients for the surface-associated water were found to be in good agreement with published experimental results. CLAYFF was also applied to study the water structure in nanopores of brucite finding again highly structured water at the surface (Wang et al., 2004). One of the recent improvements in the CLAYFF is related to the vibrational modes of brucite. Zeitler et al. (2014) have modified the Mg–O–H interaction to gain a better comparison with the quantum mechanical counterpart, which also provided an improved description of the edge surfaces of the mineral and again demonstrates the application of DFT for refining potential models.

Durability is an important property of cement-based material that determines the long-term behaviour of the material. The transport of water and ions in the nanopores of the calcium silicate hydrate (C-S-H) gel influences the durability of the cement. Hou et al. used jennite as a model to investigate the structural and dynamical properties of water and Na and Cl ions in a C-S-H system finding that the uptake of Cl is due to the formation of aggregates with Na and providing nanoscale interpretation of the ^{35}Cl NMR and isotherm adsorption experimental studies (Hou and Li, 2014).

Hydrated oxides are another class of minerals that has been often modelled. Shahsavari et al. (2011) compared the performances of a core–shell model (Pellenq et al., 2009) and CSH-FF, a modification of the CLAYFF, to simulate hydrated calcio-silicate materials; both gave good representation of the two modifications of tobermorite, but the rigid ion approach resulted clearly considerably less computational intensive than the core–shell model. The INTERFACE FF was, for example, used to further the understanding of the initial hydration and cohesive properties of cement particles made of calcium silicate, the major mineral phase in cements (Mishra et al., 2013). In this study, the authors quantified the cleavage energies, mechanical properties, the adsorption of organic additives and the agglomeration of calcium silicate cement particles, which are all related causes of the durability of cement-based materials. Finally, it is worth noting that the UFF, described earlier, has been used to simulate the interaction between ettringite and phosphate retarders on hydrating cements (Coveney and Humphries, 1996). Here again the challenge was the large amount of species involved in the system.

2.5.4 Silica and Silicates

Silica, SiO_2, is the second most abundant mineral on the crust after feldspar, another silicate, and has a structure consisting of oxygen sharing SiO_4 tetrahedra. Quartz, α and β, cristobalite and tridymite are important polymorphs of SiO_2. They are topologically identical but geometrically distinct; β quartz and cristobalite are high-temperature polymorphs. With this variety of naturally occurring polymorphs, phase transitions in silica are therefore extensively studied using simulations. The amorphisation of α quartz has been also explained by the presence of elastic instabilities (Tse and Klug, 1991; Watson and Parker, 1995a, 1995b) using the partial charge model of van Beest et al. (van Beest et al., 1990). Two related potential models, a rigid ion derived by Kramer et al. (1991) and a shell model presented by Sanders et al. (1984), were later used to study the transition between α cristobalite and β cristobalite and the disorder present in the β phase (Bourova et al., 2000). The contraction of the unit cell at high temperature was attributed to the decrease in the Si–Si distances because of the large thermal motion of oxygen atoms, which causes the neighbouring tetrahedral to come closer. The dynamical disorder in β cristobalite at high temperatures was also related to the structure to three possible modifications of the low-temperature α phase. Another example of a prediction from potential-based simulations is that zeolite-structured silicas show a negative thermal expansivity (Couves et al., 1993).

On the other hand, as the interaction of water with silica is relevant in geological (Koretsky et al., 1997; Newton and Manning, 2008), biological (Tamerler et al., 2007) and technological (Pinto et al., 2009) contexts, the application of FFs to this material has been widely investigated. Quartz, the most abundant SiO_2 polymorph in soil, is a challenging material to be studied in the presence of water. Quartz has a significant solubility in water, and with a point-of-zero charge reported between 2 and 4 (Kosmulski, 2002), it shows deprotonation of the surface silanol groups at pH above 2, leading to negatively charged surfaces. Different FFs have been used to model surfaces of quartz and silica and their interaction with water, including both shell model (de Leeuw et al., 1999), rigid ion models (Leung et al., 2006), potential parameters compatible with the CHARMM empirical FF by Lopes et al. (2006), the CHARMM water contact angle (CWCA) FF (Cruz-Chu et al., 2006; Lorenz et al., 2008) and reactive FF (Fogarty et al., 2010; Lockwood and Garofalini, 2009). However, one of the major drawbacks of is how to define the accuracy of the FF to describe the water structure above the silica surfaces. Comparison between results from FF and experiments and *ab initio* calculations is therefore crucial to assess the performances and ultimately improve the FF parameters if needed. Skelton et al. (2011) have performed a comparison between the water structure on the quartz (101-1) surface using three different FFs (CLAYFF, Lopes et al. and CWCA), *ab initio* technique and X-ray reflectivity. Overall, better agreement was found between the CLAYFF and the experimental results.

Attention has also focused on silicates, which are the most abundant mineral in the subsurface and upper mantle. Peridotite is a rock mostly consisting of olivine $(Mg, Fe)_2SiO_4$ and pyroxene $(Mg, Fe, Ca)_2(Si, Al)_2O_6$. Forsterite and fayalite are the Mg and Fe endmembers of olivine. In olivine, the silica tetrahedra are connected by divalent cations in contrast to silica polymorphs where the tetrahedral are polymerised. This structural feature causes weak mechanical resistance to weathering. The high cation content and the ease of breaking the oxygen–metal bond make this class of minerals highly reactive and particularly attractive to the carbon sequestration industry (Matter and Kelemen, 2009). The reactivity is not limited to CO_2. Water is another relevant and topical molecule, and its transport through porous rocks has been investigated. For example, recently the structure and dynamics of forsterite $-scCO_2/H_2O$ interfaces as a function of water content were studied by Kerisit et al. (2012). Their MD simulations suggested that, in the presence of sufficient water, $H_xCO_3^{(2-x)-}$ formation occurs in the water films and not via direct reaction of CO_2 with the forsterite surface.

2.5.5 Iron-Based Minerals

Amongst many varieties of iron oxides, magnetite (Fe_3O_4) and hematite $(\alpha\text{-}Fe_2O_3)$ are the most important. Hematite contains only Fe^{3+}, unlike magnetite that comprises a mixture of Fe^{2+} and Fe^{3+}. While hematite forms in oxygen-rich environments, magnetite forms in anoxic environments, one amongst the other is magnetotactic bacteria, and alters to hematite in oxic environments.

(a)   (b)

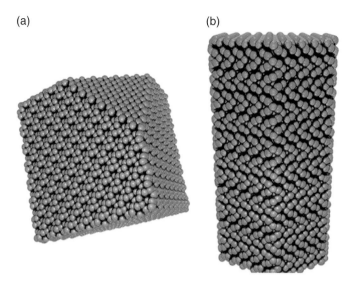

Figure 2.10 *Hematite (Fe$_2$O$_3$) nanoparticles comprising (a) the {01.2} (sides) and an oxygen terminated (00.1) (truncated corner) surfaces and (b) the {10.0} (sides) and iron-terminated (00.1) (top) surfaces.*

Weathering of this Fe-rich mineral usually enhances oxidation and produces iron oxyhydroxide minerals goethite and lepidocrocite amongst others (see Chapter 7). Magnetite is also ferromagnetic while hematite is antiferromagnetic leading to many studies on the magnetism of these minerals.

Bulk properties of the minerals are of extreme importance not only from a geochemical but also from a technological prospective. Chicot et al. (2011) have applied potential-based molecular dynamics and instrumented indentation to calculate the hardness and elastic properties of pure and complex (corrosion products) iron oxides, magnetite (Fe$_3$O$_4$), hematite (α-Fe$_2$O$_3$) and goethite (α-FeOOH) indicating good agreement between the calculated and the measured properties. Using the shell models of Lewis and Catlow (1985) and Woodley et al. (1999), a modification of these potential parameters was used also to provide the calculated infrared and Raman spectra of the minerals (Chamritski and Burns, 2005). Of particular note is the study of Kerisit and Rosso, who performed a comparative study of the rate exchange of charge transfer, applying Marcus theory (Marcus and Sutin, 1985) to wüstite (Kerisit and Rosso, 2005). Wüstite is FeO, and it forms in highly reducing conditions. Accounting for the polarisability of the system via shell models was essential in reproducing the *ab initio* charge transfer rates, compared to RIM counterpart (Bush et al., 1994).

Surfaces have also been extensively studied. For example, Kundu et al. (2006) studied the inverse spinel magnetite and its bare, hydroxylated and hydrated surfaces, to determine the predominant faces in the crystal morphology, while Spagnoli et al. have studied pure-water and salt solutions on the hematite (001) surface (Spagnoli et al., 2006) and nanoparticles (Spagnoli et al., 2009) (Figure 2.10). There are also an increasing number of studies using computer simulations in combination with experiments; adsorption of water vapour on magnetite nanoparticles was simulated, using a combination of UFF and TIP4P, to be a nucleation-like process and forming well-distinguished hydration layers (Tombácz et al., 2009). Static FFs have been widely used, but with the advent of reactive FFs, the dynamical evolution of the mineral–water interfaces can be explored, for instance, to calculate the populations of the surface functional groups, the distribution of electrolyte in the solution and the surface hydrogen bonding on the magnetite (001) surface in contact with a pure and a NaClO$_4$ water solutions (Rustad et al., 2003).

Fe-based minerals in the Earth's subsurface are also involved in the 'mineral iron respiration', a process that a diversity of microorganisms use to obtain energy from a substrate (Richardson et al., 2013) (see Chapter 11). This anaerobic process involves electron transfer (see Chapter 7) between the organic metabolism and the inorganic Fe metabolite, which is reduced from the ferric to the ferrous state; however, the mechanisms underpinning the transfer are not yet fully explained. Magnetosome membrane proteins are responsible for the direct interaction with the mineral substrate; therefore the interaction between amino acids and magnetite has attracted much attention (Buerger et al., 2013).

Iron oxyhydroxides are found in soils and are major sorption substrates for toxins, contaminants and heavy metals (Bargar et al., 1997; Waychunas et al., 2005). Perhaps goethite is the most abundant Fe oxyhydroxide, and its solubility depends on the pH and the oxidation state of Fe (Schwertmann, 1991). Explicit treatment of the solvent molecules in capturing the effect of the surface on the liquid phase and the sorption and dynamics of monovalent ions in modelling goethite–water interactions was highlighted by using a combination of static potential models (Kerisit et al., 2005b, 2006). Reactive FFs have been also heavily applied to goethite mineral surfaces to understand the proton equilibria at the surfaces (Aryanpour et al., 2010; Rustad et al., 1996). The work by Rustad and co-workers on the goethite–water interface used a water potential of Halley et al. (1993). This water potential is a polarisable and dissociating water model based on the work of Stillinger and David (1980). The advantage of using a water model that has the ability to dissociate gives the possibility of describing the acid–base chemistry at the goethite–water interface. Recently, Aryanpour and co-workers (2010) have developed a reactive FF, within the ReaxFF methodology, to describe the goethite–water interface. The structure of the solution on the goethite surface shows very good agreement with experimental data and provides the framework for future applications in other iron oxide systems. The examples given earlier highlight the fact that FFs can reproduce very complex systems with high degrees of accuracy. As the level of understanding from experimental evidence increases on processes such as crystal growth, adsorption and dissolution, the advancement of complexity in the FFs developed has had to occur to mirror that understanding. These examples also show that FFs can be used as a complement to experiment to aid in the understanding very complex geochemical systems.

2.6 Concluding Remarks

We have shown how the development of FFs has been successful for important minerals and their properties. With the availability of high-quality electronic structure simulations, there is much more data available to derive more sophisticated potentials and hence increased interest in deriving many-body potentials, such as the multipoles and EAM-based potentials (Cooper et al., 2014). Furthermore, it is worth mentioning the 'on the fly' potentials, where the potential parameters are evaluated and updated between quantum-derived structures during the course of the simulation.

However, challenges remain and can be divided in terms of models and methodologies. The widespread usage of FF-based techniques is intrinsically limited by the lack of a UFF that can be used to simulate the entire periodic table with high accuracy. The FF should be able to simulate all phases, materials, minerals, organic and biological molecules without any problem related to transferability. It should capture the differences in the interactions between species with different hybridisation, oxidation state and environment. It should deal with polarisability with a smart inclusion of all multipole interactions. It should be intrinsically reactive to deal with the change of the pH and ionic strength. Finally, each level of complexity should switchable; the FF should include parameters to turn on and off the different level of complexity. The development of the models has to be followed by the development of methodologies associated to each level of complexity in the models, such as more efficient ways to include the multipoles and their interactions and reliable ways of accounting for reactivity, without the drawback of over-parameterised FFs. Furthermore, novel methods of deriving and testing FFs need to be developed into more automatic procedures including using machine learning approaches such as artificial neural networks and multilayer perceptrons that could help determining and representing the potential energy surface (Handley and Popelier, 2010). Another important area to exploit is the development of tools that link the different levels of theory to be able to span between different length and time scales, such as the hybrid methodologies, described earlier, but as discussed, it is still in its infancy. The challenge of modelling still larger systems for longer times will ensure that the testing and development of reliable FFs will continue to be required for the foreseeable future.

References

Abascal, J.L.F., et al., A potential model for the study of ices and amorphous water: TIP4P/ice. *Journal of Chemical Physics*, 2005. **122**(23): p. 9.

Abrams, C.F. and K. Kremer, Combined coarse-grained and atomistic simulation of liquid bisphenol A-polycarbonate: liquid packing and intramolecular structure. *Macromolecules*, 2003. **36**(1): p. 260–267.

Aguado, A., et al., Multipoles and interaction potentials in ionic materials from planewave-DFT calculations. *Faraday Discussions*, 2003. **124**: p. 171–184.

Aguzzi, C., et al., Use of clays as drug delivery systems: possibilities and limitations. *Applied Clay Science*, 2007. **36**(1–3): p. 22–36.

Aicken, A.M., et al., Simulation of layered double hydroxide intercalates. *Advanced Materials*, 1997. **9**(6): p. 496–500.

Alexandrov, V. and K.M. Rosso, Insights into the mechanism of Fe(II) adsorption and oxidation at Fe–clay mineral surfaces from first-principles calculations. *The Journal of Physical Chemistry C*, 2013. **117**(44): p. 22880–22886.

Allen, M.P. and D.J. Tildesley, *Computer Simulation of Liquids*. 1989, Oxford: Clarendon Press. p. 385.

Allinger, N.L., Conformational-analysis. 130. MM2: hydrocarbon force-field utilizing V1 and V2 torsional terms. *Journal of the American Chemical Society*, 1977. **99**(25): p. 8127–8134.

Allinger, N.L., Y.H. Yuh and J.H. Lii, Molecular mechanics: the MM3 force-field for hydrocarbons. 1. *Journal of the American Chemical Society*, 1989. **111**(23): p. 8551–8566.

Allinger, N.L., K. Chen and J.-H. Lii, An improved force field (MM4) for saturated hydrocarbons. *Journal of Computational Chemistry*, 1996. **17**(5–6): p. 642–668.

Anderson, R.L., et al., Clay swelling: a challenge in the oilfield. *Earth-Science Reviews*, 2010. **98**(3–4): p. 201–216.

Applequist, J., A multipole interaction theory of electric polarization of atomic and molecular assemblies. *Journal of Chemical Physics*, 1985. **83**(2): p. 809–826.

Arcos, D., et al., Long-term geochemical evolution of the near field repository: Insights from reactive transport modelling and experimental evidences. *Journal of Contaminant Hydrology*, 2008. **102**(3–4): p. 196–209.

Aryanpour, M., A.C.T. van Duin and J.D. Kubicki, Development of a reactive force field for iron-oxyhydroxide systems. *The Journal of Physical Chemistry A*, 2010. **114**(21): p. 6298–6307.

Attard, P., Electrolytes and the Electric Double Layer, in *Advances in Chemical Physics*, ed. I. Prigogine and S.A. Rice. 2007, Chichester: John Wiley & Sons, Inc. p. 1–159.

Austen, K.F., et al., Electrostatic versus polarization effects in the adsorption of aromatic molecules of varied polarity on an insulating hydrophobic surface. *Journal of Physics. Condensed Matter*, 2008. **20**(3): p. 035215.

Bader, R.F.W., Atoms in molecules. *Accounts of Chemical Research*, 1985. **18**(1): p. 9–15.

Bains, A.S., et al., Molecular modelling of the mechanism of action of organic clay-swelling inhibitors. *Molecular Simulation*, 2001. **26**(2): p. 101−145.

Bargar, J.R., G.E. Brown and G.A. Parks, Surface complexation of Pb(II) at oxide-water interfaces. 2. XAFS and bond-valence determination of mononuclear Pb(II) sorption products and surface functional groups on iron oxides. *Geochimica et Cosmochimica Acta*, 1997. **61**(13): p. 2639–2652.

Barlow, S., et al., Molecular mechanics study of oligomeric models for poly(ferrocenylsilanes) using the extensible systematic forcefield (ESFF). *Journal of the American Chemical Society*, 1996. **118**(32): p. 7578–7592.

Bayly, C.I., et al., A well-behaved electrostatic potential based method using charge restraints for deriving atomic charges: the RESP model. *Journal of Physical Chemistry*, 1993. **97**(40): p. 10269–10280.

Berendsen, H.J.C., J.R. Grigera and T.P. Straatsma, The missing term in effective pair potentials. *Journal of Physical Chemistry*, 1987. **91**(24): p. 6269–6271.

Berthelot, D., Sur le Mélange des Gaz. *Comptes rendus de l'Academie des Sciences*, 1889. **126**: p. 1703–1706.

Bhowmik, R., K.S. Katti and D. Katti, Molecular dynamics simulation of hydroxyapatite-polyacrylic acid interfaces. *Polymer*, 2007. **48**(2): p. 664–674.

Bickmore, B.R., et al., Bond-valence constraints on liquid water structure. *The Journal of Physical Chemistry A*, 2009. **113**(9): p. 1847–1857.

Boek, E.S., P.V. Coveney and N.T. Skipper, Monte Carlo molecular modeling studies of hydrated Li-, Na-, and K-smectites: understanding the role of potassium as a clay swelling inhibitor. *Journal of the American Chemical Society*, 1995. **117**(50): p. 12608–12617.

Bourova, E., S.C. Parker and P. Richet, Atomistic simulation of cristobalite at high temperature. *Physical Review B*, 2000. **62**(18): p. 12052–12061.

Brancato, G. and M.E. Tuckerman, A polarizable multistate empirical valence bond model for proton transport in aqueous solution. *Journal of Chemical Physics*, 2005. **122**(22): p. 224507.

Brenner, D.W., Empirical potential for hydrocarbons for use in simulating the chemical vapor-deposition of diamond films. *Physical Review B*, 1990. **42**(15): p. 9458–9471.

Brenner, D.W., et al., A second-generation reactive empirical bond order (REBO) potential energy expression for hydrocarbons. *Journal of Physics. Condensed Matter*, 2002. **14**(4): p. 783–802.

Bresme, F., Equilibrium and nonequilibrium molecular-dynamics simulations of the central force model of water. *The Journal of Chemical Physics*, 2001. **115**(16): p. 7564–7574.

Brini, E., V. Marcon and N.F.A. van der Vegt, Conditional reversible work method for molecular coarse graining applications. *Physical Chemistry Chemical Physics*, 2011. **13**(22): p. 10468–10474.

Brini, E., et al., Thermodynamic transferability of coarse-grained potentials for polymer-additive systems. *Physical Chemistry Chemical Physics*, 2012. **14**(34): p. 11896–11903.

Brini, E., et al., Systematic coarse-graining methods for soft matter simulations: a review. *Soft Matter*, 2013. **9**(7): p. 2108–2119.

Brooks, B.R., et al., CHARMM: a program for macromolecular energy, minimization, and dynamics calculations. *Journal of Computational Chemistry*, 1983. **4**(2): p. 187–217.

Broukhno, A., et al., Depletion and bridging forces in polymer systems: Monte Carlo simulations at constant chemical potential. *Journal of Chemical Physics*, 2000. **113**(13): p. 5493–5501.

Broukhno, A., et al., Polyampholyte-induced repulsion between charged surfaces: Monte Carlo simulation studies. *Langmuir*, 2002. **18**(16): p. 6429–6436.

Brukhno, A.V., T. Akesson and B. Jonsson, Phase behavior in suspensions of highly charged colloids. *The Journal of Physical Chemistry B*, 2009. **113**(19): p. 6766–6774.

Buerger, A., U. Magdans and H. Gies, Adsorption of amino acids on the magnetite-(111)-surface: a force field study. *Journal of Molecular Modeling*, 2013. **19**(2): p. 851–857.

Bulo, R.E., et al., Toward a practical method for adaptive QM/MM simulations. *Journal of Chemical Theory and Computation*, 2009. **5**(9): p. 2212–2221.

Bush, T.S., et al., Self-consistent interatomic potentials for the simulation of binary and ternary oxides. *Journal of Materials Chemistry*, 1994. **4**(6): p. 831–837.

Carlson, W.D., The polymorphs of $CaCO_3$ and the aragonite-calcite transformation *Reviews in Mineralogy and Geochemistry*, 1983. **11**: p. 191–225.

Carof, A., et al., Coarse graining the dynamics of nano-confined solutes: the case of ions in clays. *Molecular Simulation*, 2014. **40**(1–3): p. 237–244.

Carrier, B., et al., Elastic properties of swelling clay particles at finite temperature upon hydration. *The Journal of Physical Chemistry C*, 2014. **118**(17): p. 8933–8943.

Catlow, C.R.A., et al., Simulating silicate structures and the structural chemistry of pyroxenoids. *Nature*, 1982. **295**(5851): p. 658–662.

Celis, R., M.C. Hermosin and J. Cornejo, Heavy metal adsorption by functionalized clays. *Environmental Science & Technology*, 2000. **34**(21): p. 4593–4599.

Chamritski, I. and G. Burns, Infrared- and Raman-active phonons of magnetite, maghemite, and hematite: a computer simulation and spectroscopic study. *The Journal of Physical Chemistry B*, 2005. **109**(11): p. 4965–4968.

Chenoweth, K., A.C.T. van Duin and W.A. Goddard, ReaxFF reactive force field for molecular dynamics simulations of hydrocarbon oxidation. *Journal of Physical Chemistry A*, 2008. **112**(5): p. 1040–1053.

Chicot, D., et al., Mechanical properties of magnetite (Fe_3O_4), hematite (α-Fe_2O_3) and goethite (α-FeO·OH) by instrumented indentation and molecular dynamics analysis. *Materials Chemistry and Physics*, 2011. **129**(3): p. 862–870.

Churakov, S.V., et al., Resolving diffusion in clay minerals at different time scales: combination of experimental and modeling approaches. *Applied Clay Science*, 2014. **96**(0): p. 36–44.

Collins, D.R. and C.R.A. Catlow, Computer-simulation of structures and cohesive properties of MICAS. *American Mineralogist*, 1992. **77**(11–12): p. 1172–1181.

Cooke, D.J. and J.A. Elliott, Atomistic simulations of calcite nanoparticles and their interaction with water. *Journal of Chemical Physics*, 2007. **127**(10): p. 104706.

Cooper, M.W.D., M.J.D. Rushton and R.W. Grimes, A many-body potential approach to modelling the thermomechanical properties of actinide oxides. *Journal of Physics. Condensed Matter*, 2014. **26**(10): 105401.

Couves, J.W., et al., Experimental-verification of a predicted negative thermal expansivity of crystalline zeolites. *Journal of Physics. Condensed Matter*, 1993. **5**(27): p. L329–L332.

Coveney, P.V. and W. Humphries, Molecular modelling of the mechanism of action of phosphonate retarders on hydrating cements. *Journal of the Chemical Society, Faraday Transactions*, 1996. **92**(5): p. 831–841.

Cox, S.R. and D.E. Williams, Representation of the molecular electrostatic potential by a net atomic charge model. *Journal of Computational Chemistry*, 1981. **2**(3): p. 304–323.

Cruz-Chu, E.R., A. Aksimentiev and K. Schulten, Water−silica force field for simulating nanodevices. *The Journal of Physical Chemistry B*, 2006. **110**(43): p. 21497–21508.

Cygan, R.T., et al., Atomistic models of carbonate minerals: bulk and surface structures, defects, and diffusion. *Molecular Simulation*, 2002. **28**(6–7): p. 475–495.

Cygan, R.T., J.J. Liang and A.G. Kalinichev, Molecular models of hydroxide, oxyhydroxide, and clay phases and the development of a general force field. *Journal of Physical Chemistry B*, 2004. **108**(4): p. 1255–1266.

Cygan, R.T., V.N. Romanov and E.M. Myshakin, Molecular simulation of carbon dioxide capture by montmorillonite using an accurate and flexible force field. *The Journal of Physical Chemistry C*, 2012. **116**(24): p. 13079–13091.

Dauberosguthorpe, P., et al., Structure and energetics of ligand-binding to proteins: *Escherichia coli* dihydrofolate reductase trimethoprim, a drug-receptor system. *Proteins: Structure, Function, and Genetics*, 1988. **4**(1): p. 31–47.

de Leeuw, N.H., F.M. Higgins and S.C. Parker, Modeling the surface structure and stability of α-quartz. *The Journal of Physical Chemistry B*, 1999. **103**(8): p. 1270–1277.

Deer, W.A., *An Introduction to the Rock-Forming Minerals/W.A. Deer, R.A. Howie, J. Zussman*, ed. J. Zussman and R.A. Howie. 1992, Harlow, England/New York, NY: Longman Scientific & Technical/Wiley.

Delgado-Buscalioni, R., K. Kremer and M. Praprotnik, Concurrent triple-scale simulation of molecular liquids. *Journal of Chemical Physics*, 2008. **128**(11): p. 114110.

Delhommelle, J. and P. Millie, Inadequacy of the Lorentz-Berthelot combining rules for accurate predictions of equilibrium properties by molecular simulation. *Molecular Physics*, 2001. **99**(8): p. 619–625.

Demichelis, R., et al., Stable prenucleation mineral clusters are liquid-like ionic polymers. *Nature Communications*, 2011. **2**: p. 1–8.

Demiralp, E., T. Cagin and W.A. Goddard, Morse stretch potential charge equilibrium force field for ceramics: application to the quartz-stishovite phase transition and to silica glass. *Physical Review Letters*, 1999. **82**(8): p. 1708–1711.

Derjaguin, B. and L. Landau, Theory of the stability of strongly charged lyophobic sols and of the adhesion of strongly charged-particles in solutions of electrolytes. *Progress in Surface Science*, 1993. **43**(1–4): p. 30–59.

di Pasquale, N., D. Marchisio and P. Carbone, Mixing atoms and coarse-grained beads in modelling polymer melts. *Journal of Chemical Physics*, 2012. **137**(16): p. 164111–164119.

Dick, B.G. and A.W. Overhauser, Theory of the dielectric constants of alkali halide crystals. *Physical Review*, 1958. **112**(1): p. 90.

Dixon, D.A., M.N. Gray and A.W. Thomas, A study of the compaction properties of potential clay-sand buffer mixtures for use in nuclear-fuel waste-disposal. *Engineering Geology*, 1985. **21**(3–4): p. 247–255.

Downing, C.A., A.A. Sokol and C.R.A. Catlow, The reactivity of CO_2 on the MgO(100) surface. *Physical Chemistry Chemical Physics*, 2014. **16**(1): p. 184–195.

Duque-Redondo, E., et al., Molecular forces governing shear and tensile failure in clay-dye hybrid materials. *Chemistry of Materials*, 2014. **26**(15): 4338–4345.

Dykstra, C.E., Electrostatic interaction potentials in molecular-force fields. *Chemical Reviews*, 1993. **93**(7): p. 2339–2353.

Ercolessi, F. and J.B. Adams, Interatomic potentials from 1st-principles calculations: the force-matching method. *Europhysics Letters*, 1994. **26**(8): p. 583–588.

Ewald, P.P., Die Berechnung optischer und elektrostatischer Gitterpotentiale. *Annalen der Physik*, 1921. **369**(3): p. 253–287.

Faux, I.D., Polarization catastrophe in defect calculations in ionic crystals. *Journal of Physics Part C: Solid State Physics*, 1971. **4**(10): p. L211–L216.

Fenter, P., et al., Is the calcite–water interface understood? Direct comparisons of molecular dynamics simulations with specular X-ray reflectivity data. *The Journal of Physical Chemistry C*, 2013. **117**(10): p. 5028–5042.

Ferguson, D.M., Parameterization and evaluation of a flexible water model. *Journal of Computational Chemistry*, 1995. **16**(4): p. 501–511.

Finnis, M., *Interatomic Forces in Condensed Matter*. 2003, Oxford: Oxford University Press.

Fischer, J., et al., Modeling of aqueous poly(oxyethylene) solutions. 2. Mesoscale simulations. *Journal of Physical Chemistry B*, 2008. **112**(43): p. 13561–13571.

Flekkoy, E.G., P.V. Coveney and G. De Fabritiis, Foundations of dissipative particle dynamics. *Physical Review E*, 2000. **62**(2): p. 2140–2157.

Fogarty, J.C., et al., A reactive molecular dynamics simulation of the silica-water interface. *The Journal of Chemical Physics*, 2010. **132**(17): p. 174704.

Freeman, C.L., et al., A new potential model for barium titanate and its implications for rare-earth doping. *Journal of Materials Chemistry*, 2011. **21**(13): p. 4861–4868.

Frenkel, D. and B. Smith, *Understanding Molecular Simulation: From Algorithms to Applications*. 1996, San Diego: Academic Press, Inc. p. 443.

Fu, Y.-T. and H. Heinz, Structure and cleavage energy of surfactant-modified clay minerals: influence of CEC, head group and chain length. *Philosophical Magazine*, 2010. **90**(17–18): p. 2415–2424.

Fu, C.-C., et al., A test of systematic coarse-graining of molecular dynamics simulations: thermodynamic properties. *Journal of Chemical Physics*, 2012. **137**(16): p. 164106.

Gale, J.D., Empirical potential derivation for ionic materials. *Philosophical Magazine B: Physics of Condensed Matter: Statistical Mechanics, Electronic, Optical and Magnetic Properties*, 1996. **73**(1): p. 3–19.

Gale, J.D. and A.L. Rohl, The general utility lattice program (GULP). *Molecular Simulation*, 2003. **29**(5): p. 291–341.

Gale, J.D., P. Raiteri and A.C.T. van Duin, A reactive force field for aqueous-calcium carbonate systems. *Physical Chemistry Chemical Physics*, 2011. **13**(37): p. 16666–16679.

Gebauer, D., A. Voelkel and H. Coelfen, Stable prenucleation calcium carbonate clusters. *Science*, 2008. **322**(5909): p. 1819–1822.

Gilbert, T.L., Soft-sphere model for closed-shell atoms and ions. *Journal of Chemical Physics*, 1968. **49**(6): p. 2640–2642.

Greathouse, J.A. and R.T. Cygan, Molecular dynamics simulation of uranyl(VI) adsorption equilibria onto an external montmorillonite surface. *Physical Chemistry Chemical Physics*, 2005. **7**(20): p. 3580–3586.

Greathouse, J.A. and R.T. Cygan, Water structure and aqueous uranyl(VI) adsorption equilibria onto external surfaces of beidellite, montmorillonite, and pyrophyllite: results from molecular simulations. *Environmental Science & Technology*, 2006. **40**(12): p. 3865–3871.

Greathouse, J.A., et al., Molecular dynamics study of aqueous uranyl interactions with quartz (010). *Journal of Physical Chemistry B*, 2002. **106**(7): p. 1646–1655.

Greathouse, J.A., et al., Uranyl surface complexes in a mixed-charge montmorillonite: Monte Carlo computer simulation and polarized XAFS results. *Clays and Clay Minerals*, 2005. **53**(3): p. 278–286.

Guilbaud, P. and G. Wipff, Force field representation of the UO_2^{2+} cation from free energy MD simulations in water. Tests on its 18-crown-6 and NO_3^- adducts, and on its calix 6 arene(6-) and CMPO complexes. *Journal of Molecular Structure. (THEOCHEM)*, 1996. **366**(1–2): p. 55–63.

Guillot, B., A reappraisal of what we have learnt during three decades of computer simulations on water. *Journal of Molecular Liquids*, 2002. **101**(1–3): p. 219–260.

Guldbrand, L., et al., Electrical double-layer forces: a Monte-Carlo study. *Journal of Chemical Physics*, 1984. **80**(5): p. 2221–2228.

Halgren, T.A., Representation of van der Waals (vdW) interactions in molecular mechanics force-fields: potential form, combination rules, and VDW parameters. *Journal of the American Chemical Society*, 1992. **114**(20): p. 7827–7843.

Halgren, T.A., Merck molecular force field .1. Basis, form, scope, parameterization, and performance of MMFF94. *Journal of Computational Chemistry*, 1996. **17**(5–6): p. 490–519.

Halley, J.W., J.R. Rustad and A. Rahman, A polarizable, dissociating molecular-dynamics model for liquid water. *Journal of Chemical Physics*, 1993. **98**(5): p. 4110–4119.

Handley, C.M. and P.L.A. Popelier, Potential energy surfaces fitted by artificial neural networks. *The Journal of Physical Chemistry. A*, 2010. **114**(10): p. 3371–3383.

Harmandaris, V.A., et al., Hierarchical modeling of polystyrene: from atomistic to coarse-grained simulations. *Macromolecules*, 2006. **39**(19): p. 6708–6719.

Hauptmann, S., et al., Potential energy function for apatites. *Physical Chemistry Chemical Physics*, 2003. **5**(3): p. 635–639.

Heinz, H., et al., Force field for mica-type silicates and dynamics of octadecylammonium chains grafted to montmorillonite. *Chemistry of Materials*, 2005. **17**(23): p. 5658–5669.

Heinz, H., et al., Self-assembly of alkylammonium chains on montmorillonite: effect of chain length, head group structure, and cation exchange capacity. *Chemistry of Materials*, 2006. **19**(1): p. 59–68.

Heinz, H., et al., Thermodynamically consistent force fields for the assembly of inorganic, organic, and biological nanostructures: the INTERFACE force field. *Langmuir*, 2013. **29**(6): p. 1754–1765.

Henderson, R.L., Uniqueness theorem for fluid pair correlation-functions. *Physics Letters A*, 1974. **49**(3): p. 197–198.

Hermans, J., et al., A consistent empirical potential for water-protein interactions. *Biopolymers*, 1984. **23**(8): p. 1513–1518.

Hess, B., C. Holm and N. van der Vegt, Modeling multibody effects in ionic solutions with a concentration dependent dielectric permittivity. *Physical Review Letters*, 2006. **96**(14): p. 147801.

Hofmann, D.W.M., L. Kuleshova and B. D'Aguanno, A new reactive potential for the molecular dynamics simulation of liquid water. *Chemical Physics Letters*, 2007. **448**(1–3): p. 138–143.

Holesova, S., et al., Antibacterial kaolinite/urea/chlorhexidine nanocomposites: experiment and molecular modelling. *Applied Surface Science*, 2014. **305**: p. 783–791.

Hou, D. and Z. Li, Molecular dynamics study of water and ions transported during the nanopore calcium silicate phase: case study of jennite. *Journal of Materials in Civil Engineering*, 2014. **26**(5): p. 930–940.

Huang, D.M., et al., Coarse-grained computer simulations of polymer/fullerene bulk heterojunctions for organic photovoltaic applications. *Journal of Chemical Theory and Computation*, 2010. **6**(2): p. 526–537.

Hwang, S., et al., The MS-Q force field for clay minerals: application to oil production. *The Journal of Physical Chemistry B*, 2001. **105**(19): p. 4122–4127.

Hynninen, A.P. and A.Z. Panagiotopoulos, Phase diagrams of charged colloids from thermodynamic integration. *Journal of Physics. Condensed Matter*, 2009. **21**(46): p. 465104.

Israelachvili, J.N., *Intermolecular and Surface Forces* (3rd Edition). 2011, Burlington: Academic Press. p. 1–674.

Izvekov, S. and G.A. Voth, Multiscale coarse graining of liquid-state systems. *Journal of Chemical Physics*, 2005. **123**(13): p. 134105.

Jones, J.E., On the determination of molecular fields. I. From the variation of the viscosity of a gas with temperature. *Proceedings of the Royal Society of London, Series A*, 1924a. **106**(738): p. 441–462.

Jones, J.E., On the determination of molecular fields. II. From the equation of state of a gas. *Proceedings of the Royal Society of London, Series A*, 1924b. **106**(738): p. 463–477.

Jonsson, B., et al., Depletion and structural forces in confined polyelectrolyte solutions. *Langmuir*, 2003. **19**(23): p. 9914–9922.

Jorgensen, W.L. and C.J. Swenson, Optimized intermolecular potential functions for amides and peptides: hydration of amides. *Journal of the American Chemical Society*, 1985. **107**(6): p. 1489–1496.

Jorgensen, W.L., et al., Comparison of simple potential functions for simulating liquid water. *Journal of Chemical Physics*, 1983. **79**(2): p. 926–935.

Jorgensen, W.L., D.S. Maxwell and J. TiradoRives, Development and testing of the OPLS all-atom force field on conformational energetics and properties of organic liquids. *Journal of the American Chemical Society*, 1996. **118**(45): p. 11225–11236.

Kalinichev, A.G., J. Wang and R.J. Kirkpatrick, Molecular dynamics modeling of the structure, dynamics and energetics of mineral–water interfaces: Application to cement materials. *Cement and Concrete Research*, 2007. **37**(3): p. 337–347.

Kamerlin, S.C.L., et al., On unjustifiably misrepresenting the EVB approach while simultaneously adopting it. *Journal of Physical Chemistry B*, 2009. **113**(31): p. 10905–10915.

Kerisit, S. and S.C. Parker, Free energy of adsorption of water and metal ions on the {1014} calcite surface. *Journal of the American Chemical Society*, 2004. **126**(32): p. 10152–10161.

Kerisit, S. and K.M. Rosso, Charge transfer in FeO: a combined molecular-dynamics and ab initio study. *The Journal of Chemical Physics*, 2005. **123**(22): p. 224712.

Kerisit, S., et al., Molecular dynamics simulations of the interactions between water and inorganic solids. *Journal of Materials Chemistry*, 2005a. **15**(14): p. 1454–1462.

Kerisit, S., et al., Atomistic simulation of charged iron oxyhydroxide surfaces in contact with aqueous solution. *Chemical Communications*, 2005b. (**24**): p. 3027–3029.

Kerisit, S., E.S. Ilton and S.C. Parker, Molecular dynamics simulations of electrolyte solutions at the (100) goethite surface. *The Journal of Physical Chemistry B*, 2006. **110**(41): p. 20491–20501.

Kerisit, S., J.H. Weare and A.R. Felmy, Structure and dynamics of forsterite–scCO$_2$/H$_2$O interfaces as a function of water content. *Geochimica et Cosmochimica Acta*, 2012. **84**(0): p. 137–151.

Khan, A., K. Fontell and B. Lindman, Phase-equilibria of some ionic surfactant systems with divalent counterions. *Colloids and Surfaces*, 1984a. **11**(3–4): p. 401–408.

Khan, A., K. Fontell and B. Lindman, Liquid crystallinity in systems of magnesium and calcium surfactants: phase-diagrams and phase structures in binary aqueous systems of magnesium and calcium di-2-ethylhexylsulfosuccinate. *Journal of Colloid and Interface Science*, 1984b. **101**(1): p. 193–200.

Kong, C.L., Combining rules for intermolecular potential parameters. 2. Rules for Lennard-Jones (12–6) potential and Morse potential. *Journal of Chemical Physics*, 1973. **59**(5): p. 2464–2467.

Konowalow, D.D., Interatomic potentials for HeNe, HeAr, and NeAr. *Journal of Chemical Physics*, 1969. **50**(1): p. 12–16.

Koretsky, C.M., et al., Detection of surface hydroxyl species on quartz, γ-alumina, and feldspars using diffuse reflectance infrared spectroscopy. *Geochimica et Cosmochimica Acta*, 1997. **61**(11): p. 2193–2210.

Korkut, A. and W.A. Hendrickson, A force field for virtual atom molecular mechanics of proteins. *Proceedings of the National Academy of Sciences of the United States of America*, 2009. **106**(37): p. 15667–15672.

Kosmulski, M., The pH-dependent surface charging and the points of zero charge. *Journal of Colloid and Interface Science*, 2002. **253**(1): p. 77–87.

Kramer, G.J., et al., Interatomic force fields for silicas, aluminophosphates, and zeolites: derivation based on ab initio calculations. *Physical Review B*, 1991. **43**(6): p. 5068–5080.

Kundu, T.K., K.H. Rao and S.C. Parker, Atomistic simulation studies of magnetite surface structures and adsorption behavior in the presence of molecular and dissociated water and formic acid. *Journal of Colloid and Interface Science*, 2006. **295**(2): p. 364–373.

Labbez, C., et al., Surface charge density and electrokinetic potential of highly charged minerals: experiments and Monte Carlo simulations on calcium silicate hydrate. *Journal of Physical Chemistry B*, 2006. **110**(18): p. 9219–9230.

Leach, A., *Molecular Modelling: Principles and Applications* (2nd Edition). 2001, Harlow, England/New York: Prentice Hall.

Lee, H., et al., A coarse-grained model for polyethylene oxide and polyethylene glycol: conformation and hydrodynamics. *Journal of Physical Chemistry B*, 2009. **1613**(40): p. 13186–13194.

Lenart, P.J., A. Jusufi and A.Z. Panagiotopoulos, Effective potentials for 1: 1 electrolyte solutions incorporating dielectric saturation and repulsive hydration. *Journal of Chemical Physics*, 2007. **126**(4): p. 044509.

Leung, K., S.B. Rempe and C.D. Lorenz, Salt permeation and exclusion in hydroxylated and functionalized silica pores. *Physical Review Letters*, 2006. **96**(9): p. 095504.

Lewis, G.V. and C.R.A. Catlow, Potential models for ionic oxides. *Journal of Physics C: Solid State Physics*, 1985. **18**(6): p. 1149–1161.

Liang, T., et al., Reactive potentials for advanced atomistic simulations. *Annual Review of Materials Research*, 2013. **43**: p. 109–129.

Lifson, S., A.T. Hagler and P. Dauber, Consistent force-field studies of inter-molecular forces in hydrogen-bonded crystals. 1. Carboxylic-acids, amides, and the C=O…H-hydrogen-bonds. *Journal of the American Chemical Society*, 1979. **101**(18): p. 5111–5121.

Lin, F.H., et al., A study of purified montmorillonite intercalated with 5-fluorouracil as drug carrier. *Biomaterials*, 2002. **23**(9): p. 1981–1987.

Liu, P. and G.A. Voth, Smart resolution replica exchange: an efficient algorithm for exploring complex energy landscapes. *Journal of Chemical Physics*, 2007. **126**(4): p. 045106.

Lobaugh, J. and G.A. Voth, The quantum dynamics of an excess proton in water. *Journal of Chemical Physics*, 1996. **104**(5): p. 2056–2069.

Lockwood, G.K. and S.H. Garofalini, Bridging oxygen as a site for proton adsorption on the vitreous silica surface. *Journal of Chemical Physics*, 2009. **131**(7): p. 074703.

Lopes, P.E.M., et al., Development of an empirical force field for silica. Application to the quartz–water interface. *The Journal of Physical Chemistry B*, 2006. **110**(6): p. 2782–2792.

Lopes, P.E.M., B. Roux and A.D. MacKerell, Jr., Molecular modeling and dynamics studies with explicit inclusion of electronic polarizability: theory and applications. *Theoretical Chemistry Accounts*, 2009. **124** (1–2): p. 11–28.

Lorentz, H.A., Ueber die Anwendung des Satzes vom Virial in der kinetischen Theorie der Gase. *Annalen der Physik*, 1881. **248**(1): p. 127–136.

Lorenz, C.D., et al., Molecular dynamics of ionic transport and electrokinetic effects in realistic silica channels. *The Journal of Physical Chemistry C*, 2008. **112**(27): p. 10222–10232.

Lyman, E., F.M. Ytreberg and D.M. Zuckerman, Resolution exchange simulation. *Physical Review Letters*, 2006. **96**(2): p. 028105.

Lyubartsev, A.P., Multiscale modeling of lipids and lipid bilayers. *European Biophysics Journal with Biophysics Letters*, 2005. **35**(1): p. 53–61.

Lyubartsev, A.P. and A. Laaksonen, Calculation of effective interaction potentials from radial-distribution functions: a reverse Monte-Carlo approach. *Physical Review E*, 1995. **52**(4): p. 3730–3737.

Lyubartsev, A.P., et al., On coarse-graining by the inverse Monte Carlo method: dissipative particle dynamics simulations made to a precise tool in soft matter modeling. *Soft Materials*, 2003. **1**(1): p. 121–137.

MacKerell, A.D., et al., All-atom empirical potential for molecular modeling and dynamics studies of proteins. *Journal of Physical Chemistry B*, 1998. **102**(18): p. 3586–3616.

Madden, P.A., et al., From first-principles to material properties. *Journal of Molecular Structure (THEOCHEM)*, 2006. **771**(1–3): p. 9–18.

Mahadevan, T.S. and S.H. Garofalini, Dissociative water potential for molecular dynamics simulations. *Journal of Physical Chemistry B*, 2007. **111**(30): p. 8919–8927.

Mahoney, M. and W. Jorgensen, A five-site model for liquid water and the reproduction of the density anomaly by rigid, nonpolarizable potential functions. *The Journal of Chemical Physics*, 2000. **112**(20): p. 8910–8922.

Marcus, R.A. and N. Sutin, Electron transfers in chemistry and biology. *Biochimica et Biophysica Acta (BBA) – Reviews on Bioenergetics*, 1985. **811**(3): p. 265–322.

Markutsya, S., et al., Evaluation of coarse-grained mapping schemes for polysaccharide chains in cellulose. *Journal of Chemical Physics*, 2013. **138**(21): p. 214108.

Marra, J., Direct measurement of the interaction between phosphatidylglycerol bilayers in aqueous-electrolyte solutions. *Biophysical Journal*, 1986. **50**(5): p. 815–825.

Marrink, S.J., A.H. de Vries and A.E. Mark, Coarse grained model for semiquantitative lipid simulations. *Journal of Physical Chemistry B*, 2004. **108**(2): p. 750–760.

Marrink, S.J., et al., The MARTINI force field: coarse grained model for biomolecular simulations. *Journal of Physical Chemistry B*, 2007. **111**(27): p. 7812–7824.

Martin, P., et al., Application of molecular dynamics DL_POLY codes to interfaces of inorganic materials. *Molecular Simulation*, 2006. **32**(12–13): p. 1079–1093.

Mason, E.A., Forces between unlike molecules and the properties of gaseous mixtures. *Journal of Chemical Physics*, 1955. **23**(1): p. 49–56.

Matter, J.M. and P.B. Kelemen, Permanent storage of carbon dioxide in geological reservoirs by mineral carbonation. *Nature Geoscience*, 2009. **2**(12): p. 837–841.

Matysiak, S., et al., Modeling diffusive dynamics in adaptive resolution simulation of liquid water. *Journal of Chemical Physics*, 2008. **128**(2): p. 024503.

Mayo, S.L., B.D. Olafson and W.A. Goddard, Dreiding: a generic force-field for molecular simulations. *Journal of Physical Chemistry*, 1990. **94**(26): p. 8897–8909.

Meakin, P. and Z. Xu, Dissipative Particle Dynamics and other particle methods for multiphase fluid flow in fractured and porous media. *Progress in Computational Fluid Dynamics*, 2009. **9**(6–7): p. 399–408.

Metropolis, N., et al., Equation of state calculations by fast computing machines. *Journal of Chemical Physics*, 1953. **21**(6): p. 1087–1092.

Milani, A., et al., Coarse-grained simulations of model polymer nanofibres. *Macromolecular Theory and Simulations*, 2011. **20**(5): p. 305–319.

Mirzoev, A. and A.P. Lyubartsev, Effective solvent mediated potentials of Na+ and Cl- ions in aqueous solution: temperature dependence. *Physical Chemistry Chemical Physics*, 2011. **13**(13): p. 5722–5727.

Mirzoev, A. and A.P. Lyubartsev, Systematic implicit solvent coarse graining of dimyristoylphosphatidylcholine lipids. *Journal of Computational Chemistry*, 2014. **35**(16): p. 1208–1218.

Mishra, R.K., R.J. Flatt and H. Heinz, Force field for tricalcium silicate and insight into nanoscale properties: cleavage, initial hydration, and adsorption of organic molecules. *The Journal of Physical Chemistry C*, 2013. **117**(20): p. 10417–10432.

Moreno, N., P. Vignal, J. Li and V. M. Calo, Multiscale modeling of blood flow: coupling finite elements with smoothed dissipative particle dynamics. *Procedia Computer Science*, 2013. **18**: p. 2565–2574.

Mueller, J.E., A.C.T. van Duin and W.A. Goddard, Development and validation of ReaxFF reactive force field for hydrocarbon chemistry catalyzed by nickel. *Journal of Physical Chemistry C*, 2010. **114**(11): p. 4939–4949.

Mulliken, R.S., Electronic population analysis on LCAO-MO molecular wave functions .1. *Journal of Chemical Physics*, 1955. **23**(10): p. 1833–1840.

Myshakin, E.M., et al., Molecular dynamics simulations of turbostratic dry and hydrated montmorillonite with intercalated carbon dioxide. *The Journal of Physical Chemistry A*, 2014. **118**(35): 7454–7468.

Neri, M., et al., Coarse-grained model of proteins incorporating atomistic detail of the active site. *Physical Review Letters*, 2005. **95**(21): p. 218102.

Newman, S.P., et al., Interlayer arrangement of hydrated MgAl layered double hydroxides containing guest terephthalate anions: comparison of simulation and measurement. *The Journal of Physical Chemistry B*, 1998. **102**(35): p. 6710–6719.

Newton, R.C. and C.E. Manning, Thermodynamics of SiO_2–H_2O fluid near the upper critical end point from quartz solubility measurements at 10 kbar. *Earth and Planetary Science Letters*, 2008. **274**(1–2): p. 241–249.

Nielsen, S.O., et al., Recent progress in adaptive multiscale molecular dynamics simulations of soft matter. *Physical Chemistry Chemical Physics*, 2010. **12**(39): p. 12401–12414.

Nielsen, M.H., S. Aloni and J.J. De Yoreo, In situ TEM imaging of $CaCO_3$ nucleation reveals coexistence of direct and indirect pathways. *Science*, 2014. **345**(6201): p. 1158–1162.

Noid, W.G., Perspective: coarse-grained models for biomolecular systems. *Journal of Chemical Physics*, 2013. **139**(9): p. 090901.

Noid, W.G., et al., The multiscale coarse-graining method. I. A rigorous bridge between atomistic and coarse-grained models. *Journal of Chemical Physics*, 2008. **128**(24): p. 244114.

Parry, D.E., Electrostatic potential in surface region of an ionic-crystal. *Surface Science*, 1975. **49**(2): p. 433–440.

Parry, D.E., Correction. *Surface Science*, 1976. **54**(1): p. 195–195.

Paul, D.R. and L.M. Robeson, Polymer nanotechnology: nanocomposites. *Polymer*, 2008. **49**(15): p. 3187–3204.

Pavese, A., et al., Modelling of the thermal dependence of structural and elastic properties of calcite, $CaCO_3$. *Physics and Chemistry of Minerals*, 1996. **23**(2): p. 89–93.

Pearlman, D.A., et al., AMBER, a package of computer-programs for applying molecular mechanics, normal-mode analysis, molecular-dynamics and free-energy calculations to simulate the structural and energetic properties of molecules. *Computer Physics Communications*, 1995. **91**(1–3): p. 1–41.

Pedone, A., et al., A new self-consistent empirical interatomic potential model for oxides, silicates, and silica-based glasses. *Journal of Physical Chemistry B*, 2006. **110**(24): p. 11780–11795.

Pellenq, R.J.-M., et al., A realistic molecular model of cement hydrates. *Proceedings of the National Academy of Sciences*, 2009. **106**(38): p. 16102–16107.

Pinto, E.M., et al., Electrochemical and surface characterisation of carbon-film-coated piezoelectric quartz crystals. *Applied Surface Science*, 2009. **255**(18): p. 8084–8090.

Plassard, C., et al., Investigation of the surface structure and elastic properties of calcium silicate hydrates at the nanoscale. *Ultramicroscopy*, 2004. **100**(3–4): p. 331–338.

Plassard, C., et al., Nanoscale experimental investigation of particle interactions at the origin of the cohesion of cement. *Langmuir*, 2005. **21**(16): p. 7263–7270.

Ponder, J.W., et al., Current status of the AMOEBA polarizable force field. *The Journal of Physical Chemistry B*, 2010. **114**(8): p. 2549–2564.

Pouget, E.M., et al., The initial stages of template-controlled $CaCO_3$ formation revealed by cryo-TEM. *Science*, 2009. **323**(5920): p. 1455–1458.

Praprotnik, M., L. Delle Site and K. Kremer, A macromolecule in a solvent: adaptive resolution molecular dynamics simulation. *Journal of Chemical Physics*, 2007. **126**(13): p. 134902.

Praprotnik, M., L. Delle Site and K. Kremer, Multiscale simulation of soft matter: from scale bridging to adaptive resolution, *Annual Review of Physical Chemistry*, 2008. **59**: p. 545–571.

Prasitnok, K. and M.R. Wilson, A coarse-grained model for polyethylene glycol in bulk water and at a water/air interface. *Physical Chemistry Chemical Physics*, 2013. **15**(40): p. 17093–17104.

Quigley, D., et al., Sampling the structure of calcium carbonate nanoparticles with metadynamics. *Journal of Chemical Physics*, 2011. **134**(4): p. 044703.

Rai, B., et al., A molecular dynamics study of the interaction of oleate and dodecylammonium chloride surfactants with complex aluminosilicate minerals. *Journal of Colloid and Interface Science*, 2011. **362**(2): p. 510–516.

Raiteri, P. and J.D. Gale, Water is the key to nonclassical nucleation of amorphous calcium carbonate. *Journal of the American Chemical Society*, 2010. **132**(49): p. 17623–17634.

Raiteri, P., et al., Derivation of an accurate force-field for simulating the growth of calcium carbonate from aqueous solution: a new model for the calcite-water interface. *Journal of Physical Chemistry C*, 2010. **114**(13): p. 5997–6010.

Ramirez-Hernandez, A., et al., Dynamical simulations of coarse grain polymeric systems: rouse and entangled dynamics. *Macromolecules*, 2013. **46**(15): p. 6287–6299.

Rappe, A.K. and W.A. Goddard, Charge equilibration for molecular-dynamics simulations. *Journal of Physical Chemistry*, 1991. **95**(8): p. 3358–3363.

Rappe, A.K., et al., UFF, a full periodic-table force-field for molecular mechanics and molecular-dynamics simulations. *Journal of the American Chemical Society*, 1992. **114**(25): p. 10024–10035.

Ray, S.S. and M. Bousmina, Biodegradable polymers and their layered silicate nano composites: in greening the 21st century materials world. *Progress in Materials Science*, 2005. **50**(8): p. 962–1079.

Raymand, D., et al., A reactive force field (ReaxFF) for zinc oxide. *Surface Science*, 2008. **602**(5): p. 1020–1031.

Reith, D., M. Putz and F. Muller-Plathe, Deriving effective mesoscale potentials from atomistic simulations. *Journal of Computational Chemistry*, 2003. **24**(13): p. 1624–1636.

Richardson, D.J., J.N. Butt and T.A. Clarke, Controlling electron transfer at the microbe–mineral interface. *Proceedings of the National Academy of Sciences*, 2013. **110**(19): p. 7537–7538.

Rosenbluth, M.N. and A.W. Rosenbluth, Further results on Monte-Carlo equations of state. *Journal of Chemical Physics*, 1954. **22**(5): p. 881–884.

Roterman, I.K., et al., A comparison of the CHARMM, AMBER and ECEPP potentials for peptides. II. φ-ψ Maps for N-acetyl alanine N′-methyl amide: comparisons, contrasts and simple experimental tests. *Journal of Biomolecular Structure and Dynamics*, 1989. **7**(3): p. 421–453.

Ruehle, V., et al., Versatile object-oriented toolkit for coarse-graining applications. *Journal of Chemical Theory and Computation*, 2009. **5**(12): p. 3211–3223.

Rustad, J.R., A.R. Felmy and B.P. Hay, Molecular statics calculations of proton binding to goethite surfaces: a new approach to estimation of stability constants for multisite surface complexation models. *Geochimica et Cosmochimica Acta*, 1996. **60**(9): p. 1563–1576.

Rustad, J.R., A.R. Felmy and E.J. Bylaska, Molecular simulation of the magnetite-water interface. *Geochimica et Cosmochimica Acta*, 2003. **67**(5): p. 1001–1016.

Rzepiela, A.J., et al., Hybrid simulations: combining atomistic and coarse-grained force fields using virtual sites. *Physical Chemistry Chemical Physics*, 2011. **13**(22): p. 10437–10448.

Sainz-Diaz, C.I., A. Hernandez-Laguna and M.T. Dove, Modeling of dioctahedral 2: 1 phyllosilicates by means of transferable empirical potentials. *Physics and Chemistry of Minerals*, 2001. **28**(2): p. 130–141.

Sanders, M.J., M. Leslie and C.R.A. Catlow, Interatomic potentials for SiO_2. *Journal of the Chemical Society, Chemical Communications*, 1984. **19**: p. 1271–1273.

Santiso, E.E. and K.E. Gubbins, Multi-scale molecular modeling of chemical reactivity. *Molecular Simulation*, 2004. **30**(11–12): p. 699–748.

Scanlon, D.O., et al., Band alignment of rutile and anatase TiO_2. *Nature Materials*, 2013. **12**(9): p. 798–801.

Schmitt, U.W. and G.A. Voth, The computer simulation of proton transport in water. *Journal of Chemical Physics*, 1999. **111**(20): p. 9361–9381.

Schwertmann, U., Solubility and dissolution of iron-oxides. *Plant and Soil*, 1991. **130**(1–2): p. 1–25.

Shahsavari, R., R.J.M. Pellenq and F.-J. Ulm, Empirical force fields for complex hydrated calcio-silicate layered materials. *Physical Chemistry Chemical Physics*, 2011. **13**(3): p. 1002–1011.

Shan, T.-R., et al., Second-generation charge-optimized many-body potential for Si/SiO_2 and amorphous silica. *Physical Review B*, 2010a. **82**(23): p. 235302.

Shan, T.-R., et al., Charge-optimized many-body potential for the hafnium/hafnium oxide system. *Physical Review B*, 2010b. **81**(12): p. 125328.

Shapley, T.V., et al., Atomistic modeling of the sorption free energy of dioxins at clay-water interfaces. *Journal of Physical Chemistry C*, 2013. **117**(47): p. 24975–24984.

Shell, M.S., Systematic coarse-graining of potential energy landscapes and dynamics in liquids. *Journal of Chemical Physics*, 2012. **137**(8): p. 084503.

Shi, S.H., et al., An extensible and systematic force field, ESFF, for molecular modeling of organic, inorganic, and organometallic systems. *Journal of Computational Chemistry*, 2003. **24**(9): p. 1059–1076.

Singh, U.C. and P.A. Kollman, An approach to computing electrostatic charges for molecules. *Journal of Computational Chemistry*, 1984. **5**(2): p. 129–145.

Sirk, T.W., S. Moore and E.F. Brown, Characteristics of thermal conductivity in classical water models. *The Journal of Chemical Physics*, 2013. **138**(6): p. 064505.

Skelton, A.A., et al., Simulations of the quartz(1011)/water interface: a comparison of classical force fields, ab initio molecular dynamics, and X-ray reflectivity experiments. *The Journal of Physical Chemistry C*, 2011. **115**(5): p. 2076–2088.

Skipper, N.T., G. Sposito and F.R.C. Chang, Monte-Carlo simulation of interlayer molecular-structure in swelling clay-minerals .2. Monolayer hydrates. *Clays and Clay Minerals*, 1995. **43**(3): p. 294–303.

Spagnoli, D. and J.D. Gale, Atomistic theory and simulation of the morphology and structure of ionic nano-particles. *Nanoscale*, 2012. **4**(4): p. 1051–1067.

Spagnoli, D., et al., Molecular dynamics simulations of the interaction between the surfaces of polar solids and aqueous solutions. *Journal of Materials Chemistry*, 2006. **16**(20): p. 1997–2006.

Spagnoli, D., et al., Prediction of the effects of size and morphology on the structure of water around hematite nanoparticles. *Geochimica et Cosmochimica Acta*, 2009. **73**(14): p. 4023–4033.

Srivastava, K.P., Unlike molecular interactions and properties of gas mixtures. *Journal of Chemical Physics*, 1958. **28**(4): p. 543–549.

Srivastava, A. and G.A. Voth, Hybrid approach for highly coarse-grained lipid bilayer models. *Journal of Chemical Theory and Computation*, 2013. **9**(1): p. 750–765.

Stillinger, F.H. and C.W. David, Study of the water octamer using the polarization model of molecular-interactions. *Journal of Chemical Physics*, 1980. **73**(7): p. 3384–3389.

Stone, A.J., Distributed multipole analysis, or how to describe a molecular charge-distribution. *Chemical Physics Letters*, 1981. **83**(2): p. 233–239.

Stone, A.J. and M. Alderton, Distributed multipole analysis: methods and applications. *Molecular Physics*, 1985. **56**(5): p. 1047–1064.

Sun, H., COMPASS: an ab initio force-field optimized for condensed-phase applications: overview with details on alkane and benzene compounds. *Journal of Physical Chemistry B*, 1998. **102**(38): p. 7338–7364.

Sun, H., et al., An ab-initio CFF93 all-atom force-field for polycarbonates. *Journal of the American Chemical Society*, 1994. **116**(7): p. 2978–2987.

Suter, J.L., et al., Large-scale molecular dynamics study of montmorillonite clay: emergence of undulatory fluctuations and determination of material properties. *The Journal of Physical Chemistry C*, 2007. **111**(23): p. 8248–8259.

Suter, J.L., et al., Recent advances in large-scale atomistic and coarse-grained molecular dynamics simulation of clay minerals. *Journal of Materials Chemistry*, 2009. **19**(17): p. 2482–2493.

Swadling, J.B., P.V. Coveney and H.C. Greenwell, Clay minerals mediate folding and regioselective interactions of RNA: a large-scale atomistic simulation study. *Journal of the American Chemical Society*, 2010. **132**(39): p. 13750–13764.

Tamerler, C., et al., Genetically engineered polypeptides for inorganics: a utility in biological materials science and engineering. *Materials Science and Engineering: C*, 2007. **27**(3): p. 558–564.

Taylor, C.D. and M. Neurock, Theoretical insights into the structure and reactivity of the aqueous/metal interface. *Current Opinion in Solid State & Materials Science*, 2005. **9**(1–2): p. 49–65.

Tenney, C.M. and R.T. Cygan, Molecular simulation of carbon dioxide, brine, and clay mineral interactions and determination of contact angles. *Environmental Science & Technology*, 2014. **48**(3): p. 2035–2042.

Teppen, B.J., et al., Molecular dynamics modeling of clay minerals .1. Gibbsite, kaolinite, pyrophyllite, and beidellite. *Journal of Physical Chemistry B*, 1997. **101**(9): p. 1579–1587.

Thuresson, A., et al., Monte Carlo simulations of parallel charged platelets as an approach to tactoid formation in clay. *Langmuir*, 2013. **29**(29): p. 9216–9223.

Tokarsky, J., P. Capkova and J.V. Burda, Structure and stability of kaolinite/TiO_2 nanocomposite: DFT and MM computations. *Journal of Molecular Modeling*, 2012. **18**(6): p. 2689–2698.

Tombácz, E., et al., Water in contact with magnetite nanoparticles, as seen from experiments and computer simulations. *Langmuir*, 2009. **25**(22): p. 13007–13014.

Tosenberger, A., et al., Particle dynamics methods of blood flow simulations. *Mathematical Modelling of Natural Phenomena*, 2011. **6**(5): p. 320–332.

Tröster, P., et al., Polarizable water models from mixed computational and empirical optimization. *The Journal of Physical Chemistry B*, 2013. **117**(32): p. 9486–9500.

Tse, J.S. and D.D. Klug, Mechanical instability of α-quartz: a molecular dynamics study. *Physical Review Letters*, 1991. **67**(25): p. 3559–3562.

Ufimtsev, I.S., et al., A multistate empirical valence bond model for solvation and transport simulations of OH- in aqueous solutions. *Physical Chemistry Chemical Physics*, 2009. **11**(41): p. 9420–9430.

van Beest, B.W.H., G.J. Kramer and R.A. van Santen, Force fields for silicas and aluminophosphates based on ab initio calculations. *Physical Review Letters*, 1990. **64**(16): p. 1955–1958.

van Duin, A.C.T., et al., ReaxFF: a reactive force field for hydrocarbons. *Journal of Physical Chemistry A*, 2001. **105**(41): p. 9396–9409.

Valero, R., et al., Perspective on diabatic models of chemical reactivity as illustrated by the gas-phase S(N)2 reaction of acetate ion with 1,2-dichloroethane. *Journal of Chemical Theory and Computation*, 2009. **5**(1): p. 1–22.

Vega, C. and J.L.F. Abascal, Simulating water with rigid non-polarizable models: a general perspective. *Physical Chemistry Chemical Physics*, 2011. **13**(44): p. 19663–19688.

Vega, C., et al., What ice can teach us about water interactions: a critical comparison of the performance of different water models. *Faraday Discussions*, 2009. **141**: p. 251–276.

Verlet, L., Computer experiments on classical fluids. I. Thermodynamical properties of Lennard-Jones molecules. *Physical Review*, 1967. **159**(1): p. 98–103.

Verlet, L., Computer experiments on classical fluids. 2. Equilibrium correlation functions. *Physical Review*, 1968. **165**(1): p. 201.

Verwey, E.J.W., J.Th.G. Overbeek and K. van Nes, *Theory of the Stability of Lyophobic Colloids*. 1948, Amsterdam: Elsevier Publishing Company Inc.

Villà, J. and A. Warshel, Energetics and dynamics of enzymatic reactions. *The Journal of Physical Chemistry B*, 2001. **105**(33): p. 7887–7907.

Vuilleumier, R. and D. Borgis, Transport and spectroscopy of the hydrated proton: a molecular dynamics study. *Journal of Chemical Physics*, 1999. **111**(9): p. 4251–4266.

Waldman, M. and A.T. Hagler, New combining rules for rare-gas van-der-Waals parameters. *Journal of Computational Chemistry*, 1993. **14**(9): p. 1077–1084.

Wallace, A.F., et al., Microscopic evidence for liquid-liquid separation in supersaturated $CaCO_3$ solutions. *Science*, 2013. **341**(6148): p. 885–889.

Wang, Z.-J. and M. Deserno, A systematically coarse-grained solvent-free model for quantitative phospholipid bilayer simulations. *Journal of Physical Chemistry B*, 2010. **114**(34): p. 11207–11220.

Wang, J., A.G. Kalinichev and R.J. Kirkpatrick, Molecular modeling of water structure in nano-pores between brucite (001) surfaces. *Geochimica et Cosmochimica Acta*, 2004. **68**(16): p. 3351–3365.

Wang, Q., D.J. Keffer and D.M. Nicholson, A coarse-grained model for polyethylene glycol polymer. *Journal of Chemical Physics*, 2011. **135**(21): p. 214903.

Wang, Q., et al., Multi-scale models for cross-linked sulfonated poly (1, 3-cyclohexadiene) polymer. *Polymer*, 2012a. **53**(7): p. 1517–1528.

Wang, Q., et al., Atomistic and coarse-grained molecular dynamics simulation of a cross-linked sulfonated poly (1,3-cyclohexadiene)-based proton exchange membrane. *Macromolecules*, 2012b. **45**(16): p. 6669–6685.

Wang, Y., et al., Molecular dynamics simulation of strong interaction mechanisms at wet interfaces in clay-polysaccharide nanocomposites. *Journal of Materials Chemistry A*, 2014. **2**(25): p. 9541–9547.

Warshel, A. and R.M. Weiss, An empirical valence bond approach for comparing reactions in solutions and in enzymes. *Journal of the American Chemical Society*, 1980. **102**(20): p. 6218–6226.

Watson, G.W. and S.C. Parker, Quartz amorphization: a dynamical instability. *Philosophical Magazine Letters*, 1995a. **71**(1): p. 59–64.

Watson, G.W. and S.C. Parker, Dynamical instabilities in alpha-quartz and alpha-berlinite: a mechanism for amorphization. *Physical Review B*, 1995b. **52**(18): p. 13306–13309.

Waychunas, G.A., C.S. Kim and J.F. Banfield, Nanoparticulate iron oxide minerals in soils and sediments: unique properties and contaminant scavenging mechanisms. *Journal of Nanoparticle Research*, 2005. **7**(4–5): p. 409–433.

Weiner, S.J., et al., An all atom force-field for simulations of proteins and nucleic-acids. *Journal of Computational Chemistry*, 1986. **7**(2): p. 230–252.

Wilson, M., et al., Molecular dynamics simulations of compressible ions. *Journal of Chemical Physics*, 1996a. **104**(20): p. 8068–8081.

Wilson, M., U. Schonberger and M.W. Finnis, Transferable atomistic model to describe the energetics of zirconia. *Physical Review B*, 1996b. **54**(13): p. 9147–9161.

Wilson, M., S. Jahn and P.A. Madden, The construction and application of a fully flexible computer simulation model for lithium oxide. *Journal of Physics. Condensed Matter*, 2004. **16**(27): p. S2795–S2810.

Wood, W.W. and F.R. Parker, Monte Carlo equation of state of molecules interacting with the Lennard-Jones potential. 1. Supercritical isotherm at about twice the critical temperature. *Journal of Chemical Physics*, 1957. **27**(3): p. 720–733.

Woodley, S.M., et al., The prediction of inorganic crystal structures using a genetic algorithm and energy minimisation. *Physical Chemistry Chemical Physics*, 1999. **1**(10): p. 2535–2542.

Woods, A.D.B., W. Cochran and B.N. Brockhouse, Lattice dynamics of alkali halide crystals. *Physical Review*, 1960. **119**(3): p. 980–999.

Wu, Y.J., H.L. Tepper and G.A. Voth, Flexible simple point-charge water model with improved liquid-state properties. *Journal of Chemical Physics*, 2006. **124**(2): p. 024503.

Wu, Y., et al., An improved multistate empirical valence bond model for aqueous proton solvation and transport. *Journal of Physical Chemistry B*, 2008. **112**(2): p. 467–482.

Wu, Z., Q. Cui and A. Yethiraj, A new coarse-grained force field for membrane-peptide simulations. *Journal of Chemical Theory and Computation*, 2011. **7**(11): p. 3793–3802.

Xie, W., et al., Development of a polarizable intermolecular potential function (PIPF) for liquid amides and alkanes. *Journal of Chemical Theory and Computation*, 2007. **3**(6): p. 1878–1889.

Zaidan, O.F., J.A. Greathouse and R.T. Pabalan, Monte Carlo and molecular dynamics simulation of uranyl adsorption on montmorillonite clay. *Clays and Clay Minerals*, 2003. **51**(4): p. 372–381.

Zeitler, T.R., et al., Vibrational analysis of brucite surfaces and the development of an improved force field for molecular simulation of interfaces. *Journal of Physical Chemistry C*, 2014. **118**(15): p. 7946–7953.

Zelko, J., et al., Effects of counterion size on the attraction between similarly charged surfaces. *Journal of Chemical Physics*, 2010. **133**(20): p. 204901.

Zeng, Q.H., et al., Molecular dynamics simulation of organic–inorganic nanocomposites: layering behavior and interlayer structure of organoclays. *Chemistry of Materials*, 2003. **15**(25): p. 4732–4738.

Zhang, G., et al., Dispersion-corrected density functional theory and classical force field calculations of water loading on a pyrophyllite(001) surface. *Journal of Physical Chemistry C*, 2012. **116**(32): p. 17134–17141.

Zhao, T. and X. Wang, Distortion and flow of nematics simulated by dissipative particle dynamics. *Journal of Chemical Physics*, 2014. **140**(18): p. 184902.

Zhu, R., et al., Sorptive characteristics of organomontmorillonite toward organic compounds: a combined LFERs and molecular dynamics simulation study. *Environmental Science & Technology*, 2011. **45**(15): p. 6504–6510.

Zhu, R., et al., Modeling the interaction of nanoparticles with mineral surfaces: adsorbed C60 on pyrophyllite. *The Journal of Physical Chemistry A*, 2013. **117**(30): p. 6602–6611.

3

Quantum-Mechanical Modeling of Minerals

Alessandro Erba[1] and Roberto Dovesi[1,2]

[1]*Dipartimento di Chimica, Università degli Studi di Torino, Torino, Italy*
[2]*NIS Centre of Excellence "Nanostructured Interfaces and Surfaces", Torino, Italy*

3.1   Introduction

The combination of growing computing resources and success of quantum-chemical methods based on the density functional theory (DFT) is rapidly widening the applicability range of first-principles techniques in the solid state to the study of minerals of geochemical interest, materials for the electronics and hydrogen storage, biomaterials, etc. (Albanese et al. 2012; Delle Piane et al. 2013; Erba et al. 2014b; Ugliengo et al. 2008). A variety of properties of materials, such as structural, electronic, vibrational, optical, elastic, magnetic, etc., can now be routinely computed with high accuracy with several periodic quantum-chemical programs. Such methods, however, in their standard formulation, describe the ground state of the system (a time-dependent formulation of the DFT would be required for describing excited states (Marques and Gross 2004)) at zero temperature and pressure, which is of course a severe limitation to their applicability in geochemical studies.

   The exact physicochemical composition of the Earth's deep interior, for instance, is still debated. Different compositional models have been proposed (Anderson and Bass 1986; Bass and Anderson 1984; Ringwood 1975) which can be mainly validated on the grounds of seismological data collected during earthquakes, provided that the individual elastic response of all possible constituents of the Earth's upper mantle, transition zone, and lower mantle is known. This characterization is an extremely difficult task from an experimental point of view (Anderson and Isaak 1995; Mainprice et al. 2013) in that high temperatures and pressures have to be simultaneously considered (typically, pressures up to 140 GPa and temperatures in the range 800–1200 K). In this respect, the predictive power of first-principles techniques could prove decisive. Nevertheless, the theoretical description of structural, thermodynamic, and elastic properties of minerals at high temperature and pressure also represents a challenge for state-of-the-art quantum-mechanical techniques for the following reasons: (i) complex algorithms have to be developed for including the effects of pressure and temperature on

Molecular Modeling of Geochemical Reactions: An Introduction, First Edition. Edited by James D. Kubicki.
© 2016 John Wiley & Sons, Ltd. Published 2016 by John Wiley & Sons, Ltd.

structural and elastic properties; if not fully automated, they would require too many separate program modules, interfaces, external units for data storage, postprocessing scripts, etc., for being routinely used; (ii) all the required algorithms have to be implemented in the same program, within the same formal framework and numerical conditions; (iii) all the algorithms have to be efficiently implemented as regards massive-parallel scalability and memory use in that the resulting calculations can become computationally rather demanding; and (iv) all the algorithms should be as general as possible in terms of dimensionality of the system under investigation (from 1-D to 2-D and 3-D), chemical composition, symmetry, electronic configuration, etc. From a more fundamental point of view, all kinds of chemical interactions must be reliably described. The DFT, in its standard formulation, is known to neglect or describe in a spurious way dispersive, London-type, interactions. In this respect, in recent years, simple *a posteriori* semi-empirical energy corrections have become extremely popular in a molecular context (Grimme 2006, 2011; Grimme et al. 2010). Even if a reparametrization of such schemes has been attempted for molecular crystals (Civalleri et al. 2008), further effort has to be put in this direction before these schemes could be fruitfully used in the simulation of several properties of the solid state.

In this chapter, we will review some of the most recent developments in this respect, as implemented in the CRYSTAL program for quantum-chemical simulations of the solid state, as we are among its developers (Dovesi et al. 2014a,b). In the last years, many efforts have been devoted to the optimization of the core algorithms of the program (calculation of the self-consistent-field (SCF) procedure and computation of atomic and cell gradients) in terms of reduced use of memory and increased efficiency in the parallel scalability (Bush et al. 2011; Dovesi et al. 2014a; Orlando et al. 2012). Several tensorial properties can now be computed automatically, such as the fourth-rank elastic tensor (Erba et al. 2014b; Perger et al. 2009), the third-rank direct and converse piezoelectric tensors (Erba et al. 2013a; Mahmoud et al. 2014b), the fourth-rank photoelastic Pockels' tensor (Erba and Dovesi 2013), the second-rank dielectric (or polarizability) tensor, and third- and fourth-rank hyperpolarizabilities (Ferrero et al. 2008a,b,c,d; Orlando et al. 2009, 2010). Generalization of the elastic and piezoelectric tensor calculation to 1-D and 2-D systems has also been performed (Baima et al. 2013; Erba et al. 2013b; Lacivita et al. 2013b). At variance with most DFT-based programs where numerical approaches are implemented, analytical infrared (IR) and Raman intensities are now available as well as the automated computation of IR and Raman spectra (Carteret et al. 2013; De La Pierre et al. 2011). More details about these spectroscopic features are given in Chapter 10. New algorithms have also been developed for the study of solid solutions and, more generally, disordered systems (D'Arco et al. 2013; Mustapha et al. 2013).

Some of these recent developments will be illustrated in this chapter, along with some application to the study of the family of silicate garnets. Garnets constitute a large class of materials of great technological and geochemical interest; they can be used as components of lasers, computer memories, and microwave optical devices and, due to high hardness and recyclability, as abrasives and filtration media (Novak and Gibbs 1971). Silicate garnets are among the most important rock-forming minerals and represent the main constituents of the Earth's lower crust, upper mantle, and transition zone. They are characterized by a cubic structure with space group $Ia\bar{3}d$ and formula $X_3Y_2(SiO_4)_3$, where the X site hosts divalent cations such as Ca^{2+}, Mg^{2+}, Fe^{2+}, and Mn^{2+} and the Y site is occupied by trivalent cations such as Al^{3+}, Fe^{3+}, and Cr^{3+}. At least 12 end-members of this family of minerals have been identified (Rickwood et al. 1968). The primitive cell contains four formula units (80 atoms), and the structure consists in alternating SiO_4 tetrahedra and YO_6 octahedra sharing corners to form a three-dimensional network. The most common end-members of the family are pyrope $Mg_3Al_2(SiO_4)_3$; almandine $Fe_3Al_2(SiO_4)_3$; spessartine $Mn_3Al_2(SiO_4)_3$; grossular $Ca_3Al_2(SiO_4)_3$; uvarovite $Ca_3Cr_2(SiO_4)_3$; and andradite $Ca_3Fe_2(SiO_4)_3$. Natural silicate garnets can be found in a wide range of chemical compositions since they form solid solutions.

The structure of this chapter is as follows: in Section 3.2 the formal framework is briefly defined within which the CRYSTAL program is developed, in terms of periodic boundary conditions and localized Gaussian-type orbital (GTO) basis sets (BSs); Section 3.3 is devoted to the discussion of a couple of techniques for the inclusion of the effect of pressure on structural properties of minerals; Section 3.4 deals with the evaluation of the elastic tensor (both at zero and nonzero pressure) and related elastic properties of minerals; a brief review on the first-principles studies that have been performed on spectroscopic properties of silicate garnets is given in Section 3.5, along with the description of the approach implemented in the CRYSTAL program for computing thermodynamic properties of solids; a brief description, along with some examples of application, of some of the techniques implemented for the study of solid solutions or disordered systems is given in Section 3.6; finally, some remarks on current developments and future challenges are given in Section 3.7, in particular as regards the inclusion of temperature on computed structural and elastic properties of minerals.

3.2 Theoretical Framework

This section is devoted to a very brief illustration of the main formal aspects of standard quantum-chemical techniques, as applied to solid-state systems. For a more detailed presentation, the reader may refer to Chapter 1. The aim is here to find the electronic ground-state solution of the static Schrodinger equation at fixed nuclei and in the absence of external fields:

$$\hat{H}_{el}\Psi_0 = E_0\Psi_0. \tag{3.1}$$

Despite the simplification introduced by translational invariance, the intrinsic periodic nature of crystals makes the solution of Equation 3.1 practically impossible unless "average-field" approximations to the electrostatic Hamiltonian are introduced, such as those represented by Hartree–Fock (HF) or Kohn–Sham (KS) approaches (Hohenberg and Kohn 1964; Kohn and Sham 1965). The nonrelativistic electrostatic Hamiltonian, $\hat{H}_{el}$, in Equation 3.1, for the N-electron system in the field of M nuclei of charge Z_A, fixed at position $\mathbf{R}_A$, is given by the following expression:

$$\hat{H}_{el} = \sum_{n=1}^{N}\frac{-\nabla_n^2}{2} + \sum_{n=1}^{N}\sum_{A=1}^{M}\frac{-Z_A}{r_{nA}} + \frac{1}{2}\sum_{n,m=1}^{N}{'}\frac{1}{r_{nm}} + \frac{1}{2}\sum_{A,B=1}^{M}{'}\frac{Z_A Z_B}{r_{AB}}. \tag{3.2}$$

Atomic units (au) are here used; the primed double sums exclude "diagonal" terms ($n = m$, $A = B$). $\hat{H}_{el}$ depends parametrically on the sets $\{\mathbf{R}\}$, $\{Z\}$ of positions and charges of the M nuclei; the same holds true for ground-state energy E_0 and wave function Ψ_0 which is an antisymmetric function of the space–spin coordinates $\mathbf{x}_n \equiv (\mathbf{r}_n, \sigma_n)$ of the N electrons.

3.2.1 Translation Invariance and Periodic Boundary Conditions

Let us consider a regular lattice of vectors $\mathbf{T_m}$ generated from D primitive linearly independent basis vectors $\mathbf{a}_i$ of ordinary space, $\mathbf{T_m} = \sum_{i=1}^{D} m_i \mathbf{a}_i$, where m_i are integers and D is the number of periodic directions (three for bulk crystals, two for slabs, one for polymeric structures). The $\mathbf{a}_i$ vectors define (though not univocally) the *unit cell* of the crystal. The coordinates and charges of all nuclei in the infinite crystal can be generated from those of a translationally irreducible finite set $\{\mathbf{R}_{A,0}; Z_{A,0}\}$ as follows: $\mathbf{R}_{A,\mathbf{m}} = \mathbf{R}_{A,0} + \mathbf{T_m}$ and $Z_{A,\mathbf{m}} = Z_{A,0}$. For solving Equation 3.1, Born–von Kàrmàn (BvK) periodic boundary conditions are adopted. They impose that Ψ_0 is cyclically periodic with respect to a superlattice of vectors $\overline{\mathbf{W}}_\mathbf{m}$ defined as $\mathbf{T_m}$, but starting from D superbasis vectors $\overline{\mathbf{A}}_i = w_i\mathbf{a}_i$:

$$\text{if:}\quad \mathbf{x}_n = (\mathbf{r}_n, \sigma_n),\ \mathbf{x}'_n = (\mathbf{r}_n + \overline{\mathbf{W}}_\mathbf{m}, \sigma_n),$$
$$\text{then:}\quad \Psi_0(\ldots, \mathbf{x}_n, \ldots) = \Psi_0(\ldots, \mathbf{x}'_n, \ldots). \tag{3.3}$$

The integers w_i define the effective number of electrons in the system: $N^{\text{eff}} = W N_0$, with $W = \prod_i w_i$, and $N_0 = \Sigma_A Z_{A,\mathbf{0}}$ the number of electrons per cell (only neutral systems are here considered). The setting of the w_i's (i.e., the so-called shrinking factor) is one of the main computational parameters of a periodic calculation.

3.2.2 HF and KS Methods

In their spin-unrestricted formulation, both HF and KS methods are intended to obtain a set of N one-electron functions, the molecular spin orbitals (MSO) (or crystalline spin orbitals (CSO) in the periodic case), $\psi_j^F(\mathbf{x}) = \phi_j^{F,\sigma}(\mathbf{r})\omega(\sigma)$ (here F stands for HF or KS and σ for α or β), which satisfy the equation:

$$\hat{h}^{F,\sigma}\phi_j^{F,\sigma}(\mathbf{r}) = \left[-\frac{\nabla^2}{2} + \sum_A \frac{-Z_A}{|\mathbf{R}_A - \mathbf{r}|} + \int \frac{\rho^F(\mathbf{r}')}{|\mathbf{r}-\mathbf{r}'|}d\mathbf{r}' + \widetilde{V}^{F,\sigma} \right]\phi_j^{F,\sigma}(\mathbf{r}) = \in_j^{F,\sigma}\phi_j^{F,\sigma}(\mathbf{r}). \tag{3.4}$$

The effective Hamiltonian $\hat{h}^{F,\sigma}$ which acts on the individual MSO contains, apart from the kinetic, nuclear attraction and *Hartree* operators (the last one expressing the Coulomb repulsion with all the electrons in the system), a *corrective potential* $\widetilde{V}^{F,\sigma}$, which differs in the two schemes. A *single-determinant N-electron function*, Ψ_0^F, can be defined, after assigning the N electrons to the N MSOs corresponding to the lowest eigenvalues $\in_j^{F,\sigma}$ of Equation 3.4 and antisymmetrizing their product. In the rest of this section we shall assume, for simplicity, that we are describing a closed-shell system so that the spin index can be dropped from the effective Hamiltonian and from the corrective potential. The position density matrix (DM) and the electron density (ED) associated with Ψ_0^F are then simply

$$P^F(\mathbf{r};\mathbf{r}') = 2\sum_{n=1}^{N/2}\phi_n^F(\mathbf{r})\left(\phi_n^F(\mathbf{r}')\right)^*;\quad \rho^F(\mathbf{r}) = P^F(\mathbf{r};\mathbf{r}) = 2\sum_{n=1}^{N/2}|\phi_n^F(\mathbf{r})|^2. \tag{3.5}$$

In the HF scheme, the corrective potential $\widetilde{V}^{HF}$ is defined by imposing that the HF energy E_0^{HF} is a minimum with respect to any other single-determinant N-electron wave function. To achieve this goal, $\widetilde{V}^{HF}$ must take the form of the *exact-exchange operator*. We have, correspondingly,

$$E_0^{HF} \equiv \left\langle \Psi_0^{HF}|\hat{H}_{el}|\Psi_0^{HF}\right\rangle = -\int\left[\frac{\nabla^2}{2}P^{HF}(\mathbf{r};\mathbf{r}')\right]_{(\mathbf{r}'=\mathbf{r})}d\mathbf{r} - \sum_A Z_A\int\frac{\rho^{HF}(\mathbf{r})}{|\mathbf{R}_A - \mathbf{r}|}d\mathbf{r}$$
$$+ \frac{1}{2}\int\frac{\rho^{HF}(\mathbf{r})\rho^{HF}(\mathbf{r}')}{|\mathbf{r}-\mathbf{r}'|}d\mathbf{r}\,d\mathbf{r}' - \frac{1}{4}\int\frac{|P^{HF}(\mathbf{r};\mathbf{r}')|^2}{|\mathbf{r}-\mathbf{r}'|}d\mathbf{r}\,d\mathbf{r}' + \frac{1}{2}\sum_{A,B=1}^{M}{}'\frac{Z_A Z_B}{r_{AB}} \geq E_0.$$

The KS scheme, formulated in the frame of DFT (Hohenberg and Kohn 1964; Kohn and Sham 1965), introduces, for any given N-electron ED, $\rho(\mathbf{r})$, two *universal functionals*: $\varepsilon_{xc}(\mathbf{r}; [\rho])$ and its functional derivative $V_{xc}(\mathbf{r}; [\rho])$. When the *exchange-correlation potential* $V_{xc}(\mathbf{r}; [\rho]^{KS})$ is used for $\widetilde{V}^{KS}$ as a multiplicative operator in Equation 3.4, the density from Equation 3.5 coincides with the *exact* ground-state ED: $\rho^{KS}(\mathbf{r}) = \rho(\mathbf{r})$. The functional $\varepsilon_{xc}(\mathbf{r};[\rho])$ allows the *exact* ground-state energy to be calculated, again with reference to the occupied KS manifold:

$$E_0^{KS} = -\int \left[\frac{\nabla^2}{2} P^{KS}(\mathbf{r};\mathbf{r}') \right]_{(\mathbf{r}'=\mathbf{r})} d\mathbf{r} - \sum_A Z_A \int \frac{\rho(\mathbf{r})}{|\mathbf{R}_A - \mathbf{r}|} d\mathbf{r}$$

$$+ \frac{1}{2} \int \frac{\rho(\mathbf{r})\rho(\mathbf{r}')}{|\mathbf{r}-\mathbf{r}'|} d\mathbf{r} d\mathbf{r}' + \int \rho(\mathbf{r}) \varepsilon_{xc}(\mathbf{r};[\rho]) d\mathbf{r} + \frac{1}{2} \sum_{A,B=1}^{M} {}' \frac{Z_A Z_B}{r_{AB}} = E_0.$$

Equation 3.4 must be solved self-consistently in both cases, because the Hartree and the corrective potential are defined in terms of the occupied manifold. Let us note that in the HF case, the *nonlocal* corrective potential is perfectly defined. On the contrary, no exact formula exists for the *local* exchange-correlation potential $V_{xc}(\mathbf{r};[\rho])$; an enormous amount of proposals has been formulated so far. The two schemes, HF and KS, can be combined together in so-called hybrid functionals which introduce a fraction of exact HF exchange into the exchange-correlation DFT functional. Hybrid functionals correct for some of the main deficiencies of pure LDA and GGA functionals, that is, they reduce the self-interaction error, and they generally widen the electronic band gap of solids (Corá et al. 2004). Among others, popular hybrid functionals are B3LYP (Becke 1993), PBE0 (Adamo and Barone 1999), and HSE06 (Heyd and Scuseria 2004).

3.2.3 Bloch Functions and Local BS

Given that the one-electron effective Hamiltonian $\hat{h}^F$ commutes with all operations of the space group $\mathcal{G}$, in particular of the subgroup $\mathcal{T}$ of pure translations, its eigenfunctions, the crystalline orbitals (CO), can be classified according to the irreducible representations (irrep) of that group. As is shown in standard textbooks (Tinkham 1964), they are then characterized by an index κ, a vector of reciprocal space, such that the corresponding COs are *Bloch functions* (BF), $\phi_n^F(\mathbf{r};\kappa)$, which satisfy the property

$$\phi_n^F(\mathbf{r}+\mathbf{T_m};\kappa) = \phi_n^F(\mathbf{r};\kappa)\exp(\iota\kappa \cdot \mathbf{T_m}). \tag{3.6}$$

Clearly, κ's differing by a reciprocal lattice vector $\mathbf{G}$ define the same irreducible representation. Among all equivalent κ's one can choose the one closest to the origin of the reciprocal space; this "minimal-length" set fills the so-called *(first)* Brillouin zone (BZ). The COs must also satisfy BvK conditions $[\phi_n^F(\mathbf{r}+\overline{\mathbf{W}}_\mathbf{m};\kappa) = \phi_n^F(\mathbf{r};\kappa)]$, which means that $\exp(\iota\kappa \cdot \overline{\mathbf{W}}_\mathbf{m}) = 1$.

In practical calculations, the orbitals must be expanded into some suitably chosen BS. In periodic systems, it is convenient to use a BS of BFs $f_\mu(\mathbf{r};\kappa)$ so that determining the COs $\phi_n^F(\mathbf{r};\kappa)$ reduces to a secular problem that involves only basis functions of that given κ. The choice of the BS is crucial and determines the algorithms and numerical methods used in the actual solution. Most calculations use either plane waves (PWs) or atom-centered local functions, AOs. PWs are the traditional choice in solid-state physics, reflecting the delocalized nature of valence and conduction electron states in crystals. Localized BSs formed by AOs, $\chi_\mu(\mathbf{r})$, are the traditional choice in quantum chemistry, reflecting the atomic composition of matter. In periodic systems, one uses Bloch sums of AOs:

$$f_\mu(\mathbf{r};\kappa) = \frac{1}{\sqrt{W}} \sum_\mathbf{T} \exp[\iota\kappa \cdot \mathbf{T}]\chi_\mu(r-\mathbf{T}). \tag{3.7}$$

The C$_{RYSTAL}$ program, on which we focus our attention, shares with standard molecular codes the use of GTOs as local basis functions. Each atom A carries p_A GTOs, each resulting from a "contraction" of M_{iA} Gaussian "primitives" of angular momentum components ℓ, m centered in $\mathbf{R}_A$:

$$\chi_{iA}(\mathbf{r}_A) = \sum_{j=1}^{M_{iA}} c_{iA,j} N^{\ell,m}(\alpha_{iA,j}) X^{\ell,m}(\mathbf{r}_A) \exp\left[-\alpha_{iA,j}\mathbf{r}_A^2\right].$$

Here $\mathbf{r}_A = \mathbf{r} - \mathbf{R}_A$, $X^{\ell,m}$ are real solid harmonics and $N^{\ell,m}$ normalization coefficients; $c_{iA,j}$ are known as "coefficients" $\alpha_{iA,j}$ as "exponents" of the GTO. See Appendix A of Erba and Pisani (2012) for more details.

The evaluation of GTO integrals in C$_{RYSTAL}$ entails problems related to the periodically infinite character of the system. Sophisticated techniques have been implemented which permit the truncation or the accurate approximation of lattice sums: Ewald techniques, multipolar treatment of nonoverlapping distributions, bipolar expansion, etc. (see Pisani et al. (1988) for a detailed description of such techniques). An attractive feature related to the *local* character of the basis functions is that not only 3-dimensional crystals but also structures periodic in 2 (slabs), 1 (polymers), and 0 (molecules) dimensions are treated by C$_{RYSTAL}$ with the same basic technology without any need of artificial replication of the subunits. Another major advantage of AOs is that point symmetry of the crystal can be fully exploited at any stage of the calculation, which, given the usual high symmetry of crystals, results in large savings of computational resources (Orlando et al. 2014; Zicovich-Wilson and Dovesi 1998).

3.3 Structural Properties

A good description of the structure of minerals generally relies on the effectiveness of minimization algorithms in exploring the potential energy surface (PES) of the system. Depending on the specific structural feature to be explored, one might be interested in characterizing global or relative minima and saddle points (for details on the implementation of transition-state search techniques in the C$_{RYSTAL}$ program, see Rimola et al. (2010)) of the PES. The energy of a crystal can be minimized both in terms of atomic coordinates within the cell and in terms of the lattice parameters of the cell; the corresponding energy gradients are implemented analytically in the C$_{RYSTAL}$ program (Doll 2001; Doll et al. 2001). A quasi-Newton technique combined with the Broyden–Fletcher–Goldfarb–Shanno algorithm for Hessian updating is used (Broyden 1970; Fletcher 1970; Goldfarb 1970; Shanno 1970) in the automated implementation of the geometry optimizer and convergence checked on both gradient components and nuclear displacements (Civalleri et al. 2001).

As an example, in Table 3.1, we report selected structural parameters of three $X_3Y_2(SiO_4)_3$ silicate garnets, as obtained by fully optimizing their structure at the B3LYP hybrid level of theory and as experimentally determined with accurate low-temperature X-ray diffraction experiments (Zicovich-Wilson et al. 2008). The lattice parameter, a; the fractional coordinates of the oxygen atom, O_i; and selected bond lengths are reported for pyrope, $Mg_3Al_2(SiO_4)_3$; grossular, $Ca_3Al_2(SiO_4)_3$; and andradite, $Ca_3Fe_2(SiO_4)_3$. The overall agreement between computed and measured structural features is quite remarkable, the lattice parameter being systematically overestimated by about 1% and, correspondingly, the YO and SiO distances being overestimated by about $0.01 - 0.02$ Å.

When structural properties of minerals at geophysical conditions have to be determined, constant-volume and constant-pressure optimizations prove extremely useful. In the next subsections,

Table 3.1 *Lattice parameter, a (in Å), fractional coordinates of the oxygen atom, O_i, and selected distances (in Å) of three $X_3Y_2(SiO_4)_3$ silicate garnets: pyrope (X = Mg and Y = Al), grossular (X = Ca and Y = Al), and andradite (X = Ca and Y = Fe).*

	Pyrope		Grossular		Andradite	
	Calc.	Exp.	Calc.	Exp.	Calc.	Exp.
a	11.5447	11.4390	11.9368	11.8450	12.1960	12.0510
O_x	0.03214	0.03291	0.03740	0.03823	0.03893	0.03914
O_y	0.04971	0.05069	0.04515	0.04528	0.04838	0.04895
O_z	0.65344	0.65331	0.65156	0.65137	0.65617	0.65534
X_1O	2.2052	2.1959	2.3301	2.3218	2.3780	2.3584
X_2O	2.3648	2.3335	2.5058	2.4865	2.5331	2.4953
YO	1.8987	1.8850	1.9398	1.9255	2.0497	2.0186
SiO	1.6496	1.6337	1.6627	1.6459	1.6612	1.6492

Zicovich-Wilson et al. (2008). Reproduced with permission of Wiley.
Experimental data are from low-temperature X-ray diffraction experiments; see Zicovich-Wilson et al. (2008) for further details.

a brief account will be given of these two alternative approaches for including the effect of pressure on computed structural properties of minerals, with particular reference to the case of silicate garnets. The inclusion of the effect of temperature on such properties will be briefly discussed in Section 3.7.

3.3.1 P–V Relation Through Analytical Stress Tensor

The stress tensor $\boldsymbol{\sigma}$ is a symmetric second-rank tensor that can be computed analytically from the total energy density derivatives with respect to strain:

$$\sigma_{ij} = \frac{1}{V}\frac{\partial E}{\partial \epsilon_{ij}} = \frac{1}{V}\sum_{\kappa=1}^{3}\frac{\partial E}{\partial a'_{\kappa i}}a_{\kappa j}, \tag{3.8}$$

with ϵ second-rank symmetric pure strain tensor and $i,j,k = x,y,z$. In the second equality, $\partial E/\partial \epsilon_{ij}$ has been expressed in terms of analytical energy gradients with respect to lattice parameters, with a_{ij} elements of a 3×3 matrix, **A**, where Cartesian components of the three lattice vectors $\mathbf{a}_1$, $\mathbf{a}_2$, and $\mathbf{a}_3$ are inserted by rows [$V = \mathbf{a}_1(\mathbf{a}_2 \times \mathbf{a}_3)$ is the cell volume]; when a distortion is applied to the cell, the lattice parameters transform as

$$a'_{ij} = \sum_{k=1}^{3}\left(\delta_{jk} + \epsilon_{jk}\right)a_{ik}, \tag{3.9}$$

where δ_{jk} is the Kronecker delta. The difficult part of the calculation of the stress tensor in Equation 3.8 is the evaluation of the analytical energy gradients with respect to the cell parameters, which have been implemented in the CRYSTAL program about 10 years ago by Doll et al. for 1-D, 2-D, and 3-D periodic systems (Doll et al. 2004, 2006).

An external "prestress" in the form of a hydrostatic pressure P,

$$\sigma_{ij}^{\text{pre}} = P\delta_{ij}, \tag{3.10}$$

can be added to that of Equation 3.8. Given that the optimizer works in terms of analytical cell gradients, in order to perform a pressure-constrained geometry optimization, the total stress tensor has to be back-transformed to obtain the corresponding constrained gradients:

$$\frac{\partial H}{\partial a_{ij}} = \frac{\partial E}{\partial a_{ij}} + PV\left(\mathbf{A}^{-1}\right)_{ji}. \tag{3.11}$$

Let us note that, with the inclusion of a hydrostatic pressure, the function to be minimized becomes the enthalpy $H = E + PV$ (Souza and Martins 1997). The geometry optimizer under an external hydrostatic pressure has been implemented by Doll (2010) in the CRYSTAL program: the optimized volume V of any crystal at a given hydrostatic pressure P can then be computed analytically.

In a couple of recent studies (Erba et al. 2014a; Mahmoud et al. 2014a), the $P–V$ relation of six silicate garnet end-members has been computed using B3LYP hybrid first-principles simulations; results are summarized in Figure 3.1 and compared with available experimental data. For synthetic pyrope, a single-crystal X-ray diffraction study up to 33 GPa by Zhang et al. (1998) was available (solid circles); a subsequent study by Zhang et al. (1999) on synthetic single-crystal grossular, andradite, and almandine up to 12, 14, and 22 GPa, respectively, was taken as a reference (solid triangles) for these three systems. Computed values for almandine are reported up to 6 GPa only because for higher pressures the SCF electronic structure calculation of the system is not converging within the numerical accuracy required by such calculations. For spessartine and uvarovite, the high-pressure

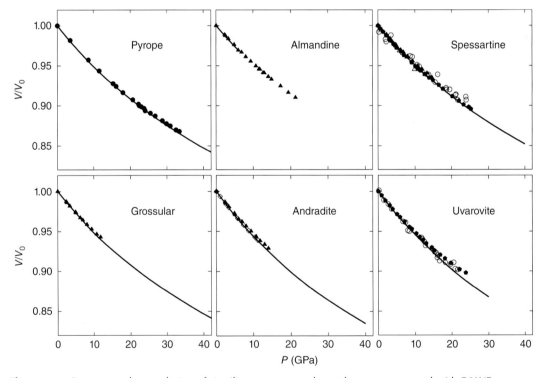

Figure 3.1 *Pressure–volume relation of six silicate garnet end-members as computed with B3LYP first-principles simulations (continuous lines) and measured experimentally (symbols); see text for details on the experiments.*

X-ray synchrotron diffraction study by Diella et al. (2004) up to 25 and 35 GPa, respectively, is taken as a reference (solid pentagons). We also report the values obtained by Leger et al. (1990) in their powder X-ray diffraction study as a function of pressure for spessartine and uvarovite up to 25 GPa (open circles). For spessartine, other two compressional experiments are considered: a recent pressure–volume–temperature study by Gréaux and Yamada (2012) up to 13 GPa (open triangles) and the single-crystal X-ray diffraction study by Zhang et al. (1999) up to 15 GPa (solid triangles).

The agreement between the computed P–V relation and the experimentally determined one is overall very satisfactory, as can be inferred from insight of Figure 3.1. As we will discuss in Section 3.4, the reliable description of the P–V relation constitutes an essential prerequisite to the study of elastic properties of minerals under geophysical pressures.

3.3.2 *P–V* Relation Through Equation of State

An alternative and commonly adopted approach for establishing the P–V relation of a mineral is using so-called equations of state (EOSs). "Cold" EOSs are energy–volume or pressure–volume relations which describe the behavior of a solid under compression and expansion, at $T = 0$ K, that is the case of standard quantum-mechanical simulations. Here we are interested in *universal* EOSs (i.e., not specific of particular materials) that are generally expressed as analytical functions of a limited set of parameters (equilibrium energy E_0, equilibrium volume V_0, equilibrium bulk modulus $K_0 = -V\partial P/\partial V$, and pressure derivative of equilibrium bulk modulus $K_0' = \partial K_0/\partial P$) for ease of interpolation, extrapolation, and differentiation and are quite used in solid-state physics and geophysics (Alchagirov et al. 2001; Cohen et al. 2000).

EOSs have experienced a large success in theoretical simulations, in that, in principle, they would allow for passing from (few) energy–volume data in the vicinity of the equilibrium volume to the P–V relation and, possibly, to high-pressure properties. To do so, energy–volume data are numerically fitted to the analytical $E(V)$ functional form of the EOS. From $P = -\partial E/\partial V$, the P–V connection is established. Let us stress, however, that the analytical expression of the $E(V)$ relation is generally obtained as a series which is truncated to some order. By taking derivatives of increasing order of this expression (for computing pressure P, bulk modulus K, and its pressure derivative K'), the error introduced by that truncation increases.

Many universal EOSs have been proposed so far (Alchagirov et al. 2001; Birch 1947, 1978; Holzapfel 1996; Murnaghan 1944; Poirier and Tarantola 1998; Vinet et al. 1986). They are all phenomenological and can behave quite differently from each other as regards extrapolation at high pressures. Comprehensive reviews and comparisons of different EOSs are available in the literature (Anderson 1995; Angel 2000; Duffy and Wang 1998; Hama and Suito 1996; Stacey et al. 1981). Four EOSs are currently implemented in the CRYSTAL program: the original third-order Murnaghan's (1944), the third-order Birch's (1947, 1978), the logarithmic Poirier–Tarantola's (1998), and the exponential Vinet's (1986).

The analytical stress tensor approach illustrated in Section 3.3.1 provides an extremely satisfactory description of the P–V relation of the six silicate garnets here considered (see Figure 3.1). One may wonder about the accuracy that could be reached in computing the same P–V relation by following the alternative scheme based on the EOSs. When extrapolating to high pressure, the four considered EOSs slightly deviated from each other, still remaining relatively close to the stress tensor reference (with maximum differences of 0.4% for pyrope, at 60 GPa, with the Poirier–Tarantola logarithmic EOS; 0.6% for grossular, at 60 GPa, with the Murnaghan EOS; 0.5% for andradite, at 40 GPa, with the Murnaghan EOS, for instance). Since differences among the four EOSs are very small, it is difficult to tell which one is providing the best description as regards the P–V relation of this family of garnets: Birch–Murnaghan for pyrope, Vinet for grossular, and Poirier–Tarantola for andradite, for instance. All of them are essentially providing an acceptable description of the compressibility of these minerals. On the contrary, large deviations from the analytical reference have been observed as regards the pressure dependence of the bulk modulus, the third-order Birch–Murnaghan one providing the best description among them (Erba et al. 2014a; Mahmoud et al. 2014a).

3.4   Elastic Properties

The physicochemical composition of the Earth's deep interior still has to be properly determined. In this respect, seismological data collected during earthquakes constitute the main source of information. In order to correctly interpret these outcomes and, possibly, to distinguish among different compositional models, to understand how seismic waves propagate during an earthquake and to trace plate dynamics, the individual elastic response of all possible constituents of the Earth's upper mantle, transition zone, and lower mantle should be fully characterized (i.e., the corresponding elastic tensors determined). This kind of characterization is particularly challenging from an experimental point of view in that simultaneous high temperatures and pressures have to be considered (Anderson and Isaak 1995; Mainprice et al. 2013).

Experimentally, compression/expansion studies at relatively low pressures/temperatures have extensively been used to fit data to various EOSs and to extrapolate at geophysical conditions (see Section 3.3.2); first derivatives of the bulk modulus K with respect to pressure and temperature could be determined which, however, showed large discrepancies among each other. As regards pressure, for instance, the comprehensive work by Knittle (1995) documents how, at ambient pressure, different experimental determinations of K agree relatively well to each other, while at high pressures disagreements up to 50% are commonly reported on K'. High pressure or high-temperature single-crystal experimental elastic studies are now becoming feasible, but still, measurements of the elastic constants of a mineral single crystal at simultaneous high pressure and temperature are rare.

Computational material science does represent, in principle, a powerful alternative, especially so if first-principles simulations are considered. DFT constitutes, indeed, an accurate theoretical framework for simulating elastic properties of minerals (Karki et al. 2001). Pressure and temperature are very different thermodynamic variables to be properly accounted for within quantum-mechanical calculations. The effect of pressure can be introduced in a relatively simple way into the picture by an EOS approach or by an analytical stress tensor approach (see Section 3.3). In both cases, the equilibrium volume (and, more generally, structure) of the crystal at any given pressure can be computed quite accurately and at relatively low computational cost: one essentially needs to optimize, under given constraints, the crystal structure at different pressures/volumes. According to this scheme, high-pressure elastic properties of many minerals of the Earth's mantle have been computed from first principles by Karki and collaborators (Karki et al. 2001). The inclusion of temperature effects on computed elastic properties is by far more complex and requires to go somehow beyond the harmonic approximation to the lattice potential according to which the lattice thermal expansion would, indeed, be zero. We will briefly address this issue in Section 3.7.

3.4.1   Evaluation of the Elastic Tensor

In the absence of any finite prestress, elastic constants can be defined as second energy density derivatives with respect to pairs of infinitesimal Eulerian strains:

$$C_{ijkl} = \frac{1}{V_0} \left(\frac{\partial^2 E}{\partial \epsilon_{ij} \partial \epsilon_{kl}} \right)_{\epsilon = 0}, \tag{3.12}$$

where V_0 is the equilibrium volume. These constants do represent the link between stress and strain via Hooke's law. In the limit of zero temperature, typical of quantum-mechanical simulations, they are also referred to as athermal elastic constants. The elastic tensor $\mathbb{C}$ is a fourth-rank tensor which, in principle, should be characterized by 81 components. Given that the pure strain tensor ϵ is symmetric, $\mathbb{C}$, in general, exhibits only 21 independent elements due to the following possible permutations among its indices: $(i \leftrightarrow j)$, $(k \leftrightarrow l)$, and $(ij \leftrightarrow kl)$ coming from the invariance with respect to the order at which the two derivatives in Equation 3.12 are performed. The point symmetry of the different lattices can further reduce the number of independent constants down to 3 (for cubic crystals). A fully automated implementation of the elastic tensor calculation is available in the CRYSTAL program (Erba et al. 2014b; Perger et al. 2009).

If a finite prestress σ^{pre} is applied in the form of a hydrostatic pressure P, as in Equation 3.10, within the frame of finite Eulerian strain, the corresponding elastic stiffness constants read (Karki et al. 1997, 2001; Wallace 1965, 1972; Wang et al. 1995):

$$B_{ijkl} = C_{ijkl} + \frac{P}{2}\left(2\delta_{ij}\delta_{kl} - \delta_{il}\delta_{jk} - \delta_{ik}\delta_{jl}\right), \qquad (3.13)$$

provided that V_0 in Equation 3.12 becomes the equilibrium volume $V(P)$ at pressure P. In the fully automated implementation in the CRYSTAL program of the calculation of the stiffness tensor $\mathbf{B}$ (and of $S = \mathbf{B}^{-1}$, the compliance tensor) under pressure (Erba et al. 2014a), $V(P)$ is obtained from the analytical stress tensor described in Section 3.3.1. An option exists for using the $V(P)$ relation obtained from a given EOS, as discussed in Section 3.3.2. Since both ϵ and δ are symmetric tensors, we can rewrite equality (3.13) as

$$B_{vu} = C_{vu} + \begin{pmatrix} 0 & P & P & 0 & 0 & 0 \\ P & 0 & P & 0 & 0 & 0 \\ P & P & 0 & 0 & 0 & 0 \\ 0 & 0 & 0 & \dfrac{-P}{2} & 0 & 0 \\ 0 & 0 & 0 & 0 & \dfrac{-P}{2} & 0 \\ 0 & 0 & 0 & 0 & 0 & \dfrac{-P}{2} \end{pmatrix}, \qquad (3.14)$$

where Voigt's notation (Nye 1957) has been used, according to which $v, u = 1, \ldots, 6$ ($1 = xx$, $2 = yy$, $3 = zz$, $4 = yz$, $5 = xz$, $6 = xy$).

Following the procedure described previously, we have computed the elastic stiffness constants B_{vu} of pyrope and grossular up to 60 GPa and those of andradite up to 40 GPa at B3LYP level of theory using all-electron BSs and the CRYSTAL program (Erba et al. 2014a). The three symmetry-independent constants, B_{11}, B_{12}, and B_{44}, are reported in Figure 3.2 as black lines; the corresponding dependence on pressure is rather similar for the three garnets, and it is quasilinear, B_{12} showing a slightly more linear behavior than B_{11} and B_{44}. Available experimental data, as obtained from Brillouin scattering measurements, are also reported in the figure.

For pyrope, two measurements are reported: one by Sinogeikin and Bass (2000) who reported values at ambient pressure and at $P = 14$ GPa (filled symbols) and one by Conrad et al. (1999) who reported values at three different pressures (empty symbols). From inspection of the left panel of Figure 3.2, one can clearly observe that (i) the two experimental datasets agree relatively well with

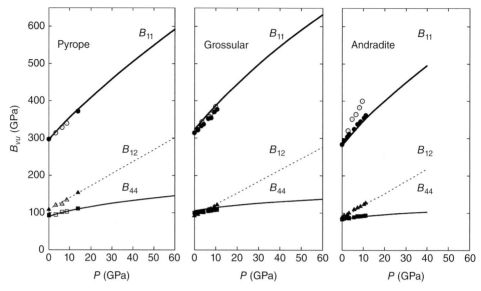

Figure 3.2 *Elastic stiffness constants* B_{vu} *of pyrope (left panel), grossular (central panel), and andradite (right panel), as a function of pressure* P. *Black lines represent computed values. All experimental values are obtained from Brillouin scattering measurements (see text for details). Erba et al. (2014a). Reproduced with permission of AIP.*

each other, (ii) absolute computed values of B_{vu} at zero pressure are extremely close to their measured counterparts (Erba et al. 2014b,c), and (iii) the computed pressure dependence of the elastic stiffness constants satisfactorily matches available experimental data in the low-pressure regime and is thus expected to be rather reliable in the high-pressure predictions. Also for grossular, two experiments are available to compare with: results by Jiang et al. (2004a) that refer to a 87% grossular-rich garnet (at eight different pressures up to 11 GPa) are reported as filled symbols; values obtained by Conrad et al. (1999) at five different pressures up to 10 GPa for the B_{11} constant are also reported as empty symbols. We observe that (i) the absolute values of the elastic constants at ambient pressure agree with the experiments as regards B_{12} and B_{44}, while B_{11} is slightly overestimated in this case (let us recall that the experiment by Jiang et al. was performed on a 87% grossular-rich garnet with 9% of andradite which exhibits lower elastic constants than grossular) and (ii) the pressure dependence of all elastic constants nicely compares with the experimental behavior; in particular, the low-pressure crossing of B_{12} and B_{44} is perfectly reproduced. For andradite, two experimental datasets are available: Jiang et al. (2004b) reported values at nine pressures up to 11 GPa (filled symbols). Again, for B_{11} we also report, as empty symbols, the less accurate results by Conrad et al. (1999) (two crystal directions were considered at each pressure with respect to 36 in Jiang et al. (2004b)) who measured the elastic constants at five pressures up to 10 GPa. Some considerations are as follows: (i) the two experiments describe a very different pressure dependence of B_{11}, (ii) both the absolute values at ambient pressure and the pressure dependence of computed constants are in good agreement with data by Jiang et al. (2004b), and (iii) given the reliable description of the pressure dependence of computed elastic constants of pyrope and grossular, our results for andradite confirm the higher accuracy of the measurements by Jiang et al. (2004b) with respect to those by Conrad et al. (1999).

3.4.2 Elastic Tensor-Related Properties

Several elastic properties of isotropic polycrystalline aggregates can be computed from the elastic stiffness and compliance constants defined previously via the Voigt–Reuss–Hill averaging scheme (Hill 1963). In particular, for cubic crystals (as in the case of silicate garnets that we will discuss in the following), the adiabatic bulk modulus K is simply defined as

$$K = \frac{1}{3}(B_{11} + 2B_{12}) \equiv \frac{1}{3}(S_{11} + 2S_{12})^{-1}. \qquad (3.15)$$

The average shear modulus $\bar{G}$ can be expressed as

$$\bar{G} = \frac{1}{10}(B_{11} - B_{12} + 3B_{44}) + \frac{5}{2}(4(S_{11} - S_{12}) + 3S_{44})^{-1}. \qquad (3.16)$$

From the bulk modulus and the average shear modulus defined earlier, Young's modulus E and Poisson's ratio σ can be defined as well:

$$E = \frac{9K_0\bar{G}}{3K_0 + \bar{G}} \quad \text{and} \quad \sigma = \frac{3K_0 - 2\bar{G}}{2(3K_0 + \bar{G})}. \qquad (3.17)$$

General expressions for crystals of any symmetry of all these quantities in terms of elastic and compliance constants can be found, for instance, in Ottonello et al. (2010). The spatial anisotropy of Young's modulus, linear compressibility, shear modulus, and Poisson's ratio can also be evaluated from the computed elastic tensor and fully characterized (Marmier et al. 2010; Nye 1957).

3.4.3 Directional Seismic Wave Velocities and Elastic Anisotropy

Single-crystal Brillouin scattering experiments allow for the accurate determination of directional seismic wave velocities and elastic anisotropy of a crystal. The manifestations of elastic anisotropy might be not so evident from the sole inspection of the elastic tensor and include (i) shear-wave birefringence, that is, the two polarizations of transverse waves travel with different velocities, and (ii) azimuthal anisotropy, that is, the seismic wave velocities depend on propagation direction. If not properly recognized, anisotropic effects are generally interpreted as due to inhomogeneities such as layering or gradients and can lead to validation of wrong compositional models (Anderson 1989). Despite their relevance to the compositional analysis of the Earth's interior, only few directional studies have been performed so far for silicate garnets (Jiang et al. 2004a,b; Sinogeikin and Bass 2000), while most studies report average seismic wave velocities (Chen et al. 1999; Gwanmesia et al. 2014, 2006; Jiang et al. 2004a,b; Kono et al. 2010; Sinogeikin and Bass 2000).

According to the elastic continuum model, the three acoustic wave velocities, along any general crystallographic direction represented by unit wave vector $\hat{\mathbf{q}}$, can be related to the elastic constants by Christoffel's equation which can be given an eigenvalue/eigenvector form as follows (Auld 1973; Musgrave 1970):

$$\mathbf{A}^{\hat{\mathbf{q}}}\mathbf{U} = \mathbf{V}^2\mathbf{U} \quad \text{with} \quad A_{jk}^{\hat{\mathbf{q}}} = \frac{1}{\rho}\sum_{il}\hat{q}_i C_{ijkl}\hat{q}_l, \qquad (3.18)$$

where $A_{jk}^{\hat{\mathbf{q}}}$ is Christoffel's matrix, ρ the crystal density, $i,j,k,l = x,y,z$ represent Cartesian directions, $\hat{q}_i$ is the ith element of the unit vector $\hat{\mathbf{q}}$, $\mathbf{V}$ is a 3×3 diagonal matrix whose three elements give the acoustic velocities, and $\mathbf{U} = (\hat{\mathbf{u}}_1, \hat{\mathbf{u}}_2, \hat{\mathbf{u}}_3)$ is the eigenvector 3×3 matrix where each column represents the polarization $\hat{u}$ of the corresponding eigenvalue. The three acoustic wave velocities, also referred to as seismic velocities, can be labeled as quasilongitudinal v_p, slow quasitransverse v_{s1}, and fast quasitransverse v_{s2}, depending on the polarization direction $\hat{u}$ with respect to wave vector $\hat{q}$ (Karki et al. 2001).

As anticipated in the preceding, the elastic anisotropy of a crystal can be fully characterized from directional seismic wave velocities. The azimuthal anisotropy for quasilongitudinal and quasitransverse seismic wave velocities can be defined as follows (Karki et al. 2001):

$$A_X = \frac{v_{X_{max}} - v_{X_{min}}}{\bar{v}_X}, \tag{3.19}$$

where $X = p$, s labels longitudinal and shear waves, and $\bar{v}_X$ is the polycrystalline isotropic average velocity obtained from the Voigt–Reuss–Hill scheme (Hill 1963). Elastic anisotropy would be zero for an ideal isotropic material; even cubic crystals, such as silicate garnets, however, show a nonzero elastic anisotropy (Karki et al. 2001). For cubic crystals, the elastic anisotropy can be given a simple expression in terms of a single anisotropy index computed from the elastic constants (Authier and Zarembowitch 2006):

$$A = \left(\frac{2B_{44} + B_{12}}{B_{11}} - 1 \right) \times 100. \tag{3.20}$$

In Figure 3.3, computed directional seismic wave velocities are reported for grossular, uvarovite, spessartine, pyrope, andradite, and almandine along an azimuthal angle θ which spans the (110) crystallographic plane of the lattice by exploring all the high-symmetry crystallographic directions: $\theta = 0°$ corresponds to the crystallographic direction [110], $\theta = 45°$ to [111] direction, $\theta = 90°$ to [001] direction, etc. Computed velocities are reported as continuous lines of increasing thickness as a function of pressure. Available directional experimental data are also reported in the figure: for andradite, data from an accurate single-crystal Brillouin scattering experiment by Jiang et al. (2004b) are reported at ambient pressure (full squares) and at 8.7 GPa (full circles); a subsequent study by Jiang et al. (2004a) on a single-crystal grossular-rich garnet at 4.3 GPa (full circles) is also taken as a reference; for pyrope, data from the study by Sinogeikin and Bass (2000) are reported. From inspection of the figure, the accuracy of the theoretical description of angular dependence, oscillation amplitudes, and pressure shift of the seismic wave velocities can be clearly seen.

From Figure 3.3, the six end-members can be sorted according to increasing propagation velocity, at zero pressure, as follows: Alm < And < Spe < Uva < Pyr < Gro; this sequence does not change under increasing pressure. Almandine shows the slowest v_p, while pyrope and grossular allow for the fastest propagation. This behavior can be rationalized in terms of the elemental composition of the end-members taking into account that seismic wave velocities are inversely proportional to the density of the material. Fe-bearing phases such as andradite and almandine are the most dense, followed by the Mn- and Cr-bearing phases, such as spessartine and uvarovite, whereas pyrope and grossular contain the lightest elements (Mg, Ca) thus being the least dense.

Seismic wave velocities increase as pressure increases in all cases. In the absence of external pressure, the six silicate garnet end-members can be sorted in terms of increasing elastic anisotropy, as follows: Spe < Pyr < Alm < Gro < And ≪ Uva. Spessartine and pyrope show very low anisotropy, while uvarovite is by far the most anisotropic among them. The elastic anisotropy of grossular, uvarovite, spessartine, and andradite increases as a function of pressure. In particular, grossular and spessartine are the most affected by pressure, with anisotropies varying from −5.6 to −13.2% and from −2.5 to −5.1%, respectively, when passing from 0 to 40 GPa; the elastic anisotropy of andradite increases from −10 to −14.5% in the same pressure range, while the anisotropy of

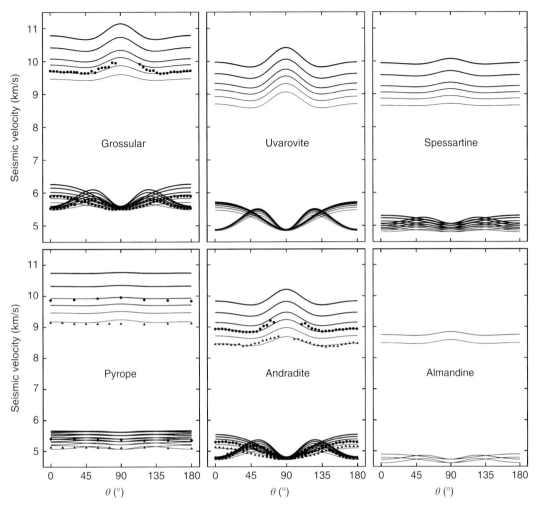

Figure 3.3 *Directional quasitransverse and quasilongitudinal seismic wave velocities of single-crystal grossular, uvarovite, spessartine, pyrope, andradite, and almandine along an azimuthal angle θ (defined in the text). Computed data at different pressures (0 GPa, 4 GPa, 8 GPa, 12 GPa, 20 GPa, 30 GPa) are reported as continuous lines of increasing thickness. Experimental data are reported when available (see text for details). Mahmoud et al. (2014a). Reproduced with permission of AIP.*

uvarovite only slightly increases from −15.8 to −16.5% passing from 0 up to 30 GPa (Mahmoud et al. 2014a). Pyrope and almandine show a different behavior under pressure: their elastic anisotropy decreases. If at ambient pressure pyrope shows a larger anisotropy than spessartine, as soon as pressure increases, the anisotropy of spessartine becomes larger than pyrope.

3.5 Vibrational and Thermodynamic Properties

A wealth of information about the chemical composition, structure, and thermodynamics of minerals can be inferred from vibrational spectroscopic measurements. In this respect, quantum-mechanical computational spectroscopy has proven to be an extremely useful complementary tool

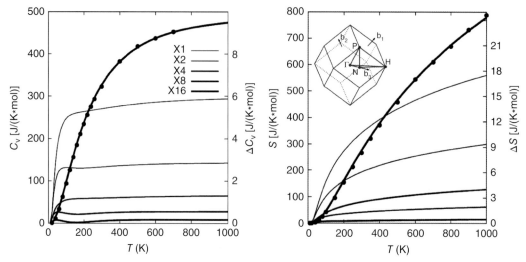

Figure 3.4 *Constant-volume specific heat C_V and entropy S of pyrope as a function of temperature, as computed at B3LYP level of theory (thick continuous line) with the largest SC considered (viz., X27) and compared with experimental data (full circles) from Haselton and Westrum (1980) and Tequi et al. (1991). On the right scale of the two panels, $\Delta C_V = C_V^{X27} - C_V^{Xn}$ and $\Delta S = S^{X27} - S^{Xn}$ are reported that show the convergence of computed thermodynamic properties on the size of the adopted SC (n = 1, 2, 4, 8, 16). The inset of the right panel shows the shape of the first Brillouin zone of silicate garnets. (a) Haselton and Westrum (1980). Reproduced with permission of Elsevier. (b) Tequi et al. (1991). Reproduced with permission of Elsevier.*

for achieving full characterization of IR and Raman spectra (see Chapter 10) in terms of peak positions, intensities, and band classifications.

In the CRYSTAL program, the calculation of vibration frequencies at the Γ point ($\mathbf{k} = 0$, at the center of the first Brillouin zone (FBZ) in reciprocal space; for a graphical representation of the shape of the FBZ of silicate garnets, see the inset of Figure 3.4), within the harmonic approximation, is available since 2003 (Pascale et al. 2004b; Zicovich-Wilson et al. 2004). The vibration frequencies at the center of the FBZ (directly comparable with the outcomes of IR and Raman measurements) are obtained from the diagonalization of the mass-weighted Hessian matrix of the second derivatives of the total energy per cell with respect to atomic displacements u:

$$W_{ai,bj}^{\Gamma} = \frac{H_{ai,bj}^{0}}{\sqrt{M_a M_b}} \quad \text{with} \quad H_{ai,bj}^{0} = \left(\frac{\partial^2 E}{\partial u_{ai}^0 \partial u_{bj}^0} \right), \tag{3.21}$$

where atoms a and b (with atomic masses M_a and M_b) in the reference cell are displaced along the ith and jth Cartesian directions. The first derivatives of the total energy per cell $\left(g_i^a = \partial E / \partial u_{ai} \right)$ with respect to atomic displacements from the equilibrium configuration $\mathcal{R}^{eq}$ are computed analytically, whereas second derivatives numerically, using a two-point formula:

$$\frac{\partial^2 E}{\partial u_{ai} \partial u_{bj}} \approx \frac{g_i^a \left(\mathcal{R}^{eq}, u_{bj} = +\bar{u} \right) - g_i^a \left(\mathcal{R}^{eq}, u_{bj} = -\bar{u} \right)}{2\bar{u}},$$

where $\bar{u} = 0.003$ Å, a value 10–50 times smaller than that usually used in other solid-state programs (Kresse and Furthmüller 1996a,b; Soler et al. 2002).

Chapter 10 is entirely devoted to the presentation of the many tools that have been developed and implemented into the public CRYSTAL program as regards the analysis of spectroscopical properties of mineral calculation of IR and Raman analytical intensities, simulation of IR and Raman spectra, isotopic substitution effect, longitudinal optical (LO)/transverse optical (TO) splitting, anharmonic corrections to hydrogen stretching modes, graphical analysis of the normal modes of vibration, scanning of energy along normal mode coordinates, etc. Several examples of applications of these techniques to minerals of geochemical interest are also given there. In the next part of this section, we will briefly review the main studies about first-principles investigations of vibrational spectroscopic properties of silicate garnets and then move to discussing how thermodynamic properties of minerals can be simulated with standard quantum-mechanical techniques.

The IR and Raman vibration frequencies of pyrope and andradite have been first computed quantum mechanically, using the B3LYP hybrid functional, in 2005 (Pascale et al. 2005a,b); all vibration modes have been characterized by evaluating the effect of the isotopic substitution. In the following years, the same methodology has been applied to other members of the silicate garnet family: grossular (Dovesi et al. 2009), spessartine (Valenzano et al. 2009), uvarovite (Valenzano et al. 2010), and almandine (Ferrari et al. 2009). In 2008, the LO/TO splitting, requiring the information about the dielectric response of the system, of pyrope, grossular, and andradite has been computed (Zicovich-Wilson et al. 2008). The complete IR reflectance spectra of the six end-members have been simulated by Dovesi et al. (2011). The 17 IR-active F_{1u} TO and LO frequencies, the corresponding oscillator strengths, the high-frequency and static dielectric constants, and the reflectance spectra have been computed. The agreement with experiments for the TO and LO peaks has been documented to be always rather satisfactory, the mean absolute difference for the whole set of data (178 peaks in total) being 5 cm^{-1}. The reflectance spectra, simulated through the classical dispersion relation, had reproduced the experimental curves extremely well. For this class of systems, the B3LYP hybrid functional has been shown to significantly improve over LDA or GGA ones (Maschio et al. 2011). Only recently, the implementation of analytical Raman intensities in the CRYSTAL14 version of the program has made the quantum-mechanical simulation of Raman spectra of crystalline materials possible (Dovesi et al. 2014a). A couple of recent studies have reported about these kinds of simulations for pyrope and grossular (Maschio et al. 2014, 2013). An example will be briefly illustrated in Section 3.6 as regards the simulation of the IR spectroscopical properties of the grossular–andradite solid solution series $Ca_3Fe_{2-2x}Al_{2x}(SiO_4)_3$ over the whole chemical composition range $0 \leq x \leq 1$ (De La Pierre et al. 2013).

3.5.1 Solid-State Thermodynamics

The calculation of the thermodynamic properties of crystals is generally more demanding with respect to the sole spectroscopic characterization as it requires the knowledge of phonon modes over the complete FBZ; phonons at points different from Γ can be obtained by building a supercell (SC) of the original unit cell, following a direct-space approach (Parlinski et al. 1997; Togo et al. 2008). The lattice vectors $\mathbf{g} = \sum_t l_t^g \mathbf{a}_t$ identify the general crystal cell where $\{\mathbf{a}_t\}$ are the direct lattice basis vectors, with $t = 1,\ldots, D$ (where D is the dimensionality of the system: 1, 2, 3 for 1-D, 2-D, 3-D periodic systems): within periodic boundary conditions the integers l_t^g run from 0 to L_t–1. The parameters $\{L_t\}$ define size and shape of the SC in direct space. Let us label with **G** the general superlattice (i.e., whose reference cell is the SC) vector, and let us introduce the $L = \Pi_t L_t$ Hessian matrices $\{\mathbf{H^g}\}$ whose elements are $H_{ai,bj}^{\mathbf{g}} = \partial^2 E / \left(\partial u_{ai}^0 \partial u_{bj}^{\mathbf{g}} \right)$ where, at variance with Equation 3.21, atom b is displaced in cell **g**, along with all its periodic images in the crystal (i.e., in cells **g** + **G**). The set of L Hessian matrices $\{\mathbf{H^g}\}$ can be Fourier transformed into a set of *dynamical matrices* $\{\mathbf{W^k}\}$, each one associated with a wave vector $\mathbf{k} = \sum_t (k_t/L_t)\mathbf{b}_t$ where $\{\mathbf{b}_t\}$ are the reciprocal lattice vectors and the integers k_t run from 0 to L_t-1:

$$\mathbf{W}^{\mathbf{k}} = \sum_{\mathbf{g}=0}^{L-1} \mathbf{M}^{-(1/2)} \mathbf{H}^{\mathbf{g}} \mathbf{M}^{-(1/2)} \exp(\imath \mathbf{k} \cdot \mathbf{g}), \tag{3.22}$$

where $\mathbf{M}$ is the diagonal matrix with the masses of the nuclei associated with the $3M$ atomic coordinates where M is the number of atoms per cell. The solution is then obtained through the diagonalization of the L matrices $\{\mathbf{W}^{\mathbf{k}}\}$:

$$\left(\mathbf{U}^{\mathbf{k}}\right)^{\dagger} \mathbf{W}^{\mathbf{k}} \mathbf{U}^{\mathbf{k}} = \mathbf{\Lambda}^{\mathbf{k}} \quad \text{with} \quad \left(\mathbf{U}^{\mathbf{k}}\right)^{\dagger} \mathbf{U}^{\mathbf{k}} = \mathbf{I}. \tag{3.23}$$

The elements of the diagonal $\mathbf{\Lambda}^{\mathbf{k}}$ matrix provide the *vibrational frequencies*, $v_i^{\mathbf{k}} = \sqrt{\lambda_i^{\mathbf{k}}}$ (au are adopted), while the columns of the $\mathbf{U}^{\mathbf{k}}$ matrix contain the corresponding *normal coordinates*. To each $\mathbf{k}$-point in the FBZ, $3M$ harmonic oscillators (i.e., phonons) are associated which are labeled by a phonon band index i ($i = 1,\ldots,3M$) and whose energy levels are given by the usual harmonic expression:

$$\varepsilon_m^{i,\mathbf{k}} = \left(m + \frac{1}{2}\right) 2\pi v_i^{\mathbf{k}}, \tag{3.24}$$

where m is an integer. The overall vibrational canonical partition function, $Q_{\text{vib}}(T)$, at a given temperature T, can be expressed as follows:

$$Q_{\text{vib}}(T) = \sum_{\mathbf{k}=0}^{L-1} \sum_{i=1}^{3M} \sum_{m=0}^{\infty} \exp\left[-\frac{\varepsilon_m^{i,\mathbf{k}}}{k_B T}\right], \tag{3.25}$$

where k_B is Boltzmann's constant. According to standard statistical mechanics, thermodynamic properties of crystalline materials such as entropy S and thermal contribution to the internal energy U can be expressed as

$$S(T) = k_B T \left(\frac{\partial \log(Q_{\text{vib}})}{\partial T}\right) + k_B \log(Q_{\text{vib}}) \quad \text{and} \quad U(T) = k_B T^2 \left(\frac{\partial \log(Q_{\text{vib}})}{\partial T}\right). \tag{3.26}$$

From the previous expression for U, the constant-volume specific heat C_V can also be computed according to $C_V = \partial U/\partial T$.

In the first-principles simulation of the thermodynamic properties of minerals, convergence of computed properties has to be carefully checked with respect to the number of $\mathbf{k}$-points considered for the phonon dispersion (or, equivalently in the direct-space approach, with respect to the size of the adopted SC). Here, we consider the theoretical description of the evolution with temperature of the constant-volume specific heat, C_V, and of the entropy, S, of the most abundant among silicate garnets: pyrope, $Mg_3Al_2(SiO_4)_3$. We compute phonon frequencies and, via Equations 3.24–3.26, thermodynamic properties with SC of increasing size (corresponding to an increasing number of $\mathbf{k}$-points in the previous formalism). We start from the primitive cell of pyrope, containing 80 atoms and corresponding to 1 $\mathbf{k}$-point, that we label $X1$; then, five other SCs are built that we label Xn as they are n times larger than the primitive one, where $n = 2, 4, 8, 16$, and 27 (i.e., containing up to 2160 atoms for $X27$ and corresponding to 27 $\mathbf{k}$-points in the FBZ). All the considered SCs are cubic so that their high symmetry can be exploited for reducing the computational cost of all the calculations (in all cases only nine SCF calculations, plus computation of the atomic gradients, are required).

In the two panels of Figure 3.4, we report the constant-volume specific heat C_V and entropy S of pyrope as computed at B3LYP level of theory (thick continuous line) with the largest SC considered (viz., X 27), and we compare with available experimental data (full circles) from Haselton and Westrum (1980) and Tequi et al. (1991). On the right scale of each of the two panels, $\Delta C_V = C_V^{X27} - C_V^{Xn}$ and $\Delta S = S^{X27} - S^{Xn}$ are reported that show the convergence of computed thermodynamic properties with respect to the size of the adopted SC ($n = 1, 2, 4, 8, 16$). It turns out that, already with an $X8$ SC, results are converged within 0.2% to the $X27$ ones for both specific heat and entropy. The agreement with experimental measurements is rather satisfactory and better than previously reported (Hofmeister and Chopelas 1991).

3.6 Modeling Solid Solutions

Solid solutions of different chemical compositions are of crucial relevance to the study of geochemical properties of minerals of the Earth's mantle. Natural silicate garnets, for instance, can be found in a wide range of chemical compositions since they form solid solutions. Solid solutions between pyralspites (pyrope–almandine–spessartine) and ugrandites (uvarovite–grossular–andradite) seldom occur in natural garnets (Hensen 1976), whereas aluminosilicate garnets with different X cations show complete solid solubility among them at high pressure (O'Neill et al. 1989). Linear composition-bulk modulus trends have been observed for garnets in the pyralspite series (Duffy and Anderson 1989; Isaak and Graham 1976; Yeganeh-Haeri et al. 1990). On the contrary, by combining the few available experimental bulk moduli of grandite (grossular–andradite) solid solutions (Babuška et al. 1978; Bass 1986; O'Neill et al. 1989), a significant deviation from linearity seems to turn out, as discussed by O'Neill et al. (1989). However, uncertainty and nonhomogeneity of the measurements still leave room for further insights (see later).

New algorithms have recently been developed and implemented into the CRYSTAL14 program for the study of solid solutions and, more generally, disordered systems (D'Arco et al. 2013; Mustapha et al. 2013). As far as solid-state solutions are concerned, the computational scheme is as follows: (i) the periodic nature of the system is preserved; (ii) in general, an SC is built in order to increase the number of atomic sites involved in the substitution; and (iii) for any chemical composition within a given series, the program finds the total number of atomic configurations and determines the symmetry-irreducible ones to be explicitly considered in order to define a statistical average.

In this section, we present the results of a first-principles theoretical study of the elastic properties of the grandite solid solution, $Ca_3Fe_{2-2x}Al_{2x}(SiO_4)_3$, as a function of its chemical composition x. Reference is made to the primitive unit cell of the end-members (cubic space group $\mathcal{G} \equiv Ia\bar{3}d$), which counts $|\mathcal{G}| = 48$ symmetry operators and four formula units $Ca_3Y_2(SiO_4)_3$. There are eight Y sites involved for substitution. Solid solutions are obtained from andradite by progressively replacing Fe^{3+} with Al^{3+} cations. Apart from the two end-members, andradite ($x = 0$) and grossular ($x = 1$), other seven compositions are explicitly considered: $x = 0.125, 0.25, 0.375, 0.5, 0.625, 0.75$, and 0.875. For each composition x, $n_{Al} = 8x$ aluminum atoms are present that correspond to $8\,!/[n_{Al}!\,(8 - n_{Al})\,!]$ different substitutional configurations, that is, cation distributions among the Y sites. There is a total of 256 possible atomic configurations over the whole range of compositions. Following the symmetry analysis recently proposed by Mustapha et al. (2013), these configurations can be partitioned into 23 distinct symmetry-independent classes (SIC). Each class L consists of $\mathcal{M}_L = |\mathcal{G}|/|\mathcal{H}_L|$ configurations that belong to a symmetry subgroup of order $|\mathcal{H}_L|$ of

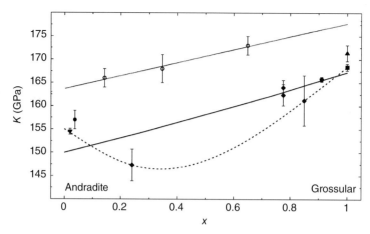

Figure 3.5 *Bulk modulus K of the grandite solid solution, $Ca_3Fe_{2-2x}Al_{2x}(SiO_4)_3$, as a function of its chemical composition x. Experimental data are reported as full and empty symbols (see text for details). When available, error bars are also shown. The solid line shows the quasilinear trend of our calculated values, whereas the thick dashed curve is drawn to provide an approximate fit to the experimental data, as suggested by Bass (1986) and O'Neill et al. (1989). Reproduced with permission of Wiley.*

the aristotype symmetry. Since all the configurations of a given SIC are equivalent to each other, the number of calculations to be actually performed reduces to one *per* SIC, that is, to a total of 23. $\mathcal{M}_L$ can then be interpreted as the multiplicity of class L (see Lacivita et al. (2014) for details on the symmetry properties of the 23 configurations). Only the highest spin ferromagnetic configurations will be considered, as the difference between ferromagnetic and antiferromagnetic energies was shown to be extremely small (Meyer et al. 2010).

In Figure 3.5, available experimental determinations of the bulk modulus K of grandite as a function of its chemical composition x are reported as full and empty symbols (Babuška et al. 1978; Bass 1986, 1989; Halleck 1973; Isaak and Graham 1976; Jiang et al. 2004a,b). For the grossular endmember, two values are reported (at $x \approx 1$, natural single-crystal samples pure to 97 and 99%, respectively): an ultrasonic measurement by Halleck (1973) (triangle) and a Brillouin scattering experiment performed by Bass (1989) (square). For andradite, two values are given as determined by Brillouin spectroscopy by Bass (1986) (circle) and Jiang et al. (2004b) (inverted triangle). From inspection of Figure 3.5, where error bars are also reported, we can clearly see how different experimental determinations of the bulk moduli of the two end-members show finite discrepancies between each other. Experimental uncertainties become even larger for most of the intermediate compositions, as we shall discuss in the following.

The main experimental investigation of the elastic properties of intermediate compositions of the grandite solid solution has been performed by Babuška et al. (1978). Four specimens of three different compositions were analyzed (rhombi in the figure). More recently, Jiang et al. (2004a) performed Brillouin spectroscopy on a grossular-rich garnet (pentagon). The thick dashed line in Figure 3.5 represents the bulk modulus trend as a function of composition x as proposed by Bass (1986) and O'Neill et al. (1989) on the grounds of available experimental data at that time. More recently, an *in situ* X-ray diffraction experiment has been performed by Fan et al. (2011) on three specimens of intermediate chemical compositions in the grandite binary. The corresponding bulk moduli have been evaluated and, though significantly overestimated with respect to all previous determinations, seem to exhibit a linear behavior (see empty circles in Figure 3.5).

The results of theoretical simulations explicitly accounting for the effect of configurational disorder are reported in Figure 3.5 as a solid line (Lacivita et al. 2014). They clearly show a quasilinear behavior of the elastic response of the grandite solid solution as a function of its chemical composition. For each composition x and each SIC, the EOS of a representative atomic configuration was computed, according to the procedure described in Section 3.3.2. As regards the two end-members, computed values are found to be in agreement with experimental data within 1% for grossular and 3% for andradite. An approximately linear dependence of volume on chemical composition is observed in the computed data (De La Pierre et al. 2013), that is, no significant excess volume of mixing is found at any intermediate composition, which confirms the ideal character of the grossular–andradite solid solution, as already suggested by several studies (Bird and Helgeson 1980; Ganguly 1976; Holdaway 1972; McAloon and Hofmeister 1995; Perchuk and Aranovich 1979) and in contrast to other investigations (Liou 1973; Meagher 1975).

These findings contribute to the definition of a homogeneous frame according to which all solid solutions of the most abundant silicate garnets (pyralspite and grandite) exhibit linear elastic properties as a function of their chemical composition. Linearity of pyralspite is well known since long, whereas the presumed nonlinearity of the grandite system has been demonstrated to be an artifact due to scarcity and heterogeneity of the available experimental measurements.

Optical (Lacivita et al. 2013a) and spectroscopical (De La Pierre et al. 2013) properties of the grossular–andradite, $Ca_3Fe_{2-2x}Al_{2x}(SiO_4)_3$, solid solution series have recently been investigated at the quantum-mechanical level of theory. Refractive indices vary quite regularly between the andradite (1.860) and grossular (1.671) end-members. The high-frequency SiO stretching modes also show a rather linear behavior of both frequencies and IR intensities as a function of x. The frequencies of the low-energy bands show an almost linear dependence on composition as well; on the contrary, the behavior of the corresponding intensities is less linear. When considering different possible atomic configurations at fixed composition, some spectral features display a clear dependence upon short-range Al/Fe cation ordering.

Garnets are among the most promising candidates as possible hydrogen storage media for the Earth's mantle, given their abundance and stability (Milman et al. 2000; Nobes et al. 2000a). A lot of attention is being devoted to the understanding of the incorporation of hydrogen into nominally anhydrous minerals (NAM) because of the remarkable effect it has on their technological and geophysical properties (Aines and Rossman 1984a,b; Beran and Putnis 1983; Freund and Oberheuser 1986; Griggs 1967; Rossman 1988; Wilkins and Sabine 1973). In particular, NAMs are of great geological interest in that they may potentially introduce large amount of "water" in the Earth mantle thus significantly modifying its elastic properties (Knittle et al. 1992; Mackwell et al. 1985; O'Neill et al. 1993).

The hydrogarnet substitution ($SiO_4 \leftrightarrow O_4H_4$) in grossular has received special attention in that it represents an effective mechanism for including hydrogen into silicate garnets (Lager et al. 1987, 2002, 2005; Lager and Von Dreele 1996; Nobes et al. 2000a,b; Olijnyk et al. 1991; Orlando et al. 2006; Pascale et al. 2004a; Pertlik 2003). Hydrogrossular can be represented by the general formula $Ca_3Al_2(SiO_4)_{3-x}(OH)_{4x}$; when $0 < x < 1.5$ it is called hibschite, and when $1.5 < x < 3$ it is called katoite (Pertlik 2003). At low temperature there is complete solid solubility of the two end-members (grossular and silicon-free katoite). Hydrogarnets are known to be stable over the whole Earth's mantle pressure range (Knittle et al. 1992); natural hydrogarnets equilibrated at 180 km depth have been characterized (O'Neill et al. 1993). In a couple of recent studies, first-principles simulations have been applied to the investigation of the behavior of katoite under pressure (Erba et al. 2015c) and of the explicit effect of chemical composition x on structural and energetic properties of hydrogrossular (Lacivita et al. 2015).

3.7   Future Challenges

Most of the techniques that we are currently developing and implementing in the CRYSTAL program, relevant to the quantum-mechanical simulation of geochemical properties of minerals, are meant to include the effect of temperature on computed properties. Standard quantum-chemical methods based on the DFT represent, indeed, a powerful tool for the accurate determination of a variety of properties of materials, such as structural, electronic, vibrational, optical, elastic, magnetic, etc. (Dovesi et al. 2005; Dronskowski 2005; Grimme et al. 2010; Pisani 1996). The growing parallel computing resources are rapidly widening the range of applicability of such schemes which can now be routinely used for studying minerals of geochemical interest. If the inclusion of pressure can be modeled in a relatively simple way (with techniques like those presented in Section 3.3), that of temperature is a much more difficult task for the solid state. The most effective technique for taking into account temperature effects (including anharmonic terms) on computed properties of materials, in particular as regards thermal nuclear motion, would be *ab initio* molecular dynamics which, however, remains rather computationally demanding (Buda et al. 1990; Car and Parrinello 1985; Vila et al. 2012).

Within the frame of standard quantum-chemical methods, temperature can be modeled by explicitly treating the lattice dynamics. A number of techniques have already been implemented in the CRYSTAL program for including the effect of harmonic thermal nuclear motion on one-electron properties such as electron charge and momentum densities (Erba et al. 2013c; Madsen et al. 2013; Pisani et al. 2012). When the lattice dynamics of a crystal is solved within the purely harmonic approximation, however, vibration frequencies are described as independent of interatomic distances, and the corresponding vibrational contribution to the internal energy of the crystal turns out to be independent of volume. It follows that, within such an assumption, a variety of physical properties would be wrongly described: thermal expansion would be null, elastic constants would not depend on temperature, constant-pressure and constant-volume specific heats would coincide with each other, thermal conductivity would be infinite as well as phonon lifetimes, etc. (Ashcroft and Mermin 1976; Baroni et al. 2010). An explicit account of anharmonic effects would require the calculation of phonon–phonon interaction coefficients with techniques such as vibrational configuration interaction (VCI), vibrational self-consistent field (VSCF), vibrational perturbation theory (VPT), transition-optimized shifted Hermite (TOSH), etc. (Jung and Gerber 1996; Lin et al. 2008; Neff and Rauhut 2009). An alternative and much simpler approach for correcting most of the aforementioned deficiencies of the harmonic approximation is the so-called quasiharmonic approximation (QHA) (Allen and De Wette 1969); according to which, the equilibrium volume, $V(T)$, at any temperature T can be deduced by minimizing Helmholtz's free energy $F(V; T)$. This approach also allows for the natural combination of temperature and pressure effects. A fully automated implementation of the QHA has recently been developed in the CRYSTAL program, which relies on computing and fitting harmonic vibration frequencies at different volumes after having performed volume-constrained geometry optimizations (Erba 2014; Erba et al. 2015d). This strategy has already been successfully applied to the investigation of thermal structural and average elastic properties of diamond, periclase, lime, Al_2O_3 corundum, and Mg_2SiO_4 forsterite (Erba 2014; Erba et al. 2015a,b,d), and investigations on molecular crystals and alkali halides are currently in progress. The quasiharmonic approximation is also being generalized to the calculation of the elastic tensor of crystals at finite temperatures.

References

Adamo C and Barone V 1999 Toward reliable density functional methods without adjustable parameters: the PBE0 model. *J. Chem. Phys.* **110**, 6158.

Aines RD and Rossman GR 1984a The hydrous component in garnets: pyralspites. *Am. Mineral.* **69**(11–12), 1116–1126.

Aines RD and Rossman GR 1984b Water in minerals? A peak in the infrared. *J. Geophys. Res. Solid Earth* **89**(B6), 4059–4071.

Albanese E, Civalleri B, Ferrabone M, Bonino F, Galli S, Maspero A, and Pettinari C 2012 Theoretical and experimental characterization of pyrazolato-based Ni(II) metal-organic frameworks. *J. Mater. Chem.* **22**, 22592–22602.

Alchagirov AB, Perdew JP, Boettger JC, Albers RC, and Fiolhais C 2001 Energy and pressure versus volume: equations of state motivated by the stabilized jellium model. *Phys. Rev. B* **63**, 224115.

Allen RE and De Wette FW 1969 Calculation of dynamical surface properties of noble-gas crystals. I. The quasiharmonic approximation. *Phys. Rev.* **179**, 873–886.

Anderson DL 1989 *Theory of the Earth*. Blackwell Scientific Publications, Boston.

Anderson OL 1995 *Equations of State of Solids for Geophysicists and Ceramic Science*. Oxford University Press, New York.

Anderson OL and Bass JD 1986 Transition region of the earth's upper mantle. *Nature* **320**, 321–328.

Anderson OL and Isaak DG 1995 *Elastic Constants of Mantle Minerals at High Temperature*. American Geophysical Union, Washington, D.C. pp. 64–97.

Angel RJ 2000 Equations of state In *High Temperature and High Pressure Crystal Chemistry*, Rev. Mineral. Geochem. 41 (ed. Hazen R and Downs R) Mineralogical Society of America, Washington, D.C. pp. 35–59.

Ashcroft NW and Mermin ND 1976 *Solid State Physics*. Saunders College, Philadelphia, PA.

Auld BA 1973 *Acoustic Fields and Waves in Solids*. Krieger Publishing Company, Malabar, FL.

Authier A and Zarembowitch A 2006 Elastic properties In *International Tables for Crystallography, Vol. D* (ed. Authier A) John Wiley & Sons, Inc. p. 72.

Babuška V, Fiala J, Kumazawa M, Ohno I, and Sumino Y 1978 Elastic properties of garnet solid-solution series. *Phys. Earth Planet. In.* **16**(2), 157–176.

Baima J, Erba A, Orlando R, Rérat M, and Dovesi R 2013 Beryllium oxide nanotubes and their connection to the flat monolayer. *J. Phys. Chem. C* **117**, 12864–12872.

Baroni S, Giannozzi P, and Isaev E 2010 Density-functional perturbation theory for quasi-harmonic calculations. *Rev. Mineral. Geochem.* **71**(1), 39–57.

Bass JD 1986 Elasticity of uvarovite and andradite garnets. *J. Geophys. Res.* **91**(B7), 7505–7516.

Bass JD 1989 Elasticity of grossular and spessartite garnets by Brillouin spectroscopy. *J. Geophys. Res.* **94**, 7621–7628.

Bass JD and Anderson OL 1984 Composition of the upper mantle: geophysical tests of two petrological models. *Geophys. Res. Lett.* **11**, 229–232.

Becke AD 1993 Density-functional thermochemistry. III. The role of exact exchange. *J. Chem. Phys.* **98**, 5648.

Beran A and Putnis A 1983 A model of the OH positions in olivine, derived from infrared-spectroscopic investigations. *Phys. Chem. Miner.* **9**(2), 57–60.

Birch F 1947 Finite elastic strain of cubic crystals. *Phys. Rev.* **71**, 809–824.

Birch F 1978 Finite strain isotherm and velocities for single-crystal and polycrystalline NaCl at high pressures and 300 K. *J. Geophys. Res.* **83**(B3), 1257–1268.

Bird DK and Helgeson HC 1980 Chemical interaction of aqueous solutions with epidote-feldspar mineral assemblages in geologic systems; 1, thermodynamic analysis of phase relations in the system CaO-FeO-Fe_2O_3-Al_2O_3-SiO_2-H_2O-CO_2. *Am. J. Sci.* **280**, 907–941.

Broyden CG 1970 The convergence of a class of double-rank minimization algorithms 1. General considerations. *IMA J. Appl. Math.* **6**(1), 76.

Buda F, Car R, and Parrinello M 1990 Thermal expansion of *c*-Si via ab initio molecular dynamics. *Phys. Rev. B* **41**, 1680.

Bush IJ, Tomic S, Searle BG, Mallia G, Bailey CL, Montanari B, Bernasconi L, Carr JM, and Harrison NM 2011 Parallel implementation of the ab initio CRYSTAL program: electronic structure calculations for periodic systems. *Proc. R. Soc. A: Math. Phys. Eng. Sci* **467**, 2112.

Car R and Parrinello M 1985 Unified approach for molecular dynamics and density-functional theory. *Phys. Rev. Lett.* **55**, 2471.

Carteret C, De La Pierre M, Dossot M, Pascale F, Erba A, and Dovesi R 2013 The vibrational spectrum of $CaCO_3$ aragonite: a combined experimental and quantum-mechanical investigation. *J. Chem. Phys.* **138**, 014201.

Chen G, Cooke JA, Jr., Gwanmesia GD, and Liebermann RC 1999 Elastic wave velocities of $Mg_3Al_2Si_3O_{12}$-pyrope garnet to 10 GPa. *Am. Mineral.* **84**, 384–388.

Civalleri B, D'Arco P, Orlando R, Saunders VR, and Dovesi R 2001 Hartree-Fock geometry optimization of periodic system with the CRYSTAL code. *Chem. Phys. Lett.* **348**, 131.

Civalleri B, Zicovich-Wilson C, Valenzano L, and Ugliengo P 2008 B3LYP augmented with an empirical dispersion term (B3LYP-D∗) as applied to molecular crystals. *CrystEngComm* **10**, 405.

Cohen RE, Gülseren O, and Hemley RJ 2000 Accuracy of equation-of-state formulations. *Am. Mineral.* **85**, 338–344.

Conrad PG, Zha CS, Mao HK, and Hemley RJ 1999 The high-pressure, single-crystal elasticity of pyrope, grossular, and andradite. *Am. Mineral.* **84**, 374.

Corá F, Alfredsson M, Mallia G, Middlemiss D, Mackrodt W, Dovesi R, and Orlando R 2004 The performance of hybrid density functionals in solid state chemistry In *Principles and Applications of Density Functional Theory in Inorganic Chemistry II* vol. 113 of *Structure and Bonding* Springer, Berlin Heidelberg. pp. 171–232.

D'Arco Ph, Mustapha S, Ferrabone M, Noël Y, De La Pierre M, and Dovesi R 2013 Symmetry and random sampling of symmetry independent configurations for the simulation of disordered solids. *J. Phys. Condens. Matter* **25**, 355401.

De La Pierre M, Noel Y, Mustapha S, D'Arco P, and Dovesi R 2013 The infrared vibrational spectrum of andradite-grossular solid solutions: a quantum mechanical simulation. *Am. Mineral.* **98**, 966–976.

De La Pierre M, Orlando R, Maschio L, Doll K, Ugliengo P, and Dovesi R 2011 Performance of six functionals (LDA, PBE, PBESOL, B3LYP, PBE0 and WC1LYP) in the simulation of vibrational and dielectric properties of crystalline compounds. The case of forsterite Mg_2SiO_4. *J. Comp. Chem.* **32**, 1775–1784.

Delle Piane M, Corno M, and Ugliengo P 2013 Does dispersion dominate over H-bonds in drug-surface interactions? The case of silica-based materials as excipients and drug-delivery agents. *J. Chem. Theory Comput.* **9**(5), 2404–2415.

Diella V, Sani A, Levy D, and Pavese A 2004 High-pressure synchrotron x-ray diffraction study of spessartine and uvarovite: a comparison between different equation of state models. *Am. Mineral.* **89**, 371–376.

Doll K 2001 Implementation of analytical Hartree-Fock gradients for periodic systems. *Comput. Phys. Commun.* **137**, 74.

Doll K 2010 Analytical stress tensor and pressure calculations with the CRYSTAL code. *Mol. Phys.* **108**(3–4), 223–227.

Doll K, Dovesi R, and Orlando R 2004 Analytical Hartree-Fock gradients with respect to the cell parameter for systems periodic in three dimensions. *Theor. Chem. Acc.* **112**(5–6), 394–402.

Doll K, Dovesi R, and Orlando R 2006 Analytical Hartree-Fock gradients with respect to the cell parameter: systems periodic in one and two dimensions. *Theor. Chem. Acc.* **115**(5), 354–360.

Doll K, Harrison NM, and Saunders VR 2001 Analytical Hartree-Fock gradients for periodic systems. *Int. J. Quantum Chem.* **82**, 1.

Dovesi R, Civalleri B, Roetti C, Saunders VR, and Orlando R 2005 Ab initio quantum simulation in solid state chemistry. *Rev. Comput. Chem.* **21**, 1.

Dovesi R, De La Pierre M, Ferrari AM, Pascale F, Maschio L, and Zicovich-Wilson CM 2011 The IR vibrational properties of six members of the garnet family: a quantum mechanical ab initio study. *Am. Mineral.* **96**, 1787–1798.

Dovesi R, Orlando R, Erba A, Zicovich-Wilson CM, Civalleri B, Casassa S, Maschio L, Ferrabone M, De la Pierre M, D'Arco Ph, Noël Y, Causá M, Rérat M, and Kirtman B 2014a CRYSTAL14: a program for the ab initio investigation of crystalline solids. *Int. J. Quantum Chem.* **114**, 1287–1317.

Dovesi R, Saunders VR, Roetti C, Orlando R, Zicovich-Wilson CM, Pascale F, Doll K, Harrison NM, Civalleri B, Bush IJ, D'Arco Ph, Llunell M, Causá M, and Noël Y 2014b *CRYSTAL14 User's Manual Università di Torino, Torino.* http://www.crystal.unito.it. Accessed December 9, 2015.

Dovesi R, Valenzano L, Pascale F, Zicovich-Wilson CM, and Orlando R 2009 Ab initio quantum-mechanical simulation of the Raman spectrum of grossular. *J. Raman Spectrosc.* **40**(4), 416–418.

Dronskowski R 2005 *Computational Chemistry of Solid State Materials.* John Wiley & Sons, Inc, Weinheim.

Duffy TS and Anderson DL 1989 Seismic velocities in mantle minerals and the mineralogy of the upper mantle. *J. Geophys. Res.* **94**, 1895–1912.

Duffy TS and Wang Y 1998 Pressure-volume-temperature equations of state. *Mineral. Soc. Am. Rev. Mineral.* **37**, 425–458.

Erba A 2014 On combining temperature and pressure effects on structural properties of crystals with standard ab initio techniques. *J. Chem. Phys.* **141**, 124115.

Erba A and Dovesi R 2013 Photoelasticity of crystals from theoretical simulations. *Phys. Rev. B* **88**, 045121.

Erba A and Pisani C 2012 Evaluation of the electron momentum density of crystalline systems from ab initio linear combination of atomic orbitals calculations. *J. Comput. Chem.* **33**, 822.

Erba A, El-Kelany KE, Ferrero M, Baraille I, and Rérat M 2013a Piezoelectricity of $SrTiO_3$: an ab initio description. *Phys. Rev. B* **88**, 035102.

Erba A, Ferrabone M, Baima J, Orlando R, Rérat M, and Dovesi R 2013b The vibration properties of the $(n, 0)$ boron nitride nanotubes from ab initio quantum chemical simulations. *J. Chem. Phys.* **138**, 054906.

Erba A, Ferrabone M, Orlando R, and Dovesi R 2013c Accurate dynamical structure factors from ab initio lattice dynamics: the case of crystalline silicon. *J. Comput. Chem.* **34**, 346.

Erba A, Mahmoud A, Belmonte D, and Dovesi R 2014a High pressure elastic properties of minerals from ab initio simulations: the case of pyrope, grossular and andradite silicate garnets. *J. Chem. Phys.* **140**, 124703.

Erba A, Mahmoud A, Orlando R, and Dovesi R 2014b Elastic properties of six silicate garnet end-members from accurate ab initio simulations. *Phys. Chem. Miner.* **41**, 151–160.

Erba A, Mahmoud A, Orlando R, and Dovesi R 2014c Erratum to: elastic properties of six silicate garnet end-members from accurate ab initio simulations. *Phys. Chem. Miner.* **41**, 161–162.

Erba A, Maul J, De La Pierre M, and Dovesi R 2015a Structural and elastic anisotropy of crystals at high pressures and temperatures from quantum mechanical methods: the case of Mg_2SiO_4 forsterite. *J. Chem. Phys.* **142**, 204502.

Erba A, Maul J, Demichelis R, and Dovesi R 2015b Assessing thermochemical properties of materials through ab initio quantum-mechanical methods: the case of α-Al_2O_3. *Phys. Chem. Chem. Phys.* **17**(17), 11670–11677.

Erba A, Montserrat Navarrete López A, Zicovich-Wilson C, and Dovesi R 2015c Katoite under pressure: an ab initio investigation of its structural, elastic and vibrational properties sheds light on the phase transition. *Phys. Chem. Chem. Phys.* **17**, 2660–2669.

Erba A, Shahrokhi M, Moradian R, and Dovesi R 2015d On how differently the quasi-harmonic approximation works for two isostructural crystals: thermal properties of MgO and CaO. *J. Chem. Phys.* **142**, 044114.

Fan D, Wei S, Liu J, Li Y, and Xie H 2011 High pressure X-ray diffraction study of a grossular–andradite solid solution and the bulk modulus variation along this solid solution. *Chin. Phys. Lett.* **28**(7), 076101.

Ferrari AM, Valenzano L, Meyer A, Orlando R, and Dovesi R 2009 Quantum-mechanical ab initio simulation of the Raman and IR spectra of $Fe_3Al_2Si_3O_{12}$ almandine. *J. Phys. Chem. A* **113**(42), 11289–11294.

Ferrero M, Rérat M, Kirtman B, and Dovesi R 2008a Calculation of first and second static hyperpolarizabilities of one- to three-dimensional periodic compounds. Implementation in the CRYSTAL code. *J. Chem. Phys.* **129**, 244110.

Ferrero M, Rérat M, Orlando R, and Dovesi R 2008b Coupled perturbed Hartree-Fock for periodic systems: the role of symmetry and related computational aspects. *J. Chem. Phys.* **128**, 014110.

Ferrero M, Rérat M, Orlando R, and Dovesi R 2008c The calculation of static polarizabilities of periodic compounds. The implementation in the CRYSTAL code for 1D, 2D and 3D systems. *J. Comput. Chem.* **29**, 1450.

Ferrero M, Rérat M, Orlando R, Dovesi R, and Bush I 2008d Coupled Perturbed Kohn-Sham calculation of static polarizabilities of periodic compound. *J. Phys. Chem. Solids* **117**, 12016.

Fletcher R 1970 A new approach to variable metric algorithms. *Comput. J.* **13**, 317.

Freund F and Oberheuser G 1986 Water dissolved in olivine: a single-crystal infrared study. *J. Geophys. Res. Solid Earth* **91**(B1), 745–761.

Ganguly J 1976 The energetics of natural garnet solid solution. *Contrib. Mineral. Petrol.* **55**(1), 81–90.

Goldfarb D 1970 A family of variable-metric methods derived by variational means. *Math. Comput.* **24**, 23.

Gréaux S and Yamada A 2012 PVT equation of state of $Mn_3Al_2Si_3O_{12}$ spessartine garnet. *Phys. Chem. Miner.* **41**, 141–149.

Griggs D 1967 Hydrolytic weakening of quartz and other silicates. *Geophys. J. Roy. Astron. Soc.* **14**(1–4), 19–31.

Grimme S 2006 Semiempirical GGA-type density functional constructed with a long-range dispersion correction. *J. Comput. Chem.* **27**(15), 1787–1799.

Grimme S 2011 Density functional theory with London dispersion corrections. *Wiley Interdiscip. Rev. Comput. Mol. Sci.* **1**(2), 211–228.

Grimme S, Antony J, Ehrlich S, and Krieg H 2010 A consistent and accurate ab initio parametrization of density functional dispersion correction (DFT-D) for the 94 elements H-Pu. *J. Chem. Phys.* **132**, 154104.

Gwanmesia GD, Wang L, Heady A, and Liebermann RC 2014 Elasticity and sound velocities of polycrystalline grossular garnet ($Ca_3Al_2Si_3O_{12}$) at simultaneous high pressures and high temperatures. *Phys. Earth Planet. In.* **228**, 80–87.

Gwanmesia GD, Zhang J, Darling K, Kung J, Li B, Wang L, Neuville D, and Liebermann RC 2006 Elasticity of polycrystalline pyrope ($Mg_3Al_2Si_3O_{12}$) to 9 GPa and 1000 C. *Phys. Earth Planet. In.* **155**, 179–190.

Halleck P 1973 *The Compression and Compressibility of Grossular Garnet: A Comparison of X-Ray and Ultrasonic Methods.* Ph.D. thesis, University of Chicago, Department of the Geophysical Sciences, Chicago.

Hama J and Suito K 1996 The search for a universal equation of state correct up to very high pressures. *J. Phys. Condens. Matter* **8**(1), 67.

Haselton HT and Westrum EF 1980 Low-temperature heat capacities of synthetic pyrope, grossular, and pyrope$_{60}$ grossular$_{40}$. *Geochim. Cosmochim. Acta* **44**, 701–709.

Hensen BJ 1976 The stability of pyrope-grossular garnet with excess silica. *Contrib. Mineral. Petrol.* **55**, 279–292.

Heyd J and Scuseria GE 2004 Efficient hybrid density functional calculations in solids: assessment of the Heyd-Scuseria-Ernzerhof screened Coulomb hybrid functional. *J. Chem. Phys.* **121**(3), 1187–1192.

Hill R 1963 Elastic properties of reinforced solids: some theoretical principles. *J. Mech. Phys. Solids* **11**, 357–372.

Hofmeister AM and Chopelas A 1991 Thermodynamic properties of pyrope and grossular from vibrational spectroscopy. *Am. Mineral.* **76**, 880–891.

Hohenberg P and Kohn W 1964 Inhomogeneous electron gas. *Phys. Rev.* **136**, B864–B871.

Holdaway M 1972 Thermal stability of Al-Fe epidote as a function of f_{O2} and Fe content. *Contrib. Mineral. Petrol.* **37**(4), 307–340.

Holzapfel WB 1996 Physics of solids under strong compression. *Rep. Prog. Phys.* **59**(1), 29.

Isaak DG and Graham EK 1976 The elastic properties of an almandine-spessartine garnet and elasticity in the garnet solid solution series. *J. Geophys. Res.* **81**, 2483–2489.

Jiang F, Speziale S, and Duffy TS 2004a Single-crystal elasticity of grossular- and almandine-rich garnets to 11 GPa by Brillouin scattering. *J. Geophys. Res.* **109**, B10210.

Jiang F, Speziale S, Shieh SR, and Duffy TS 2004b Single-crystal elasticity of andradite garnet to 11 GPa. *J. Phys. Condens. Matter* **16**, S1041.

Jung JO and Gerber RB 1996 Vibrational wave functions and energy levels of large anharmonic clusters: a vibrational SCF study of (Ar)13. *J. Chem. Phys.* **105**(24), 10682–10690.

Karki BB, Ackland GJ, and Crain J 1997 Elastic instabilities in crystals from ab initio stress—strain relations. *J. Phys. Condens. Matter* **9**(41), 8579.

Karki BB, Stixrude L, and Wentzcovitch RM 2001 High-pressure elastic properties of major materials of earth's mantle from first principles. *Rev. Geophys.* **39**, 507–534.

Knittle E 1995 Static compression measurements of equations of state In *A Handbook of Physical Constants: Mineral Physics and Crystallography* (ed. Ahrens TJ) vol. 2, AGU, Washington D.C. pp. 98–142.

Knittle E, Hathorne A, Davis M, and Williams Q 1992 A spectroscopic study of the high-pressure behavior of the O_4H_4 substitution in garnet In *High-Pressure Research: Application to Earth and Planetary Sciences* (ed. Syono Y and Manghnani MH) American Geophysical Union, Washington D.C. pp. 297–304.

Kohn W and Sham LJ 1965 Self-consistent equations including exchange and correlation effects. *Phys. Rev.* **140**, A1133.

Kono Y, Gréaux S, Higo Y, Ohfuji H, and Irifune T 2010 Pressure and temperature dependences of elastic properties of grossular garnet up to 17 GPa and 1650 K. *J. Earth Sci.* **21**, 782–791.

Kresse G and Furthmüller J 1996a Efficiency of ab initio total energy calculations for metals and semiconductors using a plane-wave basis set. *Comput. Mater. Sci.* **6**, 15.

Kresse G and Furthmüller J 1996b Efficient iterative schemes for ab initio total-energy calculations using a plane-wave basis set. *Phys. Rev. B* **54**, 11169.

Lacivita V, D'Arco Ph, Orlando R, Dovesi R, and Meyer A 2013a Anomalous birefringence in andradite-grossular solid solutions: a quantum-mechanical approach. *Phys. Chem. Miner.* **40**, 781–788.

Lacivita V, Erba A, Dovesi R, and D'Arco P 2014 Elasticity of grossular-andradite solid solution: an ab initio investigation. *Phys. Chem. Chem. Phys.* **16**, 15331–15338.

Lacivita V, Erba A, Noël Y, Orlando R, D'Arco Ph, and Dovesi R 2013b Zinc oxide nanotubes: an ab initio investigation of their structural, vibrational, elastic, and dielectric properties. *J. Chem. Phys.* **138**, 214706.

Lacivita V, Mahmoud A, Erba A, D'Arco P, and Mustapha S 2015 Hydrogrossular, $Ca_3Al_2(SiO_4)_{3-x}(H_4O_4)_x$: an ab initio investigation of its structural and energetic properties. *Am. Mineral.* **100**(11–12), 2637–2649.

Lager G, Armbruster T, and Faber J 1987 Neutron and X-ray diffraction study of hydrogarnet $Ca_3Al_2(O_4H_4)_3$. *Am. Mineral.* **72**(7–8), 756–765.

Lager GA and Von Dreele RB 1996 Neutron powder diffraction study of hydrogarnet to 9.0 GPa. *Am. Mineral.* **81**(9–10), 1097–1104.

Lager GA, Downs RT, Origlieri M, and Garoutte R 2002 High-pressure single-crystal X-ray diffraction study of katoite hydrogarnet: evidence for a phase transition from *Ia3d* to *I43d* symmetry at 5 GPa. *Am. Mineral.* **87**(5–6), 642–647.

Lager GA, Marshall WG, Liu Z, and Downs RT 2005 Re-examination of the hydrogarnet structure at high pressure using neutron powder diffraction and infrared spectroscopy. *Am. Mineral.* **90**(4), 639–644.

Leger JM, Redon AM, and Chateau C 1990 Compressions of synthetic pyrope, spessartine and uvarovite garnets up to 25 GPa. *Phys. Chem. Miner.* **17**, 161–167.

Lin C, Gilbert AT, and Gill PM 2008 Calculating molecular vibrational spectra beyond the harmonic approximation. *Theor. Chem. Acc.* **120**(1–3), 23–35.

Liou JG 1973 Synthesis and stability relations of epidote, $Ca_2Al_2FeSi_3O_{12}(OH)$. *J. Petrol.* **14**, 381–413.

Mackwell S, Kohlstedt D, and Paterson M 1985 The role of water in the deformation of olivine single crystals. *J. Geophys. Res. Solid Earth (1978–2012)* **90**(B13), 11319–11333.

Madsen AO, Civalleri B, Ferrabone M, Pascale F, and Erba A 2013 Anisotropic displacement parameters for molecular crystals from periodic Hartree-Fock and density functional theory calculations. *Acta Crystallogr. Sec. A* **69**, 309.

Mahmoud A, Erba A, Doll K, and Dovesi R 2014a Pressure effect on elastic anisotropy of crystals from ab initio simulations: the case of silicate garnets. *J. Chem. Phys.* **140**, 234703.

Mahmoud A, Erba A, El-Kelany KE, Rérat M, and Orlando R 2014b Low-temperature phase of $BaTiO_3$: piezoelectric, dielectric, elastic, and photo elastic properties from ab initio simulations. *Phys. Rev. B* **89**, 045103.

Mainprice D, Barruol G, and Ben Ismail W 2013 *The Seismic Anisotropy of the Earth's Mantle: From Single Crystal to Polycrystal*. American Geophysical Union, Washington, D.C.

Marmier A, Lethbridge ZA, Walton RI, Smith CW, Parker SC, and Evans KE 2010 Elam: a computer program for the analysis and representation of anisotropic elastic properties. *Comput. Phys. Commun.* **181**(12), 2102–2115.

Marques M and Gross E 2004 Time-dependent density functional theory. *Annu. Rev. Phys. Chem.* **55**(1), 427–455.

Maschio L, Demichelis R, Orlando R, De La Pierre M, Mahmoud A, and Dovesi R 2014 The Raman spectrum of grossular garnet: a quantum mechanical simulation of wavenumbers and intensities. *J. Raman Spectrosc.* **45**, 710–715.

Maschio L, Ferrabone M, Meyer A, Garza J, and Dovesi R 2011 The infrared spectrum of spessartine: an ab initio all electron simulation with five different functionals (LDA, PBE, PBEsol, B3LYP and PBE0). *Chem. Phys. Lett.* **501**, 612–618.

Maschio L, Kirtman B, Salustro S, Zicovich-Wilson CM, Orlando R, and Dovesi R 2013 Raman spectrum of pyrope garnet: a quantum mechanical simulation of frequencies, intensities, and isotope shifts. *J. Phys. Chem. A* **117**(45), 11464–11471.

McAloon BP and Hofmeister AM 1995 Single-crystal IR spectroscopy of grossular-andradite garnets. *Am. Mineral.* **80**, 1145–1156.

Meagher EP 1975 Crystal-structures of pyrope and grossularite at elevated-temperatures. *Am. Mineral.* **60**, 218–228.

Meyer A, Pascale F, Zicovich-Wilson CM, and Dovesi R 2010 Magnetic interactions and electronic structure of uvarovite and andradite garnets. An ab initio all-electron simulation with the CRYSTAL06 program. *Int. J. Quantum Chem.* **110**, 338.

Milman V, Winkler B, Nobes R, Akhmatskaya E, Pickard C, and White J 2000 Garnets: structure, compressibility, dynamics, and disorder. *JOM* **52**(7), 22–25.

Murnaghan FD 1944 The compressibility of media under extreme pressures. *Proc. Natl. Acad. Sci. U. S. A.* **30**, 244.

Musgrave MJP 1970 *Crystal Acoustics*. Holden-Day, San Francisco, CA.

Mustapha S, D'Arco Ph, De La Pierre M, Noël Y, Ferrabone M, and Dovesi R 2013 On the use of symmetry in configurational analysis for the simulation of disordered solids. *J. Phys. Condens. Matter* **25**, 105401.

Neff M and Rauhut G 2009 Toward large scale vibrational configuration interaction calculations. *J. Chem. Phys.* **131**(12), 124129.

Nobes R, Akhmatskaya E, Milman V, White J, Winkler B, and Pickard C 2000a An ab initio study of hydrogarnets. *Am. Mineral.* **85**(11–12), 1706–1715.

Nobes R, Akhmatskaya E, Milman V, Winkler B, and Pickard C 2000b Structure and properties of alumino-silicate garnets and katoite: an ab initio study. *Comput. Mat. Sci.* **17**(2), 141–145.

Novak GA and Gibbs GV 1971 The crystal chemistry of the silicate garnets. *Am. Mineral.* **56**, 791–825.

Nye JF 1957 *Physical Properties of Crystals*. Oxford University Press, Oxford.

Olijnyk H, Paris E, Geiger C, and Lager G 1991 Compressional study of katoite [$Ca_3Al_2 (O_4H_4)_3$] and grossular garnet. *J. Geophys. Res. Solid Earth* **96**(B9), 14313–14318.

O'Neill B, Bass JD, and Rossman GR 1993 Elastic properties of hydrogrossular garnet and implications for water in the upper mantle. *J. Geophys. Res. Solid Earth* **98**(B11), 20031–20037.

O'Neill B, Bass JD, Smyth JR, and Vaughan MT 1989 Elasticity of a grossular-pyrope-almandine garnet. *J. Geophys. Res.* **94**, 17819–17824.

Orlando R, De La Pierre M, Zicovich-Wilson CM, Erba A, and Dovesi R 2014 On the full exploitation of symmetry in periodic (as well as molecular) self-consistent-field ab initio calculations. *J. Chem. Phys.* **141**, 104108.

Orlando R, Delle Piane M, Bush IJ, Ugliengo P, Ferrabone M, and Dovesi R 2012 A new massively parallel version of crystal for large systems on high performance computing architectures. *J. Comput. Chem.* **33**, 2276–2284.

Orlando R, Ferrero M, Rérat M, Kirtman B, and Dovesi R 2009 Calculation of the static electronic second hyperpolarizability or χ(3) tensor of three-dimensional periodic compounds with a local basis set. *J. Chem. Phys.* **131**, 184105.

Orlando R, Lacivita V, Bast R, and Ruud K 2010 Calculation of the first static hyperpolarizability tensor of three-dimensional periodic compounds with a local basis set: a comparison of LDA, PBE, PBE0, B3LYP, and HF results. *J. Chem. Phys.* **132**, 244106.

Orlando R, Torres F, Pascale F, Ugliengo P, Zicovich-Wilson C, and Dovesi R 2006 Vibrational spectrum of katoite $Ca_3Al_2[(OH)_4]_3$: a periodic ab initio study. *J. Phys. Chem. B* **110**(2), 692–701.

Ottonello G, Civalleri B, Ganguly J, Perger WF, Belmonte D, and Vetuschi Zuccolini M 2010 Thermo-chemical and thermo-physical properties of the high-pressure phase anhydrous B ($Mg_{14}Si_5O_{24}$): an ab-initio all-electron investigation. *Am. Mineral.* **95**, 563–573.

Parlinski K, Li ZQ, and Kawazoe Y 1997 First-principles determination of the soft mode in cubic ZrO_2. *Phys. Rev. Lett.* **78**, 4063–4066.

Pascale F, Catti M, Damin A, Orlando R, Saunders VR, and Dovesi R 2005a Vibration frequencies of Ca_3Fe_2-Si_3O_{12} andradite: an ab initio study with the CRYSTAL code. *J. Phys. Chem. B* **109**(39), 18522–18527.

Pascale F, Ugliengo P, Civalleri B, Orlando R, D'Arco P, and Dovesi R 2004a The katoite hydrogarnet Si-free $Ca_3Al_2([OH]_4)_3$: a periodic Hartree-Fock and B3LYP study. *J. Chem. Phys.* **121**(2), 1005–1013.

Pascale F, Zicovich-Wilson CM, Gejo FL, Civalleri B, Orlando R, and Dovesi R 2004b The calculation of the vibrational frequencies of the crystalline compounds and its implementation in the crystal code. *J. Comp. Chem.* **25**, 888–897.

Pascale F, Zicovich-Wilson CM, Orlando R, Roetti C, Ugliengo P, and Dovesi R 2005b Vibration frequencies of $Mg_3Al_2Si_3O_{12}$ pyrope. An ab initio study with the CRYSTAL code. *J. Phys. Chem. B* **109**(13), 6146–6152.

Perchuk L and Aranovich L 1979 Thermodynamics of minerals of variable composition: andradite-grossularite and pistacite-clinozoisite solid solutions. *Phys. Chem. Miner.* **5**(1), 1–14.

Perger WF, Criswell J, Civalleri B, and Dovesi R 2009 Ab-initio calculation of elastic constants of crystalline systems with the CRYSTAL code. *Comput. Phys. Commun.* **180**, 1753–1759.

Pertlik F 2003 Bibliography of hibschite, a hydrogarnet of grossular type. *GeoLines* **15**, 113–119.

Pisani C 1996 *Quantum-Mechanical Ab-Initio Calculation of the Properties of Crystalline Materials* vol. 67 of *Lecture Notes in Chemistry Series*. Springer Verlag, Berlin.

Pisani C, Dovesi R, and Roetti C 1988 *Hartree-Fock Ab Initio Treatment of Crystalline solids* vol. 48 of *Lecture Notes in Chemistry Series*. Springer Verlag, Berlin.

Pisani C, Erba A, Ferrabone M, and Dovesi R 2012 Nuclear motion effects on the density matrix of crystals: an ab initio Monte Carlo harmonic approach. *J. Chem. Phys.* **137**, 044114.

Poirier JP and Tarantola A 1998 A logarithmic equation of state. *Phys. Earth Planet. In.* **109**, 1–8.

Rickwood PC, Mathias M, and Siebert JC 1968 A study of garnets from eclogite and peridotite xenoliths found in kimberlite. *Contrib. Mineral. Petrol.* **19**, 271–301.

Rimola A, Zicovich-Wilson CM, Dovesi R, and Ugliengo P 2010 Search and characterization of transition state structures in crystalline systems using valence coordinates. *J. Chem. Theory Comput.* **6**(4), 1341–1350.

Ringwood AE 1975 *Composition and Petrology of the Earth's Mantle*. McGraw-Hill, New York.

Rossman GR 1988 Vibrational spectroscopy of hydrous components. *Rev. Mineral.* **18**, 193–206.

Shanno DF 1970 Conditioning of quasi-Newton methods for function minimization. *Math. Comput.* **24**, 647.

Sinogeikin SV and Bass JD 2000 Single-crystal elasticity of pyrope and MgO to 20 GPa by Brillouin scattering in the diamond cell. *Phys. Earth Planet. Inter.* **120**, 43–62.

Soler JM, Artacho E, Gale JD, García A, Junquera J, Ordejón P, and Sánchez-Portal D 2002 The siesta method for ab initio order-n materials simulation. *J. Phys. Condens. Matter* **14**, 2745–2779.

Souza I and Martins J 1997 Metric tensor as the dynamical variable for variable-cell-shape molecular dynamics. *Phys. Rev. B* **55**, 8733–8742.

Stacey F, Brennan B, and Irvine R 1981 Finite strain theories and comparisons with seismological data. *Geophys. Surv.* **4**(3), 189–232.

Tequi C, Robie RA, Hemingway BS, Neuville D, and Richet P 1991 Melting and thermodynamic properties of pyrope ($Mg_3Al_2Si_3O_{12}$). *Geochim. Cosmochim. Acta* **55**, 10051010.

Tinkham M 1964 *Group Theory and Quantum Mechanics*. McGraw-Hill, New York.

Togo A, Oba F, and Tanaka I 2008 First-principles calculations of the ferroelastic transition between rutile-type and $CaCl_2$-type SiO_2 at high pressures. *Phys. Rev. B* **78**, 134106.

Ugliengo P, Sodupe M, Musso F, Bush IJ, Orlando R, and Dovesi R 2008 Realistic models of hydroxylated amorphous silica surfaces and MCM-41 mesoporous material simulated by large-scale periodic B3LYP calculations. *Adv. Mater.* **20**(23), 4579–4583.

Valenzano L, Meyer A, Demichelis R, Civalleri B, and Dovesi R 2009 Quantum-mechanical ab initio simulation of the Raman and IR spectra of $Mn_3Al_2Si_3O_{12}$ spessartine. *Phys. Chem. Miner.* **36**(7), 415–420.

Valenzano L, Pascale F, Ferrero M, and Dovesi R 2010 Ab initio quantum-mechanical prediction of the IR and Raman spectra of $Ca_3Cr_2Si_3O_{12}$ uvarovite garnet. *Int. J. Quantum Chem.* **110**(2), 416–421.

Vila FD, Lindahl VE, and Rehr JJ 2012 X-ray absorption Debye-Waller factors from ab initio molecular dynamics. *Phys. Rev. B* **85**, 024303.

Vinet P, Ferrante J, Smith JR, and Rose JH 1986 A universal equation of state for solids. *J. Phys. C.* **19**(20), 467.

Wallace DC 1965 Lattice dynamics and elasticity of stressed crystals. *Rev. Mod. Phys.* **37**, 57–67.

Wallace DC 1972 *Thermodynamics of Crystals*. John Wiley & Sons, Inc, New York.

Wang J, Li J, Yip S, Phillpot S, and Wolf D 1995 Mechanical instabilities of homogeneous crystals. *Phys. Rev. B* **52**, 12627–12635.

Wilkins R and Sabine W 1973 Water-content of some nominally anhydrous silicates. *Am. Mineral.* **58**(5–6), 508–516.

Yeganeh-Haeri A, Weidner DJ, and Ito E 1990 Elastic properties of the pyrope-majorite solid solution series. *Geophys. Res. Lett.* **17**, 2453–2456.

Zhang L, Ahsbahs H, and Kutoglu A 1998 Hydrostatic compression and crystal structure of pyrope to 33 GPa. *Phys. Chem. Miner.* **25**(4), 301–307.

Zhang L, Ahsbahs H, Kutoglu A, and Geiger CA 1999 Single-crystal hydrostatic compression of synthetic pyrope, almandine, spessartine, grossular and andradite garnets at high pressures. *Phys. Chem. Miner.* **27**(1), 52–58.

Zicovich-Wilson C and Dovesi R 1998 On the use of symmetry-adapted crystalline orbitals in SCF-LCAO periodic calculations. I. The construction of the symmetrized orbitals. *Int. J. Quantum Chem.* **67**, 299–309.

Zicovich-Wilson CM, Pascale F, Roetti C, Saunders VR, Orlando R, and Dovesi R 2004 The calculation of the vibration frequencies of α-quartz: the effect of Hamiltonian and basis set. *J. Comput. Chem.* **25**, 1873–1881.

Zicovich-Wilson CM, Torres FJ, Pascale F, Valenzano L, Orlando R, and Dovesi R 2008 Ab initio simulation of the IR spectra of pyrope, grossular, and andradite. *J. Comput. Chem.* **29**(13), 2268–2278.

4

First Principles Estimation of Geochemically Important Transition Metal Oxide Properties

Structure and Dynamics of the Bulk, Surface, and Mineral/Aqueous Fluid Interface

Ying Chen,[1] Eric Bylaska,[2] and John Weare[1]

[1]*Chemistry and Biochemistry Department, University of California, San Diego, La Jolla, CA, USA*
[2]*Environmental Molecular Sciences Laboratory, Pacific Northwest National Laboratory, Richland, WA, USA*

4.1 Introduction

Reactions in the mineral surface/reservoir fluid interface control many geochemical processes such as the dissolution and growth of minerals (Yanina and Rosso 2008), heterogeneous oxidation/reduction (Hochella 1990, Brown 2001, Hochella et al. 2008, Navrotsky et al. 2008), and inorganic respiration (Newman 2010). Key minerals involved in these processes are the transition metal oxides and oxyhydroxides (e.g., hematite, Fe_2O_3, and goethite, FeOOH) (Brown et al. 1999, Brown 2001, Hochella et al. 2008, Navrotsky et al. 2008). To interpret and predict these processes, it is necessary to have a high level of understanding of the interactions between the formations containing these minerals and their reservoir fluids. However, these are complicated chemical events occurring under a wide range of T, P, and X conditions, and the interpretation is complicated by the highly heterogeneous nature of natural environments (Hochella 1990, Hochella et al. 2008, Navrotsky et al. 2008) and the electronic and structural complexity of the oxide materials involved (Cox 1992, Kotliar and Vollhardt 2004, Navrotsky et al. 2008). In addition, also because of the complexity of the minerals involved and the heterogeneous nature of natural systems, the direct observation

Molecular Modeling of Geochemical Reactions: An Introduction, First Edition. Edited by James D. Kubicki.
© 2016 John Wiley & Sons, Ltd. Published 2016 by John Wiley & Sons, Ltd.

of these reactions at the atomic level is experimentally extremely difficult. Theoretical simulations will provide important support for analysis of the geochemistry of the mineral surface/fluid region as well as provide essential tools to extrapolate laboratory measurements to the field environment.

The atomic-level interpretation of the chemical events occurring in the surface and fluid regions has recently been supported by the rapid development of new high-resolution spectroscopic measurements of highly ordered systems (e.g., clean well-ordered surfaces with "best possible" ordered layers) using synchrotron light sources (Renaud 1998, Fenter and Sturchio 2004, Park et al. 2005, 2010, Lo et al. 2007, Tanwar et al. 2007, Catalano et al. 2010a, 2010b, Fenter et al. 2010a, 2010b, Ghose et al. 2010, Fulton et al. 2012, Huang et al. 2014, Massey et al. 2014, Stubbs et al. 2015, Yuan et al. 2015) and high-resolution NMR (Anovitz et al. 2009, 2010). As well as providing new insights into the surface structure of these complex materials, these observations provide a rich and challenging database for quantification of theoretical predictions.

The objective of this chapter is to describe our efforts to use first principles dynamical simulation methods (based on direct solution to the electronic structure Schrödinger equations (Car and Parrinello 1985, Remler and Madden 1990, Marx and Hutter 2012)) to analyze and interpret these data (atom and electronic structure). With further development, these methods will provide a means to extrapolate observations from highly ordered materials to much more heterogeneous natural environments. The target of our recent calculations is the structure of the surface interface and the development of methods to simulate this dynamic region. Because of the weak interactions of the species (H_2O molecules, solutes, etc.) in the interface at finite temperature, atoms are in constant motion. The measurements that are available (e.g., X-ray diffraction (Brown and Sturchio 2002, Fenter 2002)) are of the average (equilibrium) structure of these fluctuating species. In order to interpret the data, a dynamical theory of the atomic motion in the interface region must be used. The need to seamlessly model the transition metal bulk, solid surface, fluid interface, and fluid bulk poses a difficult problem for simulation because the bonding character changes from ionic and covalent bonding in the solid material to polarized closed-shell hydrogen bonding (H-bonding) in the interface region and to closed-shell/H-bonding water–water and water–solute interaction in the bulk fluid region.

The simulation methods discussed in this article efficiently carry out dynamical simulations with forces calculated directly from first principles. There have been several previous efforts to calculate the properties of transition metal oxide surface properties for geochemical applications using first principles methods (Becker et al. 1996, Rosso and Rustad 2001, Kubicki et al. 2008), but most of this work has focused on the static structure of the mineral–water interface. These methods avoid the problem of defining empirical force fields for conventional molecular dynamics (CMD) simulations by calculating the interactions between atoms on the fly directly using various levels of approximation to the electronic Schrödinger equation (herein called *ab initio* molecular dynamics (AIMD)) (Car and Parrinello 1985, Remler and Madden 1990, Marx and Hutter 2012)) in the three regions (mineral, interface, and solution bulk). There have been a number of efforts to model the dynamical behavior of these systems using CMD (Rustad et al. 1996, 2003, Shroll and Straatsma 2003, Kerisit 2011, Kerisit et al. 2012). Although these calculations have led to important insights, detailed interpretation of data is limited by the difficulty of defining empirical force fields that reflect changes in bonding and electronic structure in the three regions (Kerisit et al. 2012). On the other hand, although AIMD methods provide a more reliable description of the changing interactions in the system, the calculation of the forces for these complex systems is much more computationally demanding. Therefore, the limitations as to the equilibration of the system and the number of atomic species that can be considered are more restrictive for AIMD methods than CMD methods. The results presented here are close to the limit of what can be practically calculated with presently available computational platforms (Bylaska et al. 2011b).

The transition metal oxide and oxyhydroxide minerals considered in this chapter and their interface properties are important to other technological applications such as solar hydrogen production, heterogeneous catalysis, and magnetic materials applications (Renaud 1998, Valdes et al. 2012). First principles calculations for these systems have been reported (Huda et al. 2010, Pozun and Henkelman 2011, Valdes et al. 2012). The calculations reported here use the NWChem software, which can be downloaded from http://www.nwchem-sw.org/index.php/Download. These application packages have been designed for implementation on highly parallel computers (up to 100 K cores; http://www.nwchem-sw.org/index.php/Benchmarks) and with special emphasis on providing tools for interpreting chemical environments.

4.2 Overview of the Theoretical Methods and Approximations Needed to Perform AIMD Calculations

In order to model interactions of transition metal oxide minerals with solution interfaces, it is necessary to have an accurate and realistic representation of the bulk/surface (10–15 Å), the interface water–solute, and the bulk solution regions. For the transition metal oxides, the computational complexity is greatly increased by the highly correlated electronic behavior of these materials and the complexity of the unit cells. The possibility of electron localization in the unit cell due to local character of the 3d atomic orbitals in these materials leads to local spin ordering within the unit cell further complicating the electronic structure calculation. These effects are extremely difficult to capture theoretically and remain a current topic of theoretical research in condensed matter physics (Kotliar and Vollhardt 2004).

As will be discussed in more detail in the following, there are significant problems with the application of orbital density functional theory (DFT) (Hohenberg and Kohn 1964, Kohn and Sham 1965, Parr and Yang 1995) to these problems. However, a minimum requirement to achieve spin ordering is to enforce single electron orbital occupation (e.g., the number of orbitals equals the number of electrons, roughly 1000 valence electrons in the calculations reported in the following). For surfaces interacting with a loosely bound fluid layer, the observed structure (e.g., CTR measurements (Fenter and Sturchio 2004)) represents the equilibrium average of the positions of the solution species (H_2O and dissolved solutes). The strength of interaction and the nature of the bonding of the species (e.g., highly polarized and H-bonding for H_2O) vary with the position of the molecule in the mineral/fluid interface. This effect is incorporated in the AIMD simulation, but the number of particles required to capture changes in bonding due to local and long-range interactions greatly affects the time required for simulation. Typically, dynamical simulation time scales that can be practically reached are up to few hundred picoseconds (Atta-Fynn et al. 2013), but most simulations are significantly shorter.

The calculation of the electronic structure of a highly correlated system is a problem of current interest in condensed matter physics (Kotliar and Vollhardt 2004). Fortunately, there is an approximate approach, the DFT of Hohenberg, Kohn, and Sham (Hohenberg and Kohn 1964, Kohn and Sham 1965, Parr and Yang 1995) that provides estimates of many properties at a practical computational cost. Though not expected to provide more than qualitative accuracy for spin-dependent properties, essentially all AIMD methods implement this approach. To clarify the following discussion, it will be useful to briefly outline some of the aspects of this theory that affect the accuracy possible in dynamical simulations and its application to highly correlated systems. (For more detail on DFT, refer to Chapter 1.)

Hohenberg and Kohn demonstrated that the total electronic energy of a many-electron system can be written as an orbital-based functional of the electron density (Hohenberg and Kohn 1964, Kohn and Sham 1965, Parr and Yang 1995). That is,

$$E[\rho, \{\boldsymbol{R}_I\}] = \sum_{i=1}^{\text{occupied}} \left\langle \psi_i^{\text{KS}} \left| -\frac{1}{2}\nabla^2 \right| \psi_i^{\text{KS}} \right\rangle + \sum_{I=1}^{\text{nion}} \int V_{\text{ext}}(r, \boldsymbol{R}_I)\rho(r)dr + \frac{1}{2}\int\int \frac{\rho(r)\rho(r')}{|r-r'|}drdr' + E_{\text{xc}}[\rho]$$

(4.1)

In Equation 4.1 $E[\rho, \{\boldsymbol{R}_I\}]$ is the total electronic energy and is a function of the electron density,

$$\rho(r) = \sum_{i=1}^{\text{occupied}} \left| \psi_i^{\text{KS}}(r) \right|^2$$

(4.2)

and the atomic positions $\boldsymbol{R}_I$. $E_{\text{xc}}[\rho]$ is the exchange–correlation energy defined in Hohenberg and Kohn (Hohenberg and Kohn 1964, Parr and Yang 1995).

The existence of the functional $E_{\text{xc}}[\rho]$ is demonstrated by the Hohenberg Kohn theorem, but the form as a function of density is still a topic of much research (Burke 2012, Becke 2014). In Equation 4.1 $\psi_i^{\text{KS}}(r)$ are the Kohn–Sham orbital wave functions (Kohn and Sham 1965) found from the constrained variation of the total energy as (Parr and Yang 1995)

$$\left[-\frac{1}{2}\nabla^2 + \sum_{I=1}^{\text{nion}} V_{\text{ext}}(r, \boldsymbol{R}_I) + \int \frac{\rho(r')}{|r-r'|}dr' + v_{\text{xc}}(\rho(r)) \right] \psi_i^{\text{KS}}(r) = \varepsilon_i \psi_i^{\text{KS}}(r)$$

(4.3)

The orbital Kohn–Sham functions must be constrained to be orthonormal:

$$\left\langle \psi_i^{\text{KS}} | \psi_j^{\text{KS}} \right\rangle = \delta_{i,j}$$

(4.4)

$v_{\text{xc}}(\rho(r))$ is the exchange–correlation function defied by functional variation of Equation 4.1. Our objective is to calculate the dynamics of the system from the solution to the Kohn–Sham equations (Eq. 4.3). Given the forces

$$F_I(\boldsymbol{R}_I) = -\left.\frac{\partial E}{\partial \boldsymbol{R}_I}\right|_{\frac{\delta E}{\delta \rho(r)}=0}$$

(4.5)

we can write Newton's equations of motions as

$$M_I \ddot{\boldsymbol{R}}_I = \boldsymbol{F}_I$$

(4.6)

Time integration of Equation 4.6 provides the first principles dynamics of the system of particles (AIMD) by the Kohn–Sham equations (Eq. 4.3). Car and Parrinello (CP) (Car and Parrinello 1985) have introduced a modification of Equations 4.3–4.6 that can lead to a more efficient algorithm. These equations are used in most of the dynamical calculations reported in the following. In this algorithm the dynamical variables are expanded to include the Kohn–Sham wave functions and fictitious dynamical equations. Lagrange multipliers, Λ_{ij}, maintain orthogonality between the Kohn–Sham orbitals. The dynamical equations for this approach are modified to be

$$\mu \ddot{\psi}_i^{KS}(\mathbf{r}) = \left[-\frac{1}{2}\nabla^2 + \sum_{I=1}^{nion} V_{ext}(\mathbf{r}, \mathbf{R}_I) + \int \frac{\rho(\mathbf{r}')}{|\mathbf{r}-\mathbf{r}'|}d\mathbf{r}' + v_{xc}(\rho(\mathbf{r})) \right]\psi_i^{KS}(\mathbf{r}) - \sum_{j=1}^{occupied}\Lambda_{ij}\psi_i^{KS}(\mathbf{r}) \qquad (4.7)$$

In Equation 4.6, $\ddot{\psi}_i^{KS}(\mathbf{r})$ is the fictitious acceleration of the Kohn–Sham wave function (as defined in the Car–Parrinello approach (Car and Parrinello 1985, Remler and Madden 1990)). A corresponding fictitious kinetic energy, $KE(\psi)$, of the wave function degree of freedom may also be computed. This propagation must be carried out with the orthonormality constraints, Equation 4.4, held tight (Car and Parrinello 1985, Remler and Madden 1990, Marx and Hutter 2012). The magnitude of $KE(\psi)$ is controlled by the fictitious mass μ and must be kept small to obtain realistic dynamics for the real position variables, $\mathbf{R}_I(t)$. To further define the application of Equations 4.1–4.6 to model systems, choices must be made that affect the accuracy and efficiency of the calculation. These are discussed in the following. More detailed discussions of some of the more complex of these issues are given in the appendices.

The first consideration that must be made is the number of atoms (number of electrons, N) to be included in the simulation. The numerical work in the calculation scales as N^3. For the problems we will discuss in the following, the target systems are periodic bulk materials (3-D periodicity, test calculations; see following text) or perfect surface terminations with disordered fluid interfaces (2-D periodicity). For the bulk phases the periodic unit cell defines the minimum size of the calculation with appropriate Brillouin zone sampling (Ziman 1972, Ashcroft and Mermin 1976). (In some cases we use a large unit cell (e.g., $X \times Y \times Z\,Å^3$) to avoid Brillouin sampling (Ziman 1972, Remler and Madden 1990).) For interface problems, we use a slab construction (limited number of bulk layers). For some problems where required in order to model changes in the surface fluid structure, we will use an enlarged surface unit cell. This will be indicated in the calculations discussed in the application sections in the following.

The nonlinear partial differential equations (Eqs. 4.3 and 4.6) are not defined until the external potential, $V_{ext}(\mathbf{r}, \mathbf{R}_I)$, the exchange–correlation function, $v_{xc}(\rho(\mathbf{r}))$, and the fictitious mass, μ, are defined. μ is a parameter that controls how tightly the Kohn–Sham DFT equation constraint, Equation 4.6, is maintained. As $\mu \to 0$, the Kohn–Sham equations are closely constrained, but the fictitious dynamics of the wave functions are very fast, implying that the time integration step must be kept very short (e.g., 0.12 fs) or the system will not stay close to the Born–Oppenheimer surface defined by the solution to Equation 4.3. Also as μ increases, the rate of transfer of energy into the fictitious kinetic energy of the electronic degrees of freedom (Remler and Madden 1990) increases. As this occurs, the total energy in real structural coordinates will not be conserved. Generally a compromise must be made between the efficiency of the CP algorithm and the retention of the kinetic energy in the particle degrees of freedom.

The exchange–correlation potential, $v_{xc}(\rho(\mathbf{r}))$, is an unknown function and must be considered as part of the phenomenology of DFT. There have been many efforts to develop functions that will represent accurately a large number of systems (Zhao and Truhlar 2008, Burke 2012, Becke 2014, Luo et al. 2014). These include purely local functions (LDA) (Kohn and Sham 1965), semilocal functions (e.g., PBE (Perdew et al. 1997)), and many others (Zhao and Truhlar 2008). In systems that contain localized d states, exchange is particularly important and exact exchange corrected functionals (hybrid functionals, e.g., PBE0 (Adamo and Barone 1999)) and DFT+U corrections (Anisimov et al. 1993, Liechtenstein et al. 1995, Dudarev et al. 1998, Rollmann et al. 2004, Zhou et al. 2004) are frequently used. In the following, we will present computational results that evaluate the use of several of the most popular versions of exchange correlation (e.g., PBE (Perdew et al. 1997) and PBE0 (Adamo and Barone 1999)) for calculations applicable to the transition metal oxide fluid interface.

To solve Equations 4.3 and 4.6, the Kohn–Sham functions must be expanded in a basis. Two common choices are local functions (atom-like orbital functions typically used in molecular calculations) (Szabo and Ostlund 1996) and plane waves (typically used in condensed matter calculations) (Ashcroft and Mermin 1976, Car and Parrinello 1985, Marx and Hutter 2012). To achieve the efficiency necessary for dynamical simulations, the calculations that we report here utilize plane waves. There are efficient local basis dynamical methods as well (Marx and Hutter 2012). In either choice, the basis must be large enough so that the solutions to the KS PDEs (e.g., Eqs. 4.3 and 4.6) are close to the intrinsic accuracy of the DFT approximations.

The external potential, $V_{ext}(r, R_I)$, representing the attractive interaction between the electrons and the atomic nuclei centers is the remaining function in Equations 4.3 and 4.7 that must be specified before calculation. In an all-electron calculation, this potential is defined in terms of the nuclear charge and position, Z_I and R_I. The core electrons and other valence electron self-consistently screen these potentials. However, the effective atom–electron potentials experienced by the valence electrons are still fast varying near the nuclear centers. In addition, the orthogonality of the valence solutions to the core atom-like states in heavy atoms creates nodes in the valence orbital functions also resulting in their fast variation. To expand these rapid changes in plane waves would require an impractically large basis. (There are similar problems even when local basis functions are used.) In addition, the magnitude of the computational problem is a function of the number of Kohn–Sham orbital functions retained in the calculation (leading to an approximately N^3 scaling). With the expectation that bond formation between atoms in a condensed system is accounted for by the valence electrons (wave functions), it is common to develop potentials (pseudopotentials) that operate only on the valence wave functions. The idea behind these potentials is that a much smoother potential may be developed for the valence electrons while still providing accurate bond energies and wave function variations in the bonding region. In addition, the introduction of these potentials removes the need to include core orbitals. There is an extensive literature in this area (see, e.g., Pickett 1989). The general theory is discussed in more detail in Appendix A.1.

Pseudopotentials are developed from atomic calculations performed at the same level as proposed for the full many-atom condensed phase calculation. Therefore, the use of pseudopotentials does not represent a parameterization of the target condensed phase problem. However, there are issues that must be addressed that affect the transferability of the pseudopotential from the atomic problem to the condensed phase problem. Possibly the most important decision that has to be made is how many valence orbitals on a particular atom will contribute to the structure of the valence band. As we have mentioned, the time to solution for any DFT solver increases dramatically as the number of valence orbitals increases. Hence, the number of orbitals retained in the calculation is a critical decision. For the second row elements (see the α-Al_2O_3 (corundum) calculations in the following), this is fairly straightforward. The $n_p = 1$ (n_p is the principal quantum number) 1s orbital function in the core is well separated from the $n_p = 2$, 2s, and 2p orbital functions. On the other hand, for second row elements, the separation between the 2s and the 2p functions in the $n_p = 2$ shell is not sufficient (even on the right-hand side of the periodic table) to suppress hybridization in condensed materials (Kawai and Weare 1990). For these materials, the 2s and 2p orbital pseudopotentials are needed.

For the third row and higher periods, this decision is tricky. For example, for the 3d Fe atom (an important component of the transition metals of interest to this work), the 3s, 3p, 3d, and 4s are all candidates for hybridization. For efficiency, the five 3d orbitals and the fact that there may be high-spin and low-spin configurations create a difficult problem. For these atoms, we have found that the 3s and 3p orbitals that are filled in the isolated atom may play a role in the bulk bonding problem (Fulton et al. 2010, 2012). Including these orbitals is straightforward but does increase the cost of the calculation. In the next section, we will discuss this issue by direct comparison of calculated results to observed bulk properties. These calculations will illustrate the level of DFT theory that is necessary to describe (even qualitatively) transition metal oxide minerals and also set the stage for the discussion of the effects of adding a fluid interface in the section following.

The projector augmented plane wave method (PAW) of Blochl removes many of the problems of the somewhat *ad hoc* nature of the pseudopotential approach (Blochl 1994, Holzwarth et al. 1997, Kresse and Joubert 1999, Valiev and Weare 1999, Bylaska et al. 2001, 2002, Valiev et al. 2003). Similar to pseudopotential methods, PAW retains the use of a plane wave basis and all the advantages associated with it. However, in the PAW approach instead of discarding the rapidly varying parts of the electronic functions, these are projected onto a local basis set (e.g., a basis of atomic functions). By carefully choosing the local basis, the convergence of the plane wave problem can be systematically improved. The PAW method is an all-electron method, and no part of the electron density is removed from the problem. In addition, the norm-conservation condition can be relaxed in PAW resulting in smaller plane wave cutoffs. Both of these features can offer a significant advantage over the pseudopotentials, in particular for systems containing elements with hard pseudopotentials (e.g., oxygen, fluorine, first row transition metals, and lanthanides).

4.3   Accuracy of Calculations for Observable Bulk Properties

A first consideration in the development of an atomic-level model is to establish the expected accuracy of the methods of calculation that you intend to use. In this section, we will evaluate the accuracy of the application of the approaches that we have discussed when used to calculate bulk properties (a relatively inexpensive calculation). Similar calculations and comparisons to solution properties can be found in our recent efforts to simulate X-ray structures of aqueous solutions (Fulton et al. 2010, 2012). In the following sections, we will discuss the application to surface (Section 4.4) and mineral/fluid interface problems (Section 4.5).

The transition metal oxide hematite (α-Fe_2O_3) and the oxyhydroxide goethite (α-FeOOH) are central to many geochemical applications and are a significant challenge to the computational methods that we are using. The structures of the unit cells of these solids are given in Figure 4.1. The *d* electrons in these materials are localized about the Fe atoms and exhibit strong correlation. Because of the complicated physics associated with modeling the correlation of five *d* electrons in confined orbits, the development of a computationally tractable model for their electronic structure leads to some uncertainty (difficult choices) in the development of pseudopotentials and in the level of exchange used. In order to compare the accuracy of these calculations to a very well-determined mineral, we include similar calculations of corundum (α-Al_2O_3). This mineral has a structure similar to hematite, but the pseudopotential is more straightforward to develop and the electronic structure calculation is relatively easy because it is closed shell.

4.3.1   Bulk Structural Properties

Tables 4.1 and 4.3 summarize our calculations of the bulk lattice structures given in Figure 4.1 using the various levels of approximation discussed in Section 4.2. The estimates of the bond lengths illustrated in Figure 4.1 are given in Tables 4.2 and 4.4.

4.3.1.1   Calculated Structures of Corundum

The conventional unit cell lattice parameters for this mineral are reported at various levels of electronic structure calculation (LDA, PBE, and PBE0) as well as observed values in Table 4.1. The

(a)   (b)

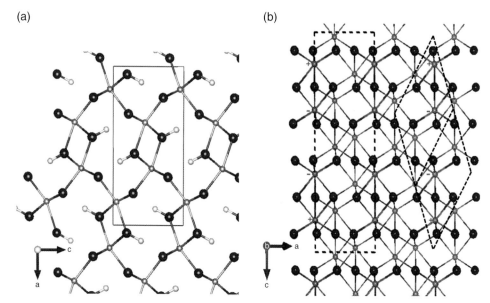

Figure 4.1 *Unit cells for bulk goethite structure (a) and corundum structure (b) (left, 30-atom cell; right, 10-atom cell). Hematite and corundum have same crystal structure (left, 30-atom cell; right, 10-atom cell). Fe and Al, blue; O, red; H, white. The "a," "b," and "c" are the lattice vectors.*

Table 4.1 *Lattice parameters of the conventional cell of corundum (Å for a, b, c; ° for α, β, γ) calculated using LDA, PBE, and PBE0 plane wave DFT calculations.*

Lattice parameters	LDA	PBE	PBE0	Experiment[a]
Supercell	1 × 1 × 1	1 × 1 × 1	1 × 1 × 1	
MP	2 × 2 × 1	2 × 2 × 1	1 × 1 × 1	
a	4.675	4.767	4.681	4.758
b	4.675	4.767	4.681	4.758
c	12.736	12.999	12.909	12.990
α	90.0	90.0	90.0	90
β	90.0	90.0	90.0	90
γ	120.0	120.0	119.9	120

100 Ry (2721 eV) and 200 Ry (5442 eV) were used for the wave function and density cutoff energies. The corundum conventional cell contains 12 Al and 18 O atoms.
[a] Kirfel and Eichhorn (1990).

Table 4.2 *Bond lengths and atom center distances (in Å) of corundum calculated using LDA, PBE, and PBE0 plane wave DFT calculations.*

Distances	LDA	PBE	PBE0	Experiment[a]
Supercell	1 × 1 × 1	1 × 1 × 1	1 × 1 × 1	
MP	2 × 2 × 1	2 × 2 × 1	1 × 1 × 1	
Al–Al-a	2.611	2.666	2.638	2.654
Al–Al-b	3.757	3.833	3.816	3.840
Al–O1	1.817	1.853	1.818	1.854
Al–O2	1.941	1.980	1.962	1.971

100 Ry (2721 eV) and 200 Ry (5442 eV) were used for the wave function and density cutoff energies.
[a] Kirfel and Eichhorn (1990).

Table 4.3 *Lattice parameters for the conventional cells of hematite and goethite (Å for a, b, c; ° for α, β, γ) calculated using LDA, PBE, PBE+U, and PBE0 plane wave DFT calculations.*

Lattice parameters	LDA PSP1	LDA PSP2	PBE PSP1	PBE PSP2	PBE+U PSP1	PBE0 PSP1	Experiment
Hematite							
Supercell	$1 \times 1 \times 1$	$1 \times 1 \times 1$	$1 \times 1 \times 1$	$1 \times 1 \times 1$	$1 \times 3 \times 2$	$1 \times 1 \times 1$	
MP	$2 \times 2 \times 1$	$2 \times 2 \times 1$	$2 \times 2 \times 1$	$2 \times 2 \times 1$	$1 \times 1 \times 1$	$1 \times 1 \times 1$	
a	4.683	4.498	5.154	4.947		5.122	5.036[a]
b	4.681	4.499	5.155	4.947		5.122	5.036[a]
c	13.515	13.228	13.820	13.602		13.63	13.747[a]
α	90.68	90.65	89.98	89.93		87.84	90[a]
β	89.27	89.37	90.02	90.07		92.16	90[a]
γ	119.1	119.3	119.9	120.0		119.6	120[a]
Goethite							
Supercell	$1 \times 1 \times 1$	$1 \times 1 \times 1$	$1 \times 1 \times 1$	$1 \times 1 \times 1$	$2 \times 2 \times 1$	$1 \times 1 \times 1$	
MP	$1 \times 3 \times 2$	$1 \times 3 \times 2$	$1 \times 3 \times 2$	$1 \times 3 \times 2$	$1 \times 1 \times 1$	$1 \times 1 \times 1$	
a	9.662	9.285	10.107	9.858		9.812	9.951–9.956[b]
b	2.586	2.644	3.024	2.976		3.168	3.018–3.025[b]
c	4.383	4.220	4.611	4.515		4.240	4.598–4.616[b]
α	90.0	90.1	90.0	90.0		90.0	90[b]
β	90.0	90.0	90.0	90.0		90.0	90[b]
γ	90.0	89.9	90.0	90.0		90.0	90[b]

100 Ry (2721 eV) and 200 Ry (5442 eV) were used for the wave function and density cutoff energies. The hematite conventional cell contains 12 Fe and 18 O atoms, and the goethite conventional cell contains 4 Fe, 8 O, and 4 H atoms. U = 4 eV for all PBE+U calculations in this paper. The PSP1 and PSP2 pseudopotentials for Fe atom have [Ar]3s3p and [Ar] as core. The largest calculation here is PBE+U calculation using $2 \times 2 \times 1$ cell that contains 408 spin-up and 408 spin-down electrons.
[a] Maslen et al. (1994). Reproduced with permission from Wiley.
[b] Alvarez et al. (2008).

relatively simple LDA calculations provide reasonable results comparing to experiments, with a difference around 2.0%. Using this level of approximation, the bonds are usually shorter than experimental observations. The semilocal GGA calculations, PBE, which can be done for roughly the same computational cost as LDA give much better agreement with experiment, within about 0.2%. The higher-level electronic structure calculation, PBE0, provides slightly less accuracy of 1.6%. PBE0 is about the highest-level calculation that can practically be performed for these minerals. It is discouraging that the accuracy of geometry decreases slightly in going from PBE to PBE0. However, the excellent results from the more efficient PBE calculations are quite positive. Problems of this sort are common in structural DFT calculations (Khein et al. 1995). There is much more support for the PBE0-type calculations in the electronic structure calculations for the spin-ordered systems hematite and goethite. The bond lengths for corundum given in Table 4.3 show the same trends. Again, the LDA bond lengths are shorter than observed. PBE calculations provide predictions that are in better agreement with the structural data. In this case as well, PBE0 calculations are not as accurate as the PBE calculations. Although changes can be made to the hybrid functional (Pozun and Henkelman 2011) or other functionalities introduced (Zhao and Truhlar 2008), this is probably the best level of agreement experiment one can expect from a present-day DFT calculation. (You might want to read and refer to DeMichelis et al. (2010). They use a variety of methods to model silicates which may be an even more tractable system.)

Table 4.4 *Bond lengths and atom center distances (in Å) of hematite and goethite calculated using LDA, PBE, PBE+U, and PBE0 plane wave DFT calculations.*

Distances	LDA PSP1	LDA PSP2	PBE PSP1	PBE PSP2	PBE+U PSP1	PBE0 PSP1	Experiment
Hematite							
Supercell	1 × 1 × 1	1 × 1 × 1	1 × 1 × 1	1 × 1 × 1	1 × 3 × 2	1 × 1 × 1	
MP	2 × 2 × 1	2 × 2 × 1	2 × 2 × 1	2 × 2 × 1	1 × 1 × 1	1 × 1 × 1	
Fe—O1	1.878	1.795	1.994	1.887	1.996	1.964	1.946[a]
Fe—O2	1.971	1.930	2.132	2.122	2.187	2.137	2.116[a]
Fe—Fe-a	2.646	2.595	2.906	2.933	2.922	2.904	2.900[a]
Fe—Fe-b							
Goethite							
Supercell	1 × 1 × 1	1 × 1 × 1	1 × 1 × 1	1 × 1 × 1	2 × 2 × 1	1 × 1 × 1	
MP	1 × 3 × 2	1 × 3 × 2	1 × 3 × 2	1 × 3 × 2	1 × 1 × 1	1 × 1 × 1	
O1—H	1.093	1.094	0.973	0.968	0.982	0.963	
O2—H	1.302	1.306	1.743	1.808	1.651	1.637	
Fe—O1	1.915	1.867	2.134	2.100	2.123	2.079	2.077–2.103[b]
Fe—O2	1.906	1.846	1.973	1.911	1.992	1.949	1.937–1.954[b]
Fe—Fe-a	2.890	2.802	3.363	3.317	3.323	3.245	3.283–3.311[b]
Fe—Fe-b	3.415	3.316	3.456	3.352	3.493	3.381	3.432–3.465[b]

100 and 200 Ry were used for the wave function and density cutoff energies.
[a] Maslen et al. (1994). Reproduced with permission from Wiley.
[b] Alvarez et al. (2008).

4.3.1.2 Calculated Structures for Hematite and Goethite

In Tables 4.3 and 4.4 structure calculations at the various levels are reported for hematite and goethite. The LDA results are significantly worse than the PBE results. The average agreement for unit cell parameters is ~0.1 Å (~1.5%) for the PBE and PBE0, whereas the LDA result is ~0.3 Å (~3.0%). Due to the strongly correlated nature of d electrons for these systems, GGA calculations are expected to be less accurate than the hybrid calculations (Luo et al. 2014). From Table 4.4, the difference from experiment of the bond lengths (Fe—O bond) within the unit cell is <0.02 Å (~1%) for the PBE0 calculation, which is more accurate than the 0.05 Å (~2%) difference observed for the PBE calculation.

Note on the accuracy and consistency of XRD measurements: Considering the dispersion of an X-ray source and the error of the aligned instrument, the expected precision of lattice parameters and bond lengths in a single observation can be up to 0.00001–0.0001 Å (Herbstein 2000). However, there are issues with the reproducibility of reported structures from similar experimental setups. For example, the difference of hematite lattice parameter between Blake's measurement (Blake et al. 1966) and Finger's (Finger and Hazen 1980) work is ~0.03 Å. These differences between intrinsic error and reported differences could be due to disorder of crystal samples, thermal, and other effects.

An additional problem for application of DFT to these materials is the accuracy of the pseudopotential representation of the atomic potential. If the Fe^{3+} 3s and 3p play a role in the hybridization of the d orbitals, they need to be included in the pseudopotential of the valence band (see Appendix A.1). While the atomic energies of these orbitals are well below the 3d states (around 30 eV below bottom of d states in DFT), we have found in prior aqueous solution simulations that their inclusion improves the M—O bond lengths in the solution phase (Fulton et al. 2010, 2012). Results including the 3s and 3p in the valence structure are included in Table 4.3. In these applications, we found that the results generated from the two pseudopotentials have similar accuracy. This suggests that without losing much accuracy in goethite and hematite surface simulations, we can use a Fe pseudopotential, which puts the 3s and 3p orbitals in the core. This leads to a much more efficient calculation.

Hybrid DFT calculations are still extremely expensive for dynamical simulations. As another approach to add some model-based exchange correction in the calculation of these minerals that contain localized d states, we implemented the computationally efficient DFT+U method (Anisimov et al. 1993, Liechtenstein et al. 1995, Dudarev et al. 1998, Rollmann et al. 2004, Zhou et al. 2004). By providing an approximate exact exchange correction (based on a Hubbard-type model), this kind of calculation is able to provide some improvement over GGA at a small computational cost. The DFT+U framework is discussed in Appendix A.2 DFT+U. The parameters U and J for Fe atoms in this work are set to be 4 eV and 1 eV, respectively, for Fe in hematite and goethite (Rollmann et al. 2004, Tunega 2012). As will be discussed in the next section, this method provides a reasonable estimation of the localized *d* electronic structure properties of these materials while at the same time producing structural results that are as good as a DFT GGA calculation (see Tables 4.3 and 4.4).

4.3.1.3   Comparison of PAW and Pseudopotential Calculations for Corundum

While the norm-conserving pseudopotential DFT method produces structures that agree very well with experimental observations, a drawback of this method is that fairly large cutoff energies (i.e., plane wave basis set size) are needed for these calculations to be reliable (e.g., 100 Ry or 2721 eV). The PAW method and the related Vanderbilt ultrasoft pseudopotentials (Vanderbilt 1990) have a potential advantage of being able to use smaller cutoff energies. To demonstrate this, we show in Table 4.5 the effect cutoff energy has on the lattice parameters for corundum in the PAW method and the norm-conserving pseudopotential method. As can be seen in this table, the lattice parameters with the PAW method are accurate down to 40 Ry, whereas for the norm-conserving pseudopotential method cutoff energies below 80 Ry produce questionable results. In principle, reducing the

Table 4.5 *Effect of cutoff energy on the lattice parameters of the conventional cell of corundum (Å for a, b, c; ° for α, β, γ) in PAW and norm-conserving pseudopotential (PSPW) calculations.*

Lattice parameters	PAW PBE	PAW PBE	PAW PBE	PAW PBE	PAW PBE	Experiment[a]
Cutoff	20 Ry	40 Ry	60 Ry	80 Ry	100 Ry	
a	4.14	4.74	4.73	4.74	4.75	4.758
b	4.14	4.74	4.73	4.74	4.75	4.758
c	11.32	13.02	13.00	13.03	13.07	12.990
α	90	90	90	90	90	90
β	90	90	90	90	90	90
γ	90	120	120	120	120	120
Lattice parameters	PSPW PBE	PSPW PBE	PSPW PBE	PSPW PBE	PSPW PBE	Experiment[a]
Cutoff	20 Ry	40 Ry	60 Ry	80 Ry	100 Ry	
a	3.57	4.42	4.78	4.75	4.73	4.758
b	3.57	4.42	4.78	4.75	4.73	4.758
c	11.55	13.33	13.13	13.06	13.01	12.990
α	90	90	90	90	90	90
β	90	90	90	90	90	90
γ	120	120	120	120	120	120

The R-3c space group was used to optimize the unit cell. The conventional cell of corundum contains 12 Al and 18 O atoms. The PAW PBE and PSPW PBE calculations are for a 1 × 1 × 1 cell with just Γ-point sampling. Plane wave density cutoffs in these calculations are twice the wave function cutoff energies.
[a] Kirfel and Eichhorn (1990).

cutoff energy by a factor of two results in a calculation that is approximately three times faster. This is a significant reduction in computational time, but unfortunately there are extra costs associated the PAW method that can significantly reduce these speedups. The largest extra cost in a standard PAW method is the calculation of atomic exchange–correlation terms. To reduce these costs to an acceptable level, the one-dimensional integrations used in the calculations have to be highly optimized. These optimizations have been implemented in the NWChem software, but further optimization is an ongoing project.

4.3.2   Bulk Electronic Structure Properties

4.3.2.1   Band Gaps of Corundum, Hematite, and Goethite

The calculated direct band gaps (defined in these calculations to be the energy difference between the highest-filled and the lowest-unfilled band states) for the three minerals are reported in Table 4.6. For corundum (a closed electronic shell system), the band gap calculated using PBE is 7.0 eV. The hybrid functional PBE0 gives 9.33 eV in good agreement with the experimental range of 8.8–10.8 eV (Tews and Gründler 1982, Perevalov et al. 2007). We note that the inexpensive DFT+U method cannot be used in this mineral because it does not have localized d states.

For the transition metal oxides and oxyhydroxides, the band gap is sensitive to level of calculation showing large differences as illustrated in Table 4.6. DFT+GGA gives band gaps that are far too small, and HF gives band gap estimates that are very large. This is consistent with the results in Table 4.6. We note that PBE+U and PBE0 give similar estimates. Both are much better than produced by PBE. PBE+U produces the best results. To some extent, this excellent agreement results from the choice of U and J values (U = 4 eV, J = 1 eV), which are found by comparing the DFT+U calculated properties using different values (band gap, local spin moment, unit cell volume, etc.) with experimental results (Rollmann et al. 2004). However, for some materials choosing U and J can be less straightforward (Kumar et al. 2013). As discussed in the appendix, the DFT+U level of the theory is computationally much more efficient (roughly similar to the DFT calculation). The agreement illustrated in Table 4.6 supports the use of this level of theory in the much larger surface + interface problems reported in the following. A general conclusion is that some level of exact exchange mixing (hybrid DFT, DFT+U) must be included in order to have any agreement with observations when calculating electronic structures of transition metal oxide or oxyhydroxide. For the present, DFT+U is the only theory with performance that will allow significant dynamical simulation. Similar conclusions are derived from the local spin estimation and projected density of states in the following subsection.

Table 4.6 *Band gaps (in eV) for corundum, hematite, and goethite calculated using PBE, PBE+U, and PBE0 plane wave DFT calculations.*

Crystal	PBE	PBE+U (U = 4 eV)	PBE0	HF	Experiment
Corundum	7.00		9.33	24.52[a]	8.8–10.8[b]
Hematite	0.20	1.63	5.80	15.58[a]	2.0[c]
Goethite	0.48	1.84	4.13	17.32[a]	2.1–2.5[d]

100 Ry (2721 eV) and 200 Ry (5442 eV) were used for the wave function and density cutoff energies (PBE+U calculation were not done for corundum because it does not have localized d states).
[a] Perevalov et al. (2007).
[b] Tews and Gründler (1982).
[c] Mochizuki (1977).
[d] Cornell (2003).

4.3.2.2 Spin Configuration in Hematite and Goethite

As a result of the strongly local behavior of the 3d band, the Fe^{3+} centers are qualitatively described as a high-spin transition metal ion (Cox 1992). In an ionic lattice model of a solid (Cox 1992), each ion would have a spin of $5/2\ \mu_B$. Goethite and hematite are antiferromagnetic materials with a total unit cell spin of 0. However, since there are several Fe^{3+} in the unit cell of these minerals ($4Fe^{3+}$ in goethite 16-atom cell and $12Fe^{3+}$ in the hematite 30-atom hexagonal unit cell), there may be several possible arrangements of the localized spin within the unit cell. Not surprisingly, this alignment of the local spins will affect the energy and several spin structures need to be tested to predict the lowest-energy ordering for such systems.

Because of electron delocalization, the electron spin localized on a single atom in a bulk system may be different from that of the isolated ion. In the delocalized electron picture we are using, the minimum energy of the unit cell is used to identify the localization of the orbital function. The localization will therefore depend on the level of calculation we use and on the distribution (arrangement) of spins in the unit cell. To allow spin order in the bulk unit cell, each electron occupies a single spin orbital. Consistent with the antiferromagnetic property of the minerals, the total spin of the unit cell is held to be zero by equal occupation of spin-up and spin-down orbitals. Even though the total spin in the unit cell is constrained to be zero, the local spin projection on various Fe ions may be different because of the breaking of translational symmetry of the 3d band states within the unit cell. In order to quantify the localized spin, a localized representation of the d band, atomic occupation must be projected out of the calculated d band. In the calculations reported here, local projectors are constructed from the localized atomic functions used to produce the Fe pseudopotentials (see Appendix A.1).

To generate a minimum-energy, spin-ordered state, we first equally fill the up and down orbital states. This ensures that the system will reach an antiferromagnetic state in the unit cell. Despite the great flexibility we have in the basis set (i.e., many plane waves), DFT solvers have a difficult time finding spatially localized states. To facilitate this procedure, we begin the calculation by generating a spin-ordered state in the unit cell by giving weight (via a spin penalty function also based on the atomic projectors; see Appendix A.1) to the pseudopotential on ion sites. Spin localization (i.e., breaking the symmetry) will lead to a more stable energy and break the local symmetry of the unit cell. This will create spin localization with the penalty potential turned on. If the ordered spin state is stable, the spin structure will be retained on further optimization after the constraint is removed. We then use this procedure to generate spin-localized states with different arrangements within the unit cell. With further optimization, these states have different energies. We select the lowest state as the potential ground state.

The ordering and magnitude of the local spin vary with level of calculation (e.g., PBE, PBE+U, or PBE0). Results for the local spin moments (the calculated number difference between spin-up and spin-down electrons on each Fe center) for hematite and goethite using different levels of calculation are included in Table 4.7. Note that the PBE+U and PBE0 levels of calculation are closer to the experimental values reported. Similar to band gap, results from DFT+U and hybrid DFT agree better with experiments, especially for goethite. For hematite, the values from those two methods are lower than experiments but still much closer comparing to pure DFT result. Again, some level of exact exchange must be included in order to approach agreement with observations.

In Table 4.8, we report calculated energy differences between possible local spin orderings for Fe atoms in each layer in hematite using DFT and DFT+U. As mentioned previously, hematite is anti-ferromagnetic, so the sum of local spin moment is 0 in one unit cell. We calculated the energies for three possible spin orderings $(+--+)(++--)(+-+-)$ along [001] in the hexagonal conventional cell (Figure 4.1B left). The calculated energy ordering of DFT and DFT+U for different spin

Table 4.7 *Local magnetic moment (μ (μB/Fe atom)) for each Fe site in hematite and goethite calculated using PBE, PBE+U, and PBE0 plane wave DFT calculations.*

Crystal	PBE	PBE+U (U = 4 eV)	PBE0	HF	Experiment
Hematite	3.60[a]	4.10[a]	4.39[a]	4.60[a]	4.64[b]
Goethite	3.54[c]	3.75[c]	3.90[c]	4.56[c]	3.8[d]

100 Ry (2721 eV) and 200 Ry (5442 eV) were used for the wave function and density cutoff energies.
[a] Chen (2015).
[b] Coey and Sawatzky (1971).
[c] Chen (2015), Coey and Sawatzky (1971).
[d] Bocquet and Kennedy (1992).

Table 4.8 *Energy difference (in meV/Fe atom) between spin configurations of the hematite cell from PBE and PBE+U plane wave DFT calculations.*

Calculation	+−−+	+−+−	++−−	++++
PBE (meV/Fe atom)	0	228.0	278.6	704.9
PBE+U (meV/Fe atom)	0	192.6	192.8	433.1

100 and 200 Ry were used for the wave function and density cutoff energies. There are two types of Fe atoms pair in (001) and [111] directions. Fe—Fe-a in Figure 4.1 means Fe—Fe pair with short distance, and Fe—Fe-b means Fe—Fe pair with long distance. +−−+ ground state means, in both directions, Fe atoms in short-distance pair have same spin while Fe atoms in long-distance pair have different spin. Spin-up (+) and spin-down (−) have been marked to the atoms in Figure 4.1.

configurations is the same (+−−+ along the [111] direction of the primitive cell). The system in the (+−−+) spin ordering has the lowest energy. The lowest spin-ordering configuration is also (+−−+) along [111] in the rhombohedral primitive cell (Figure 4.1B right). For comparison, the fully aligned ferromagnetic state (++++) is also calculated. The energy is higher than all three antiferromagnetic configurations.

The results included in Tables 4.6 and 4.7 emphasize that electronic structure calculations for these highly correlated materials are still uncertain. In general, it is expected that the PBE results tend to be overdelocalized (low-spin moments). Adding exact exchange as in DFT+U and PBE0 produces larger spin moments. Hartree–Fock (HF) tends to overlocalize and the spin moments for this theory are too large as compared to observations. However, consistent with the band gap calculations, it is necessary to include some level of exact exchange to achieve reasonable agreement with the electronic structure data.

4.3.2.3 Projected Densities of States (PDOS)

The distribution of states as a function of energy (density of states (DOS)) is a property important to the prediction of conduction, bonding, and spectroscopic properties of materials (Ziman 1972, Ashcroft and Mermin 1976). To provide more detail about the electronic structure in the unit cell, we project out of the DOS the local variation around individual atoms (a local or projected density of states (PDOS)). The PDOS used in our calculations is defined in Appendix A.1. In Figures 4.2 and 4.3, the PDOS for the various bulk minerals are reported.

Corundum is an insulator with a band gap near 7.0 eV (PBE96), as seen in the PDOS, Figure 4.2a. The densities near the top of the valence band (i.e., Fermi level) are primarily from O-2p part of the band but also contain a small Al-3s and Al-3p component. Lower-energy valance bands (left peak) contain Al-3s, Al-3p, and O-2s component. The peak of conduction band originates from the local

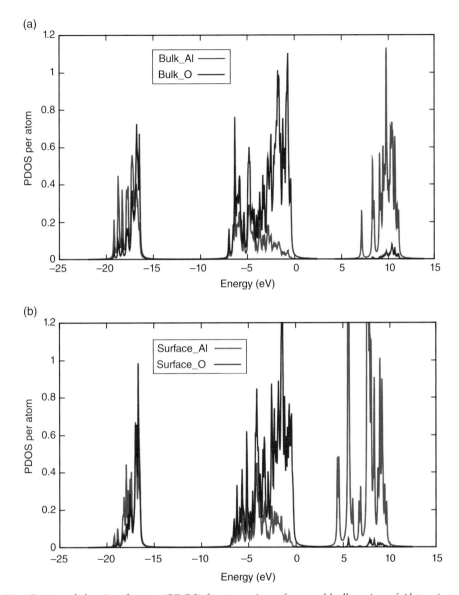

Figure 4.2　*Projected density of states (PDOS) for atoms in surface and bulk region of Al-terminated corundum (001) using PBE96. (a) Bulk corundum and (b) Al-terminated corundum (001) surface.*

Al-3s and Al-3p interaction. There is considerable 3s–3p hybridization of Al atom as expressed in the PDOS.

The PDOS near the Fermi level for bulk hematite using different methods has been plotted in Figure 4.3. Note the very narrow band gap in the PBE calculation. The valence (occupied) states just below the Fermi level have an atomic Fe-3d and O-2p components. Conduction bands above the band gap belong to the minority spin on the Fe^{3+} ion (In Figure 4.3, $y > 0$ are spin-up band functions, and $y < 0$ are spin-down). Most of the Fe-d component in valence bands is spin-down and the Fe-3d component in the conduction bands is primarily spin-up. The most important difference

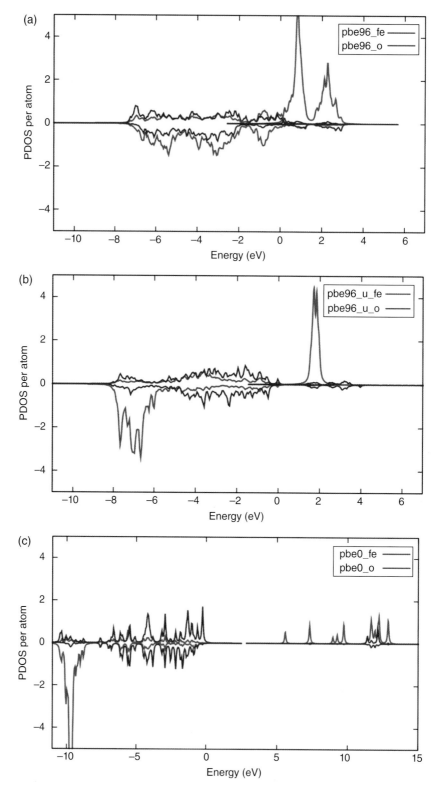

Figure 4.3 *Projected density of states for bulk hematite using different DFT levels. The pink line represents PDOS for the Fe atoms, and the red line represents PDOS for the O atoms. The line above the middle line (y = 0) is spin-up density line and below y = 0 is spin-down density line. The upper, middle, and lower figures are from (a) PBE, (b) PBE+U, and (c) PBE0, respectively.*

between the various levels of calculations is that in the DFT+U and the PBE0 calculation, the occupied d band is much narrower and moved to lower energy than in the PBE calculation reflecting the expected localization in a more correlated theory. This means that the states just below the band gap in the PBE+U and PBE0 calculation are primarily O-2sp states, suggesting that these materials are O-2p → Fe-3d charge transfer insulators (near ion lattices, which is consistent with X-ray absorption and emission spectra measurements) (Fujimori et al. 1986, Lad and Henrich 1989, Ciccacci et al. 1991, Cox 1992, Dräger et al. 1992). When using the PBE functional, more Fe-3d and O-2p hybridization is obtained leading to the high Fe-3d → Fe-3d character in states at the top of the band in hematite. The widths of the O-2sp bands in all the PBE0 and DFT+U calculations are similar. These calculations again show that some level of exchange beyond PBE is necessary to provide a reasonable interpretation for these minerals. However, in the surface and interface calculations that we report in the following section, the PBE0 calculation is extremely expensive, so we do most of these calculations at the DFT+U level.

4.4   Calculation of Surface Properties

4.4.1   Surface Structural Properties

4.4.1.1   Surface Termination

To initiate calculations of the surface and interface regions, unreconstructed surfaces cut from the bulk (see Figures 4.4 and 4.5) are taken as starting structures to be optimized. Because of the loss of three-dimensional symmetry, the calculations can become expensive, and it is not feasible to search all possible configurations. Electronic structure calculations have not been applied directly to more

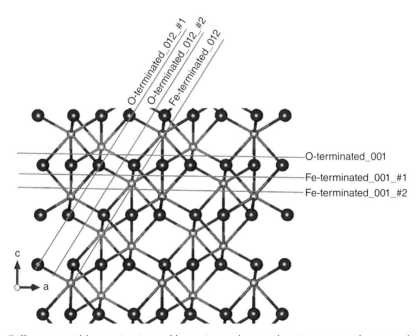

Figure 4.4   *Different possible terminations of hematite and corundum (oxygen, red; iron and aluminum, blue). The "a" and "c" are the lattice vectors.*

(a)

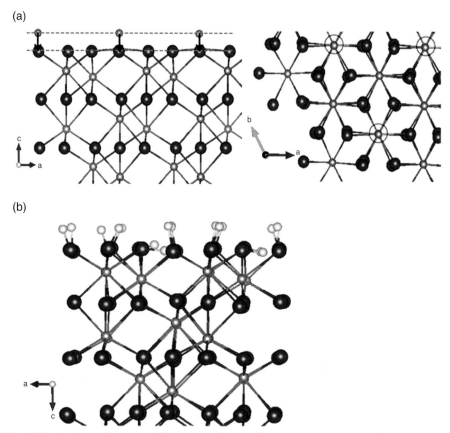

(b)

Figure 4.5 *Al_2O_3 (001) surfaces: (a) Al-terminated corundum (001) surface (top Al layer moves down to O layer). (b) O-terminated corundum (or hematite) (001) protonated surface. The "a," "b," and "c" are the lattice vectors.*

complicated models (such as roughness, steps) that have been considered in experiments but will need many more atoms computationally. As illustrated in Figures 4.4 and 4.5, selecting different surface cleavage planes (terminations) results in different surface structures and compositions. In forming the cleaved surface structure, bonds are broken, leaving the surface atoms in the newly terminated surface with unsaturated bonds. These dangling bonds may be passivated (terminated) by surface reconstruction, by reaction with surrounding phases, or by protonation. These changes affect the local electronic structure and nature of bond formation in the surface region. Different sample preparations and observational environments can result different termination structures with the same surface planes and atomic compositions. For the transition metal oxide and oxyhydroxide minerals, determining this surface structure and the composition in a particular experimental setup may be difficult (Kerisit 2011). Additional information may be obtained from computational estimates of the relative stability of various surfaces (*ab initio* thermodynamics (Wang et al. 2000, Reuter and Scheffler 2002, Lo et al. 2007)). In principle, temperature corrections to the surface free energy may be calculated (Esler et al. 2010). However, for minerals at low temperatures and pressures, it is reasonable to use enthalpy differences, neglecting the zero-point and temperature effects in the free energy. One needs to keep in mind, however, that neglect of entropic effects can make determination of the equilibrium configuration problematic when the enthalpy differences between configurations are relatively small (Wesolowski et al. 2012).

4.4.1.2 Corundum Terminations

The closed-shell mineral corundum (α-Al_2O_3) has the same structure as one of our principal mineral targets, hematite (α-Fe_2O_3), but is an easier computational problem allowing us to explore the accuracy of further choices that affect the accuracy of our calculations, for example, the number of mineral layers in the calculation required for reasonable accuracy and the loss of 3-D symmetry in the surface region (Rustad et al. 2003, Liu et al. 2010, Kumar et al. 2013).

The changes in structure created by expanding the surface unit shell from (1×1) to (2×2) or (2×3) surfaces are shown in Tables 4.1–4.3. These results suggest that at a unit cell least around 10×10 Å is required to retain the accuracy of the DFT calculation. In our plane wave basis calculations, we use a unit with cell with a large dimension perpendicular to the surface to include at least four layers of metal atoms (roughly 10–15 Å) to accurately represent the transition from the bulk to the surface region. This means that at least a few hundred of atoms are needed to model the surface structure accurately.

4.4.1.3 Surface Relaxation Corundum (001)

For this mineral, the (001) surface termination obtained in an experimental setup is dependent on the method generating of the surface and the measurement environment (e.g., for corundum, high vacuum (Renaud 1998), aqueous overlayers (Catalano 2011)). For the (001) surface in vacuum, the Al termination (Figure 4.5a) appears to be the most stable (Renaud 1998), whereas for surfaces in contact with absorbed water layer the O-termination (Figure 4.5b) agrees well with interpretation (Catalano 2011) and the Al-terminated surface has not been reported.

Considerable surface reconstruction has been observed for the Al-terminated (001) surface shown in Figure 4.5A (Guenard et al. 1998, Renaud 1998). Our calculations show that the unsaturated Al atoms in the terminated top layer move down into the surface forming shorter bonds with the surface O atoms. These PBE results agree with prior DFT (Manassidis et al. 1993) and qualitatively with empirical potential results (Catlow et al. 1982, Mackrodt et al. 1987). Our locally optimized calculations show the top Al layer has been moved down by 84.8%; this is in qualitative agreement with experimental estimates of 51.0% (Guenard et al. 1996). The Al atoms in the top layer are bonded with the three surface Al atoms, while the bulk Al atoms are sixfold coordinated. The Al—O bond length in the reconstructed surface is 1.678 Å (Al—O bond lengths in the solid are 1.853 and 1.980 Å).

The surface termination shown in Figure 4.5b agrees most closely with experiment (Catalano 2011). In the calculations for this surface, the excess negative charge created in the termination is passivated by protonation of the three dangling O bonds. As opposed to the Al-terminated surface, the protonation of the O surface bonds reduces the excess valence of the exposed O atoms and reduces the need to form shortened bonds (Brown and Shannon 1973, Brown 2014). Therefore, there is much less reconstruction in this surface than in the Al termination (see Figure 4.3). For the calculated O-terminated hydrated surface of corundum (001), the O layer on top moves up by 3.5%. The two types of bond lengths of Al—O are 1.87 and 2.01 Å versus the bulk bond lengths of 1.85 and 1.98 Å.

4.4.1.4 Protonation of the Goethite (100) Surfaces

In Figure 4.6, we show results of the calculation of the cleavage of goethite to produce the two possible (100) surfaces with different surface protonation schemes (see also Kubicki et al. (2008)).

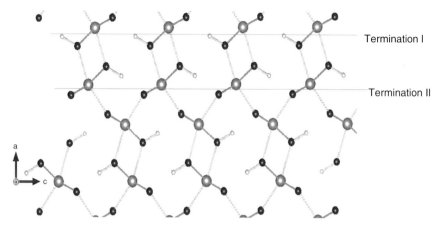

Figure 4.6 *Bulk structure and two types of (100) surface terminations of goethite. The "a," "b," and "c" are the lattice vectors.*

We estimated the surface energies of these two terminations with full reconstruction after terminating the unsaturated O atoms with protons to produce a neutral surface. In this calculation, the termination I produced a surface energy of 156.7 meV/$Å^2$ versus 130.57 meV/$Å^2$ for termination II. This predicts termination II is more stable which agrees with the interpretation of the experimental observation of Ghose et al. (2010).

4.4.1.5 The Protonation of Hematite Surfaces

For the O-terminated hematite (012) surface, Catalano and Fenter (Catalano et al. 2007) proposed O-terminated_012_#1, Figure 4.4, after investigating this surface using X-ray reflectivity (crystal truncation rod (CTR) experiments (Fenter and Sturchio 2004)). On the other hand, Tanwar et al. (2007) proposed the O-terminated_012_#2 in Figure 4.4 to interpret X-ray CTR diffraction data. We used Catalano's termination for this surface in the calculations reported here. In these calculations, the O atoms are capped with protons.

Tanwar et al. (2009) proposed both single-domain and double-domain models for hematite (001) in vacuum. Trainor et al. (2004) studied hydrated hematite (001) surface and suggested two hydroxyl moieties on surface. For the O—O—Fe—O—Fe—O—O—Fe—R (012) cleavage of hematite, there are two possible protonation models (Kerisit 2011), producing a neutral surface (OH)—(OH)—Fe—O—Fe—O—O—Fe—R (Figure 4.7a, Model 1) (Yin and Ellis 2009) and (OH$_2$)—O—Fe—O—Fe—O—O—R (Figure 4.7b, Model 2). In both models, there are singly Fe-coordinated (O^{II}) and triply Fe-coordinated (O^{I}) oxygen atoms. The O^{II} is deeper in the surface layer. (In the bulk O atoms are tetrahedrally coordinated with Fe atoms (Figure 4.1b).) Both these models require the expansion of the 2d surface unit cell to 2×2.

To determine the most stable surface protonation scheme, we optimized the structures of the two different (2×2) protonated surface models at the PBE+U level. (PBE+U is used for all hematite surface and interface calculations in the following.) The Model 1 protonation scheme (OH)—(OH)—Fe—O—Fe—O—O—Fe—R was found to be 11.7 meV/A^2 lower in surface energy than Model 2. After structural optimization the deeper O^{II}H hydroxyl bridges to the O^{I}H further stabilize this surface termination by forming an H-bond (2.73 Å, 154.6°) (Figure 4.4a and b, H-bond 1). The stability of the Model 1 structure may be supported by bond valence analysis (Brown and Shannon 1973, Brown 2009). H-bond 1 reduces the bond valence of (O^{II}) producing slight oversaturated

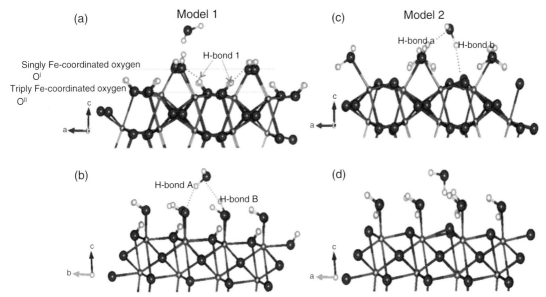

Figure 4.7 *Water molecule absorbed on different hydration models for hematite (012) surface. (a) and (b) are side views of Model 1, and (c) and (d) are side views of Model 2. The "a," "b," and "c" are the lattice vectors.*

valences. The protonation of O^I contributes bond valence to the O^I oxygen, which is only singly Fe coordinated. Even with this protonation, the O atoms in the terminated surface still are considerably under saturated (in the bond valence analysis) and will be sites for absorption of electron-donating species.

4.4.2 Electronic Structure in the Surface Region

4.4.2.1 Corundum (001) Al Terminated

The PDOS for surface atoms of the Al-terminated (001) (projection on the Al atom) and bulk corundum (001) (also projected at the Al atom) are illustrated in Figure 4.2. While valence bands look similar in the PDOS, the biggest difference between bulk and surface corundum is the existence of surface states in the edge of conduction bands, which narrow the band gap from 7.0 to 4.0 eV. The reason for this is the large change in the position of the surface Al and the resulting surface-localized states created by the (001) reconstruction (see Figure 4.5). Similar band gap changes have been reported in other materials (Lazic et al. 2013, Caban-Acevedo et al. 2014) and are important to solar energy conversion.

4.4.2.2 Oxygen-Terminated Hematite (001) Surface

In Figure 4.8a, the PDOS for Fe atoms in bulk and on the (001) surface have been plotted. The PDOS widens slightly in the surface region; however, the bandwidths, band gaps, and local spin moments change very little on approaching the surface. The surface geometry of the (001) termination is illustrated in Figure 4.5b. Each Fe atom in hematite is hexagonally coordinated connecting two adjacent O layers. This coordination is similar to the bulk so the chemical environment of

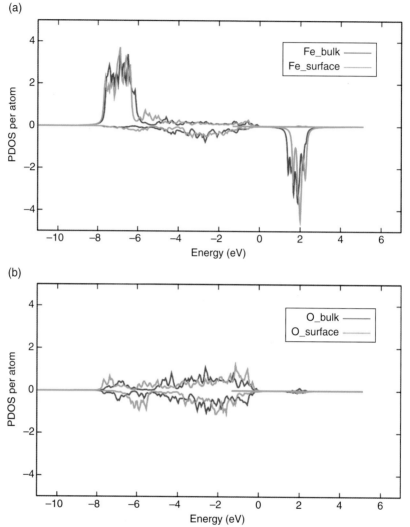

Figure 4.8 *Projected density of states for Fe atoms (a) and O atoms (b) in surface and bulk region of O-terminated hematite (001) using PBE+U.*

surface Fe layer is similar. We show elsewhere (Chen 2015) that the spin ordering does not change in going from the bulk to the surface.

The PDOS projected from the surface O is illustrated in Figure 4.8b. The spin-up ($y > 0$) and spin-down ($y < 0$) DOS are the same for O atoms in the bulk. However, in the surface region as is illustrated in Figure 4.8b, there is a substantial difference between the spin-up and spin-down PDOS. As we have shown in solution simulations of transition metal ions (Bogatko et al. 2010), the local spin moment of transition metal atom can influence the spin polarization of connected O atom. The transition metal oxide hematite has spin moments locally on metal atoms in the bulk. However, O atoms in hematite are not spin polarized because up-spin and down-spin are connected to adjacent metal atoms with opposite spin moment directions in the symmetrical bulk structure. This suggests that the presence of a surface termination removes metal atoms in certain spin direction and breaks this symmetry leading to the polarized spin states of the surface O atoms.

4.4.3 Water Adsorption on Surface

Even with protonation to remove excess charge, the valence of the O atoms in the surface may remain under saturated (i.e., able to accept more positive charge) (Brown and Shannon 1973, Brown 2014). This means that proton interaction between water molecules outside the surface O atoms and the H-bonds between surface atoms can stabilize the surface region. This will play a significant role in determining the water structure on surface.

Recently major advances have been made in the observation of water layer structures on the surface region using synchrotron-generated X-ray diffraction. Ghose et al. (2010) employed CTR techniques to study the 3-D structure of water layers on the goethite (100) surface. Tanwar et al. (2007) also used this technique to investigate water adsorption on hematite (012) surface. Structural models identified in this work can be investigated with the computational tools. *Ab initio* (DFT+U) methods have been used by Kubicki et al. (2008) to study water layer structure on goethite (100) surface. More recent calculations by Chen (2015) have investigated water adsorption for the same surface. Lo et al. (2007) concluded the surface water can stabilize surface structure by hydrogen bonding when studying water adsorption on hematite (012) surface.

As a starting point for adsorption optimizations of the interface structure on hematite, we studied the single water adsorption process on the protonated hematite (012) surface by placing one water molecule in contact with the (2×2) surfaces of Model 1 and Model 2 shown in Figure 4.7. The adsorption energy for one water molecule on both protonated models can be calculated as

$$E_{\text{adsorption}} = E_{\text{water}} + E_{\text{surface}} - E_{\text{water-surface}}$$

Our calculated water adsorption energy for Model 1 is 28.05 kJ/mol. This is slightly greater than the adsorption energy 23.15 kJ/mol calculated for Model 2. In Model 1, the absorbed water forms an H-bond with an (O^I) atom as a donor (2.71 Å, 160.4°, H-bond A in Figure 4.7b) and an H-bond with another (O^I) atom as an acceptor (2.70 Å, 168.9°, H-bond B in Figure 4.7b). In Model 2 (see Figure 4.7c and d), the absorbed water forms an H-bond with O^{II} donating electrons (2.90 Å, 166.0°, H-bond b in Figure 4.7c). It also forms an H-bond with an $O^I - H$ hydroxyl where its proton points toward the surface (2.75 Å, 146.4°, H-bond a in Figure 4.7c). The side views of water adsorption, hydrated models, and H-bonds are shown in Figure 4.7.

Combining the result from surface protonation section, Model 1 is slightly more stable energetically as a bare protonated surface. This structure also has slightly greater water adsorption energy when one water molecule is absorbed on the (2×2) surface. However, the energy difference between Model 1 and Model 2 is small, suggesting that the coexistence of Model 1 and Model 2 is possible. Catalano and Fenter (Catalano et al. 2007) in analyzing their CTR data on this surface proposed water layer, which is regarded forming hydrogen bonds with (O^{II}) atoms agreeing more with Model 2. In our Model 1, the $O^{II}H$ group is oversaturated and forms a donor hydrogen bond with O^I (Figure 4.4a), suggesting that it will not form hydrogen bonds to additional waters. On the other hand, in Model 2, O^{II} atoms are undersaturated, and it is possible to attract water molecules, which may form a water layer closer to the surface and more in agreement with the CTR data interpretation of Catalano and Fenter.

4.5 Simulations of the Mineral–Water Interface

The interactions of the solution species in the interface region are relatively weak and at finite temperature the moles are in constant motion. For example, CTR measurements of surface regions (Fenter 2002, Fenter and Sturchio 2004, Park et al. 2005, Catalano et al. 2007, 2010a, 2010b, Fenter et al. 2010a, 2010b, 2013) are observing the thermally averaged structure of the system. For such systems the dynamics of the system must be taken into account. This may be done with CMD (Shroll and Straatsma 2003, Kerisit 2011, Boily 2012, Kerisit et al. 2012) in which classical force fields are used or with much more computer-intensive models in which the forces between atoms are calculated from first principles (Kubicki et al. 2008, Kumar et al. 2013, Huang et al. 2014).

Kerisit (2011) used CMD simulations with four different force fields to determine the atomic-level structure of three hematite–water interfaces, (001), (110), and (012). The simulation results were compared with experimental electron density profile obtained from CTR data (Catalano et al. 2007). In CMD, large-scale models containing many thousands of atoms can be used due to the simplicity of the empirical potentials. However, these interaction models cannot account for the changes in bonding conditions that occur in the interface region. These many-body effects are especially difficult to treat for the transition metal oxides and oxyhydroxides. The first principles dynamical methods we use avoid these problems by calculating the interactions directly from the solution to the electronic structure problem. This produces a parameter-free model of the interface region.

The theoretical background of AIMD has been discussed in Section 4.2. Similar to the bulk and surface optimization calculation, DFT and DFT+U are the most common electronic structure methods used in AIMD. Huang et al. (2014) ran AIMD simulation, structural properties, and infrared spectra of the corundum (001)/water interface. Kumar et al. used AIMD to study H-bonds and vibrations of water on (110) rutile. In the calculations reported here, AIMD simulations using the Car–Parrinello molecular dynamics (CPMD) method were used (see Section 4.2). We have reported AIMD studies of the goethite (100)–water interface (Chen 2015), comparing types of surface bonds, types of water molecules, and dynamics process on the surface. Hematite (001) and (012) surface–water interface properties as well as Fe^{2+} ion adsorption processes are also in preparation for publication (Chen 2015). Some of these results for the (012) termination of hematite are briefly discussed in the following.

4.5.1 CPMD Simulations of the Vibrational Structure of the Hematite (012)–Water Interface

Similar to our slab in the surface section previously, we include four Fe layers and employ Model 1 (Figure 4.7), which can be represented by $(OH)_2$-X-$(OH)_2$-Fe_2-O_2-Fe_2-O_2-O_2-Fe_2-O_2-Fe_2-$(OH)_2$-X-$(OH)_2$ in c axis. Between two surfaces of the slab in the simulation cell (periodic boundary condition employed), we add 54 water molecules ($c = 25.0$ Å). The density of water is set to 1.0 g/cm^3. The total simulated cell is electrically neutral; it has 1104 valence electrons and 290 atoms, which include 32 Fe, 118 O, and 140 H. Equation of motions in CPMD were integrated using Verlet algorithm, with a time step of 0.12 fs and fictitious orbital mass of 600.0 a.u. All H atoms have been replaced by deuterium to allow longer time steps. For very accurate thermodynamics and time correlation functions, one needs to be careful about using different masses. This can affect the entropy and other chemical behaviors such as water dissociation constants (K_w) (Mesmer and Herting 1978). For the most part these effects are small, and in any case to obtain a very accurate thermodynamic description of H vibrations will require that they be treated using

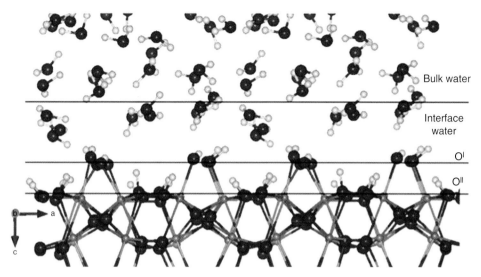

Figure 4.9 *Interfacial and bulk water regions of hematite (012)–water interface. The "a," "b," and "c" are the lattice vectors.*

path integral quantum dynamic techniques (Allen and Tildesley 1987). The canonical ensemble was chosen for this simulation. Nose–Hoover thermostat was used to control the temperature for ions (300 K) and electrons (1200 K). CPMD simulations have been performed 12 ps for each system, and 1 ps equilibration was performed after which the trajectory data was collected.

As discussed in the Section 4.4, the $O^I D$ and $O^{II} D$ bonds (Figure 4.7) play significant roles in surface structural and dynamical properties. The changes in structure as the interface layer goes from the surface interface to the water bulk are illustrated in Figure 4.9 and discussed in more detail in other publications (Chen 2015). In the following, we present the power spectrum analysis using generating trajectory from CPMD.

4.5.1.1 Calculation of Power Spectrum

The power spectrum can be calculated by taking Fourier transform of the velocity autocorrelation functions (Kohanoff 1994, Thomas et al. 2013, Huang et al. 2014):

$$P(\omega) = m \int \left\langle \dot{\boldsymbol{R}}(\tau) | \dot{\boldsymbol{R}}(t+\tau) \right\rangle_\tau e^{-i\omega t} dt$$

$\left\langle \dot{\boldsymbol{R}}(\tau) | \dot{\boldsymbol{R}}(t+\tau) \right\rangle$ denotes the velocity autocorrelation of the velocity $\dot{\boldsymbol{R}}$. Deuterium atom positions (all H atoms have been replaced by deuterium) associated with water molecules in the trajectories of the species in the hematite (012)–water interface are used to calculate power spectrum. The results are shown in Figure 4.10.

The peaks located in the range of 0–800 cm^{-1} in both Figures 4.10a and b, corresponding to the intermolecular librational and translational motions, are difficult to interpret. In Figure 4.10b, there are peaks around 1185 cm^{-1} which are not shown in Figure 4.10a. That peak is due to the bending motion of water molecules, so that's not observed in vibration of the OD group spectrum. The value

(a)   (b)

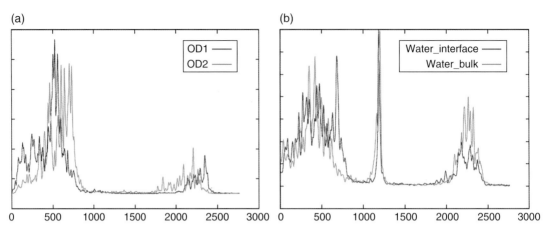

Figure 4.10 *Computed power spectrum of two types of OD groups on hematite (012) surface, water on interfacial region and bulk region. (a) O^ID and $O^{II}D$ surface group (cm^{-1}) and (b) bulk water molecules (cm^{-1}).*

$1185\,\mathrm{cm}^{-1}$ is in good agreement with the experimental value $1209\,\mathrm{cm}^{-1}$ in liquid water. The peaks in the range from 1800 to $2500\,\mathrm{cm}^{-1}$ are the contribution of OD stretching motion modes. In Figure 4.8a, the direct water surface bond with O^ID has a blueshift compared to $O^{II}D$; this shows $O^{II}D$ forms a relatively stronger H-bond with O^I, while O^ID likely has a weakened or H-bond in the interfacial region. In Figure 4.10b, in higher-frequency region, there is one peak $2180\,\mathrm{cm}^{-1}$, which is closer to experimental bulk ice frequency $2190\,\mathrm{cm}^{-1}$ and another peak $2287\,\mathrm{cm}^{-1}$ that is closer to experimental bulk liquid frequency $2260\,\mathrm{cm}^{-1}$. This indicates water molecules in interfacial region have both liquid-like water and icelike vibrational properties as has been identified in the corundum calculations of (Huang et al. 2014).

4.5.2 CPMD Simulations of Fe^{2+} Species at the Mineral–Water Interface

Fe^{2+}–Fe^{3+} redox cycling associated with iron (hydr)oxides is important to many geochemical and environmental processes including CO_2 sequestration (DePaolo and Orr 2008), inorganic respiration, confinement of toxic materials (Brown et al. 1999, Brown 2001, Hochella et al. 2008, Navrotsky et al. 2008), dissolution and secondary mineral precipitation (Yanina and Rosso 2008), and cycling of biological nutrients (Newman 2010). Probably the simplest reaction in Fe^{2+}–Fe^{3+} cycling is the oxidation of aqueous Fe^{2+}(aq) with Fe^{3+} (oxyhydr)oxides such as hematite and goethite. Recently experiments by Rosso, Scherer, and others (Williams and Scherer 2004, Larese-Casanova and Scherer 2007, Cwiertny et al. 2008, Gorski and Scherer 2008, 2009, 2011, Yanina and Rosso 2008, Scherer et al. 2010, Schaefer et al. 2011, Katz et al. 2012, Latta et al. 2012) have put into question the classical view of absorption Fe^{2+} occurring at static surface sites (e.g., surface complexation modeling) (Stumm et al. 1976, Davis and Kent 1990, Dzombak 1990, Sverjensky 1993, Katz and Hayes 1995) suggesting a much more fundamentally based framework (Gorski and Scherer 2011) in which the adsorbed Fe^{2+} oxidizes to Fe^{3+} with the electron transferring into the conduction or polaron (Rosso et al. 2003) band of the Fe^{3+} oxide, followed either by electron trapping in the solid and participating in a redox reaction with an environmental contaminant near the iron oxide surface (Bylaska et al. 2011a) or by electron transport leading to another surface Fe^{3+} to reductively dissolve (Yanina and Rosso 2008).

Accurate modeling of these interface processes for the iron oxide minerals presents a difficult electronic structure problem because their properties are heavily dependent on strongly localized d electrons, complex hydration processes, disorder of the surface/solution interface, the interaction of the solution phase with the highly charged mineral surface, etc. A previous electronic structure study using the DFT+U method has called into question whether or not Fe^{2+} spontaneously oxidizes upon bonding to a Fe^{3+} oxide surface (Russell et al. 2009). In this study they did not find evidence for electron transfer from a Fe^{2+} hexaqua complex into the (021) surface of goethite. However, they did find that the U correction changed the character of the small amount of electron sharing present from O-2p orbitals to Fe-3d orbitals. To handle this apparent discrepancy, it was suggested that defects might be playing a critical role in the spontaneous oxidation of absorbed Fe^{2+}. Because of the limitations of their approach at the time, these simulations did not contain waters of solvation and a search for lowest-energy conformations of the adsorbed Fe^{2+} ions was not possible at the time.

For this chapter, we revisited this problem performing studies of Fe^{2+} + hematite–water interfaces using the PBE+U method with full solvation and limited equilibrium using CPMD. Both (012) and (001) surfaces were simulated. As shown in Figure 4.11, we also found very little electron transfer resulting from the Fe^{2+} bidentate bond to the (012) hematite surface. However, we did find a significant amount of electron transfer with a tridentate bond to the (001) surface, which is somewhat consistent with the assertions of Rosso and Scherer (Rosso et al. 2003, Williams and Scherer 2004, Kerisit and Rosso 2006, Larese-Casanova and Scherer 2007, Cwiertny et al. 2008, Gorski and Scherer 2008, 2009, 2011, Yanina and Rosso 2008, Catalano et al. 2010a, 2010b, Scherer et al. 2010, Schaefer et al. 2011, Katz et al. 2012, Latta et al. 2012). While the PBE+U approach can lead to nontrivial amount of charge delocalization, a full conversion of a near surface Fe^{3+} to Fe^{2+} was not observed at this level of theory for this surface.

(a)   (b)

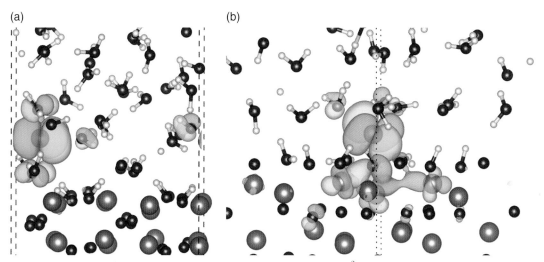

Figure 4.11   *(a) Highest occupied molecular orbital (HOMO) of Fe^{2+} bidentate bond to the 012 hematite surface. Note little charge transfer. (b) HOMO of Fe^{2+} tridentate bond to the 001 hematite surface. Note charge transfer.*

4.6 Future Perspectives

The development of realistic molecular models of geochemical processes is becoming possible because of the availability of ever-increasing computational capabilities. In this chapter, we discuss key aspects associated with simulating the structure and dynamics of the mineral–water interface using plane wave DFT. To illustrate these types of simulations, we showed a variety of results for the bulk, surface, and mineral–water interface properties of the well-characterized minerals corundum, hematite, and goethite. Good results are now obtainable using this level of theory, and further improvements in accuracy and efficiency of these methods will make these methods even more accessible to geochemical researchers in the future.

While density functional calculations provide remarkable structural predictions, they can be qualitatively incorrect in predicting properties such as band gaps, conductivity, and spin states. These problems are partially treated by adding exact exchange. However, the inclusion of exchange in this way is not justifiable and is an *ad hoc* correction. To make further progress along these lines, other higher-level methods that go beyond DFT-based methods, such as dynamical mean field theory (DMFT), GW, and variational and diffusion Monte Carlo methods, will need to be carefully tested on geochemical systems.

Besides the well-known issues of modern electronic structure theory, certainly the most important unsolved problem in simulation technology is the inability of present methods to efficiently search phase space and identify reaction mechanisms. This is an especially difficult problem for first principles simulation methods because the typical time scales of these processes far exceed the time scales that can be achieve even by CMD. This means that much more efficient sampling methods for identifying relevant structure and mechanisms must be developed. Lastly, it is important to recognize that the new computers that are appearing are based on the inclusion of many processor and fast communication. The development of algorithms that will capture the performance of these emerging computers is a difficult problem. To continue to make progress in the use of first principles methods for complex geochemical systems, scalability to very large numbers of processor cores will be needed. This will not only require novel software development and performance analysis tools but also a reevaluation of mathematical and algorithmic approaches.

Acknowledgments

This research was supported by the BES Geosciences program and the BES Heavy element program of the US Department of Energy, Office of Science—DE-AC06-76RLO 1830. Additional support from ASCR petascale tools program and EMSL operations. EMSL operations are supported by the DOE's Office of Biological and Environmental Research. We wish to thank the Scientific Computing Staff, Office of Energy Research, and the US Department of Energy for a grant of computer time at the National Energy Research Scientific Computing Center (Berkeley, CA). Some of the calculations were performed on the Chinook and Cascade computing systems at the Molecular Science Computing Facility in the William R. Wiley Environmental Molecular Sciences Laboratory (EMSL) at PNNL.

Appendix

In the following sections, we discuss briefly some of the details of the approximations that are necessary to define plane wave DFT. This discussion supports the discussions of accuracy that we report in Sections 4.2–4.4 in the main text.

A.1   Short Introduction to Pseudopotentials

The variation/strength of the atom center electron potential, V_{ext}, must be reduced in order for KS wave functions to be expanded in a reasonable number of terms. In order to do this, pseudopotentials have been developed (Pickett 1989). These are widely used; nevertheless, it is typical to have to modify these functions to obtain the accuracy required for a particular calculation. A significant advance in the development of the pseudopotential method was made by Hamann, Schluter, and Chiang (HSC) (Hamann et al. 1979) with the introduction of norm-conserving pseudopotentials. While there are differences between various approaches, all popular pseudopotentials adopt the basic prescription of HSC. These methods have been highly developed in the condensed matter community and are well explained and reviewed (see, e.g., the detailed review of Pickett (1989) and the original papers cited therein). The basic idea of pseudopotential is that the core region of the atomic potential is replaced by a much slower varying function designed to specifically reproduce the behavior of the valence wave functions in regions outside the core (presumed to be the bonding region). The smoothed potential has a nodeless solution that can be expanded by a smaller plane wave basis. It can be shown that with proper care, replacing the atomic potential with a pseudopotential will produce the same solutions beyond the region of replacement while also maintaining the normalization of the orbital function.

Pseudopotentials are derived from first principles single-atom DFT calculations at the same level of approximation (GGA or hybrid exchange) as used in the full many-atom condensed system simulation. These potentials are precomputed before use in the condensed matter calculation so the simulation remains parameter-free (no parameters adjusted in the simulation). On the other hand, there are issues such as the contribution of the atomic valence structure to the bonding that must be decided before the development of the pseudopotential. These choices will determine the transferability of the pseudopotential from the atomic to the condensed environment. Such issues are carefully discussed in the Picket review. In the calculations reported in the next section, the effects of various assumptions will be tested by comparison of calculations to bulk structural observations.

In the HSC approach, given the selection of valence orbitals (corresponding to various l to be included in the active space), a pseudopotential for each total angular momentum is found from the direct inversion of the Schrödinger equation (with a selected DFT functional; see Hamann 1989). This produces a nonlocal pseudopotential of the form

$$V^{\text{pseud}} = V_M^{\text{val}}(\boldsymbol{r}) + \hat{V}_{\text{ps}}(\boldsymbol{r},\boldsymbol{r}') = V_M^{\text{val}}(\boldsymbol{r}) + \sum_{l,m} Y_{lm}(\hat{\boldsymbol{r}}) V_l(r)\delta(r-r') Y_{lm}^*(\hat{\boldsymbol{r}}')$$

where $V_M^{\text{val}}(\boldsymbol{r})$ is the Coulomb and exchange potential due to the (nonactive) valence electrons, $Y_{lm}(\hat{\boldsymbol{r}})$ is the spherical harmonic defined by the angular momentum, l, and magnetic quantum, m, numbers, $\hat{\boldsymbol{r}}$ is a unit vector in the $\boldsymbol{r}$ direction, and $V_l(r)$ is the radial potential found from the inversion of the DFT solution to the radial Schrödinger equation for the equivalent atomic problem (see HSC (Hamann et al. 1979, Hamann 1989)). The operator $\hat{V}_{\text{ps}}(\boldsymbol{r},\boldsymbol{r}')$ acts on function of $\boldsymbol{r}$ as

$$\hat{V}_{\text{ps}}\psi(\boldsymbol{r}) = \int \left[\sum_{l,m} Y_{lm}(\hat{\boldsymbol{r}}) V_l(r)\delta(r-r') Y_{lm}^*(\hat{\boldsymbol{r}}')\psi(\boldsymbol{r}')dr' \right.$$

This potential has a semilocal form, neither just local (radial) nor fully separable (see KB (Bylander and Kleinman 1984)). In this semilocal form, the pseudopotential is computationally difficult to calculate with a plane wave basis set, because the kernel integration is not separable in $\boldsymbol{r}$ and $\boldsymbol{r}'$ (see KB (Bylander and Kleinman 1984)). To produce a more efficient calculation while retaining as much of the atomic form as possible, Kleinman and Bylander approximated the form by

$$\hat{V}_{ps}^{KB}(\boldsymbol{r},\boldsymbol{r}') = V_{local}(\boldsymbol{r}) + \sum_{l,m} P_{lm}(\boldsymbol{r})h_l P_{lm}^*(\boldsymbol{r}')$$

where the atom-centered projectors $P_{lm}(\boldsymbol{r})$ are of the form

$$P_{lm}(\boldsymbol{r}) = [V_l(r) - V_{local}(r)]\widetilde{\varphi}_l(r)Y_{lm}(\hat{\boldsymbol{r}}).$$

and the coefficient h_l is

$$h_l = \left\{ 4\pi \int_0^{\infty} \widetilde{\varphi}_l(r)[V_l(r) - V_{local}(r)]\widetilde{\varphi}_l(r)r^2 dr \right\}^{-1}$$

where $\widetilde{\varphi}_l(r)$ are the zero radial node pseudo-wave functions of the potentials, $V_l(r)$, calculated in the atomic environment. Note that $\hat{V}_{ps}^{KB}|\widetilde{\varphi}_l Y_{lm}\rangle = V_l|\widetilde{\varphi}_l Y_{lm}\rangle$, that is, that the fully nonlocal KB form preserves the form of the potential in the atomic problem. The choice of the local potential $V_{local}(r)$ is somewhat arbitrary, but for transition metals it is often chosen to be the $V_{l=0}(r)$ potential. A larger series expansion in pseudo-wave functions can be used to improve the fully local description of the semilocal form. This leads to the general form

$$\hat{V}_{ps}(\boldsymbol{r},\boldsymbol{r}') = V_{local}(\boldsymbol{r}) + \sum_{l,m}\sum_{n,n'} P_{nlm}(\boldsymbol{r})h_l^{n,n'} P_{n'lm}^*(\boldsymbol{r}')$$

There is a large body of literature describing pseudopotential methods and illustrating the accuracy and efficiency of this approach (Pickett 1989). For use in the structural calculations, the pseudopotentials are developed entirely from fitting atomic calculations and, therefore, should not be considered as part of the data fitting process. Nevertheless, there are questions about accuracy of the pseudopotential approximation, for example, how many $\widetilde{\varphi}_{nl}(r)$ are required to accurately represent the valence structure of the condensed system and how much of the unscreened atomic potential is assigned as the core region (roughly speaking the region removed). This is a function of the atomic structure of the particular element (e.g., separation of the highest-filled lowest excited states, highest valence states, etc.). In this work we found that including the 3s and 3p functions in the active space of the pseudopotential of the Fe^{3+} and Cr^{3+} ions considerably improved the agreement with the scattering data. The default pseudopotential included only the 3d orbitals. Additional issues that have to be considered in the pseudopotential representation include the functional form for the $V_l(r)$ potentials selected and the evaluation of the parameters in these potentials by comparison to atomic calculations at the same level of electronic structure calculation.

The radius of the region of replacement of the external with the pseudopotential form determines to a large extent the smoothness of the pseudopotential (the larger the region, the smoother the pseudopotential). However, if this region is too large, the bond formation will be affected and the pseudopotential representation will produce incorrect bonding results. An example of the derived smooth pseudopotential and nodeless pseudo-wave functions is given in Figure A.1 for the Fe^{3+} ion:

$$E_{psp} = \sum_{\sigma=\uparrow,\downarrow}\sum_{i=1}^{n_{elc}^{\sigma}}\sum_{I=1}^{n_{ion}} \left(\langle \psi_i^{\sigma} | V_{local}^I | \psi_i^{\sigma} \rangle + \sum_{l=0}^{l_{max}^I}\sum_{m=-l}^{l}\sum_{n=1}^{n_{max}^I}\sum_{n'=1}^{n'_{max}^I} \langle \psi_i^{\sigma} | P_{nlm}^I \rangle h_l^{I,n,n'} \langle P_{n'lm}^I | \psi_i^{\sigma} \rangle \right)$$

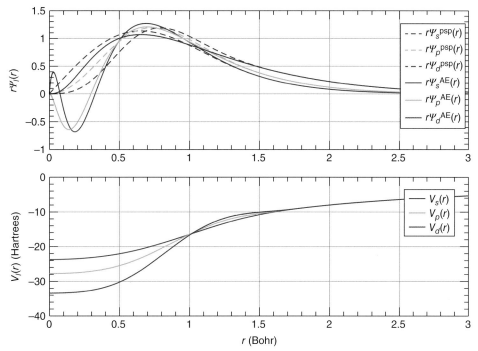

Figure A.1 *Comparison of the pseudo-wave functions (dashed lines) with the full-core atomic valence wave functions (solid lines) for Fe^{3+}. The lower panel shows the corresponding pseudopotentials.*

In the next section, we will illustrate the accuracy of the various choices discussed previously by direct application to the calculation of observed bulk properties of three minerals: goethite (FeOOH), hematite (Fe_2O_3), and corundum (Al_2O_3).

A.1.1 The Spin Penalty Pseudopotential

As discussed in the text, in order to reliably calculate spin-ordered systems in DFT, it is often necessary to develop a spin-ordered initial state prior to full optimization or dynamical simulation. In fact there may be a number of competing spin orderings in the unit cell (see Figure 4.1). A convenient way to generate these states is to introduce terms in the external potential (e.g., a pseudopotential) that will stabilize the electronic wave function at a selected set of sites, that is, add a spin penalty function to the pseudopotential, This will break the symmetry of the wave function and create spin localization. If the state is an approximation to the DFT spin eigenfunctions, further optimization with the spin penalty turned off will lead to a spin-ordered DFT solution. The expectation of the spin penalty energy, $E_{psp + penalty}$, is

$$E_{psp + penalty} = \sum_{\sigma = \uparrow, \downarrow} \sum_{i=1}^{n_{elc}^{\sigma}} \sum_{I=1}^{n_{ion}} \left(\langle \psi_i^{\sigma} | V_{local}^I | \psi_i^{\sigma} \rangle \right.$$

$$\left. + \sum_{l=0}^{l_{max}^I} \sum_{m=-l}^{l} \sum_{n=1}^{n_{max}^I} \sum_{n'=1}^{n'_{max}^I} (1 - \delta_{l,l^{\sigma}} \delta_{I,ionlist} (\xi^{\sigma} - 1)) \langle \psi_i^{\sigma} | P_{nlm}^I \rangle h_l^{I,n,n'} \langle P_{n'lm}^I | \psi_i^{\sigma} \rangle \right).$$

A.1.2   Projected Density of States from Pseudo-Atomic Orbitals

The projected density of states (PDOS) are calculated from

$$\rho_I(\varepsilon) = \sum_{i,i'} |\langle \widetilde{\varphi}_{I,i'} | \psi_i \rangle|^2 \delta(\varepsilon - \varepsilon_i)$$

where I is the atom index for the PDOS, n is the index of valence states in the system, and $\widetilde{\varphi}_{I,i'}$ is the ith pseudo-atomic orbital centered in the Ith atom. The pseudo-atomic orbital is generated by solving Kohn–Sham Equation 4.3 using pseudopotential approach to single atoms.

A.2   Hubbard-Like Coulomb and Exchange (DFT+U)

In DFT framework, most exchange–correlation functionals are generated from expansions around homogeneous electron gas limit. When DFT is used to model these systems that have localized electrons, the predicted electronic states can be significantly away from the localized states.

DFT+U is inspired by the Hubbard model, which can be used to model strongly correlated systems. It adds a Hubbard term (on-site Coulomb and exchange) to the DFT functional for strongly correlated electrons (d, f electrons…) and uses a regular DFT functional on other valence electrons.

The Hubbard term is expressed using Slater integrals and local occupation matrix:

$$E_{\text{Hubbard}} = \frac{1}{2} \sum_{I,\sigma} \sum_{m,n,x,y} \rho_{mn}^{I,\sigma} \rho_{xy}^{I,-\sigma} \left\langle \chi_m^I \chi_x^I | V_{ee} | \chi_n^I \chi_y^I \right\rangle$$

$$+ \frac{1}{2} \sum_{I,\sigma} \sum_{m,n,x,y} \rho_{mn}^{I,\sigma} \rho_{xy}^{I,\sigma} \left(\left\langle \chi_m^I \chi_x^I | V_{ee} | \chi_n^I \chi_y^I \right\rangle - \left\langle \chi_m^I \chi_n^I | V_{ee} | \chi_x^I \chi_y^I \right\rangle \right)$$

$$\rho_{mn}^{I,\sigma} = \sum_k f_k \left\langle \chi_m^I | \varphi_k^\sigma \right\rangle \left\langle \varphi_k^\sigma | \chi_n^I \right\rangle$$

However, the double-counting problem appears, because the Hartree term and exchange–correlation term in DFT functional both include some fragments of Coulomb interactions. The double-counting term has been derived and been subtracted from the total energy functional:

$$E_{\text{DFT}+U} = E_{\text{DFT}} + E_{\text{Hubbard}} \left[\{ \rho_{mm}^{I,\sigma} \} \right] - E_{\text{DC}} \left[\{ \rho^{I,\sigma} \} \right]$$

$$E_{\text{DC}} = \frac{1}{2} \bar{U} \sum_I N_\sigma^I N_{-\sigma}^I + \frac{1}{2} (\bar{U} - \bar{J}) \sum_I \sum_\sigma N_\sigma^I (N_{-\sigma}^I - 1)$$

N is the trace of local occupation matrix. $\bar{U}$ and $\bar{J}$ are the average Coulomb and exchange parameters. There are different approaches to calculate those two numbers. Dudarev et al. (1998) used those two average parameters to replace Slater integrals in equation, and DFT+U functionals have been simplified. In this approach, only $(\bar{U} - \bar{J})$ is meaningful:

$$E_{\text{DFT}+U} = E_{\text{DFT}} + \frac{1}{2} (\bar{U} - \bar{J}) \sum_{I,\sigma} \left(\sum_j \rho_{jj}^{I,\sigma} - \sum_{j,l} \rho_{jl}^{I,\sigma} \rho_{lj}^{I,\sigma} \right)$$

There are different choices of projectors χ_m^I, which can be used in the calculation (i.e., projectors from pseudopotential or PAW), and the number of projectors for each calculation is limited, so the cost of DFT+U method is approximately equivalent to DFT methods which is much cheaper than adding exact exchange.

A.3 Overview of the PAW Method

The main idea in the PAW method (Blochl 1994) is to project out the high-frequency components of the wave function in the atomic sphere region. Effectively this splits the original wave function into two parts:

$$\psi_n(\mathbf{r}) = \widetilde{\psi}_n(\mathbf{r}) + \sum_I \psi_n^I(\mathbf{r})$$

The first part $\widetilde{\psi}_n(\mathbf{r})$ is smooth and can be represented using a plane wave basis set of practical size. The second term is localized with the atomic spheres and is represented on radial grids centered on the atoms as

$$\psi_n^I(\mathbf{r}) = \sum_\alpha \left(\varphi_\alpha^I(\mathbf{r}) - \widetilde{\varphi}_\alpha^I(\mathbf{r}) \right) c_{n\alpha}^I$$

where the coefficients $c_{n\alpha}^I$ are given by

$$c_{n\alpha}^I = \left\langle \widetilde{p}_\alpha^I \middle| \widetilde{\psi}_n \right\rangle$$

This decomposition can be expressed using an invertible linear transformation, T, which relates the stiff one-electron wave functions ψ_n to a set of smooth one-electron wave functions $\widetilde{\psi}_n$:

$$\widetilde{\psi}_n = T\psi_n$$

$$\psi_n = T^{-1}\widetilde{\psi}_n$$

which can be represented by fairly small plane wave basis. The transformation T is defined using a local PAW basis, which consists of atomic orbitals, $\varphi_\alpha^I(\mathbf{r})$, smooth atomic orbitals, $\widetilde{\varphi}_\alpha^I(\mathbf{r})$ which coincide with the atomic orbitals outside a defined atomic sphere, and projector functions, $p_\alpha^I(\mathbf{r})$, where I is the atomic index and α is the orbital index. The projector functions are constructed such that they are localized within the defined atomic sphere and in addition are orthonormal to the atomic orbitals. Blochl defined the invertible linear transformations by

$$T = 1 + \sum_I \sum_\alpha \left(\middle| \widetilde{\varphi}_\alpha^I \right\rangle - \left| \varphi_\alpha^I \right\rangle \right) \left\langle p_\alpha^I \middle|$$

$$T^{-1} = 1 + \sum_I \sum_\alpha \left(\middle| \varphi_\alpha^I \right\rangle - \left| \widetilde{\varphi}_\alpha^I \right\rangle \right) \left\langle \widetilde{p}_\alpha^I \middle|$$

$$\left| \widetilde{p}_\alpha^I \right\rangle = \sum_\beta \left[\left\langle \widetilde{p}^I \middle| \varphi^I \right\rangle \right]_{\alpha\beta}^{-1} \left| p_\beta^I \right\rangle$$

The main effect of the PAW transformation is that the fast variations of the valence wave function in the atomic sphere region are projected out using local basis set, thereby producing a smoothly varying wave function that may be expanded in a plane wave basis set of a manageable size.

In order for a PAW calculation to be manageable, it is important for the PAW basis not to be too large and for the smooth atomic orbitals and smooth projector functions to be adequately described by fairly small plane wave basis. However, for the PAW calculation to be accurate, the basis must also accurately describe the regions near the atomic centers. Procedures used to determine the PAW basis can be found in several places, and for the most part these procedures are the same as the ones used to generate Vanderbilt pseudopotentials (Vanderbilt 1990), with additional Gram–Schmidt steps to enforce the orthonormality relations.

The expression for the total energy in PAW method can be separated into the following 15 terms:

$$E_{\mathrm{PAW}} = \widetilde{E}_{\mathrm{kinetic-pw}} + \widetilde{E}_{\mathrm{vlocal-pw}} + \widetilde{E}_{\mathrm{Coulomb-pw}} + \widetilde{E}_{\mathrm{xc-pw}} + E_{\mathrm{ion-ion}} + E_{\mathrm{kinetic-atom}} + E_{\mathrm{local-atom}} + E_{\mathrm{xc-atom}}$$

$$+ E_{\mathrm{cmp-vloc}} + E_{\mathrm{Hartree-atom}} + E_{\mathrm{cmp-cmp}} + E_{\mathrm{cmp-pw}} + E_{\mathrm{valence-core}} + E_{\mathrm{kinetic-core}} + E_{\mathrm{ion-core}}$$

The first five terms are essentially the same as for a standard pseudopotential plane wave program, minus the nonlocal pseudopotential, where

$$\widetilde{E}_{\mathrm{kinetic-pw}} = \sum_i \sum_G \frac{|G|^2}{2} \widetilde{\psi}^*(G) \widetilde{\psi}(G)$$

$$\widetilde{E}_{\mathrm{Coulomb-pw}} = \frac{\Omega}{2} \sum_{G \neq 0} \left(\frac{4\pi}{|G|^2} \right) \widetilde{\rho}^*(G) \widetilde{\rho}(G)$$

$$\widetilde{E}_{\mathrm{xc-pw}} = \frac{\Omega}{N_1 N_2 N_3} \sum_r \widetilde{\rho}(r) \epsilon_{\mathrm{xc}}(\widetilde{\rho}(r))$$

$$E_{\mathrm{ion-ion}} = \frac{1}{2\Omega} \sum_{G \neq 0} \left(\frac{4\pi}{|G|^2} \right) e^{-\frac{|G|^2}{4\epsilon}} \sum_{I,J} Z_I e^{-iG \cdot R_I} Z_J e^{-iG \cdot R_J} + \frac{1}{2} \sum_a \sum_{I,J \in |R_I - R_J + a|} Z_I Z_J \frac{\mathrm{erf}(\epsilon |R_I - R_J + a|)}{|R_I - R_J + a|}$$

$$- \frac{\epsilon}{\sqrt{\pi}} \sum_I Z_I^2 - \frac{\pi}{2\epsilon^2 \Omega} \left(\sum_I Z_I \right)^2$$

$$\widetilde{E}_{\mathrm{vlocal-pw}} = \sum_G \widetilde{\rho}(G) V_{\mathrm{local}}(G)$$

The local potential in the $\widetilde{E}_{\mathrm{vlocal-pw}}$ term is the Fourier transform of

$$V_{\mathrm{local}}(r) = -\sum_I Z_I \frac{\mathrm{erf}(|r - R_I|/\sigma_I)}{|r - R_I|} + v_{\mathrm{ps}}^I(|r - R_I|)$$

It turns out that for many atoms σ_I needs to be fairly small. This results in $V_{\mathrm{local}}(r)$ being stiff. However, since in the aforementioned integral this function is multiplied by a smooth density $\widetilde{\rho}(G)$, the expansion of $V_{\mathrm{local}}(G)$ only needs to be the same as the smooth density. The auxiliary pseudopotential $v_{\mathrm{ps}}^I(|r - R_I|)$ is defined to be localized within the atomic sphere and is introduced to remove ghost states due to local basis set incompleteness.

The next four terms are atomic based and they essentially take into account the difference between the true valence wave functions and the pseudo-wave functions:

$$E_{\text{kinetic}-\text{atom}} = \sum_I \sum_i \sum_{\alpha\beta} \langle \tilde{\psi}_i | \tilde{p}_\alpha^I \rangle \left(t_{\text{atom}}^I \right)_{\alpha\beta} \langle \tilde{p}_\beta^I | \tilde{\psi}_i \rangle$$

$$E_{\text{local}-\text{atom}} = \sum_I \sum_i \sum_{\alpha\beta} \langle \tilde{\psi}_i | \tilde{p}_\alpha^I \rangle \left(u_{\text{atom}}^I \right)_{\alpha\beta} \langle \tilde{p}_\beta^I | \tilde{\psi}_i \rangle$$

$$E_{\text{xc}-\text{atom}} = \sum_I \sum_{\theta,\varphi} w_{\theta\varphi} \int_0^{r_{\text{cut}}^I} r^2 \left(\rho^I(r,\theta,\varphi)\epsilon_{\text{xc}}\left(\rho^I(r,\theta,\varphi)\right) - \tilde{\rho}^I(r,\theta,\varphi)\epsilon_{\text{xc}}\left(\tilde{\rho}^I(r,\theta,\varphi)\right) \right) dr$$

$$E_{\text{Hartree}-\text{atom}} = \sum_I W_{\text{atom}}^I$$

$$= \frac{1}{2} \sum_I \sum_i \sum_{\alpha\beta} \langle \tilde{\psi}_i | \tilde{p}_\alpha^I \rangle \langle \tilde{p}_\beta^I | \tilde{\psi}_i \rangle \sum_j \sum_{\mu\nu} \langle \tilde{\psi}_j | \tilde{p}_\mu^I \rangle \langle \tilde{p}_\nu^I | \tilde{\psi}_j \rangle \sum_{lm} \tau_{l_\alpha m_\alpha, l_\beta m_\beta}^{lm} \tau_{l_\mu m_\mu, l_\nu m_\nu}^{lm} \left(V_{\text{Heff}}^I \right)_{\alpha\beta\mu\nu}^l$$

The next three terms are the terms containing the compensation charge densities:

$$E_{\text{cmp}-\text{vloc}} = \sum_G \left[\rho_{\text{cmp}}(G)\tilde{V}_{\text{local}}(G) + \tilde{\rho}_{\text{cmp}}(G)\left(V_{\text{local}}(G) - \tilde{V}_{\text{local}}(G) \right) \right]$$

$$+ \int \left(\rho_{\text{cmp}}(r) - \tilde{\rho}_{\text{cmp}}(r) \right)\left(V_{\text{local}}(r) - \tilde{V}_{\text{local}}(r) \right) dr$$

$$E_{\text{cmp}-\text{cmp}} = \Omega \sum_{G \neq 0} \left(\frac{4\pi}{|G|^2} \right) \left[\rho_{\text{cmp}}(G)\tilde{\rho}_{\text{cmp}}(G) - \frac{1}{2}\tilde{\rho}_{\text{cmp}}(G)\tilde{\rho}_{\text{cmp}}(G) \right]$$

$$+ \frac{1}{2} \int\int \frac{\left(\rho_{\text{cmp}}(r) - \tilde{\rho}_{\text{cmp}}(r) \right)\left(\rho_{\text{cmp}}(r') - \tilde{\rho}_{\text{cmp}}(r') \right)}{|r - r'|} dr dr'$$

$$E_{\text{cmp}-\text{pw}} = \Omega \sum_{G \neq 0} \left(\frac{4\pi}{|G|^2} \right) \rho_{\text{cmp}}(G)\tilde{\rho}(G)$$

In the first two formulas, the first terms are computed using plane waves and the second terms are computed using Gaussian two-center integrals. The smooth local potential in the $E_{\text{cmp}-\text{vloc}}$ term is the Fourier transform of

$$\tilde{V}_{\text{local}}(r) = - \sum_I Z_I \frac{\text{erf}(|r - R_I|/\tilde{\sigma}_I)}{|r - R_I|}$$

The stiff and smooth compensation charge densities in the aforementioned formula are

$$\rho_{\text{cmp}}(r) = \sum_I \sum_{lm} Q_{lm}^I g_{lm}^{\sigma_I}(r - R_I)$$

$$\tilde{\rho}_{\text{cmp}}(r) = \sum_I \sum_{lm} Q_{lm}^I g_{lm}^{\tilde{\sigma}_I}(r - R_I)$$

where

$$Q_{lm}^I = \sum_i \sum_{\alpha\beta} \langle \widetilde{\psi}_i | \widetilde{p}_\alpha^I \rangle \langle \widetilde{p}_\beta^I | \widetilde{\psi}_i \rangle \tau_{l_\alpha m_\alpha, l_\beta m_\beta}^{lm} \left(q_{\text{comp}}^I \right)_{\alpha\beta}^l$$

The decay parameter σ_I is defined the same as aforementioned, and $\widetilde{\sigma}_I$ is defined to be smooth enough in order that $\widetilde{\rho}_{\text{cmp}}(\mathbf{r})$ and $\widetilde{V}_{\text{local}}(\mathbf{r})$ can readily be expanded in terms of plane waves. The final three terms are the energies that contain the core densities:

$$E_{\text{valence}-\text{core}} = \sum_i \sum_I \sum_{\alpha\beta} \sum_I \sum_{\alpha\beta} \langle \widetilde{\psi}_i | \widetilde{p}_\alpha^I \rangle V_{\text{valence}-\text{core}\,\alpha\beta}^I \langle \widetilde{p}_\beta^I | \widetilde{\psi}_i \rangle$$

$$E_{\text{kinetic}-\text{core}} = \sum_c \int_0^\infty \left\{ \left(\varphi_{n_c l_c}^I(r) \right)' \left(\varphi_{n_c l_c}^I(r) \right)' - + l_c (l_c + 1) \frac{\varphi_{n_c l_c}^I(r) \varphi_{n_c l_c}^I(r)}{r^2} \right\} dr$$

$$E_{\text{ion}-\text{core}} = \sum_I \frac{1}{2} \int \int \frac{\rho_c^I(\mathbf{r}) \rho_c^I(\mathbf{r}')}{|\mathbf{r} - \mathbf{r}'|} d\mathbf{r} d\mathbf{r}' - \int \frac{\rho_c^I(\mathbf{r})}{|\mathbf{r}|} (Z_I + Z_I^{\text{core}})$$

The matrix elements contained in the aforementioned formula are

$$\left(t_{\text{atom}}^I \right)_{\alpha\beta} = \frac{\delta_{m_\alpha m_\beta} \delta_{l_\alpha l_\beta}}{2} \int_0^{r_{\text{cut}}^I} \left\{ \left(\varphi_{n_\alpha l_\alpha}^I(r) \right)' \left(\varphi_{n_\beta l_\beta}^I(r) \right)' - \left(\widetilde{\varphi}_{n_\alpha l_\alpha}^I(r) \right)' \left(\widetilde{\varphi}_{n_\beta l_\beta}^I(r) \right)' \right.$$

$$\left. + l_\alpha (l_\alpha + 1) \frac{\varphi_{n_\alpha l_\alpha}^I(r) \varphi_{n_\beta l_\beta}^I(r) - \widetilde{\varphi}_{n_\alpha l_\alpha}^I(r) \widetilde{\varphi}_{n_\beta l_\beta}^I(r)}{r^2} \right\} dr$$

$$\left(u_{\text{atom}}^I \right)_{\alpha\beta} = \frac{Z_I}{4\pi} \left(V_{\text{comp}}^I \right)_{\alpha\beta}^{l=0} + \frac{2Z_I}{\sqrt{2\pi}\sigma_I} \left(q_{\text{comp}}^I \right)_{\alpha\beta}^{l=0}$$

$$+ \delta_{m_\alpha m_\beta} \delta_{l_\alpha l_\beta} \int_0^{r_{\text{cut}}^I} \left\{ \varphi_{n_\alpha l_\alpha}^I(r) \varphi_{n_\beta l_\beta}^I(r) \left(\frac{-Z_I}{r} \right) + \widetilde{\varphi}_{n_\alpha l_\alpha}^I(r) \widetilde{\varphi}_{n_\beta l_\beta}^I(r) \left(-v_{\text{ps}}^I(r) \right) \right\} dr$$

$$\left(V_{\text{Heff}}^I \right)_{\alpha\beta\mu\nu}^l = \left(V_{\text{H}}^I \right)_{\alpha\beta\mu\nu}^l - 2 \left(V_{\text{comp}}^I \right)_{\alpha\beta}^l \left(q_{\text{comp}}^I \right)_{\mu\nu}^l - \left(v_g^I \right)'^l \left(q_{\text{comp}}^I \right)_{\alpha\beta}^l \left(q_{\text{comp}}^I \right)_{\mu\nu}^l$$

$$\left(V_{\text{H}}^I \right)_{\alpha\beta\mu\nu}^l = \frac{4\pi}{2l+1} \int_0^{r_{\text{cut}}^I} \int_0^{r_{\text{cut}}^I} \left(\frac{r_<^l}{r_>^{l+1}} \right) \left(\varphi_{n_\alpha l_\alpha}(r) \varphi_{n_\beta l_\beta}(r) \varphi_{n_\mu l_\mu}(r) \varphi_{n_\nu l_\nu}(r) \right.$$

$$\left. - \widetilde{\varphi}_{n_\alpha l_\alpha}(r) \widetilde{\varphi}_{n_\beta l_\beta}(r) \widetilde{\varphi}_{n_\mu l_\mu}(r) \widetilde{\varphi}_{n_\nu l_\nu}(r) \right) dr dr'$$

$$\left(V_{\text{comp}}^I \right)_{\alpha\beta}^l = \frac{4\pi}{2l+1} \int_0^{r_{\text{cut}}^I} \int_0^{r_{\text{cut}}^I} \widetilde{\varphi}_{n_\alpha l_\alpha}(r) \widetilde{\varphi}_{n_\beta l_\beta}(r) \left(\frac{r_<^l}{r_>^{l+1}} \right) g_l^I(r') r'^2 dr dr'$$

$$\left(q^I_{\text{comp}}\right)^l_{\alpha\beta} = \int_0^\infty r^l \left(\varphi_{n_\alpha l_\alpha}(r)\varphi_{n_\beta l_\beta}(r) - \widetilde{\varphi}_{n_\alpha l_\alpha}(r)\widetilde{\varphi}_{n_\beta l_\beta}(r)\right) dr$$

$$\left(v^l_g\right)^l = \frac{4\sqrt{2\pi}}{(2l+1)(2l+1)!!\sigma_I^{2l+1}}$$

$$\tau^{lm}_{l_\alpha m_\alpha, l_\beta m_\beta} = \int_0^{2\pi}\int_0^\pi T_{lm}(\theta,\phi)T_{l_\alpha m_\alpha}(\theta,\phi)T_{l_\beta m_\beta}(\theta,\phi)\sin\theta d\theta d\phi$$

References

Adamo, C. and V. Barone (1999). "Toward reliable density functional methods without adjustable parameters: The PBE0 model." *Journal of Chemical Physics* **110**(13): 6158–6170.

Allen, M. P. and D. Tildesley (1987). *Computer Simulation of Liquids*. Oxford: Clarendon Press.

Alvarez, M., E. E. Sileo and E. H. Rueda (2008). "Structure and reactivity of synthetic Co-substituted goethites." *American Mineralogist* **93**: 584–590.

Anisimov, V. I., I. V. Solovyev, M. A. Korotin, M. T. Czyżyk and G. A. Sawatzky (1993). "Density-functional theory and NiO photoemission spectra." *Physical Review B* **48**(23): 16929–16934.

Anovitz, L. M., D. R. Cole, G. Rother, J. W. Valley and A. Jackson (2010). "Analysis of nano-porosity in the St. Peter Sandstone using (ultra) small angle neutron scattering." *Geochimica et Cosmochimica Acta* **74**(12): A26.

Anovitz, L. M., G. W. Lynn, D. R. Cole, G. Rother, L. F. Allard, W. A. Hamilton, L. Porcar and M. H. Kim (2009). "A new approach to quantification of metamorphism using ultra-small and small angle neutron scattering." *Geochimica et Cosmochimica Acta* **73**(24): 7303–7324.

Ashcroft, N. W. and N. D. Mermin (1976). *Solid State Physics*. New York: Holt, Rinehart and Winston.

Atta-Fynn, R., E. J. Bylaska and W. A. de Jong (2013). "Importance of counteranions on the hydration structure of the curium ion." *Journal of Physical Chemistry Letters* **4**(13): 2166–2170.

Becke, A. D. (2014). "Perspective: Fifty years of density-functional theory in chemical physics." *Journal of Chemical Physics* **140**(18): 18A301.

Becker, U., M. F. Hochella and E. Apra (1996). "The electronic structure of hematite{001} surfaces: Applications to the interpretation of STM images and heterogeneous surface reactions." *American Mineralogist* **81**(11–12): 1301–1314.

Blake, R. L., R. E. Hessevick, T. Zoltai and L. W. Finger (1966). "Refinement of the hematite structure." *American Mineralogist* **51**(1966): 123–129.

Blochl, P. E. (1994). "Projector augmented-wave method." *Physical Review B* **50**(24): 17953–17979.

Bocquet, S. and S. J. Kennedy (1992). "The Néel temperature of fine particle goethite." *Journal of Magnetism and Magnetic Materials* **109**: 260–264.

Bogatko, S. A., E. J. Bylaska and J. H. Weare (2010). "First principles simulation of the bonding, vibrational, and electronic properties of the hydration shells of the high-spin Fe^{3+} ion in aqueous solutions." *The Journal of Physical Chemistry A* **114**(5): 2189–2200.

Boily, J.-F. (2012). "Water structure and hydrogen bonding at goethite/water interfaces: Implications for proton affinities." *The Journal of Physical Chemistry C* **116**(7): 4714–4724.

Brown, G. E. (2001). "Surface science—how minerals react with water." *Science* **294**(5540): 67–69.

Brown, G. E., V. E. Henrich, W. H. Casey, D. L. Clark, C. Eggleston, A. Felmy, D. W. Goodman, M. Gratzel, G. Maciel, M. I. McCarthy, K. H. Nealson, D. A. Sverjensky, M. F. Toney and J. M. Zachara (1999). "Metal oxide surfaces and their interactions with aqueous solutions and microbial organisms." *Chemical Reviews* **99**(1): 77–174.

Brown, G. E. and N. C. Sturchio (2002). "An overview of synchrotron radiation applications to low temperature geochemistry and environmental science." In Fenter, P. A., Rivers, M. L., Sturchio, N. C. and Sutton,

S. R. (Eds.) *Applications of Synchrotron Radiation in Low-Temperature Geochemistry and Environmental Sciences*, Vol. **49**. Washington, DC: Mineralogical Society of America, pp. 1–115.

Brown, I. D. (2009). "Recent developments in the methods and application of the bond valence." *Chemical Reviews* **109**: 6858–6919.

Brown, I. D. (2014). "Bond valence theory." In Brown, I. D. and Poeppelmeier, K. R. (Eds.) *Bond Valences*, Vol. **158**. Heidelberg: Springer, pp. 11–58.

Brown, I. D. and R. D. Shannon (1973). "Empirical bond-strength bond-length curves for oxides." *Acta Crystallographica. Section A* **29**(3): 266–282.

Burke, K. (2012). "Perspective on density functional theory." *Journal of Chemical Physics* **136**(15): 150901.

Bylander, D. M. and L. Kleinman (1984). "Outer-core electron and valence electron pseudopotential." *Physical Review B* **29**(4): 2274–2276.

Bylaska, E. J., A. J. Salter-Blanc and P. G. Tratnyek (2011a). "One-electron reduction potentials from chemical structure theory calculations." In Tratnyek, P. G., Grundl, T. J. and Haderlein, S. B. (Eds.) *Aquatic Redox Chemistry*, Vol. **1071**. Washington, DC: American Chemical Society, pp. 37–64.

Bylaska, E. J., K. Tsemekhman, N. Govind and M. Valiev (2011b). "Large-scale plane-wave-based density functional theory: Formalism, parallelization, and applications." In Reimers, J. R. (Ed.) *Computational Methods for Large Systems: Electronic Structure Approaches for Biotechnology and Nanotechnology*. Hoboken, NJ: John Wiley & Sons, Inc., pp. 77–116.

Bylaska, E. J., M. Valiev, R. Kawai and J. H. Weare (2002). "Parallel implementation of the projector augmented plane wave method for charged systems." *Computer Physics Communications* **143**(1): 11–28.

Bylaska, E. J., M. Valiev and J. H. Weare (2001). "Comparing the efficiency and accuracy of the projector augmented wave DFT method against Gaussian based DFT methods." *Abstracts of Papers of the American Chemical Society* **221**: U435.

Caban-Acevedo, M., N. S. Kaiser, C. R. English, D. Liang, B. J. Thompson, H. E. Chen, K. J. Czech, J. C. Wright, R. J. Hamers and S. Jin (2014). "Ionization of high-density deep donor defect states explains the low photovoltage of iron pyrite single crystals." *Journal of the American Chemical Society* **136**(49): 17163–17179.

Car, R. and M. Parrinello (1985). "Unified approach for molecular-dynamics and density-functional theory." *Physical Review Letters* **55**(22): 2471–2474.

Catalano, J. G. (2011). "Weak interfacial water ordering on isostructural hematite and corundum (001) surfaces." *Geochimica et Cosmochimica Acta* **75**(8): 2062–2071.

Catalano, J. G., P. Fenter and C. Park (2007). "Interfacial water structure on the (012) surface of hematite: Ordering and reactivity in comparison with corundum." *Geochimica et Cosmochimica Acta* **71**(22): 5313–5324.

Catalano, J. G., P. Fenter, C. Park, K. M. Rosso, A. J. Frierdich and B. T. Otemuyiwa (2010a). "Fe(II)-induced structural transformations of hematite surfaces and their impact on contaminants." *Geochimica et Cosmochimica Acta* **74**(12): A150.

Catalano, J. G., P. Fenter, C. Park, Z. Zhang and K. M. Rosso (2010b). "Structure and oxidation state of hematite surfaces reacted with aqueous Fe(II) at acidic and neutral pH." *Geochimica et Cosmochimica Acta* **74**(5): 1498–1512.

Catlow, C. R. A., R. James, W. C. Mackrodt and R. F. Stewart (1982). "Defect energetics in alpha-Al_2O_3 and rutile TiO_2." *Physical Review B* **25**(2): 1006–1026.

Chen, Y. (2015). *Ab initio study of structure and dynamics of bulk, surface and the mineral/aqueous fluid interface regions*. University of California, San Diego, CA.

Ciccacci, F., L. Braicovich, E. Puppin and E. Vescovo (1991). "Empty electron states in Fe_2O_3 by ultraviolet inverse-photoemission spectroscopy." *Physical Review B* **44**(19): 10444–10448.

Coey, Y. M. D. and G. A. Sawatzky (1971). "A study of hyperfine interactions in the system $(Fe_{1-x}Rh_x)_2O_3$ using the Mossbauer effect (Bonding parameters)." *Journal of Physics C: Solid State Physics* **4**: 2386.

Cornell, R. M. S. U. (2003). *The Iron Oxides: Structure, Properties, Reactions, Occurrences, and Uses*. Weinheim: Wiley-VCH.

Cox, P. A. (1992). *Transition Metal Oxides*. Oxford: Clarendon Press.

Cwiertny, D. M., R. M. Handler, M. V. Schaefer, V. H. Grassian and M. M. Scherer (2008). "Interpreting nanoscale size-effects in aggregated Fe-oxide suspensions: Reaction of Fe(II) with goethite." *Geochimica et Cosmochimica Acta* **72**(5): 1365–1380.

Davis, J. A. and D. B. Kent (1990). "Surface complexation modeling in aqueous geochemistry." *Reviews in Mineralogy and Geochemistry* **23**: 177–260.

DeMichelis, R., B. Civalleri, M. Ferrabone and R. Dovesi (2010). "On the performance of eleven DFT functionals in the description of the vibrational properties of aluminosilicates." *International Journal of Quantum Chemistry* **110**: 406–415.

DePaolo, D. J. and F. M. Orr (2008). "Geoscience research for our energy future." *Physics Today* **61**(8): 46–51.

Dräger, G., W. Czolbe and J. A. Leiro (1992). "High-energy-spectroscopy studies of a charge-transfer insulator: X-ray spectra of alpha-Fe_2O_3." *Physical Review B* **45**(15): 8283–8287.

Dudarev, S. L., G. A. Botton, S. Y. Savrasov, C. J. Humphreys and A. P. Sutton (1998). "Electron-energy-loss spectra and the structural stability of nickel oxide: An LSDA+U study." *Physical Review B* **57**(3): 1505–1509.

Dzombak, D. A. (1990). *Surface Complexation Modeling: Hydrous Ferric Oxide*. New York: John Wiley & Sons.

Esler, K. P., R. E. Cohen, B. Militzer, J. Kim, R. J. Needs and M. D. Towler (2010). "Fundamental high-pressure calibration from all-electron quantum Monte Carlo calculations." *Physical Review Letters* **104**(18): 185702.

Fenter, P., S. Kerisit, P. Raiteri and J. D. Gale (2013). "Is the calcite–water interface understood? Direct comparisons of molecular dynamics simulations with specular X-ray reflectivity data." *The Journal of Physical Chemistry C* **117**(10): 5028–5042.

Fenter, P., S. S. Lee, C. Park, J. G. Catalano, Z. Zhang and N. C. Sturchio (2010a). "Imaging interfacial topography and reactivity with X-rays." *Geochimica et Cosmochimica Acta* **74**(12): A287.

Fenter, P., S. S. Lee, C. Park, J. G. Catalano, Z. Zhang and N. C. Sturchio (2010b). "Probing interfacial reactions with X-ray reflectivity and X-ray reflection interface microscopy: Influence of NaCl on the dissolution of orthoclase at pOH 2 and 85°C." *Geochimica et Cosmochimica Acta* **74**(12): 3396–3411.

Fenter, P. and N. C. Sturchio (2004). "Mineral-water interfacial structures revealed by synchrotron X-ray scattering." *Progress in Surface Science* **77**(5–8): 171–258.

Fenter, P. A. (2002). "X-ray reflectivity as a probe of mineral-fluid interfaces: A user guide." In Fenter, P. A., Rivers, M. L., Sturchio, N. C. and Sutton, S. R. (Eds.) *Applications of Synchrotron Radiation in Low-Temperature Geochemistry and Environmental Sciences*, Vol. **49**. Washington, DC: Mineralogical Society of America, pp. 149–220.

Finger, L. W. and R. M. Hazen (1980). "Crystal structure and isothermal compression of Fe_2O_3, Cr_2O_3, and V_2O_3 to 50 kbars." *Journal of Applied Physics* **51**(10): 5362–5367.

Fujimori, A., M. Saeki, N. Kimizuka, M. Taniguchi and S. Suga (1986). "Photoemission satellites and electronic structure of Fe_2O_3." *Physical Review B* **34**(10): 7318–7328.

Fulton, J. L., E. J. Bylaska, S. Bogatko, M. Balasubramanian, E. Cauet, G. K. Schenter and J. H. Weare (2012). "Near-quantitative agreement of model-free DFT-MD predictions with XAFS observations of the hydration structure of highly charged transition-metal ions." *Journal of Physical Chemistry Letters* **3**(18): 2588–2593.

Fulton, J. L., S. M. Kathmann, G. K. Schenter, E. J. Bylaska, S. A. Bogatko and J. H. Weare (2010). "XAFS spectroscopy and molecular dynamics: Aqueous ions and ion pairs under non-ideal conditions." *Geochimica et Cosmochimica Acta* **74**(12): A311.

Ghose, S. K., G. A. Waychunas, T. P. Trainor and P. J. Eng (2010). "Hydrated goethite (alpha-FeOOH) (100) interface structure: Ordered water and surface functional groups." *Geochimica et Cosmochimica Acta* **74**(7): 1943–1953.

Gorski, C. A. and M. M. Scherer (2008). "ENVR 87-Influence of magnetite stoichiometry on Fe(II) uptake and nitroaromatic reduction." *Abstracts of Papers of the American Chemical Society* **236**: 87-ENVR.

Gorski, C. A. and M. M. Scherer (2009). "Influence of magnetite stoichiometry on Fe-II uptake and nitrobenzene reduction." *Environmental Science & Technology* **43**(10): 3675–3680.

Gorski, C. A. and M. M. Scherer (2011). "Fe^{2+} sorption at the fe oxide-water interface: A revised conceptual framework." In Tratnyek, P. G., Grundl, T. J. and Haderlein, S. B. (Eds.) *Aquatic Redox Chemistry*, Vol. **1071**. Washington, DC: American Chemical Society, pp. 315–343.

Guenard, P., G. Renaud, A. Barbier and M. Gautier-Soyer (1996). "Determination of the α-Al_2O_3 (0001) surface relaxation and termination by measurements of crystal truncation rods." *MRS Online Proceedings Library* **437**: 15–20.

Guenard, P., G. Renaud, A. Barbier and M. Gautier-Soyer (1998). "Determination of the alpha-Al_2O_3 (0001) surface relaxation and termination by measurements of crystal truncation rods." *Surface Review and Letters* **5**(1): 321–324.

Hamann, D. R. (1989). "Generalized norm-conserving pseudopotentials." *Physical Review B* **40**(5): 2980–2987.

Hamann, D. R., M. Schluter and C. Chiang (1979). "Norm-conserving pseudopotentials." *Physical Review Letters* **43**(20): 1494–1497.

Herbstein, F. H. (2000). "How precise are measurements of unit-cell dimensions from single crystals?" *Acta Crystallographica. Section B* **56**(4): 547–557.

Hochella, M. F. (1990). "Atomic-structure, microtopography, composition, and reactivity of mineral surfaces." *Reviews in Mineralogy* **23**: 87–132.

Hochella, M. F., S. K. Lower, P. A. Maurice, R. L. Penn, N. Sahai, D. L. Sparks and B. S. Twining (2008). "Nanominerals, mineral nanoparticles, and earth systems." *Science* **319**(5870): 1631–1635.

Hohenberg, P. and W. Kohn (1964). "Inhomogeneous electron gas." *Physical Review B* **136**(3B): B864.

Holzwarth, N. A. W., G. E. Matthews, R. B. Dunning, A. R. Tackett and Y. Zeng (1997). "Comparison of the projector augmented-wave, pseudopotential, and linearized augmented-plane-wave formalisms for density-functional calculations of solids." *Physical Review B* **55**(4): 2005–2017.

Huang, P., T. A. Pham, G. Galli and E. Schwegler (2014). "Alumina(0001)/water interface: Structural properties and infrared spectra from first-principles molecular dynamics simulations." *The Journal of Physical Chemistry C* **118**(17): 8944–8951.

Huda, M. N., A. Walsh, Y. F. Yan, S. H. Wei and M. M. Al-Jassim (2010). "Electronic, structural, and magnetic effects of 3d transition metals in hematite." *Journal of Applied Physics* **107**(12): 123712.

Katz, J. E., X. Y. Zhang, K. Attenkofer, K. W. Chapman, C. Frandsen, P. Zarzycki, K. M. Rosso, R. W. Falcone, G. A. Waychunas and B. Gilbert (2012). "Electron small polarons and their mobility in iron (oxyhydr)oxide nanoparticles." *Science* **337**(6099): 1200–1203.

Katz, L. E. and K. F. Hayes (1995). "Surface complexation modeling. I. Strategy for modeling monomer complex formation at moderate surface coverage." *Journal of Colloid and Interface Science* **170**(2): 477–490.

Kawai, R. and J. H. Weare (1990). "From van der Waals to metallic bonding—the growth of Be clusters." *Physical Review Letters* **65**(1): 80–83.

Kerisit, S. (2011). "Water structure at hematite-water interfaces." *Geochimica et Cosmochimica Acta* **75**(8): 2043–2061.

Kerisit, S. and K. M. Rosso (2006). "Computer simulation of electron transfer at hematite surfaces." *Geochimica et Cosmochimica Acta* **70**(8): 1888–1903.

Kerisit, S., J. H. Weare and A. R. Felmy (2012). "Structure and dynamics of forsterite-scCO(2)/H_2O interfaces as a function of water content." *Geochimica et Cosmochimica Acta* **84**: 137–151.

Khein, A., D. J. Singh and C. J. Umrigar (1995). "All-electron study of gradient corrections to the local-density functional in metallic systems." *Physical Review B* **51**(7): 4105–4109.

Kirfel, A. and K. Eichhorn (1990). "Accurate structure analysis with synchrotron radiation. The electron density in Al_2O_3 and Cu_2O." *Acta Crystallographica Section A* **46**: 271–284.

Kohanoff, J. (1994). "Phonon spectra from short non-thermally equilibrated molecular dynamics simulations." *Computational Materials Science* **2**(2): 221–232.

Kohn, W. and L. J. Sham (1965). "Self-consistent equations including exchange and correlation effects." *Physical Review* **140**(4A): 1133.

Kotliar, G. and D. Vollhardt (2004). "Strongly correlated materials: Insights from dynamical mean-field theory." *Physics Today* **57**(3): 53–59.

Kresse, G. and D. Joubert (1999). "From ultrasoft pseudopotentials to the projector augmented-wave method." *Physical Review B* **59**(3): 1758–1775.

Kubicki, J. D., K. W. Paul and D. L. Sparks (2008). "Periodic density functional theory calculations of bulk and the (010) surface of goethite." *Geochemical Transactions* **9**: 4.

Kumar, N., P. R. C. Kent, D. J. Wesolowski and J. D. Kubicki (2013). "Modeling water adsorption on rutile (110) using van der Waals density functional and DFT+U methods." *The Journal of Physical Chemistry C* **117**(45): 23638–23644.

Lad, R. J. and V. E. Henrich (1989). "Photoemission study of the valence-band electronic structure in FexO, Fe_3O_4, and alpha-Fe_2O_3 single crystals." *Physical Review B* **39**(18): 13478–13485.

Larese-Casanova, P. and M. M. Scherer (2007). "Fe(II) sorption on hematite: New insights based on spectroscopic measurements." *Environmental Science & Technology* **41**(2): 471–477.

Latta, D. E., C. A. Gorski, M. I. Boyanov, E. J. O'Loughlin, K. M. Kemner and M. M. Scherer (2012). "Influence of magnetite stoichiometry on U-VI reduction." *Environmental Science & Technology* **46**(2): 778–786.

Lazic, P., R. Armiento, F. W. Herbert, R. Chakraborty, R. Sun, M. K. Y. Chan, K. Hartman, T. Buonassisi, B. Yildiz and G. Ceder (2013). "Low intensity conduction states in FeS_2: Implications for absorption, open-circuit voltage and surface recombination." *Journal of Physics. Condensed Matter* **25**(46): 465801.

Liechtenstein, A. I., V. I. Anisimov and J. Zaanen (1995). "Density-functional theory and strong interactions: Orbital ordering in Mott-Hubbard insulators." *Physical Review B* **52**(8): R5467–R5470.

Liu, L.-M., C. Zhang, G. Thornton and A. Michaelides (2010). "Structure and dynamics of liquid water on rutile TiO_2 (110)." *Physical Review B* **82**: 161415.

Lo, C. S., K. S. Tanwar, A. M. Chaka and T. P. Trainor (2007). "Density functional theory study of the clean and hydrated hematite (1(1)over-bar02) surfaces." *Physical Review B* **75**(7): 075425.

Luo, S. J., B. Averkiev, K. R. Yang, X. F. Xu and D. G. Truhlar (2014). "Density functional theory of open-shell systems. The 3d-series transition-metal atoms and their cations." *Journal of Chemical Theory and Computation* **10**(1): 102–121.

Mackrodt, W. C., R. J. Davey, S. N. Black and R. Docherty (1987). "The morphology of alpha-Al_2O_3 and alpha-Fe_2O_3—the importance of surface relaxation." *Journal of Crystal Growth* **80**(2): 441–446.

Manassidis, I., A. Devita and M. J. Gillan (1993). "Structure of the (0001) surface of alpha-Al_2O_3 from first principles calculations." *Surface Science* **285**(3): L517–L521.

Marx, D. and J. Hutter (2012). *Ab Initio Molecular Dynamics: Basic Theory and Advanced Methods*. Cambridge, UK: Cambridge University Press.

Maslen, E. N., V. A. Streltsov, N. R. Streltsova and N. Ishizawa (1994). "Synchrotron X-ray study of the electron-density in alpha-Fe_2O_3." *Acta Crystallographica. Section B* **50**: 435–441.

Massey, M., J. Lezama-Pacheco, M. E. Jones, E. Ilton, J. Cerrato, J. Bargar and S. Fendorf (2014). "Competing retention pathways of uranium upon reaction with Fe(II)." *Geochimica et Cosmochimica Acta* **142**: 166–185.

Mesmer, R. E. and D. L. Herting (1978). "Thermodynamics of ionization of D_2O and $D_2PO_4^-$." *Journal of Solution Chemistry* **7**: 901–912.

Mochizuki, S. (1977). "Electrical conductivity of α-Fe_2O_3." *Physica Status Solidi A* **41**: 591–594.

Navrotsky, A., L. Mazeina and J. Majzlan (2008). "Size-driven structural and thermodynamic complexity in iron oxides." *Geochimica et Cosmochimica Acta* **72**(12): A673.

Newman, D. K. (2010). "Feasting on minerals." *Science* **327**(5967): 793–794.

Park, C., P. A. Fenter, J. G. Catalano, S. S. Lee, K. L. Nagy and N. C. Sturchio (2010). "Aqueous-mineral interfaces toward extreme conditions: The potential experimental approaches with synchrotron X-ray probe." *Geochimica et Cosmochimica Acta* **74**(12): A794.

Park, C., P. A. Fenter, N. C. Sturchio and J. R. Regalbuto (2005). "Probing outer-sphere adsorption of aqueous metal complexes at the oxide-water interface with resonant anomalous X-ray reflectivity." *Physical Review Letters* **94**(7): 076104.

Parr, R. G. and W. T. Yang (1995). "Density-functional theory of the electronic-structure of molecules." *Annual Review of Physical Chemistry* **46**: 701–728.

Perdew, J. P., K. Burke and M. Ernzerhof (1997). "Generalized gradient approximation made simple (vol 77, pg 3865, 1996)." *Physical Review Letters* **78**(7): 1396.

Perevalov, T. V., A. V. Shaposhnikov, V. A. Gritsenko, H. Wong, J. H. Han and C. W. Kim (2007). "Electronic structure of α-Al_2O_3: Ab initio simulations and comparison with experiment." *JETP Letters* **85**(3): 165–168.

Pickett, W. E. (1989). "Pseudopotential methods in condensed matter applications." *Computer Physics Reports* **9**(3): 115–197.

Pozun, Z. D. and G. Henkelman (2011). "Hybrid density functional theory band structure engineering in hematite." *Journal of Chemical Physics* **134**(22): 224706.

Remler, D. K. and P. A. Madden (1990). "Molecular-dynamics without effective potentials via the Car-Parrinello approach." *Molecular Physics* **70**(6): 921–966.

Renaud, G. (1998). "Oxide surfaces and metal/oxide interfaces studied by grazing incidence X-ray scattering." *Surface Science Reports* **32**(1–2): 1.

Reuter, K. and M. Scheffler (2002). "Composition, structure, and stability of RuO_2(110) as a function of oxygen pressure." *Physical Review B* **65**(3): 035406.

Rollmann, G., A. Rohrbach, P. Entel and J. Hafner (2004). "First-principles calculation of the structure and magnetic phases of hematite." *Physical Review B* **69**(16): 165107.

Rosso, K. M. and J. R. Rustad (2001). "Structures and energies of AlOOH and FeOOH polymorphs from plane wave pseudopotential calculations." *American Mineralogist* **86**(3): 312–317.

Rosso, K. M., D. M. A. Smith and M. Dupuis (2003). "An ab initio model of electron transport in hematite (α-Fe_2O_3) basal planes." *Journal of Chemical Physics* **118**(14): 6455–6466.

Russell, B., M. Payne and L. C. Ciacchi (2009). "Density functional theory study of Fe(II) adsorption and oxidation on goethite surfaces." *Physical Review B* **79**(16): 165101.

Rustad, J. R., A. R. Felmy and E. J. Bylaska (2003). "Molecular simulation of the magnetite-water interface." *Geochimica et Cosmochimica Acta* **67**(5): 1001–1016.

Rustad, J. R., A. R. Felmy and B. P. Hay (1996). "Molecular statics calculations for iron oxide and oxyhydroxide minerals: Toward a flexible model of the reactive mineral-water interface." *Geochimica et Cosmochimica Acta* **60**(9): 1553–1562.

Schaefer, M. V., C. A. Gorski and M. M. Scherer (2011). "Spectroscopic evidence for interfacial Fe(II)-Fe(III) electron transfer in a clay mineral." *Environmental Science & Technology* **45**(2): 540–545.

Scherer, M., C. Gorski, M. Schaefer, D. Latta, E. O'Loughlin, M. Boyanov and K. Kemner (2010). "Fe(II)-Fe (III) electron transfer in Fe oxides and clays: Implications for contaminant transformations." *Geochimica et Cosmochimica Acta* **74**(12): A920.

Shroll, R. M. and T. P. Straatsma (2003). "Molecular dynamics simulations of the goethite-water interface." *Molecular Simulation* **29**(1): 1–11.

Stubbs, J., A. Chaka, E. Ilton, C. Biwer, M. Engelhard, J. Bargar and P. Eng (2015). "UO$_2$ oxidative corrosion by non-classical diffusion." *Physical Review Letters* **114**(24): 246103.

Stumm, W., H. Hohl and F. Dalang (1976). "Interaction of metal-ions with hydrous oxide surfaces." *Croatica Chemica Acta* **48**(4): 491–504.

Sverjensky, D. A. (1993). "Physical surface-complexation models for sorption at the mineral–water interface." *Nature* **364**: 776–780.

Szabo, A. and N. S. Ostlund (1996). *Modern Quantum Chemistry: Introduction to Advanced Electronic Structure Theory*. Mineola, NY: Dover Publications.

Tanwar, K. S., C. S. Lo, P. J. Eng, J. G. Catalano, D. A. Walko, G. E. Brown, G. A. Waychunas, A. M. Chaka and T. P. Trainor (2007). "Surface diffraction study of the hydrated hematite (1(1)over-bar-02) surface." *Surface Science* **601**(2): 460–474.

Tanwar, K. S., S. C. Petitto, S. K. Ghose, P. J. Eng and T. P. Trainor (2009). "Fe(II) adsorption on hematite (0001)." *Geochimica et Cosmochimica Acta* **73**(15): 4346–4365.

Tews, W. and R. Gründler (1982). "Electron-energy-loss spectroscopy of different Al$_2$O$_3$ modifications. I. Energy loss function, dielectric function, oscillator strength sum rule and the quantity $\epsilon_2 E$." *Physica Status Solidi B* **109**(1): 255–264.

Thomas, M., M. Brehm, R. Fligg, P. Vöhringer and B. Kirchner (2013). "Computing vibrational spectra from ab initio molecular dynamics." *Physical Chemistry Chemical Physics* **15**(18): 6608–6622.

Trainor, T. P., A. M. Chaka, P. J. Eng, M. Newville, G. A. Waychunas, J. G. Catalano and G. E. Brown Jr (2004). "Structure and reactivity of the hydrated hematite (0001) surface." *Surface Science* **573**(2): 204–224.

Tunega, D. (2012). "Theoretical study of properties of goethite (α-FeOOH) at ambient and high-pressure conditions." *The Journal of Physical Chemistry C* **116**(11): 6703–6713.

Valdes, A., J. Brillet, M. Gratzel, H. Gudmundsdottir, H. A. Hansen, H. Jonsson, P. Klupfel, G. J. Kroes, F. Le Formal, I. C. Man, R. S. Martins, J. K. Norskov, J. Rossmeisl, K. Sivula, A. Vojvodic and M. Zach (2012). "Solar hydrogen production with semiconductor metal oxides: New directions in experiment and theory." *Physical Chemistry Chemical Physics* **14**(1): 49–70.

Valiev, M., E. J. Bylaska and J. H. Weare (2003). "Calculations of the electronic structure of 3d transition metal dimers with projector augmented plane wave method." *Journal of Chemical Physics* **119**(12): 5955–5964.

Valiev, M. and J. H. Weare (1999). "The projector-augmented plane wave method applied to molecular bonding." *The Journal of Physical Chemistry A* **103**(49): 10588–10601.

Vanderbilt, D. (1990). "Soft self-consistent pseudopotentials in a generalized eigenvalue formalism." *Physical Review B* **41**(11): 7892.

Wang, X. G., A. Chaka and M. Scheffler (2000). "Effect of the environment on alpha-Al$_2$O$_3$ (0001) surface structures." *Physical Review Letters* **84**(16): 3650–3653.

Wesolowski, D. J., J. O. Sofo, A. V. Bandura, Z. Zhang, E. Mamontov, M. Predota, N. Kumar, J. D. Kubicki, P. R. C. Kent, L. Vlcek, M. L. Machesky, P. A. Fenter, P. T. Cummings, L. M. Anovitz, A. Skelton and J. Rosenqvist (2012). "Comment on 'structure and dynamics of liquid water on rutile TiO$_2$ (110)'." *Physical Review B* **85**: 167401.

Williams, A. G. B. and M. M. Scherer (2004). "Spectroscopic evidence for Fe(II)-Fe(III) electron transfer at the iron oxide-water interface." *Environmental Science & Technology* **38**(18): 4782–4790.

Yanina, S. V. and K. M. Rosso (2008). "Linked reactivity at mineral-water interfaces through bulk crystal conduction." *Science* **320**(5873): 218–222.

Yin, S. and D. E. Ellis (2009). "DFT studies of Cr(VI) complex adsorption on hydroxylated hematite ($1\bar{1}02$) surfaces." *Surface Science* **603**(4): 736–746.

Yuan, K., E. S. Ilton, M. R. Antonio, Z. Li, P. Cook and U. Becker (2015). "Electrochemical and spectroscopic evidence for the one-electron reduction of U(VI) to U(V) on magnetite." *Environmental Science & Technology* **49**(10): 6206–6213.

Zhao, Y. and D. G. Truhlar (2008). "Density functionals with broad applicability in chemistry." *Accounts of Chemical Research* **41**(2): 157–167.

Zhou, F., M. Cococcioni, C. A. Marianetti, D. Morgan and G. Ceder (2004). "First-principles prediction of redox potentials in transition-metal compounds with LDA+U." *Physical Review B* **70**(23): 235121.

Ziman, J. M. (1972). *Principles of the Theory of Solids*. Cambridge, UK: Cambridge University Press.

5

Computational Isotope Geochemistry

James R. Rustad

Corning Incorporated, Corning, NY, USA

This chapter is concerned with calculation of the energetics of isotope exchange reactions from electronic structure calculations (see also Chapter 7). The history of this area of research has been marked by a serendipitous simultaneous rapid evolution of analytical techniques that allowed highly precise measurement of the isotopic compositions of earth materials over a broad range of the periodic table and advances in computational chemistry software and hardware that allowed sufficiently accurate and feasible calculations of vibrational frequencies also over a broad range of the periodic table.

The discussion here focuses on the thermodynamics of equilibrium isotope exchange as derived from harmonic partition functions. In this context, the overall strategy in calculating isotope exchange equilibria is:

1. Make a guess at an atomic-level model of the two environments that will exchange isotopes. As an example, these two environments could be HCO_3^- (aq) and $CaCO_3$ exchanging carbon isotopes. The molecular models representing the environments could be the $HCO_3^-.7H_2O$ and $CCa_8O_{36}^{52-}$ embedded clusters in Figure 5.1.
2. Use an electronic structure code, implementing some representation of electron–electron interactions (this representation is often called the "model chemistry"), to adjust the atom positions (or a subset of these positions) in each environment until the forces associated with all allowed degrees of freedom are zero.
3. For each environment, use the electronic structure code to find the harmonic vibrational frequencies associated with these degrees of freedom. Two lists are needed for each site on which the isotope exchange will take place, one set of frequencies with the heavy isotope and one with the light isotope.
4. For each of these lists of frequencies, calculate the harmonic partition function ratio for the heavy and light isotopes.

Molecular Modeling of Geochemical Reactions: An Introduction, First Edition. Edited by James D. Kubicki.
© 2016 John Wiley & Sons, Ltd. Published 2016 by John Wiley & Sons, Ltd.

(a)   (b)

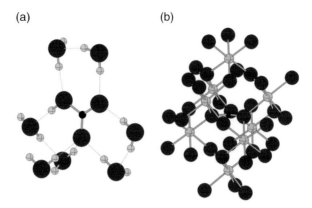

Figure 5.1   *Molecular representation of environments for carbonate ion in (a) HCO₃⁻ (aq) and (b) CaCO₃. Both representations are "core" structures to be embedded in more extended structures which are not shown so that the core structures are more easily seen.*

5. Take the logarithm of the ratio of the partition function ratios to find the free energy of isotope exchange.

This chapter focuses on how to understand the essential problem in simple terms, how to choose representative atomic environments and model chemistries to give the best possible results, and some techniques for assessing the errors in quantum chemical calculation of isotope fractionation factors. The chapter also discusses some generalizations of the "rules of thumb" often used to qualitatively understand isotope fractionation that have been more completely understood through electronic structure calculations. It is mainly aimed at people new to such calculations, but parts of the chapter, especially those parts concerned with model construction and error estimates, could also be interesting to more experienced modelers.

5.1   A Brief Statement of Electronic Structure Theory and the Electronic Problem

As reviewed in this volume and in many places elsewhere,[1] an electronic structure calculation gives the total electronic energy (total potential and electron kinetic energy) of a collection of atoms as well as the forces on each of the nuclear centers in a collection of atoms. To solve this problem one generally has to specify:

a. A set of nuclear positions. These are often represented simply by positive point charges corresponding to the atomic number and a given mass corresponding to the isotope of interest. For problems with heavy elements, the shapes of particular nuclei may also be specified.
b. A set of basis functions for expressing the electronic wave function (in molecular orbital (MO) calculations) or the electron density (in density functional theory (DFT)).

---

[1] A good source of information that is also a good read is C. J. Cramer's book *Essentials of Computational Chemistry* (2004). An older book, published before the widespread DFT revolution in the 1990s, is A. Szabo and N. Ostlund's *Modern Quantum Chemistry* (1996). Another great source of information is the slides from T. Helgaker's talks on electronic structure theory (http://folk.uio.no/helgaker/talks). The book *Molecular Electronic Structure Theory* (2000) by T. Helgaker, P. Jørgensen, and J. Olsen is authoritative but challenging for someone with a geological/geochemical background.

c. A model for the electron–electron interactions. This might come from DFT or from MO methods. The choices of (b) and (c) are often coupled and sometimes collectively referred to as the "model chemistry" or "level of theory."

d. The total electron spin angular momentum (the number of spin-up—the number of spin-down electrons, $n_{up} - n_{down} + 1$ is sometimes called the "multiplicity").

e. An initial guess for the electronic state (e.g., the electron densities of a noninteracting collection of atoms).

Calculating the properties of the "gas" of electrons whizzing around the fixed nuclei is a complex many-body problem. To a first approximation, each electron can be thought of as moving in the average field of all the other electrons (the "Hartree–Fock" approximation). But since they are quantum objects, the electrons do not have "paths." If one electron interacts with another, the two may emerge from the interaction with their identities exchanged. Of course, it cannot be said, after the interaction, which electron is which; one never knows exactly where the electrons are so there is no way to tell if they have exchanged. This effect gives rise to an interaction called the "exchange interaction." Also, there will be large instantaneous deviations from the average field as the electron swims around in the gas, due to other electrons colliding with it. For example, the interaction of a "d" electron in a ferric iron atom with the four other d electrons will deviate strongly from any "average" field seen by any "average" electron. These deviations from the mean field picture contribute to the "correlation energy."

The two main methods used in electronic structure codes are the MO methods and DFT methods. These are discussed in many places, including elsewhere in this volume. The DFT approach is to parameterize the exchange and correlation effects into a functional, called the exchange–correlation functional, that gives the energy of the gas based on the electron density at all points in the gas (local density approximation (LDA)) and may also include terms related to the gradient of the density (generalized gradient approximation (GGA)) (Kohn and Sham, 1965). The MO methods start with the Hartree–Fock solution to the electronic problem, which has exact exchange but limited correlation, and then make improved estimates of the correlation energy beyond mean field theory. The MO methods have the advantage that they can be systematically improved, but the systematic improvement is expensive. The most efficient approach to estimate the correlation is through the use of "collision" operators to give single, double, triple, and other excitations to the reference Hartree–Fock wave function and then calculate the energies associated with these excitations. This is called coupled cluster (CC) theory (Bartlett and Musiał, 2007). Another approach, called many-body perturbation theory, can be used to approximate or augment CC. The most commonly used version of CC theory explicitly deals with single and double (SD) excitations and uses a perturbative correction for triple (T) excitations (abbreviated CCSD(T)). Second-order Møller–Plesset perturbation theory (MP2; Møller and Plesset, 1934) can often be used as a sort of approximation to CCSD (note that this does not imply that CCSD predictions are necessarily closer to experiment than MP2). These methods (MP2, CCSD, and CCSD(T)) are computationally very expensive methods, in part because they require large correlation-consistent basis sets (Dunning, 1989; Kendall et al., 1992) to do their job. Because they grow with at least the fifth power of the system size (MP2), they can be applied to only very small systems. At the time of this writing, a full vibrational calculation on an ion with six water molecules around it would be a large calculation for these kinds of methods. In most geochemical applications their main utility is in benchmarking DFT methods. Unlike DFT, they have the right physics to treat weak interactions such as the van der Waals interaction.

The most basic solution to the electronic problem is to plug in a guess at the initial wave function (or electron density in the case of DFT) and run through the iterative self-consistent field (SCF) loop in the electronic structure code to converge to a stationary (i.e., time-independent) electronic state.

In simple problems, this state would normally be the lowest energy state (or ground state), and the code would yield the total energy (electron kinetic energy, electron–electron interaction energy, electron–nuclear attraction energy, and nuclear repulsion energy) as well as the forces on each of the nuclei (gradients of the energy with respect to nuclear positions). The energy depends, in the Born–Oppenheimer approximation, parametrically on the nuclear coordinates. In the Born–Oppenheimer approximation, the time scale for nuclear motion is regarded as so slow relative to the time scale of electronic motion that the electrons completely relax around the nuclei as they move and thus provide a "potential energy surface" for nuclear motion, meaning that for a given set of nuclear coordinates, the forces on the nuclei are always the same, regardless of how the system arrived at that set of nuclear coordinates.

The forces are often used to find a set of nuclear positions where all of these forces (or at least some subset of these forces) are zero. This is known as "geometry optimization" and locates a minimum on the potential energy surface. This optimization may apply to all degrees of structural freedom in the system or some subset of them (others being constrained to particular values for reasons to be discussed).

Issues associated with (a–c) are discussed in more detail in Chapter 1. Concerning the more purely practical issues (d–e), it is not at all uncommon that the iterative SCF procedure for solution of the electronic problem fails to converge. For simple cases, this problem can be solved by running the calculation with a smaller basis set, saving the result, and using it as a guess for a new calculation with the larger basis set. Sometimes additional flexibility in generating guesses for the initial electronic state is required to obtain convergence. One approach that often works for molecular systems is to put the molecule in an electric field. This usually enables the system to achieve SCF convergence. Then the converged guess can be put in a weaker field and run through another SCF cycle. Gradually the field can be reduced to zero through several of these cycles. For transition metal systems with multiple centers having unpaired electrons, it helps to build up the guess with calculations on small fragments or individual atoms, assembling these into a larger system meeting the requirements in (c). This capability is sometimes called the "fragment guess." NWChem (Valiev et al., 2010) and Gaussian (Frisch et al., 2004) codes have this capability. One should always remember the possibility of converging to excited electronic states. Excited states, of course, have different potential energy surfaces and different vibrational frequencies than ground states. Another potential problem is that it is possible to converge to a stationary state that is not even a local minimum in energy. In some codes, such as Gaussian, there are facilities to do stability analyses to check whether a true minimum has been found in the iterative SCF procedure. For systems with multiple transition metals, it is probably an excellent investment to run such an analysis.

5.2   The Vibrational Eigenvalue Problem

Having found such a stationary state at the minimum in the electronic energy, with all forces associated with the relevant degrees of freedom equal to zero (within some preselected criteria), the derivatives of these forces with respect to the nuclear positions (sometimes called the force constants) are also calculated. In some cases it may be possible to calculate this matrix analytically through perturbation theory; in others it may be necessary to physically displace each atom (perhaps taking account of symmetry) and calculate the force derivatives numerically. However they are obtained, the force derivatives form a matrix which may be written as $dF_{\alpha i}/d\beta_j/\sqrt{(m_i m_j)}$, where $dF_{\alpha i}$ is the change in the force felt by atom i in the α direction when atom j is displaced in the β direction (e.g., α and β could be either x, y, or z if Cartesian coordinates are used) and m_i and m_j are the masses of atoms i and j.

The eigenvectors of this matrix give the normal modes of the system of atoms (i.e., the atomic displacements associated with each vibrational frequency), and the square roots of the eigenvalues give the frequencies of oscillation associated with the normal modes (see Chapter 10). In the harmonic approximation, solving the eigenvalue problem turns the complicated motions of the system of atoms into $3N$ independent harmonic oscillator problems. The frequencies of these oscillators depend, in general, on the masses of all the atoms because of the $\sqrt{(m_i m_j)}$ term in $dF_{\alpha i}/d_{\beta j}/\sqrt{(m_i m_j)}$. On the other hand, if a particular vibrational mode involves no displacement of atom i, the frequency will be independent of the mass of atom i. Note that in the Born–Oppenheimer approximation, the matrix of force constants $dF_{\alpha i}/d_{\beta j}$ does not depend on the nuclear masses, so that if the eigenvalue problem needs to be solved for a different set of atomic masses, only the $\sqrt{(m_i m_j)}$ needs to be modified and there is no need to recompute the matrix of force constants. Relative to the time required to compute the $dF_{\alpha i}/d_{\beta j}$ matrix, the solution of the vibrational eigenvalue problem takes negligible computational time. So to do a problem involving isotopic substitution, the force-constant matrix $dF_{\alpha i}/d_{\beta j}$ is computed, saved, and then divided by the $\sqrt{(m_i m_j)}$ term according to the isotope substitution scheme of interest. The matrix $dF_{\alpha i}/d_{\beta j}$ is also called the "Hessian matrix," and the $dF_{\alpha i}/d_{\beta j}/\sqrt{(m_i m_j)}$ is often called the "dynamical matrix." In the simple case of a single atom vibrating harmonically in an isotropic potential well, the dynamical matrix A is very simple:

$$
\begin{matrix}
K/m & 0 & 0 \\
0 & K/m & 0 \\
0 & 0 & K/m
\end{matrix}
$$

Solving the eigenvalue problem $Ax = \lambda x$, one can see by inspection that there are three orthogonal eigenvectors (100), (010), and (001) with eigenvalues (K/m) and, hence, the frequencies are simply the familiar $\omega = \sqrt{(K/m)}$.

As a practical matter, if the reader is trying out one of these calculations, the first task is to figure out how to save the Hessian ($dF_{\alpha i}/d_{\beta j}$) matrix separately and then to do the mass substitution to get the dynamical matrix $dF_{\alpha i}/d_{\beta j}/\sqrt{(m_i m_j)}$ to find the vibrational frequencies. Because most people doing electronic structure calculations are not isotope geochemists and are just interested in getting the frequencies for the standard atomic weights, isotope substitution is not a commonly carried out procedure, and it may take some time to figure out how to save $dF_{\alpha i}/d_{\beta j}$, especially for researchers doing calculations on periodic solids.[2] Generally it is best to try it out first on a simple system. Hessian matrix evaluations on large systems can be computationally demanding. If the format of the Hessian matrix can be determined, it is easy to write small code to read it in, do the mass division, and use pass it to any suitable program (like JACOBI.f in the Numerical Recipes codes (Press et al., 1986)) to find the eigenvalues and eigenvectors. As an aside, keep in mind that a numerical evaluation of the Hessian can usually be restarted in the event of a computer crash, whereas with an analytical Hessian, one has to start over completely.[3]

---

[2] In one commonly used density functional code for solids, the facility to do simple mass substitutions to rebuild the dynamical matrix had not even been considered by the developers; the entire construction of the Hessian part of matrix had to be done all over again with the new mass!
[3] The author once had his computer tied up for 6 weeks evaluating an analytical Hessian for a particularly large system. On the last day, watching anxiously as a freak September California storm approached, he saw the last iteration on the coupled perturbed Hartree–Fock solution complete just as the power went out and the entire matrix was lost. If you do these calculations frequently on your own computers, it's a good idea to invest in a backup power system.

5.3 Isotope Exchange Equilibria

Through solving the eigenvalue problem, we have now turned our problem into $3N$ independent quantum harmonic oscillator problems. The quantum mechanical solution of the harmonic oscillator is discussed in every physical chemistry textbook. The take-home message is that there is a set of energy levels $(n + \frac{1}{2})h\omega$, where $h =$ Planck's constant, $\omega =$ frequency, and $n = 0,1,2,\ldots$, with the interesting point being the existence of a zero-point energy that is not zero. Due to the uncertainty principle, a particle cannot sit right at the bottom of its potential well in quantum mechanics. But a heavier particle gets closer to zero than a lighter particle because its frequency is lower and its zero-point energy $\frac{1}{2}h\omega$ is less.

By running the electronic structure code, we now have two lists of frequencies for each environment involved in the isotope exchange, one list of frequencies for the heavy isotope and one list for the light isotope. The basic physics of isotope exchange energetics is nicely reviewed in many places (e.g., by Schauble (2004) and Wolfsberg et al. (2009) provides a great discussion from a computational point of view). Only a brief discussion is provided here.

Geological processes fractionate isotopes for many reasons. One important reason is the difference in vibrational free energy between products and reactants in an equilibrium involving isotope exchange reactions such as

$$^{h}M_A + {}^{l}M_B = {}^{h}M_B + {}^{l}M_A \tag{5.1}$$

where M is some element in environment A (e.g., an aqueous solution) and environment B (e.g., a mineral) and ^{h}M represents the heavy isotope of M and ^{l}M represents the light isotope. An obvious point worth emphasizing is that if the isotope is the same (i.e., $h = l$), then the energy change is zero. This means that the isotope exchange reaction is a particularly simple one with the electronic contributions cancelling on each side of the reaction. That said, most of the rest of the world is focused on computing energies for reactions such as

$$2CH_4 + 2O_2 = CO_2 + 2H_2O \tag{5.2}$$

here the electronic contributions don't cancel. It is good to keep in mind that most of the efforts of the quantum chemistry community are focused on reactions like (5.2) rather than reactions like (5.1). Reaction 5.1 is sort of the ultimate isodesmic reaction, as no bonds at all are broken or formed. Theoretical improvements designed to obtain better energies for reactions such as Reaction 5.2 (i.e., improved exchange–correlation functionals for reaction energies in DFT) are not necessarily going to make great improvements in the energies for Reaction 5.1.

The simplest way to view Reaction 5.1 is to imagine that element M vibrates harmonically in the A and B environments with characteristic force constants K_A and K_B, representing the stiffness of each environment, where a higher K corresponds to a stiffer environment. It was described earlier how to evaluate this stiffness with an electronic structure code by finding the equilibrium structure, where the forces on all active atoms are zero, and then displacing the M atom by dx and calculating the force F on M which, in the harmonic approximation, is given by $-Kdx$ where $K = dF/dx$. At the bottom of an approximately parabolic well, K is the curvature of the energy as a function of distance away from the origin $(dF/dx = d^2E/dx^2)$.

Imagine that in our reaction, we take a system where $K_A = 1$, $K_B = 2$, $^{h}M = 2$, and $^{l}M = 1$. Thus on the right-hand side of Equation 5.1, we put the heavy isotope in the stiff environment and the light isotope in the soft environment. On the left-hand side of the reaction, we put the heavy isotope in the soft environment and the light isotope in the stiff environment.

What would be the equilibrium constant for Reaction 5.1 at zero temperature? In this case, the energy E of the products and reactants can be evaluated using $E = \frac{1}{2}h\omega$ (n is zero at zero temperature in $E = (n + \frac{1}{2})h\omega$). Remembering that $\omega = \sqrt{(K/M)}$ and using units such that $h = 1$, the zero-point energy of the product side of Equation 5.1 is

$$\frac{1}{2}\sqrt{(2/2)} + \frac{1}{2}\sqrt{(1/1)} = 1$$

the zero-point energy of the reactant side is

$$\frac{1}{2}\sqrt{(2/1)} + \frac{1}{2}\sqrt{(1/2)} = 3/(2\sqrt{2}) \approx 1.06066$$

and the energy change for the reaction (the energy of the products minus the energy of the reactants) is approximately -0.06066. The minus sign indicates that the product side of the reaction is more favorable; thus it is apparently better to pair the heavy isotope with the stiffer environment. This makes physical sense because the heavy isotope, having the lower-frequency spring vibration, will burrow itself lower into the vibrational potential energy well than will the light isotope $(\frac{1}{2}\sqrt{(K/M_h)} < \frac{1}{2}\sqrt{(K/M_l)})$. This burrowing is more pronounced as the spring vibrational frequency increases, being proportional to the square root of the stiffness K. The stiffer the environment, the lower the heavy isotope rides in the well relative to the light isotope. So, to achieve the lowest energy possible, the heavy isotope will tend to partition into the strongest bonding environments.

If temperature increases beyond zero kelvin, the Boltzmann law gives the populated energy levels and the zero-point energy is replaced with the slightly more complex equation:

$$E = h\omega \left[\frac{1}{2} + 1/\left(e^{h\omega/kT} - 1 \right) \right] \tag{5.3}$$

In any real system, there will be many characteristic vibrational frequencies associated with atoms in particular environments, associated with the normal modes as described earlier. But this is straightforward because through the solution of the eigenvalue problem each of these normal modes is by definition independent of the other normal modes. Since the energy of many oscillators is the sum of the energies of the individual oscillators, and the logarithm of the partition function gives the free energy, all the normal modes are multiplied together in the partition function for a single isotope exchange (Bigeleisen and Mayer, 1947; Urey, 1947):

$$\beta = (Q_h/Q_l) = \prod_{1-3N} (u_{hi}/u_{li}) \times [\exp(-u_{hi}/2)/(1 - \exp(-u_{hi}))] \times [(1 - \exp(-u_{li}))/\exp(-u_{li}/2)] \tag{5.4}$$

where $u_{h/li} = h\omega_{(h/l)}/kT$

For gas-phase molecules there are three rotations and three translations that have zero frequency that do not contribute to the β (sometimes called the reduced partition function ratio, or RPFR). The equilibrium constant for a single isotope exchange between environments A and B, often called α_{AB}, is given by β_A/β_B where β_A is the RPFR for environment A. To keep all the signs straight between the heavy and light isotopes, just remember that the larger the β associated with a particular environment, the more that environment accumulates heavy isotopes. So that if the beta for $^{13}C/^{12}C$ substitution in CO_2 is β_{CO_2} and the beta for $^{13}C/^{12}C$ substitution in CH_4 is β_{CH_4}, the heavy carbon accumulates in CO_2 and the equilibrium constant for the reaction

$$^{12}CO_2 + {}^{13}CH_4 = {}^{13}CO_2 + {}^{12}CH_4 \tag{5.5}$$

Table 5.1 *Frequencies (cm^{-1}) for CO_2 and CH_4 calculated with DFT using the exchange–correlation functional B3PW91 and basis set aug-cc-pVTZ.*

CO_2 $\beta = 1.198533$	
Heavy	Light
660.48	679.83
660.48	679.83
1374.27	1374.27
2358.26	2427.35
CH_4 $\beta = 1.118065$	
Heavy	Light
1318.47	1326.71
1318.47	1326.71
1318.47	1326.71
1551.32	1551.32
1551.32	1551.32
3033.69	3033.69
3137.72	3148.93
3137.72	3148.93
3137.72	3148.93

Masses: ^{12}C = 12.000000 amu; ^{13}C = 13.0033548 amu.

with the equilibrium constant given by

$$\alpha = \beta_{CO_2}/\beta_{CH_4} \tag{5.6}$$

that is, the β in the numerator goes with the environment having the heavy isotope on the product side of the reaction and the one in the denominator goes with the environment having the heavy isotope on the reactant side of the equation.

Generalization to periodic solids requires sums over points in the Brillouin zone. The partition function can also be generalized in terms of the vibrational density of states as authoritatively discussed by Kieffer (1982). There is a certain beauty, however, in keeping things straightforward and simple and representing environments in crystalline materials as embedded molecules (as discussed in the following) without the intellectual baggage of k-points and Brillouin zones. The molecular approach has some other practical advantages in terms of available software, which, in general, is more well developed for molecules than for periodic solids.

The main point is that, in the harmonic approximation, we can associate a single number called the RPFR with each environment and a particular pair of isotopes, h and l. The ratio of the RPFRs gives the equilibrium constant for the isotope exchange reaction between the two environments, and the environment enriched in the heaviest isotope has the highest RPFR. This is often reported as $1000 \ln(K)$ in "per mil" notation.

To illustrate these points consider the isotope exchange reaction in (5.5). We need a list of frequencies for CO_2 with ^{12}C and ^{13}C as well as one for CH_4 with ^{12}C and ^{13}C. Choosing the standard atomic weights for O and H, we obtain the lists in Table 5.1.

Fortran code for calculation of the RPFR given a list of frequencies for heavy and light masses

```
implicit none
real*8 numerator, denominator, prod
real*8 u,up,h,c,a
real*8 omegah(1000),omegal(1000)
real*8 rt,hc
```

```
      integer nfreq,i
      c=29979245800.0d0
      a=6.022141d23
      h=6.626069d-34
      rt=8.3144*298.15d0
      hc=h*c*a
56    format(i5,2f12.4,e20.10,f10.6)
      read(*,*)nfreq
      doi=1,nfreq
        read(*,*)omegah(i),omegal(i)
      enddo
      prod=1.0d0
      doi=1,nfreq
        u=hc*omegah(i)/(rt)
        up=hc*omegal(i)/(rt)
        numerator=(u*dexp(-u/2.0d0)/(1.-dexp(-u)))
        denominator=(up*dexp(-up/2.0d0)/(1.-dexp(-up)))
        prod=prod*numerator/denominator
        print 56,i,omegah(i),omegal(i),(numerator/denominator)-1.
   &,prod
      enddo
      stop
      end
```

Running the lists in Table 5.1 with the code given earlier, we obtain $\beta_{CO_2} = 1.198533$ and $\beta_{CH_4} = 1.118065$. The equilibrium constant for the reaction is then $1.198533/1.118065 = 1.071971$, which would correspond to an enrichment of ^{13}C in CO_2 of approximately 69.5‰ at 25°C. This is an equilibrium value and says nothing about the time scale to reach equilibrium, which may be very long at 25°C.

These are the main points on the subject of equilibrium isotope distributions in the harmonic approximation. Some qualitative lessons from the discussion need to be highlighted.

5.4   Qualitative Insights

Short bonds tend to have higher vibrational frequencies than long bonds. This is known as Badger's rule (Badger, 1934). Especially when considering a specific type of bond, for example, between iron and oxygen, we expect that shorter iron–oxygen bonds will be stiffer than longer iron–oxygen bonds. Any systematic chemical environmental change that affects the bond length will likely be manifested in a change in isotope fractionation. For example, any ionic compound in which Fe is in the Fe(III) oxidation state will tend to accumulate heavy iron relative to a compound in which iron is in the Fe(II) oxidation state. This is because the Fe(III)–ligand bond lengths are shorter than the Fe(II)–ligand bond lengths and therefore their vibrational frequencies are higher, and therefore it is somewhat more favorable to put heavy isotopes in the Fe(III)–ligand bonds. Elements with a lower coordination number have shorter bonds than elements with a higher coordination number, so that we can expect, for example, that silicon isotopes in enstatite, with fourfold coordinated silicon, should tend to be isotopically heavy relative to majorite garnet, with sixfold coordinated silicon.

For open-shell transition metals with multiple unpaired electrons, elements in the low-spin state will tend to have shorter, stronger bonds (if the electrons pair up, they need less room) and therefore tend to accumulate heavier isotopes than elements in high-spin states.

Besides offering quantitative estimates of such effects, electronic structure calculations have identified some interesting insights into exceptions to these generalizations. For example, take the fractionation of iron isotopes between ferropericlase $\left(Mg_{(1-x)}Fe_xO\right)$ and ferroperovskite $\left(Mg_{(1-x)}\right.$ $\left.Fe_xSiO_3\right)$ in the Earth's mantle. By the rule of thumb that the heavy isotopes fractionate into the environments with the lowest coordination numbers, it would be expected that the 6-fold coordinated iron in ferropericlase would be isotopically heavier than the 12-fold coordinated iron in ferroperovskite. In fact, DFT calculations predict the opposite. Because of its simple chemistry, the cumulative contributions to the RPFR of ferropericlase are finished by approximately $800\,cm^{-1}$ (i.e., there are no frequencies above $800\,cm^{-1}$ involving motion of iron). For ferroperovskite, on the other hand, contributions to the RPFR from coupling of iron motion to the Si—O stretching frequencies continue to be made through $900\,cm^{-1}$. These are enough to drive the RPFR for ferroperovskite above ferropericlase, despite the 6-fold versus 12-fold coordination environment. In this case, chemical composition is more important than coordination number in determining the final RPFR (see Rustad and Yin, 2009).

The ferropericlase/ferroperovskite system is also interesting from the point of view of the effect of spin state on the RPFR. For ferropericlase the room-temperature RPFR between high-spin and low-spin electronic states is about 7.6‰, with the low-spin state having the shorter bond and the larger RPFR (and, hence, a stronger tendency to be enriched in heavy iron). For ferroperovskite the difference between the RPFR for the high-spin state and the RPFR for the low-spin state is much smaller. This unexpected behavior arises from the asymmetry of the Fe—O bonds in the low-spin coordination environment. While in ferropericlase all Fe—O bonds become significantly shorter after the spin transition, in ferroperovskite the low-spin ferrous iron is too small for the 12-fold coordination environment and sits asymmetrically in the 12-fold coordinated perovskite "B" site with some short bonds and some long bonds. These tend to compensate one another in their contributions to the RPFR so that the spin transition has little effect (Rustad and Yin, 2009).

Along these lines, we have talked loosely about the differences in "stiffness" in the environment and made the observation that we should expect to find the heaviest isotopes preferentially in the stiffest bonding environments. But this is an ambiguous statement in some ways. One needs to be careful not to define "stiff environment" in terms of, for example, shear modulus. Consider the isotopic composition of magnesium dissolved in calcite in equilibrium with an aqueous solution. If asked about the distribution of ^{26}Mg and ^{24}Mg in these environments, one might, from the foregoing discussion, answer that the ^{26}Mg should be concentrated in the calcite, as the calcite environment is obviously "stiffer" than water. But what matters here is the frequencies experienced by the magnesium in each of the environments, not the fact that the shear modulus of water is zero while the shear modulus of calcite is not. It turns out, in fact, that heavy magnesium accumulates in the aqueous solution. Analogous to the ferropericlase/ferroperovskite example given previously, this is mostly because of high-frequency vibrations caused by magnesium coupling with wagging motions of water molecules (Rustad et al., 2010a). This is predicted by first-principles calculations, but it is counterintuitive, if one uses a definition of stiffness that is too rigid.

5.5   Quantitative Estimates

There is no doubt that through detailed calculations we have some better qualitative insights into the factors that influence these familiar rules of thumb in describing equilibrium isotope fractionation in different chemical environments. These types of insights are largely independent of the details of the calculations, that is, choices we make for (a–c) of Section 5.1.

If an accurate prediction is sought, then conclusions drawn will usually depend on the choices we make for (a–c) of Section 5.1. In other words, rather than asking, "Is Ca^{2+} (aq) enriched or depleted in heavy calcium relative to calcite?" one is interested in the question, "Is the $^{44}Ca - {}^{40}Ca$ separation factor between Ca^{2+}(aq) and calcite 1‰ or is it closer to 2‰?" To address quantitative problems, at least at present, effective computational chemistry depends on one's ability to choose problems that naturally result in a high degree of error cancellation. Consider the isotopic fractionation between Fe^{3+} and Fe^{2+} in aqueous solution. This is a great example of a problem with a large signal that does not depend very much on the basis set used, the model for the electron–electron interactions, or the particular molecular model used to represent the systems ($Fe(H_2O)_6^{3+/2+}$, that is, an iron ion with six water molecules around it, works fine). The reason is that beyond the first solvation shell, the environments of Fe^{2+} and Fe^{3+} are very similar. Although there is a significant change in the RPFR resulting from improving the molecular model from $Fe(H_2O)_6^{2+}$ to $Fe(H_2O)_6.12H_2O^{2+}$ (i.e., including a partial second solvation shell), there is also a very similar change in going from $Fe(H_2O)_6^{3+}$ to $Fe(H_2O)_6.12H_2O^{3+}$, and this part of the RPFR cancels in considering fractionation between the two environments.

A more difficult problem is the fractionation between, say, Fe^{2+}(aq) and hematite, α-Fe_2O_3. These environments are quite different from one another, so that one cannot rely on error cancellation in the representation of the environment. A basis set/environment combination that worked well for the Fe^{2+}/Fe^{3+} fractionation might not work well at all for this problem; one has to get closer to the "true" RPFR characteristic of the environment of interest. In other words, if there is a significant effect on the absolute RPFR by including a partial second solvation shell for Fe^{2+}(aq), this will not be compensated by a corresponding effect in Fe_2O_3, as that environment is different from the aqueous environment. In this case, it turned out that a second solvation shell, plus a continuum representation of the rest of the solvation environment, was required to get an accurate value for the individual Fe^{2+}(aq) RPFR and, hence, for the aqueous/mineral fractionation (Rustad et al., 2010b).

Addressing quantitative problems requires that due attention be paid to (a–c). The basis set needs to be large enough so that the answer does not change significantly upon making it a little larger but not so large that it makes the problem computationally intractable. Generally the investigator must experiment to see how the results depend on the basis set, picking the best affordable one and then backing down a bit to a smaller set of functions and see how the results change. Or another approach is to pick a very small system, yet still meaningful, and use that system as a benchmark. There are many kinds of basis sets, including localized functions such as Gaussians (Dunning, 1989) or Slater-type functions, as well as plane waves (Troullier and Martins, 1991). To use plane waves, the core region must be taken care of separately either with an auxiliary localized basis set (the so-called linearized augmented plane wave or LAPW method as used, e.g., in WIEN2K (Schwarz and Blaha, 2003)) or through a pseudopotential representing the repulsion of the core electrons or by smoothing the plane waves near the core in a prescribed way (this is the so-called projector augmented wave (PAW) method (Bloechl, 1992)).

Different environments will in general have different basis set requirements. For example, when looking at isotope effects arising from isotope variations in nuclear shape (Schauble, 2007), it might be advantageous to use Slater-type functions rather than Gaussian functions as these are a more faithful representation of the electron density at the core, which is where the action is in this type of problem. It is also a good idea to keep system-dependent basis set convergence issues in mind when trying to manage error cancellation. For example, the Mg^{2+}(aq) requires a more complete basis set than the vibrational spectrum of Mg^{2+} in $MgCO_3$. This is because the aquo ion includes important contributions from the relatively weak interactions between water molecules, and large basis sets with diffuse functions are required to get a converged result, even if one is using DFT and not describing the interactions accurately. Consider a problem involving fractionation between an aquo ion and a mineral. The aquo ion environment is run with a small basis set A and a large basis set B. The cluster representing the mineral environment is larger, however, and can only be run with basis set A. Should you estimate the fractionation by running basis A on both the mineral cluster and the aquo ion cluster? At first, you might think that better error cancellation could be obtained this way, but what ends up happening is that the mineral RFPR would hardly change at all in going from basis A to basis B, but the aquo ion would have a relatively large change. So a better estimate is actually obtained by running the aquo ion with the large basis set and the mineral cluster with the small basis set. Of course, the best approach would be to run them both with basis set B, but the B-aquo ion/A mineral combination will be closer to the B-aquo ion/B mineral combination than will the A-aquo ion/A mineral combination (Rustad et al., 2010b).

Concerning the model for the electron–electron interactions: for most problems of interest in geochemistry, we live in the world of DFT. The choice of the exchange–correlation potential representing the electron–electron interactions is much like the basis set, but worse, because, while we know that a larger basis set will always give a truer picture of theoretical predictions, the list of DFT exchange–correlation functionals is always growing. Many of these are highly specialized, designed to offer improvements only in small areas (i.e., the calculation of reaction barriers) without any systematic improvement over the more standard functionals in other respects. Further, the world of geochemistry is filled with water, a solvent influenced by weak interactions, such as dispersion and hydrogen bonding, which are difficult to describe using DFT.

While, in geochemical problems, we mostly work with systems sufficiently large that we have to use DFT, it's a good idea to check the DFT results with MO calculations on small systems. For example, one finds that one cannot afford to run accurate MO methods (large basis MP2 or CCSD(T)) for the $Mg(H_2O)_6.12H_2O^{2+}$ model that was put together to represent the aquo ion nor the $Mg(CO_3)_6.18Ca$ embedded cluster that was put together to represent the Mg^{2+}-bearing calcite. On the other hand, these methods might be feasible for both $Mg(H_2O)_6^{2+}$ and the $Mg-CO_3$ dimer. The B3LYP/6-311++G(2d,2p) calculation that was being contemplated for the large system can now be checked against the MP2/aug-cc-pVTZ calculation on the small system. Say, for example, that check shows that, for the small system, the fractionation factor calculated using DFT is 3.2‰ higher than the fractionation factor calculated with the MO method. First, this indicates immediately that the DFT calculation for the large system is not going to give accuracy better than 3.2‰. However, given a known offset between the DFT result and the MO result on a very similar but much smaller system, a reasonable person might accept a correction of 3.2‰ to the fractionation factor calculated with DFT on the large system as an estimate of a value that might be obtained if accurate MO calculations were possible on the large system. Putting together an effective ladder of successively cross-checked molecular models often gives good insights into the likely accuracy of the calculations and allows informed estimates to be made about the likely magnitude and direction of corrections that would be expected at higher levels of theory. Putting dispersion/van der Waals interactions into DFT in the most consistent and general way is an active area of research, likely to have an important impact on geochemical applications (see, e.g., Fornaro et al., 2014).

Finally, choosing where to put all the atoms to best represent the system of interest is a crucial task. This is in some ways the heart of what is called "molecular modeling." In gas-phase molecules and bulk solids (with sufficiently small unit cells), the representations are straightforward. Aqueous environments, however, are dynamic and may need to be represented by more than one configuration. Surface environments such as surface metal centers or sorbed surface complexes are rarely known with any certainty. Often molecular simulation methods such as molecular dynamics or Monte Carlo methods can be used to generate such configurations. Careful, imaginative construction of representative environments can be one of the most difficult (and most fun!) aspects of doing electronic structure calculations for complex problems involving interfaces or solvated species, where precise configurations are not known.

In my own work, I have tended to prefer to represent exchange sites in minerals through embedded clusters rather than as periodic solids. In part, this is because much of that work was focused on isotope fractionation between aquo ions and minerals, and I wanted to use the same methods on both the aquo ion and the mineral system to achieve the best possible level of error cancellation. The molecules-as-minerals idea goes back to Gibbs' 1982 paper (Gibbs, 1982). Then, as now, the motivation for this is that much more flexible computational methods are available for clusters than for periodic solids. If one insists on representing sites in minerals as periodic solids, using anything other than pure DFT functionals is very expensive. MO methods beyond Hartree–Fock, such as MP2 or CCSD(T), have not been implemented in accessible solid-state codes. True, there are the DFT+U (the so-called Hubbard model where an extra repulsive term is added to, e.g., d electrons in a transition metal atom which "see" each other much more explicitly than could be captured in an average, mean field theory; Anisimov et al., 1997), but such methods are mainly used to improve band gaps rather than achieve better general model chemistries.

Setting up clusters to represent crystalline and surface environments is straightforward with the right molecular modeling tools (a useful code for this is CrystalMaker™). As an example, let's set up a calculation for modeling a CO_3 group in calcite. Since there is only a single type of CO_3 group, we can imagine taking a central carbonate molecule, all eight cations attached to this central group, and then the CO_3 ions attached to those eight (other than the central carbonate). The bonds from the outer shell of CO_3 ions to the outer Ca^{2+} ions are replaced by terminating nuclear centers (sometimes called "link atoms") with a +2/6 charge to match the Pauling bond strength (charge/coordination number) coming into the oxygen atoms on the carbonate from these outer Ca^{2+} ions. The central carbonate group, the Ca^{2+} ions attached to it, and all oxygens attached to the Ca^{2+} ions are chosen as vibrationally active, and the rest of the atoms are fixed in their measured positions. The central $CO_3Ca_8O_{33}$ molecule in Figure 5.1 then vibrates within the fixed outer rind. In this way, a "molecule" has been created which ought to have fractionation characteristics very similar to the periodic solid from which the molecule was created. Within this molecule it is straightforward to use hybrid functionals or to put an aug-cc-pVTZ basis on the central carbonate. In principle, one can even treat the central carbonate unit at the MP2 level embedded within a DFT treatment of the rest of the cluster (Tuma and Sauer, 2006).

For systems investigated so far (Fe_2O_3 (Rustad and Dixon, 2009), $CaCO_3$ (Rustad et al., 2008), $Mg_{1-x}Fe_xO$ and $Mg_{1-x}Fe_xSiO_3$ (Rustad and Yin, 2009)), the embedded cluster type of approach gives excellent agreement with full lattice dynamics treatments. One issue with the use of clusters is that one has to know the mineral structure. This might be a problem if calculations are to be done at high pressure and the lattice parameters are not known. Another problem is the inconsistency in plugging the experimentally determined structure into, for example, a DFT method that wouldn't recover exactly the measured structure. DFT in the GGA tends to overestimate bond lengths and lattice parameters can be overestimated by 1–2%. A possible work-around would be to optimize the crystal structure with DFT and then construct the cluster from the DFT-optimized structure. On the other hand, it's usually a good idea to use/impose the experimental structural information if one has access to it. This is a good strategy even for highly accurate calculations on small molecules (Helgaker et al., 2008).

The requirements for clusters representing aquo ions depend strongly on whether the aquo ion is a cation or an anion. Cations usually can be represented by symmetric clusters with an explicit first and (partial) second solvation shell that are then embedded in a continuum representation of the rest of the solvent (see Figure 5.2). First-and-partial-second-shell solvent representations for cations are available in the literature for cations in tetrahedral and octahedral coordination. Anions require many explicit water molecules, at least thirty, and, in my experience, do not respond predictably to embedding in a continuum solvent. (Note that the cluster representing HCO_3^- (aq) in Figure 5.1 is only the core of a much larger cluster and would not, by itself, be sufficient for accurate computation of the fractionation factor for $^{12}C - ^{13}C$ exchange between the bicarbonate aquo ion and calcite.) Because there are no "standard" structures available for clusters of this size, the only way to generate them is through molecular dynamics or Monte Carlo methods. If there are no interaction potentials available, then one has to perform ab initio molecular dynamics studies to generate a sufficient number of representative conformers (10–20).

Surface environments may also be of interest. Although this area has not been extensively investigated, experience so far suggests that surface effects are small. For example, Rustad and Dixon (2009) looked at the RPFR for iron isotope exchange in hematite in the bulk as well as at the hematite (012) surface with both molecular and dissociative water adsorption and saw, surprisingly, almost no difference at all between the bulk RPFR and the RPFR values of both types of surface environments. More studies will be required before any general conclusions about surface isotope effects can be drawn. Of course, surface environments can have higher concentrations of elements in different oxidation states, for example, iron atoms at iron oxide surfaces may be partially reduced (Henderson et al., 1998; Williams and Scherer, 2004), and this will certainly have consequences for the isotopic composition of the surface iron atoms (Frierdich et al., 2014; Handler et al., 2009).

A major problem in application of electronic structure methods to isotope fractionation problems today is that variations in basis sets, electron–electron interaction models, and environmental representations can easily result in a random walk in combinations of these factors. In particular, reasonable choices for each of these factors can cause 1–2‰ variations in positive and negative directions,

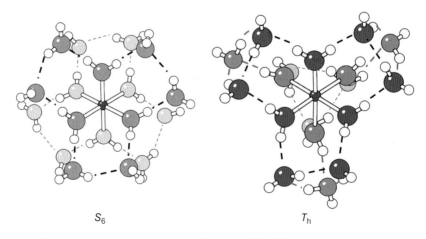

S_6 T_h

Figure 5.2   *Two different cluster representations $M(H_2O)_6 \cdot 12H_2O^{n+}$ for octahedrally coordinated aquo ions with explicit first and second shell water molecules.*

and it is often possible to find some combination that gives an "expected" answer. Individual investigators no doubt can recognize this sort of pathology—no conscientious investigator would try nine combinations of three basis sets and three exchange–correlation functionals, find the one that works, and publish the "successful" calculation without mentioning the failure of the others. If the random walk is carried out over multiple research groups, however, the combination that agrees with "experiment" is the one that tends to get published and the end result is the same as if a single investigator "cherry-picked" a certain combination without telling the scientific community about the other failures. Nothing is really learned this way, unless one can step back and find some combination that tends to work well in a variety of situations. At this point in time, even in relatively well-defined chemical environments, such as ions in solution, at aqueous–mineral interfaces, in oxide, silicate, and carbonate minerals, there seems to be no prescription for success in terms of exchange–correlation functional choices. Thus for any given problem, it is best to try a range of basis sets and exchange–correlation functionals and structural environments to generate some reasonable range of predictions before coming to any strong conclusions.

To help illustrate these issues, consider Figure 5.3 showing the relative error in the computed equilibrium constants for isotope exchange reactions involving small molecules, originally studied by Richet et al. (1977) over a wide range of DFT exchange–correlation functionals and basis sets. Clearly there are certain basis sets (cc-pVDZ) and exchange–correlation functionals (m06-hf) that tend to perform worse than others overall. However, the figure also shows that the performance is highly element specific, with the (aug)cc-pVDZ basis set being notably worse than other basis sets for oxygen exchange reactions. The DZVP(2) family of basis sets (Godbout et al., 1992), which were constructed specifically to be used in DFT calculations in the early days of application of DFT to molecules, do, in fact, appear to perform as well as the more expensive triple zeta basis sets, overall, but again, this is element specific, with the DZVP(2) family performing less well for oxygen exchange reactions. For carbon, LDA seems to perform very well with a wide range of basis sets.

It should be kept in mind that these reactions never involve hydrogen-bonded systems with weak interactions and are thus not representative of most of the reactions that would really be of interest in low-temperature geochemistry. For example, the HCTH-407 functional (Boese and Handy, 2001) is one of the better-performing DFT functionals in Figure 5.3 but does not do nearly as well when hydrogen-bonded systems are considered. Nevertheless, Figure 5.3 drives home the point that there are, at this point in time, no "magic" DFT functionals that work particularly well for computing isotope exchange equilibria over a broad range of chemical systems.

Example: Calculating the ^{11}B–^{10}B isotope fractionation factor for $B(OH)_3(aq)$ and $B(OH)_4^-(aq)$
The fractionation of ^{11}B and ^{10}B between $B(OH)_3$ (aq) and $B(OH)_4^-$ (aq) has been used in paleoclimate studies to estimate the pH of the oceans on time scales of 20 million year (Hemming and Hanson, 1992). The main assumption is that marine carbonates only incorporate $B(OH)_4^-$ and not $B(OH)_3$ (although it seems less and less likely the more is known about carbonate crystal growth mechanisms; see Demichelis et al., 2011). To make the points made in the previous discussion more concrete and to show the utility of making a ladder of multiple techniques and system size, consider the calculation of the ^{11}B $-$ ^{10}B fractionation between $B(OH)_3$ (aq) and $B(OH)_4^-$ (aq):

$$^{10}B(OH)_3(aq) + {}^{11}B(OH)_4^-(aq) = {}^{11}B(OH)_3(aq) + {}^{10}B(OH)_4^- \qquad (5.7)$$

as studied by Rustad et al. (2010a). Figure 5.4 shows the equilibrium constant for the reaction as a function of the number of solvating water molecules. For the 32-water case both $B(OH)_3$ and $B(OH)_4^-$ have 32 waters, taken from 10 independent configurations from an ab initio molecular dynamics simulation (AIMD) of $B(OH)_3$ (aq) and $B(OH)_4^-$ (aq). For the intermediate points

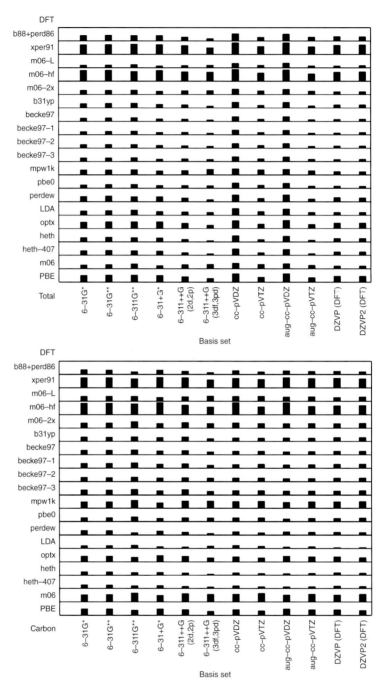

Figure 5.3 *Error over small molecules in Richet et al. (5.3). Taken from P. Zarzycki and J. R. Rustad (A review of hydrogen, carbon, nitrogen, oxygen, sulphur, and chlorine stable isotope fractionation among gaseous molecules: a quantum chemical study, unpublished).*

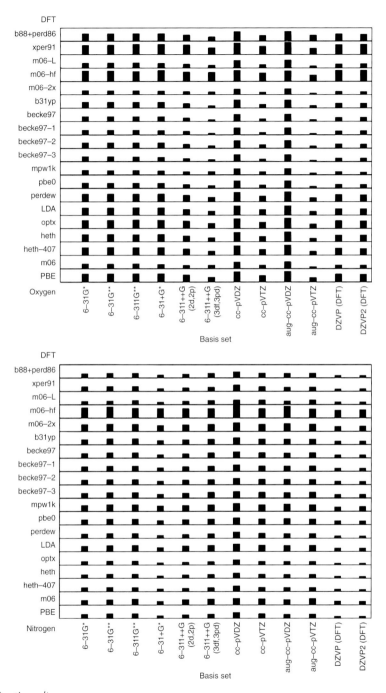

Figure 5.3 *(Continued)*

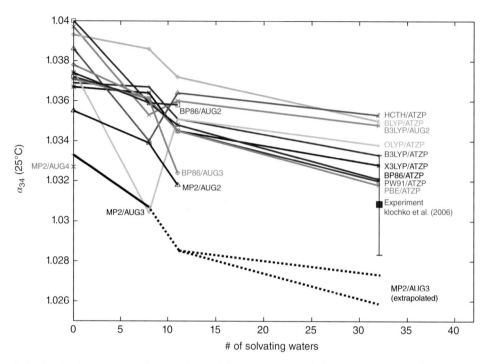

Figure 5.4 *Equilibrium constant for Reaction 5.3 for various model chemistries as a function of the number of solvating waters. Dashed lines are extrapolations.*

$B(OH)_3$ has six solvation waters because this was the hydration number determined in the AIMD calculations. Two systems are used for $B(OH)_4^-$, one having 8 waters and another having 11 waters. The $B(OH)_4 \cdot 11H_2O^-$ cluster was chosen because AIMD simulations of the borate ion in solution give a second solvation shell of approximately 11 waters. $B(OH)_4 \cdot 8H_2O^-$ was chosen because of its high symmetry (even $B(OH)_4 \cdot 11H_2O^-$ with its C_1 symmetry was too large for MP2/aug-cc-pVTZ calculation in 2008–2009 when this work was done). The conclusions that can be drawn from Figure 5.3 are extremely powerful. Based on the smaller clusters using MP2/aug-cc-pV(D,T,Q)Z, one arrives at a pretty convincing estimate of the MP2/aug-cc-pVTZ calculation for the $B(OH)_4 \cdot 11H_2O^-$ system, even though the calculation could not actually be carried out. Further, the almost uniform decrease in going from $B(OH)_4 \cdot 11H_2O^-$ to $B(OH)_4 \cdot 32H_2O^-$ using DFT gives a fair level of confidence about what would be found had it been possible to do the MP2/aug-cc-pVTZ calculations on the $B(OH)_3 \cdot 32H_2O$ and $B(OH)_4 \cdot 32H_2O^-$ clusters.

As far as the DFT calculations go, based on the BP86/AUG2 and BP86/AUG3 entries, one can anticipate a drop of 3–4‰ in going from the aug-cc-pVDZ to aug-cc-pVTZ basis. This would happen to bring the B3LYP/aug-cc-pVTZ estimate very close to the experimental value. So while it looks like the hybrid B3LYP functional is not as close to experiment as the pure functionals such as PBE, PW91, and BP86, part of the reason for that is that the 6-311+G** basis does not give converged results. In a sense, we made up for using a less reliable functional by also using a relatively poor basis set. This no doubt sounds like a lot of speculation, but this kind of activity is essential for getting some perspective on the calculations. Another result that can be anticipated is that after improving the basis sets, the B3LYP/aug-cc-pVTZ calculations are likely to be in better agreement with experiment than the MP2/aug-cc-pVTZ calculations. Because it is generally known that the

MP2/aug-cc-pVTZ calculations will be much more reliable in general for hydrogen-bonded systems, the apparently better agreement for the B3LYP calculations is an alert that there is likely something else going on that is not accounted for, such as anharmonicity. Thus, it might be found that a hybrid DFT functional with a converged basis set gives essentially the same answer as an MP2 calculation with a converged basis set plus an anharmonic correction. The bottom line is that it's easy to stumble around with these kinds of small variations, occasionally finding the experimental value at various intermediate points, and then claiming success only to find that the same combination fails on a different system. "Right for the right reason" is a common phrase in computational chemistry, and it's a good idea to keep this in mind when calculating isotope fractionation factors. The disconcerting conclusion from Figure 5.4 is that MP2/aug-cc-pVTZ, with at least 32 water molecules, is clearly what is needed, in terms of choices of model chemistries, yet these calculations are not going to be available for some time.

Again, it is important to remember that all these considerations apply to doing calculations at near chemical accuracy. The calculations of Liu and Tossell (2005) and Zeebe (2005) played a huge role in providing motivation to go back and look at the 1.19 value for the equilibrium constant of Reaction 5.7, as taken from Kotaka and Kakihana (1977) (that turned out to have the mode assignments wrong for $B(OH)_4^-$), and in providing the motivation for the key experiments of Byrne et al. (2006), which, at long last, provided a correct value as a foundation for paleoclimate studies. To my mind this is one of the premier successes of quantum chemistry in geology. Here we are actually approaching what I would call "molecular geology," as opposed to "molecular geochemistry." The latter implies the study of molecular-level processes in geochemistry, which may be interesting, but are usually far removed from saying very much about how we interpret the rock record. The term "molecular geology" is much more powerful and implies that these kinds of studies actually made a difference in how we interpret Earth history. The fact that, through the boron proxy, the community misinterpreted 20 million years of Earth history because of a wrongly assigned vibrational mode in $B(OH)_4^-$ in the original work of Kotaka and Kakihana (1977) is a cogent illustration of the "house of cards" that is built up in these kinds of efforts. These investigators, who were working in the nuclear industry, never thought, of course, that their model would be used in this way.

5.6 Relationship to Empirical Estimates

At this point, the reader may be wondering why one does not use experimentally determined frequencies in systems for which those are available. The problem with this approach is that calculated harmonic frequencies are often compared to measured anharmonic frequencies without a harmonic correction. Unless the necessary measurements to extract actual harmonic frequencies from the experimental measurements have been done (and this is extremely rare for any system of geochemical interest), we cannot learn anything quantitative by making comparisons with anharmonic spectra.

A great example is the work of Deines (2004) on the carbonate system. This paper presents a painstaking review of the literature, with the author critically compiling vibrational frequencies on a series of carbonate minerals with both the aragonite and calcite structures from multiple sources, and then using empirical force fields to estimate the isotopic substitution-induced shift for each of the frequencies, also including contributions from acoustic vibrations and external vibrations. This all must have taken years, and since the error in the measured vibrational frequencies is much smaller than the intrinsic error in the calculated frequencies, one might at first think that this is maybe a superior approach to first-principles calculations. The problem with this approach is that it is not a good idea to plug anharmonic frequencies in a harmonic partition function. Comparing

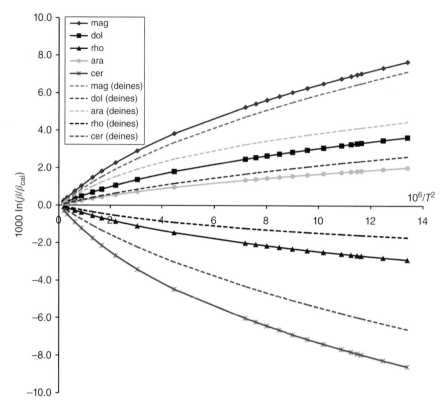

Figure 5.5   *$^{13}C/^{12}C$ fractionation relative to calcite for a series of carbonate minerals. Solid lines with markers are calculated from first-principles, and dashed lines are estimated empirically. ara, aragonite; cer, cerussite; dol, dolomite; mag, magnesite; rho, rhodochrosite. Taken from Deines (2004). Reproduced with permission of Elsevier.*

Deines' estimates to estimates from first-principles calculations in Figure 5.5, it is apparent that while there is overall reasonable qualitative agreement, there are important quantitative differences. Where there are data, such as for calcite–CO_2 and aragonite–calcite, the first-principles calculations are in better agreement with experiment. In the end, the first-principles estimates for the $^{12}C - ^{13}C$ carbonate–calcite fractionation factors are probably more reliable, even though they took much less time and effort to make and even though there are substantial disagreements between the measured anharmonic vibrational frequencies and the calculated harmonic frequencies.

Rustad and Bylaska (2007) initially tried such an approach from a computational point of view in the borate–boric acid system, using classical molecular dynamics to calculate the vibrational spectrum of these species in solution (by Fourier transforming the velocity autocorrelation function), running separate simulations of $^{10}B(OH)_3$ (aq), $^{11}B(OH)_3$ (aq), $^{11}B(OH)_4^-$ (aq), and $^{11}B(OH)_4^-$ (aq). The idea was that the simulations would reveal the vibrational spectrum very clearly, make vibrational mode assignments to determine the vibrational multiplicity, and calculate the harmonic partition function from these assignments. The advantage over other first-principles estimates for the boron isotope fractionation factor (such as carried out Liu and Tossell, 2005) would be that the results would sample many configurations for each aquo ion, not just one, as had been done in previous work. The work was done extremely carefully with sophisticated Monte Carlo uncertainty estimates, and the authors thought the results were going to be revolutionary. Instead, they were ridiculous. The fractionation factor went in the opposite direction, with $B(OH)_4^-$ being enriched in the heavy isotope. When the authors tried extracting a single configuration from the molecular dynamics simulation, optimized it in periodic boundary conditions, and calculated the frequencies as outlined previously in the harmonic approximation, a value of 1.028 was obtained with the PBE functional and a plane-wave basis with a 90 hartree cutoff, agreeing almost perfectly with the Hartree–Fock 6-31G* calculation of Liu and Tossell (2005). Figure 5.4 shows that this level of agreement was to some extent fortuitous; however, it was deflating to go through all that work only to find that all the fancy techniques were superfluous and, in the end, reproduce almost exactly the very calculation on which we were trying to improve. But the study served as an illustrative lesson reinforcing how important it is to use harmonic frequencies in a harmonic partition function. Looking at the real vibrational density of states in aqueous solution was, however, instrumental in discovering the misassigned vibrational modes in the Kotaka and Kakihana (1977) model that helped dislodge the erroneous 1.19 that was latched onto by the boron pH proxy community, so the effort was not totally fruitless.

The issue of harmonic frequencies versus measured frequencies brings up an additional point about the practice of scaling calculated frequencies to better match measured frequencies. It has long been realized that there are systematic errors in calculated frequencies and there has been much work done on finding ways to correct these errors through the use of scaling factors (Scott and Radom, 1996). In doing this for isotope exchange equilibria, one has to be sure to use scaling factors that have been designed to recover harmonic frequencies or better zero-point energies. It is probably not a good idea to, for example, find one's own scaling factors by correcting calculated harmonic frequencies to measured anharmonic frequencies and then using the "corrected" frequencies to calculate the harmonic partition function.

5.7   Beyond the Harmonic Approximation

The next big step in the ab initio computation of isotope fractionation factors is to incorporate anharmonicity. This is especially necessary for hydrogen isotopes where the predictions from harmonic theory are not very useful. One way to go beyond the harmonic approximation is to use path integral molecular dynamics (PIMD) methods to treat the quantum aspects of the dynamics. In AIMD, the forces are calculated from quantum mechanics, but the nuclei move classically in response to these forces. There are no isotope effects in classical AIMD simulations at equilibrium, as the equilibrium fractionation is strictly a quantum mechanical effect. To represent quantum delocalization, PIMD treats each nucleus as many quantum replicas connected by effective temperature- and mass-dependent springs, each of the replicas behaving as classical particles. So instead of just one, for example, H atom, one has, for example, 10 or 20 replicas all slightly displaced from one another. Quantum dynamics is the whole point of PIMD, and in these calculations the mass of the exchanging atom can be gradually perturbed from one isotope into another and free energy of this imposed perturbation can be evaluated from what are now standard molecular dynamics free energy techniques (of course, one could do this also in a classical system, gradually change an H atom into a D atom, but the energy change would be zero). This is a highly elegant and general approach, free from any assumptions about harmonicity and free of the myriad complexities of explicit anharmonic vibrational corrections, but is only now beginning to be applied to the simplest systems (Webb and Miller, 2014). While the initial results are extremely promising, time will tell whether the pitfalls associated with achieving equilibrium and statistical convergence of the free energy evaluation would ever make this a generally feasible approach.

Another way to account for anharmonicity is to make anharmonic corrections to the zero-point energies. These are very complicated but have been coded up as a "black box" add-on to the Gaussian suite of electronic structure programs. A good discussion of these methods from a geochemical perspective can be found in Liu et al. (2010).

About the only thing that can be said for certain is that one definitely must include anharmonic effects for hydrogen–deuterium isotope fractionation. For heavier isotopes, it is not possible at this time to make any general statements. It seems likely that the treatment of anharmonicity will end up being another factor in the basis set, electron–electron model, and environmental representation grid that has to be systematically explored, on a variety of system sizes, such as summarized in Figure 5.4, to draw reliable quantitative estimates from first-principles calculations.

5.8 Kinetic Isotope Effects

Kinetic isotope effects have also been widely studied with computational methods. In the harmonic approximation, no really new principles are involved. The heavy isotopes bury themselves more deeply in the potential well associated with the reactants and are thus harder to lift out and slower to react. Of course, they also bury themselves more deeply into the transition state; however, in the transition state the bonding is much weaker and the effect is much less, and so the net result is that the heavy isotopes tend to react more slowly. The barrier heights may be more challenging to predict and the transition states more challenging to locate.

In addition, there is an effect that comes about because the efficiency of the transmission across the barrier is, in principle, dependent on mass. For example, if we somehow pushed the whole system slowly up to the top of the barrier from the reactant side and then just stopped, let go, and observed which direction it would fall, the probability that it would ultimately fall to the product side of the reaction depends not only on how the system is pushed to the top but also depends, in general, on the masses of all the atoms in the system. Thus, there is an isotope effect associated with the probability of actually crossing the barrier, as well as an isotope effect on the barrier height itself. This concept was recently applied by Hofmann et al. (2012) to look at isotope effects on water exchange kinetics. In a somewhat similar vein, Kavner et al. (2008) have worked out a generalization of the Marcus theory of electron transfer to include isotope fractionation driven by redox reactions. Such process as mass dependence of barrier transmission can in principle be simulated directly using quantum dynamics methods such as PIMD discussed previously. Certainly it will be exciting to see what kinds of information can be obtained from such simulations.

5.9 Summary and Prognosis

In a work such as this, one feels compelled to make some sort of summary statement, which usually includes some view of what the future of the subject at hand might look like. One could say overall that first-principles calculations have given the community fast access to reasonably accurate models that, in the past, had to be constructed by hand in a relatively labor-intensive process that required a great deal of ingenuity on the part of the investigator (i.e., Deines, 2004; Richet et al., 1977). More importantly, the calculations allow modeling of arbitrary environments, such as surfaces, for which it would have been almost impossible to build empirical force fields from vibrational spectroscopy. Even aqueous environments are too difficult for the construction of reliable empirical models, as shown in the boron isotope example discussed previously. Moreover, the first-principles calculations often (though not always; see Richet et al., 1977) improve significantly on the empirical approach by giving harmonic vibrational frequencies directly, without requiring extraction of harmonic frequencies from anharmonic measurements.

It is clear, however, that the field has to evolve before the prediction of equilibrium fractionation factors to 0.1–1‰ accuracy can be achieved for general systems (as shown in Figure 5.4). Given the inherent difficulties in applying quantitative ab initio calculations (which grow as at least the fifth power of system size) to complex geochemical systems, it seems that improvements will come from DFT. Computational geochemists will pay close attention to the research going on in the area of improving DFT calculations on systems with hydrogen bonding and dispersion interactions, as these are likely to improve isotope fractionation calculations for mineral/aqueous systems. In addition, one could imagine a concerted program to invent a new DFT functional that was specifically constructed for the purpose of accurate calculation of isotope fractionation factors in geochemical systems. Such an effort would naturally have to be restricted, for example, to oxide-water systems (oxide here being broadly defined to include oxides, silicates, carbonates, sulfates, phosphates).

First-principles calculations have already helped advance the science of isotope geochemistry, having played an important role in helping formulate the field of clumped isotopes (Eiler, 2007; Schauble et al., 2006) as well as elucidating the role of nuclear volume effects in heavy elements (Schauble, 2007). A fascinating step in analytical work is the ability to measure position-specific isotopic compositions in molecules and minerals (Eiler, 2013). Such measurements promise an entirely new window into geochemical processes. Interpretation of position-specific isotope signatures can be greatly aided by calculations (Rustad, 2009; Rustad and Zarzycki, 2008); no doubt we can look to much more work in this area in future years.

Acknowledgments

The author would like to thank Brooke H. Dallas, Jason Boettger, and James D. Kubicki for insightful reviews and suggestions and Piotr Zarzycki for providing Figure 5.3.

References

Anisimov V. I., Aryasetiawan F., and Lichtenstein A. I. (1997) First-principles calculations of the electronic structure and spectra of strongly correlated systems: the LDA+U method. *Journal of Physics: Condensed Matter* **9**, 767–808.

Badger R. M. (1934) A relation between internuclear distances and bond force constants. *Journal of Chemical Physics* **2**, 128–132.

Bartlett R. J. and Musiał M. (2007) Coupled-cluster theory in quantum chemistry. *Reviews of Modern Physics* **79**, 291–352.

Bigeleisen J. and Mayer M. G. (1947) Calculation of equilibrium constants for isotopic exchange reactions. *Journal of Chemical Physics* **15**, 261–267.

Bloechl P. E. (1992) Projector augmented-wave method. *Physical Review B* **50**, 17953–17979.

Boese A. D. and Handy N. C. (2001) A new parametrization of exchange-correlation generalized gradient approximation functionals. *Journal of Chemical Physics* **114**, 5497–5503.

Byrne R. H., Yao W., Klochko K., Tossell J. A., and Kaufman A. J. (2006) Experimental evaluation of the isotopic exchange equilibrium $^{10}B(OH)_3 + {}^{11}B(OH)_4^- = {}^{11}B(OH)_3 + {}^{10}B(OH)_4^-$ in aqueous solution. *Deep Sea Research Part I: Oceanographic Research Papers* **53**, 684–688.

Cramer C. J. (2004) *Essentials of Computational Chemistry: Theories and Models*, 2nd Edition, John Wiley & Sons, Ltd: Chichester, UK, 618 pages.

Deines P. (2004) Carbon isotope effects in carbonate systems. *Geochimica et Cosmochimica Acta* **68**, 2659–2679.

Demichelis R., Raiteri P., Gale J. D., Quigley D., and Gebauer D. (2011) Stable prenucleation mineral clusters are liquid-like ionic polymers. *Nature Communications* **2**, 590.

Dunning T. H., Jr. (1989) Gaussian basis sets for use in correlated molecular calculations. I. The atoms boron through neon and hydrogen. *Journal of Chemical Physics* **90**, 1007–1023.

Eiler J. M. (2007) "Clumped-isotope" geochemistry—the study of naturally-occurring, multiply-substituted isotopologues. *Earth and Planetary Science Letters* **262**, 309–327.

Eiler J. M. (2013) The isotopic anatomies of molecules and minerals. *Annual Review of Earth and Planetary Sciences* **41**, 411–441.

Fornaro T., Biczysko M., Monti S., and Barone V. (2014) Dispersion-corrected DFT approaches for anharmonic vibrational frequency calculations: nucleobases and their dimers. *Physical Chemistry Chemical Physics* **16**, 10112–10128.

Frierdich A. J., Beard B. L., Reddy T. R., Scherer M. M., and Johnson C. M. (2014) Iron isotope fractionation between aqueous Fe(II) and goethite revisited: new insights based on a multi-direction approach to equilibrium and isotopic exchange rate modification. *Geochimica et Cosmochimica Acta* **139**, 383–398.

Frisch M. J., Trucks G. W., Schlegel H. B., Scuseria G. E., Robb M. A., Cheeseman J. R., Montgomery J. A., Jr., Vreven T., Kudin K. N., Burant J. C., Millam J. M., Iyengar S. S., Tomasi J., Barone V., Mennucci B., Cossi M., Scalmani G., Rega N., Petersson G. A., Nakatsuji H., Hada M., Ehara M., Toyota K., Fukuda R., Hasegawa J., Ishida M., Nakajima T., Honda Y., Kitao O., Nakai H., Klene M., Li X., Knox J. E., Hratchian H. P., Cross J. B., Bakken V., Adamo C., Jaramillo J., Gomperts R., Stratmann R. E., Yazyev O., Austin A. J., Cammi R., Pomelli C., Ochterski J. W., Ayala P. Y., Morokuma K., Voth G. A., Salvador P., Dannenberg J. J., Zakrzewski V. G., Dapprich S., Daniels A. D., Strain M. C., Farkas O., Malick D. K., Rabuck A. D., Raghavachari K., Foresman J. B., Ortiz J. V., Cui Q., Baboul A. G., Clifford S., Cioslowski J., Stefanov B. B., Liu G., Liashenko A., Piskorz P., Komaromi I., Martin R. L., Fox D. J., Keith T., Al-Laham M. A., Peng C. Y., Nanayakkara A., Challacombe M., Gill P. M. W., Johnson B., Chen W., Wong M. W., Gonzalez C., and Pople J. A. (2004) *Gaussian 03*, Revision C.02, Gaussian, Inc.: Wallingford, CT.

Gibbs G. V. (1982) Molecules as models for bonding in silicates. *American Mineralogist* **67**, 421–450.

Godbout N., Salahub D. R., Andzelm J., and Wimmer E. (1992) Optimization of Gaussian-type basis sets for local spin-density calculations: 1. Boron through neon, optimization technique and validation. *Canadian Journal of Chemistry* **70**, 560–571.

Handler R. M., Beard B. L., Johnson C. M., and Scherer M. M. (2009) Atom exchange between aqueous Fe(II) and goethite: an Fe isotope tracer study. *Environmental Science and Technology* **43**, 1102–1107.

Helgaker T., Jorgensen P., and Olsen J. (2000) *Molecular Electronic Structure Theory*, John Wiley & Sons, Inc.: New York, 944 pages.

Helgaker T., Klopper W., and Tew D. P. (2008) Quantitative quantum chemistry. *Molecular Physics* **106**, 2107–2143.

Hemming N. G. and Hanson G. N. (1992) Boron isotopic composition and concentration in modern marine carbonates. *Geochimica et Cosmochimica Acta* **56**, 537–543.

Henderson M. A., Joyce S. A., and Rustad J. R. (1998) Interaction of water with the (1×1) and (2×1) surfaces of α-Fe_2O_3 (012). *Surface Science* **417**, 66–81.

Hofmann A. E., Bourg I. C., and DePaolo D. J. (2012) Ion desolvation as a mechanism for kinetic isotope fractionation in aqueous systems. *Proceedings of the National Academy of Sciences of the United States of America* **109**, 18689–18694.

Kavner A., John S. G., Sass S., and Boyle E. A. (2008) Redox-driven stable isotope fractionation in transition metals: application to Zn electroplating. *Geochimica et Cosmochimica Acta* **72**, 1731–1741.

Kendall R. A., Dunning T. H., Jr., and Harrison R. J. (1992) Electron-affinities of the 1st-row atoms revisited—systematic basis-sets and wave-functions. *Journal of Chemical Physics* **96**, 6796–6809.

Kieffer S. W. (1982) Thermodynamics and lattice vibrations of minerals: 5. Applications to phase equilibria, isotopic fractionation, and high-pressure thermodynamic properties. *Reviews of Geophysics and Space Physics* **20**, 827–849.

Klochko K., Kaufman A. J., Yao W. S., Byrne R. H., and Tossell J. A. (2006) Experimental measurement of boron isotope fractionation in seawater. *Earth and Planetary Science Letters* **248** (1–2), 276–285. 10.1016/j.epsl.2006.05.034

Kohn W. and Sham L. J. (1965) Self-consistent equations including exchange and correlation effects. *Physical Review* **140**, A1133–A1138.

Kotaka M. and Kakihana H. (1977) Equilibrium constants for boron isotope-exchange reactions. *Bulletin of the Research Laboratory for Nuclear Reactors* **2**, 1–12.

Liu Y. and Tossell J. A. (2005) Ab initio molecular orbital calculations for boron isotope fractionations on boric acids and borates. *Geochimica et Cosmochimica Acta* **69**, 3995–4006.

Liu Q., Tossell J. A., and Liu Y. (2010) On the proper use of the Bigeleisen–Mayer equation and corrections to it in the calculation of isotopic fractionation equilibrium constants. *Geochimica et Cosmochimica Acta* **74**, 6965–6983.

Møller C. and Plesset M. S. (1934) Note on an approximation treatment for many-electron systems. *Physical Review* **46**, 618–622.

Press W. H., Flannery B. P., and Teukolsky S. A. (1986) *Numerical Recipes*, Cambridge University Press: Cambridge/New York.

Richet P., Bottinga Y., and Javoy M. (1977) A review of hydrogen, carbon, nitrogen, oxygen, sulphur, and chlorine stable isotope fractionation among gaseous molecules. *Annual Review of Earth and Planetary Sciences* **5**, 65–110.

Rustad J. R. (2009) Ab initio calculation of the carbon isotopic signatures of amino acids. *Organic Geochemistry* **40**, 720–723.

Rustad J. R. and Bylaska E. J. (2007) Ab initio calculation of isotopic fractionation in $B(OH)_3$(aq) and $B(OH)_4^-$ (aq). *Journal of the American Chemical Society* **129**, 2222–2223.

Rustad J. R. and Dixon D. A. (2009) Prediction of iron-isotope fractionation between hematite (α-Fe_2O_3) and ferric and ferrous iron in aqueous solution from density functional theory. *Journal of Physical Chemistry A* **113**, 12249–12255.

Rustad J. R. and Yin Q.-Z. (2009) Iron isotope fractionation in the Earth's lower mantle. *Nature Geoscience* **2**, 514–518.

Rustad J. R. and Zarzycki P. (2008) Calculation of site-specific carbon isotope fractionation in pedogenic oxide minerals. *Proceedings of the National Academy of Sciences of the United States of America* **105**, 10297–10301.

Rustad J. R., Bylaska E. J., Jackson V. E., and Dixon D. A. (2010a) Calculation of boron-isotope fractionation between $B(OH)_3$ (aq) and $B(OH)_4^-$ (aq). *Geochimica et Cosmochimica Acta* **74**, 2843–2850.

Rustad J. R., Casey W. H., Yin Q.-Z., Bylaska E. J., Felmy A. R., Bogatko S. A., Jackson V. E., and Dixon D. A. (2010b) Isotopic fractionation of Mg^{2+}, Ca^{2+}, and Fe^{2+} (aq) with carbonate minerals. *Geochimica et Cosmochimica Acta* **74**, 6301–6323.

Rustad J. R., Nelmes S. L., Jackson V. E., and Dixon D. A. (2008) Quantum-chemical calculations of carbon-isotope fractionation in CO_2 (g), aqueous carbonate species and carbonate minerals. *Journal of Physical Chemistry A* **112**, 542–555.

Schauble E. A. (2004) Applying stable isotope fractionation theory to new systems. *Reviews in Mineralogy and Geochemistry* **55**, 65–111.

Schauble E. A. (2007) Role of nuclear volume in driving equilibrium stable isotope fractionation of mercury, thallium, and other very heavy elements. *Geochimica et Cosmochimica Acta* **71**, 2170–2189.

Schauble E. A., Ghosh P., and Eiler J. M. (2006) Preferential formation of bonds in carbonate minerals, estimated using first-principles lattice dynamics. *Geochimica et Cosmochimica Acta* **70**, 2510–2529.

Schwarz K. and Blaha P. (2003) Solid state calculations using WIEN2K. *Computational Materials Science* **28**, 259–273.

Scott A. P. and Radom L. (1996) Harmonic vibrational frequencies: an evaluation of Hartree-Fock, Moller-Plesset, quadratic configuration interaction, density functional theory and semiempirical scale factors. *Journal of Physical Chemistry A* **100**, 16502–16513.

Szabo A. and Ostlund N. S. (1996) *Modern Quantum Chemistry: Introduction to Advanced Electronic Structure Theory*, Dover Publications: Mineola, NY, 480 pages.

Troullier N. and Martins J. L. (1991) Efficient pseudopotentials for plane-wave calculations. *Physical Review B* **43**, 1993–2006.

Tuma C. and Sauer J. (2006) Treating dispersion effects in extended systems by hybrid MP2 : DFT calculations—protonation of isobutene in zeolite ferrierite. *Physical Chemistry Chemical Physics* **8**, 3955–3965.

Urey H. C. (1947) The thermodynamic properties of isotopic substances. *Journal of the Chemical Society*, 562–581.

Valiev M., Bylaska E. J., Govind N., Kowalski K., Straatsma T. P., van Dam H. J. J., Wang D., Nieplocha J., Apra E., Windus T. L., and de Jong W. A. (2010) NWChem: a comprehensive and scalable open-source solution for large scale molecular simulations. *Computer Physics Communications* **181**, 1477–1489.

Webb M. A. and Miller T. F., III (2014) Position-specific and clumped stable isotope studies: comparison of the Urey and path-integral approaches for carbon dioxide, nitrous oxide, methane, and propane. *Journal of Physical Chemistry A* **118**, 467–474.

Williams A. G. and Scherer M. M. (2004) Spectroscopic evidence for Fe(II)-Fe(III) electron transfer at the iron oxide-water interface. *Environmental Science and Technology* **38**, 4782–4790.

Wolfsberg M., Hook W. A., Paneth P., and Rebelo L. P. N. (2009) *Isotope Effects in the Chemical Geological and Bio Sciences*, Springer: Dordrecht/New York, 466 pages.

Zeebe R. E. (2005) Stable boron isotope fractionation between dissolved $B(OH)_3$ and $B(OH)_4^-$. *Geochimica et Cosmochimica Acta* **69**, 2753–2766.

6

Organic and Contaminant Geochemistry

Daniel Tunega,[1] Martin H. Gerzabek,[1] Georg Haberhauer,[1] Hans Lischka,[2,3]
and Adelia J. A. Aquino[1,2]

[1]*Institute for Soil Research, University of Natural Resources and Life Sciences, Vienna, Austria*
[2]*Department of Chemistry and Biochemistry, Texas Tech University, Lubbock, TX, USA*
[3]*Institute for Theoretical Chemistry, University of Vienna, Vienna, Austria*

6.1 Introduction

Environmental pollution is one of the major problems of our civilization mainly due to urban, industrial and agricultural human activities. Organic and contaminant geochemistry comprises a broad range of studies of physical and chemical processes influencing behavior, distribution, and fate of organic species and contaminants on the Earth and in extraterrestrial materials. Characterization and understanding of these processes is crucial from many aspects, for example, for determination of environmental risk or for control and management of surface and ground water resources, hazardous wastes, contaminated soils and sediments, and geologic repositories (Appelo and Postma, 2005; Berkowitz et al., 2008).

Soil represents a main buffer and filter in the environment, through which pollutant substances can migrate and enter into a food chain or ground water reservoirs (Sposito, 1984; Sparks, 1999). Soil is unconsolidated, complex and heterogeneous matter consisting of inorganic and organic materials, water, air, and living organisms and the overall behavior of the soil is determined by the chemical, physical and biological properties of its components (Wolfe and Seiber, 1993; Sparks, 1995; Hornsby et al., 1996; Yaron et al., 1996). Therefore, polluting substances can undergo complex physical, chemical and biological transformation processes in the soil.

The behavior in the environmental systems also depends on their chemical nature and properties. For example, nonpolar hydrophobic polycyclic aromatic hydrocarbons (PAHs) can be sequestered in the environment due to strong interactions with carbonaceous materials often appearing in soils and sediments (e.g., black carbon) (Lohmann et al., 2005). There is a special focus on persistent organic pollutants of natural or anthropogenic origin with long half-lives. A typical example is

---

Molecular Modeling of Geochemical Reactions: An Introduction, First Edition. Edited by James D. Kubicki.
© 2016 John Wiley & Sons, Ltd. Published 2016 by John Wiley & Sons, Ltd.

dichlorodiphenyltrichloroethane (DDT) having a soil half-life in a range of 2–15 years (US Environmental Protection Agency, 1989; Augustijn-Beckers et al., 1994). These substances are usually resistant to chemical, photolytic, or biodegradation processes and exhibit typically high lipid solubility (low water solubility) leading to their bioaccumulation in fatty tissues of living organisms.

As soils are open systems, external factors such as temperature, radiation, precipitation, wetting/drying cycles, or soil management affect the behavior and transportation of pollutants in the environment. The activity of microorganisms in soils significantly contributes to the chemical transformations of polluting substances. All those physical, chemical, and biological processes may be complicated and obscured appearing at very different time and spatial scales. For example, a photochemical degradation of some pollutant can be usually a fast process but produced metabolites (eventually with a higher toxicity than the parent material) can have much higher resistance to further (bio)mineralization processes. Moreover, the simultaneous action of various processes can have a synergetic effect (positive or negative) on persistence, sequestration, bioavailability, and degradation of pollutants (Konda et al., 2002).

Although a variety of physicochemical processes affect the fate of contaminants in natural environments, the sorption to the solid soil matrix is one of the most important phenomena determining other processes such as transport, diffusion, runoff, leaching, or volatilization. Sorption/desorption processes comprise a number of subprocesses with different kinetic rates and operating over broad time scales ranging from hours through years. Owing to the complicated structure and architecture of soil matrices (e.g., existence of pores from nano- to mesopore sizes), standard adsorption processes are often accompanied by other processes such as pore condensation. Sorption/desorption characteristics of soils and sediments are of particular concern because of further contamination, for example, of groundwater systems or for a design of effective waste management and remediation techniques (Rombke et al., 1996; Weber et al., 2001). Although soils are structurally and chemically very complex heterogeneous natural geosorbents, classical adsorption models such as Langmuir or Freundlich are often used to characterize adsorption behavior of the contaminants in soils. There are more complex theories and models for simulation of sorption processes in soils, and details can be found, for example, in Sparks (1999), Sparks et al. (1999), and Weber et al. (1991).

Coexistence of inorganic mineral particles, organic soft matter (natural/soil organic matter, (NOM/SOM), and their mutual associates and/or aggregates represents a base for highly heterogeneous surfaces with numerous chemically reactive functional groups, structural defects, hydrophilic and hydrophobic sites that are reflecting different binding mechanisms of adsorbed pollutant species, and, consequently, also great differences in adsorption energies. The origin of the binding forces in soil adsorption processes is governed by the different types of interaction mechanisms such as ionic, hydrogen, and covalent bonding, charge-transfer and electron donor-acceptor mechanisms, van der Waals forces, ligand exchange, complexation, hydrophobic bonding, partitioning, or cooperative interactions among adsorbed molecules (Weber et al., 1991; Senesi, 1992; Delle Site, 2001; Konda et al., 2002). The formation, structure, and stability of adsorption complexes are strongly affected by the presence of water in soil systems. Usually, water is present in soils in unsaturated conditions and its content is strongly variable depending upon external conditions (humidity, precipitation). Water can be concentrated in hydrophilic domains of soil matrices (e.g., polar groups of NOM or mineral surfaces), and its strong solvation potential can destabilize, for example, hydrogen-bonded complexes. On the other hand, nonpolar groups and functionalities (e.g., aliphatic chains, aromatic systems) can form water repellent hydrophobic domains able to accumulate nonpolar species (Schaumann, 2005; Schaumann and LeBoeuf, 2005; Schaumann and Bertmer, 2008; Schneckenburger et al., 2012).

It is difficult to distinguish and quantify a contribution of individual (pure) components to overall sorption processes from the standard adsorption studies on natural, complex, and heterogeneous geosorbents such as soils and sediments. Therefore, there is a necessity of case studies with better defined probe materials such as pure minerals, more homogeneous organic matrices (e.g., cellulose), or organic phases extractable from soils (e.g., humic acids (HAs)). However, the traditional batch sorption studies usually provide adsorption isotherms and corresponding adsorption constants (e.g., K_L, K_F) or partitioning coefficients (e.g., K_d, K_{ow}, K_{OC}) that are macroscopic characteristics telling little about sorption mechanisms at a molecular scale. Traditional large-scale experimental studies are more and more extended and combined with specific studies using advanced experimental techniques able to descend to the molecular scale with an effort to explain fundamental processes and to understand molecular mechanisms that control disposition and interactions of chemical contaminants in the natural geosorbents. These molecular-scale investigations comprise a variety of spectroscopic, microscopic, and imagining techniques such as AFM, STM, XPS, ESEM, SIMS, or EXAFS that contribute significantly to revealing the molecular origin of interactions of contaminants that can be a solid basis for an accurate prediction of the fate of these chemicals in geosorbent systems. Several comprehensive reviews on application of spectroscopic and microscopic techniques in molecular environmental geochemistry are available in the literature (O'Day, 1999; Fenter et al., 2002; Al-Abadleh and Grassian, 2003; Henderson et al., 2014).

The complex character of interactions of contaminants with natural geosorbents and many factors that can affect them represent a great challenge for experimental investigations. In the effort to understand elementary mechanisms of these interactions at molecular scale, computational chemistry offers powerful and effective tools. They can contribute significantly to the exploration and identification of basic mechanisms controlling driving forces of kinetics and thermodynamics of contaminants in geochemical processes. Moreover, the achievements in computational geochemistry can be of more general importance for various branches of science and vice versa. For example, the statistical QSAR method, which has been used for years in the prediction of physicochemical properties or theoretical molecular descriptors (e.g., toxicity, carcinogenicity) of chemicals in organic and pharmacologic chemistry, was applied successfully in a prediction of the free energy of hydration and aqueous solubility for a set of pesticide compounds (Klamt et al., 2002, 2009). The knowledge and experiences gathered during the last decades with computer simulations in many fields of chemistry and physics (e.g., material design, biochemistry, pharmacology, mineralogy) can be successfully applied in organic and contaminant geochemistry. In addition to exploring basic mechanisms, molecular simulations can contribute to the explanation and interpretation of the experimental observations that are frequently complicated and ambiguous. Therefore, theoretical methods represent an increasingly powerful complement for experimentalists. For example, Bucheli and Gustaffson (2000, 2003) presented K_d values for the adsorption of PAHs and PBCs by soot, an important natural geosorbent in soils and sediments. These studies were complemented with quantum chemical (QC) calculations performed by Kubicki (2005, 2006) finding a good correlation between calculated adsorption energies (ΔE_{ads}) and K_d values.

Another advantage of the molecular modeling methods can be found in their capability to predict physical and chemical properties of organic contaminants in cases when experiments would be impractical or too time-consuming. Computer simulations can be used in building a variety of physical models and hypotheses and their testing and validation. One of the biggest challenges in molecular modeling is bridging different temporal and spatial scales (i.e., the size of the systems and the length of processes) in the experiment and in the molecular simulations. A typical example is the prediction of thermodynamic quantities that are classical macroscopic observables.

6.1.1 Review Examples of Molecular Modeling Applications in Organic and Contaminant Geochemistry

The previous section indicates that there is a broad field open for the application of the molecular simulation methods in the research of geochemical surfaces and interfaces. In fact, there are numerous papers on molecular simulations regarding the interactions of a variety of chemical substances (e.g., heavy metals, radionuclides, small organic molecules, amino acids, peptides, DNA) with various geosorbents available, and it is practically impossible to report all of them. This section focuses on representative examples where molecular simulation methods have been used in the investigation of mechanism of interactions between organic contaminants and geochemically relevant geosorbents such as typical minerals, NOM, and black carbon. The organic contaminants studied in the presented examples comprise pesticides, fragments and components of crude oils and oil sands, PAHs, antibiotics, highly energetic explosives, and organic toxicants. From this review our own papers are excluded as they are discussed in detail in Section 6.3.

6.1.1.1 Pesticides

The adsorption of two pesticides (atrazine and dinitro-ortho-cresol (DNOC)) on montmorillonite surfaces was modeled by means of a classical force field molecular dynamics (FF-MD) approach (see Chapter 2) (Yu et al., 2003). It was shown that in the absence of water, organic compounds form complexes of a maximal contact with the surface. In the presence of a sufficient amount of water, the adsorbed molecules are strongly destabilized and immersed in the aqueous phase.

An experimental adsorption study of dioxin by smectite clays saturated with different exchangeable cations was complemented by the FF-MD simulations (Liu et al., 2012a). The sorption of dioxin depends strongly on the hydration of the exchangeable cations, the surface charge density of the smectite clay, and the location (tetrahedral vs. octahedral) of isomorphous substitution in the clay. MD simulations confirmed the observed trends and also showed the important role of dioxin oxygen in the surface complexation.

The FF-MD simulations were used to estimate relative clay–organic interaction enthalpies for a series of nitroaromatic solutes and hydrated, K-saturated montmorillonite, for comparison with experimental adsorption isotherm data for the same clay–nitroaromatic systems (Aggarwal et al., 2007). The trend of the computed interaction enthalpies (e.g., -234 ± 17 kJ/mol for trinitrobenzene and -154 ± 16 kJ/mol for p-nitrobenzene) agreed modestly well with the trend of adsorption maxima from the experiments.

The mechanism and relatively high sorption of dibenzo-p-dioxin (DD) by Cs-saponite were studied experimentally and theoretically (Liu et al., 2009). *Ab initio* calculations explained geometrical structures of the DD molecules in the interlayer space and showed that Cs^+ interacts with dioxin ring oxygen atoms and benzene ring π-electrons.

Combined Monte Carlo (MC) and MD techniques have been used in the study of atrazine behavior in saturated sands (Cosoli et al., 2010). Diffusion coefficients, binding energies, and concentration profiles have been determined for interactions of atrazine with silica. The results confirm a moderate atrazine adhesion onto silica framework and a more favorable tendency to bind with water.

Density functional theory (DFT) (see Chapter 1) was used in simulations of adsorption of 2-methyl-4-chlorophenoxyacetic acid (MCPA) and 2-(4-chlorophenoxy)-2-methylpropanoic acid both in neutral and ionized forms on a model surface of muscovite (Ramalho et al., 2013). The ionized adsorbates interact more strongly with the surface than do their neutral forms.

Site-specific bonding of herbicides in SOM complexes was performed by Négre et al. (2001). In the work the interactions of the imidazole herbicides imazapyr, imazethapyr, and imazaquin with HAs and SOM models were investigated both experimentally and by molecular mechanics (MM) calculations. The modeling calculations showed that there are stable complex structures involving mostly van der Waals and electrostatic interactions in agreement with hydrophobic bonding and charge transfer interface as specified by the experimental findings in the work.

The formation processes of complexes between a three dimensional (3D) model of HA and a successively increasing number (1–30) of diethyl phthalate (DEP) molecules were studied by means of classical MM (Schulten et al., 2001). The intermolecular forces that stabilize the complexes were found to be mainly driven by hydrogen bonds and van der Waals interactions. It was observed that absorption of DEP molecules occurs in internal voids of the HA structure for low concentrations whereas for high concentrations, adsorption prevails on the outer surface of the HA structure.

Blotevogel et al. (2010) investigated redox-promoted degradation processes to predict potential degradation pathways by using DFT calculations. The authors identified the thermodynamic conditions required for the redox reactions taking hexamethylphosphoramide (HMPA) as a test case. HMPA is a broadly used solvent with the possibility to act as a groundwater pollutant. Later a more detailed DFT study was performed with the aim to predict the HMPA persistence in aqueous phase (Blotevogel et al., 2011). The authors concluded that hydrolysis of the P—N bond is the only thermodynamically stable way that could drive to the HMPA degradation under reducing conditions.

6.1.1.2   *Antibiotics*

Frequently used in the farming praxis, antibiotics represent a serious potential risk for a contamination of soils. In several papers molecular simulations were used in the study of interactions of antibiotics with soil components contributing to understanding their biological availability and retention in natural and engineered soil environments. The binding mechanism of one of tetracycline antibiotics (oxytetracycline) on smectite clays was studied experimentally (X-ray, IR, and NMR) accompanied by MC molecular simulations (Aristilde et al., 2010). An impact of pH was observed on the binding mechanism that involves the adsorbed moiety via the protonated dimethylamino group. MC simulations indicated that interlayer spacing and charge localization of clay layers dictate favorable binding conformations of the intercalated tetracycline molecules facilitating multiple interactions, in agreement with the spectroscopic data.

Aristilde and Sposito (2010) studied interactions between a widely used fluoroquinolone antibiotic ciprofloxacin (Cipro) and a well-tested molecular model of humic substance (HS) (Schulten and Schnitzer, 1993; Sutton et al., 2005) in a polar environment through classical MD simulations. The authors presented a detailed analysis characterizing adsorption of Cipro by both protonated and metal cation-bearing HS. The results stressed the capacity of Cipro to participate in multiple H-bonding through their polar groups demonstrating its high affinity for the HS especially if its acidic groups are fully protonated. At circumneutral pH, HS-complexed divalent metal cations could increase binding through carboxylate groups of the antimicrobial structure forming a ternary HS–metal complex.

6.1.1.3   *Explosives and Organic Toxicants*

Sarin and soman, organophosphorus compounds, are nerve agents used in chemical weapons. Their interactions with magnesium oxide and clay minerals were studied using cluster models for minerals and Møller–Plesset perturbation theory to the second order (MP2) and B3LYP methods (both combined with the 6–31G(d) basis set) (Michalková et al., 2004a, b, 2006). The cluster models of clay minerals represented tetrahedral edges. The charge of the systems and a termination of the mineral fragment determine the strength of the intermolecular interactions. In the neutral complexes, sarin and soman are physisorbed on the mineral fragments through hydrogen bonds. The covalent chemical bonding is formed between a phosphorus atom of sarin/soman and oxygen atom from the negatively charged fragments. Based on the calculated Gibbs free energies, the authors predicted that only chemically bound complexes are thermodynamically stable at room temperature.

In the work by Tsendra et al. (2014), a cluster approach extended to the quantum mechanical/molecular mechanical (QM/MM) methodology was applied using several density functionals and the MP2 method to simulate the adsorption of selected nitrogen-containing compounds, 2,4,6-trinitrotoluene (TNT), 2,4-dinitrotoluene (DNT), 2,4-dinitroanisole (DNAN), and 3-nitro-1,2,4-triazole-5-one (NTO) on the hydroxylated (100) surface of α-quartz. The structural properties were calculated using the M06-2X and PBE functionals both including the dispersion correction (D3) in the calculations. The MP2 method was used to calculate the adsorption energies. Although all molecules are physisorbed, the silica surface showed a different sorption affinity toward the chemicals as a consequence of their electronic structure. Adsorption occurs through multiple hydrogen bonds between the polar functional groups of studied chemicals and surface silanol groups. NTO was found to be the most strongly adsorbed.

6.1.1.4   Oil Components

Exploitation of crude oil and oil sand resources represents another high potential risk for the environmental pollution by various oil components. Structurally complex substances such as resins, asphaltenes, and bitumens or their simpler fragment models have already been investigated in several works.

The review by Greenfield (2011) discusses the structure and properties of resins, asphaltenes, and bitumens achieved from molecular dynamics (MD), quantum mechanics (QM), coarse graining, and thermodynamic model approaches.

In the paper by Murgich et al. (1998), FF-MD calculations were performed on the model asphaltene and resins adsorbed on neutral surfaces of kaolinite in vacuum. Van der Waals interaction provides the largest contribution (60–70%) to the binding mechanism. Smaller contributions come from Coulombic interactions (20–30%) and hydrogen bonds (10% or less). The variation in the polarity of the resin introduced only minor changes in the energy of interaction with the surfaces. It was also determined that the resin has stronger interaction with the surfaces of kaolinite than asphaltene, indicating that these surfaces may perturb the aggregates formed by the organic molecules that are found in crude oil.

Molecular-scale sorption, diffusion, and distribution of asphaltene, resin, aromatic, and saturate fractions of heavy crude oil on quartz surface were studied using FF-MD simulation by Wu et al. (2013). Similarly to the work by Murgich et al. (1998), the authors concluded that despite of the variety of the adsorbed substances, the main contribution to the interactions is represented by the van der Waals energy. Another conclusion was that the most likely oil distribution on quartz surface was those aromatics and saturates transported randomly into and out of the complex consisting of asphaltenes surrounded by resins, which was influenced by temperature.

The binding energies of benzene, *n*-hexane, pyridine, 2-propanol, and water interacting with the kaolinite surfaces were calculated at the DFT level using the exchange-hole dipole moment dispersion model (Johnson and Otero-de-la-Roza, 2012). It was documented that the hydrophilic alumina surface has a stronger affinity to all investigated molecules than the hydrophobic siloxane surface. Hydrogen bonding for pyridine, 2-propanol, and water, OH$\cdots$π interactions for benzene, and CH$\cdots$O interactions for *n*-hexane were found as the dominant noncovalent interactions.

A combination of MM and DFT methods was used in the calculation of the charge distribution and the adsorption energy on the (001) surface of hematite (α-Fe_2O_3) for a set of 12 molecules representing fragments of resins and asphaltenes (Murgich et al., 2001). The results showed that noncovalent bonding is preferential in the surface complexes. Molecules with high aromaticity and low H/C ratio show high adsorption energy that corresponds with the experimental finding for asphaltenic deposits extracted from wells (Murgich et al., 1999).

Decane, methylbenzene, pyridine, and acetic acid were selected as components in crude oils with a different polarity, and their adsorption behavior on water-wet silica surface was investigated by classical MD simulation (Zhong et al., 2013). The simulation results indicated that polar components could penetrate through water film and adsorb on silica surface, while it was difficult for nonpolar components. It was analyzed that the adsorption capability of oil components is related to three factors: interaction between oil components and silica surface, penetration in water film, and competitive adsorption with water molecules. The authors proposed a two-step adsorption process on the base of the molecular simulations. The polar oil components preferentially absorb on mineral surface promoting further adsorption of nonpolar components.

6.1.1.5   Aromatic Contaminants

The physicochemical properties and the binding characteristics of 18 PAH molecules docking to molecular structural models of fulvic acids (FA), HA, and SOM were studied by Saparpakorn et al. (2007) performing QC semiempirical calculations. The FA, HA, and SOM models were taken from Buffle et al. (1977), Stevenson (1982), and Schulten and Schnitzer (1993), respectively. From the docked conformations of PAHs onto FA, HA, and SOM structures, it was found that π–π interactions and H-bonding were significant for the binding of the PAHs. The final docked energies demonstrated that the PAHs bind to SOM more strongly than to both forms of HA and FA individually.

MM and QM studies were carried out by Kubicki and Apitz (1999) to investigate interactions of several NOM models with different organic compounds. The goal was to test different computational methodology for estimating the stability of the isolated systems and the sorption mechanism of organic compounds in NOM. Followed work by Kubicki (2000) comprised energy minimization and dynamics simulations on *n*-hexane soot model interacting with pyrene including validation of different FF approaches in providing representative soot nanoparticle structures and their properties.

Ab initio calculations were used to investigate the electronic and energetic behavior accompanying adsorption of aromatic molecules such as four polychlorinated dibenzo-*p*-dioxin molecules of different polarities onto an insulating hydrophobic (001) surface of pyrophyllite (Austen et al., 2008). The fairly weak interactions were observed to be dominated by local electrostatics rather than global multipoles or hybridization. A small transfer of electron density was observed from the molecule to the surface.

Adsorption of benzene and benzene-1,4-diol on two silica surface models with either hydrophobic or hydrophilic properties was studied by means of calculations based on local Gaussian basis functions and the B3LYP-D functional (dispersion corrected) on periodic structural models (Rimola et al., 2010). It was found that the inclusion of dispersion corrections to the functional dramatically affects both the intermolecular geometries and the adsorption energies. The adsorption of the aromatic molecules on the hydrophobic silica surface is dictated by dispersion and weak CH$\cdots$O(Si)O interactions. For hydrophilic surfaces dispersion is still large despite the fact that adsorption energies are almost doubled with respect to the hydrophobic surface due to weak hydrogen bonding through OH$\cdots\pi$ interactions.

The adsorption of benzene and several PAHs on the carbonaceous surfaces from the gas phase and water solution was investigated using several different levels of theory including DFT, MP2, and CCSD(T) (Scott et al., 2012). Both periodic and cluster approaches were used in the study. Good agreement was revealed for the theoretical and experimental adsorption energies of benzene and PAHs adsorbed on the modeled carbon surfaces. The work was later extended on nitrogen-containing aromatic compounds (e.g., pyridine) applying M06-2X and BLYP-D2 functionals

(Scott et al., 2014). The most stable systems were in a parallel orientation toward the modeled carbon surface. The calculated adsorption enthalpies, Gibbs free energies at 298.15 K, and partition coefficients were found to be in a good agreement with available experimental data (Table 1 in Scott et al. (2014)).

6.2   Molecular Modeling Methods

This section briefly summarizes molecular modeling methods, tools, and techniques relevant for organic and contaminant geochemistry. Details can be found in many excellent textbooks and reviews available in the literature. Moreover, some other chapters of this volume (Chapters 1–3) also describe particular theoretical methods and modeling techniques and their importance for geochemistry in more detailed aspects.

Generally, computational chemistry offers a broad scale of modeling methods dealing with models at a molecular scale. The molecular modeling comprises two basic groups of methods outgoing from (i) MM and (ii) QM, respectively. In the MM methods, classical particles represent a model system consisting of interacting atoms and the main task is to find proper, relatively simple analytical expressions able to describe a potential energy hypersurface of the system under study. On the other hand, in QM, systems are represented as a set of mutually interacting nuclei and electrons and the whole information about the interacting system is keyed in its wave function Ψ. Important physical properties of the system can be achieved by a solution of the Schrödinger equation. In several textbooks the reader can find a comprehensive overview of fundamentals of both QM and MM methods (Leach, 2001; Young, 2001; Frenkel and Smit, 2002; Cramer, 2004).

6.2.1   Molecular Mechanics: Brief Summary

Only basic principles of MM and some practical comments are addressed here. For more details we recommend, for example, Burkert and Allinger (1982) and Rappé and Casewit (1997). Chapter 2 also provides information on MM methods and their applications in geochemistry. MM or force field (FF) methods use analytical expressions for calculation of the potential energy, $U(\boldsymbol{R})$, for the given configuration of atoms, $\boldsymbol{R}$. The potential energy of the system is expressed usually in a simple analytical form containing empirical parameters (FF parameters) expressed in terms of interatomic potential functions of interacting atoms in the system. In the classical MM the atoms are represented as the smallest dimensionless entities (particles) having no internal structure. Thus, in this type of simulation, electrons are not explicitly present in the system, but point charges can be assigned to these particles. Complete analytical expression of the potential energy function of the many-body system is a difficult task. Such function can be expressed as a series over sum of two-, three-, and high-order terms, which are functions of geometrical variables, for example, distances and angles. In practice, such series are truncated and high-order terms are neglected and only two- and eventually also three- or four-body interactions are expressed in an analytical form. According to the nature of interactions, the total potential energy of a system is frequently partitioned to the following energy components:

$$U_{\text{Tot}} = U_{\text{Bond}} + U_{\text{Angle}} + U_{\text{Tors}} + U_{\text{Coul}} + U_{\text{VDW}} \qquad (6.1)$$

The first three terms in this equation represent bond-stretching, bond angle, and torsional terms, respectively, defining bonding energy of the system. The last two terms express nonbonding energy terms: the Coulombic, U_{Coul}, and the van der Waals energy, U_{VDW}.

There are many ways how to express individual components, and here only a frequently used approach is briefly described. For describing bond-stretching and bond angle terms, a simple approximation in a form of harmonic potential is used in a following way:

$$U_{\text{Bond}} = K_{\text{b}}(r - r_0)^2 \tag{6.2}$$

$$U_{\text{Angle}} = K_{\text{a}}(\theta - \theta_0)^2 \tag{6.3}$$

where K_{b} and K_{a} are empirical force constants, r is the separation distance between two atoms (bond), θ is a bond angle for three sequentially bonded atoms, and r_0 and θ_0 are distance and bond angle for the equilibrium state (corresponding to the lowest energy), respectively. The harmonic approximation is suitable for states, in which atoms are near the equilibrium configuration and atoms oscillate around the equilibrium positions with small deviations. In fact, the harmonic potential can describe most of the vibrations well but for larger deviations it fails. In such a case a more complex function is required to be able to describe an anharmonic nature of vibrations or even more also bond dissociation limit. Such function is, for example, Morse potential with three parameters. In practice, force constants in Equations 6.2 and 6.3 can be obtained from the analysis of the vibrational spectra.

The torsion potential, U_{Tors}, expresses rotational energy of four atoms connected in sequence (dihedral angle). This potential is periodic and requires an expansion of periodic functions such as a Fourier series. Typical expression for U_{Tors} used in many FF programs is

$$U_{\text{Tors}} = V_1/2(1 + \cos(\omega)) + V_2/2(1 + \cos(2\omega)) + V_3/2(1 + \cos(3\omega)) \tag{6.4}$$

were V_{1-3} are empirical coefficients and ω is a torsional angle. Proper torsional functions are necessary for a correct description of conformations of large flexible molecules (e.g., proteins), and particular terms are related to a chemical nature and hybridization of particular atoms involved in the torsional bonding.

From two nonbonding terms in Equation 6.1, Coulombic potential has a long-range character, whereas U_{VDW} term is of a short-range nature. From classical electrostatics, Coulombic interaction energy can be expressed in a multipole expansion. In the most common practice, only the zeroth-order expansion is used to express electrostatic interactions by using point charges assigned to the particular atoms in the model. Then, the electrostatic term is expressed as the sum of pairwise interactions of point charges where pair interactions are a function of the $1/r_{ij}$, with r_{ij} being the distance between charges q_i and q_j assigned to atoms i and j[1]:

$$U_{\text{Coul}} = \sum \frac{q_i \cdot q_j}{r_{ij}} \tag{6.5}$$

The summation runs over all atomic pairs while avoiding duplications. If the atomic charges are of opposite sign, Equation 6.5 gives a negative (attractive) energy, and if the charges are of the same sign, the result of the summation is a positive (repulsive) energy. The summation in Equation 6.5 is critical for periodic systems due to long-range nature of the electrostatic interactions and requires special mathematical methods to obtain its proper convergence (e.g., Ewald summation (Darden et al., 1993)).

The second nonbonding term in Equation 6.1 has a physical meaning in interactions of fluctuating electron densities of atoms or molecules, and it is usually decomposed onto two parts. The first part

---

[1] Atomic units are used for physical constants in equations.

is attractive and is referred as the London dispersion term that is proportional $1/r^6$. The second component describes Pauli repulsion, which rapidly increases with decreasing distance between atoms and is frequently expressed as a $1/r^{12}$ function. The combination of the attractive dispersion and repulsion terms is often expressed in the form of the Lennard–Jones function (also known as LJ or 12–6 potential):

$$U_{\text{VDW}} = \sum 4\varepsilon_{ij} \left[\left(\frac{\sigma_{ij}}{r_{ij}} \right)^{12} - \left(\frac{\sigma_{ij}}{r_{ij}} \right)^{6} \right] \tag{6.6}$$

where ε_{ij} is a depth of the LJ potential between two atoms (i and j) and σ_{ij} constant is related to a distance at the minimum of LJ potential as $r_{ij0} = 2^{(1/6)}\sigma_{ij}$. The LJ potential is used in many FF because of its simple expression (there are also alternative expressions for LJ potential). The non-bonding interactions can be expressed also by different types of potentials such as Buckingham potential (exponential function describes repulsion part) or Morse potential. In specific cases, also 9–6 potential can be used, for example, for simulations of interactions at solid surfaces (Heinz et al., 2008). Apart from the electrostatic potential (ESP), the LJ potential rapidly diminishes with the distance and in practical calculations a cutoff distance is applied (typically 5–10 Å) to reduce the summation in Equation 6.6.

The most crucial task in the FF methods is to find proper constants. There are, in principle, two basic ways how to obtain FF parameters. One way is by fitting parameters to experimental data (e.g., structural, thermodynamic, or spectroscopic) for a set of molecules. The second way comprises fitting to QM calculations (e.g., potential energy surface (PES)) on molecular models (or molecular clusters, periodic slab or bulk models if parameters for solids and surfaces are requested). In both ways bonding and also non-bonding terms can be determined.

A sensitive part of the FF potentials is the Coulombic contribution due to its long-range nature and the fact that atomic charges cannot be assigned arbitrarily. There are several ways for the assignment of atomic charges. However, the most convenient approach is based on accurate QM calculations. Calculations are usually performed on molecules, molecular clusters, or simple periodic solid systems. The most typical method for obtaining charges is their fitting to the quantum chemically calculated ESP, which is derived from the electron densities. There are various approaches such as CHELP (Chirlian and Francl, 1987), CHELPG (Breneman and Wiberg, 1990), RESP (Bayly et al., 1993), or MK (Singh and Kollman, 1984). A required condition for the determination of the FF parameters and atomic charges is a mutual balance between the Coulombic and nonbonding energy terms. Thus, the combination of the atomic charges from one type of FFs with another one has to be carefully considered.

The development of the FF parameters requires their backward validation to ensure that molecular simulations will reproduce energies and properties with a required accuracy. This validation can be performed on some set of molecules, which are similar to those used in the parameter fitting and calculated molecular properties can be compared either with experimental data and/or accurate QC calculations. Praxis showed that it is practically impossible to develop "universal" FF parameters for each element in the periodic table that would be able to reproduce satisfactorily properties of that element in different chemical situations. Instead of that FF parameters are developed for several atomic types of the same element. Carbon atom with its different hybridization states is the classical example.

Nowadays the molecular simulation methods offer a broad range of various FF programs usually developed for particular sets of molecular systems. Many of the FF parameters are developed for organic and bioorganic molecules and macromolecules (carbohydrates, peptides, proteins, nucleic acids, polymers) such as AMBER (Pearlman et al., 1995), CHARMM (MacKerell et al., 1998), GROMACS (Hess et al., 2008), MMFF (Halgren, 1996), CVFF (Sun et al., 1994), UFF (Rappé et al., 1992), and OPLS-AA (Jorgensen et al., 1996). A comprehensive overview can be found in table 2.1 in the textbook by Cramer (2004). Special attention was devoted to developing FF models for water because it is an important polar solvent and also has unusual properties. From many existing FF descriptions of water, SPC (Berendsen et al., 1981), SPC/E (Berendsen et al., 1987), and TIP3P/TIP4P (Jorgensen et al., 1983) models can be assigned to the most frequently used in the simulations.

Apart from many available FF parameters for organic molecules, a development of FF parameters for inorganic materials was not such intensive and more specialized FFs have been developed for specific types of inorganic materials (e.g., EAM-FF for metals (Daw et al., 1993) or Buckingham-type potentials for spinels (Fang et al., 2000)). Such FFs are usually less transferable and could contain additional specific energy terms. One of the main reasons for such situation is in more complex chemistry of inorganic substances (composition, bonding) comparing to organic species. Moreover, the accuracy of the FF parameters developed for bulk solids can be problematic for modeling of the surfaces and interfacial properties such as hydration energies, surface tensions, and interaction energies. In addition, nonbonding terms in the energy expressions for inorganic and organic compounds, which are necessary for the simulation of organic–inorganic interfaces, are often not compatible. In spite of that, there is an effort to develop FFs able to overcome depicted problems and using a similar platform as in the standard FF for organics to be able to simulate inorganic compounds and their interfaces with a good accuracy. For example, CLAYFF is a FF developed for simulation of hydrated crystalline compounds (e.g., clay minerals) and their interfaces with fluid phases, which comprises harmonic bond-stretching and angle-bending, Coulombic, and LJ terms (Cygan et al., 2004). Recently, a new effort in modeling of interfaces has been presented in a form of INTERFACE-FF (Heinz et al., 2013). This FF operates as an extension of common harmonic FFs using the same functional form and combination rules. The development is focused on a production of thermodynamically consistent FF parameters for simulations of inorganic–organic and inorganic–biomolecular interfaces. Currently, the INTERFACE-FF includes parameters for aluminosilicates, metals, oxides, sulfates, and apatites. Another long-standing limitation of classical FFs is inability to simulate chemical reactions (i.e., bond-breaking and bond-forming processes). This limitation can be overcome by developing new types of FF, generally called reactive FFs (Liang et al., 2013). ReaxFF is a typical representative of this class of FFs originally developed for reactions of hydrocarbons (van Duin et al., 2001).

6.2.2 Quantum Mechanics: Overview

This section only briefly summarizes QM methods. Principles can be found in many excellent books (e.g., Hehre, 1998; Lowe and Peterson, 2005; Shankar, 1994; Szabo and Ostlund, 1989), and also Chapter 1 in this book provides an overview.

The heart of the QM is the Schrödinger equation, which describes the space and time dependence of a system consisting of interacting electrons and nuclei. Mathematically, the Schrödinger equation is the second-order partial differential equation, and its nonrelativistic time-independent form is

$$\hat{H}\Psi = E\Psi \qquad (6.7)$$

where $\hat{H}$ is the differential Hamiltonian operator, E is the total energy, and Ψ is the wave function of the system. The Hamiltonian can be expressed as the sum of the kinetic and potential energy operators:

$$\hat{H} = -\frac{1}{2m}\nabla^2 + V(\boldsymbol{R},\boldsymbol{r}) \tag{6.8}$$

In Equation 6.8, m is the mass of the ith particle, ∇^2 is the Laplacian operator, and $V(\boldsymbol{R},\boldsymbol{r})$ is the operator describing the potential energy associated with the Coulombic interactions of all electrons and nuclei in the system. The existence of a wave function is one of the postulates in QM and has to satisfy certain constraints. It is interpreted in terms of a probability to find electrons in a configurational space and is used for obtaining the energy of the system. Exact solution of the Schrödinger equation for many-particle system is not possible, and for practical applications its solution requires some approximations. For example, the Born–Oppenheimer approximation separates effectively nuclear and electronic motion; thus, electronic wave function only parametrically depends on the positions of nuclei.

QM-based methods can be classified to four basic groups: Hartree–Fock (HF), *ab initio* correlation (post-HF methods), DFT, and semiempirical methods. HF and post-HF methods involve a $3N$-dimensional antisymmetric wave function for a system of N electrons. The HF approach is the simplest one as it considers exchange for a single determinant wave function. The HF method is almost universally used as the reference independent particle model. Therefore, quantum chemistry approaches are based on the HF method as a first approximation (Charlotte, 1987). The electronic correlation energy neglected in the HF method is included in the configuration interaction (CI) method (Szalay et al., 2012). This method possesses a large flexibility in computing various types of electronic structures and excitations but is computationally very expensive. Møller–Plesset (MP) perturbation theory (1934) is another post-HF *ab initio* method. It corrects the HF method by including electron correlation effects using the Rayleigh–Schrödinger perturbation theory (RS-PT), most commonly to second order (MP2). MP2 is computationally efficient and can be used for many practical applications in geochemistry. Coupled cluster (CC) theory (Čížek, 1991; Stanton and Bartlett, 1993; Paldus, 2005) formulates the electronic Schrodinger equation as a nonlinear equation, allowing the calculation of size-consistent high-precision approximations of the ground state solution. The form of CC, which includes singles and doubles and noniterative triples (CCSD(T)), is considered to be the most accurate method routinely available. Applications are available (as for MP2) for cases where the HF wave function is a good starting point.

A fundamental issue to be considered in the application of *ab initio*-based methods is the size of the system, the type of property desired to be analyzed, and the computational time required. The QC methods are computationally intensive scaling with $O(N^4)$ and more especially the correlated post-HF methods.

Therefore, in this section we will emphasize the DFT-based methods as they are by far the most frequently used methods to describe large molecular systems of the size of several hundred atoms. The popularity of DFT has roots in its economy and efficiency comparing to the standard HF and post-HF methods as scaling is $\sim O(N^3)$.

6.2.2.1 Density Functional Theory: Basic Principles

DFT defines the electronic states of atoms, molecules, and solids, in terms of the 3D electronic density of the system. It means a great simplification over the wave function-based methods (HF, CI, MP, and CC). The method is based on Hohenberg–Kohn theorems (1964), and subsequently Kohn and Sham have shown a practical way to apply it (1965).

The Hohenberg–Kohn Theorems.   The first Hohenberg–Kohn theorem establishes that the electron density exclusively determines the Hamiltonian operator and therefore all the properties of the system. This first theorem states that the external potential $V_{\text{ext}}(\vec{r})$ is (to within a constant) a unique functional of $\rho(\vec{r})$. Since in turn $V_{\text{ext}}(\vec{r})$ fixes $\hat{H}$ it is clear that the full many-particle ground state is a unique functional of $\rho(\vec{r})$. Thus, $\rho(\vec{r})$ determines the number of electrons N and $V_{ext}(\vec{r})$ and consequentially all the properties of the system such as the kinetic energy ($T[\rho]$), the potential energy $V(T[\rho])$, and the total energy ($E[\rho]$). The total energy can be written as

$$E[\rho] = T[\rho] + E_{Ne}[\rho] + E_{ee}[\rho] = \int \rho(\vec{r}) V_{Ne}(\vec{r}) d\vec{r} + F_{HK}[\rho] \qquad (6.9)$$

$$F_{HK}[\rho] = T[\rho] + E_{ee} \qquad (6.10)$$

In Equation 6.10, $F_{HK}[\rho]$ denotes a universal functional that contains the kinetic energy $T[\rho]$ and the electron–electron interaction, $E_{ee}[\rho]$. This functional is the holy grail of DFT. The subscript Ne in Equation 6.9 is regarding to the electron-nuclear interaction. It is important to emphasize that this functional is completely independent of the system; it applies well to an atom as well to large molecules. If the functional $F_{HK}[\rho]$ was known, it would be possible to solve the Schrodinger equation exactly.

The electronic interaction functional can be expressed as

$$E_{ee}[\rho] = \frac{1}{2} \int \int \frac{\rho(\vec{r_1})\rho(\vec{r_2})}{r_{12}} d\vec{r_1} d\vec{r_2} + E_{ncl}[\rho] = J[\rho] + E_{ncl}[\rho] \qquad (6.11)$$

In Equation 6.11, $E_{ncl}[\rho]$ is the nonclassical contribution to the $E_{ee}[\rho]$ term encompassing all the effects of self-interaction correction, exchange, and Coulomb correlation. $J[\rho]$ is the classical Coulomb part. To find expressions describing the $T[\rho]$ and $E_{ncl}[\rho]$ functionals is the key challenge in DFT.

The second Hohenberg–Kohn theorem establishes a variational principle, which can be expressed as

$$E_0 \le E[\widetilde{\rho}] = T[\widetilde{\rho}] + E_{Ne}[\widetilde{\rho}] + E_{ee}[\widetilde{\rho}] \qquad (6.12)$$

These two theorems lead to the major DFT statement:

$$\delta\left[E[\rho] - \mu\left(\int \rho(r)dr - N\right)\right] = 0 \qquad (6.13)$$

Equation 6.13 means that the ground state energy and density correspond to the minimum of a functional $E[\rho]$ with the constraint that the density refers to the correct number of N electrons. The Lagrange multiplier related to this constraint is the electronic *chemical potential* μ. Thus, we arrive at the interesting conclusion that there is a *universal* functional $E[\rho]$ which, if it is known, could be used in Equation 6.13 to compute the *exact* ground state density and energy.

The Kohn–Sham Approach. The equations of Kohn and Sham, published in 1965, turn DFT into a practical tool (Kohn and Sham, 1965). The Kohn–Sham approach to DFT provides an exact description of the interacting many-particle systems in terms of an effective noninteracting particle system. The effective potential of the Kohn–Sham system (noninteracting particles) can be completely described by the electron density of the interacting system. The total energy functional in the Kohn–Sham approach contains the kinetic energy, electron–electron interaction, and interaction with the external potential terms:

$$E[\rho] = T[\rho] + V_{ext}[\rho] + V_{ee}[\rho] \qquad (6.14)$$

The external potential is defined as

$$V_{ext}[\rho] = \int \hat{V}_{ext}\, \rho(r)\, dr \tag{6.15}$$

$T[\rho]$ and $V_{ee}[\rho]$ are unknown functionals. In the fictitious system of noninteracting particles idealized by Kohn and Sham, the kinetic energy functional is described by means of a single determinant wave functions in N orbitals, ϕ_i:

$$T_S[\rho] = -\frac{1}{2}\sum_i^N \langle \phi_i | \nabla^2 | \phi_i \rangle \tag{6.16}$$

Obviously T_S, the kinetic energy of the noninteracting particles, is not equal to the true kinetic energy of the real system. Kohn and Sham accounted for it by writing $F[\rho]$ as

$$F[\rho] = T_S[\rho] + J[\rho] + E_{XC}[\rho] \tag{6.17}$$

In Equation 6.17, E_{XC} is denominated as exchange–correlation energy and contains everything that is unknown. It can be defined based on Equation 6.17 as shown below.

$$E_{XC}[\rho] = (T[\rho] - T_S[\rho]) + (E_{ee}[\rho] - J[\rho]) \tag{6.18}$$

Using the variational principle in order to minimize the energy expression under the constraint $\langle \phi_i | \phi_j \rangle = \delta_{ij}$, one arrives at the final Kohn–Sham equations:

$$\left(-\frac{1}{2}\nabla^2 \left[\int \frac{\rho(\vec{r}_2)}{r_{12}} + V_{XC}(\vec{r}_1) - \sum_A^M \frac{Z_A}{r_{1A}}\right]\right)\phi_i = \left(-\frac{1}{2}\nabla^2 + V_S(\vec{r}_1)\right)\phi_i = \epsilon_i \phi_i \tag{6.19}$$

$$V_S(\vec{r}_1) = \int \frac{\rho(\vec{r}_2)}{r_{12}} d(\vec{r}_2) + V_{XC}(\vec{r}_1) - \sum_A^M \frac{Z_A}{r_{1A}} \tag{6.20}$$

This set of nonlinear Kohn–Sham equations describes the performance of noninteracting particles in an effective local potential. For the exact local potential, the "orbitals" yield the exact ground state density through Equation 6.19 and the exact ground state energy through Equation 6.20.

6.2.2.2 DFT Functionals: General Overview

The major problem in DFT theory is the fact that the exact functionals for exchange and correlation are not known except for the homogeneous electron gas. However, the success of DFT relies on the existence of approximations permitting to calculate electronic structure and physical quantities with a good accuracy. In the starting era of DFT, the local density approximation (LDA) became a frequently used approximation, especially in solid-state physics (Parr and Yang, 1989). In LDA, the exchange–correlation energy functional terms depend simply on the electron density at the coordinate, where the functional is calculated. The LDA was later extended to the local spin-density approximation (LSDA) by the inclusion of the electron-spin term (Vosko et al., 1980). The improvement of the LDA functionals led to the generalized gradient approximation (GGA) method (Langreth and Mehl, 1983; Becke, 1988; Perdew et al., 1992) by including gradients of the electron density to the functional forms. Within the frame of the GGA theory, numerous exchange–correlation functionals have been developed for chemical applications (Zhao and Truhlar, 2008a). Among them PBE (Perdew et al., 1996), PW91 (Perdew and Wang, 1992), and BLYP

(Becke, 1988; Lee et al., 1988; Miehlich et al., 1989) can be considered as frequently used functionals in many types of DFT calculations on molecules or solid-state materials. The B3LYP functional has achieved a high popularity, especially in the chemical community (Becke, 1993). This functional belongs to a hybrid type, in which HF exact exchange is mixed with DFT exchange–correlation using adjustable parameters. These parameters are fitted on a training set of molecules. The popularity of B3LYP lies in its relatively good accuracy in predicting of many properties including the thermochemical parameters for a wide range of chemical species.

From the semiempirical methods it is worthy to briefly introduce density functional tight binding (DFTB) approach as it is based on the DFT framework with the inclusion of empirical parameters (Seifert et al., 1996; Elstner et al., 1998). It introduces several approximations that reduce the calculations of complicated electronic integrals. The Hamiltonian and overlap matrices contain one- and two-center contributions only. They are calculated and tabulated in advance as functions of the distance between atomic pairs. Therefore, this method (and its self-consistent charge extension, SCC-DFTB) is orders of magnitudes faster than the standard DFT methods. The improvements known from the standard DFT methods such as dispersion corrections are also adapted in the DFTB method. Thus, this approach offers an acceptable quality in comparison to standard DFT methods and can be used for systems of thousands of atoms. The foundations of the DFTB method are reviewed by Seifert and Joswig (2012). However, the DFTB, as each parametric method, has certain limitations, for example, transferability of parameters for the same element in different chemical environments.

6.2.2.3 Dispersion Corrections to DFT

DFT represents one of the most successful theoretical methods for investigation of structure and properties of molecules and solids in various fields of material research because it holds an excellent balance of computational costs and reached accuracy. DFT methods are even faster and efficient for periodic solid systems if they combine atomic basis sets expressed in plane waves with atomic pseudopotentials. However, it has also been realized early that DFT is not accurate enough in a prediction of properties such as charge transfer, band gap, and excited states and in some specific physical situations. One big problem of the standard DFT is that it is incorrect in the description of weak nonbonding interactions such as dispersion, π–π stacking, or hydrogen bonding. The fundamental reason for this failure is the inability of the method to account for a nonlocal electron correlation effect. Therefore, the successful application of the standard DFT functionals on systems with dominating weak interactions (e.g., molecular crystals) is rather limited.

A lot of effort has been devoted in a systematic and active development of DFT functionals to overcome this problem. The development of new methods is conducted at several levels of theory—from purely empirical corrections to the standard DFT functionals to rigorous dispersion functionals derived from first principles. The van der Waals density functional (vdW-DF) method is going beyond the local and semilocal DFT approximation by employing a nonlocal functional to approximate the correlation. This first-principles approach deals with the van der Waals forces by including nonlocal correlations from the electron response to the electrodynamic field (Dion et al., 2004; Thonhauser et al., 2007; Lee et al., 2010). Another group of the improved DFT methods includes flexible hybrid meta-GGA functionals generally called Minnesota functionals. They are, for example, able to cover dispersion interactions in various nonbonding complexes for shorter distances, and the functionals such as M05-2X and M06-2X (Zhao and Truhlar, 2008b; Zhao et al., 2005) have increasing popularity in using by the QC community.

The class of methods based on a combination of conventional functionals with empirical dispersion corrections became quickly popular because of its effectivity and simple implementation.

The dispersion corrections include r^6 term (eventually also higher terms) and respective methods are generally known as DFT-D methods (Grimme, 2004, 2006, 2011; Grimme et al., 2010). Another method, based on empirical corrections to the standard DFT functionals, is known as atom-centered dispersion-correcting pseudopotential (DCP) method (von Lilienfeld et al., 2004; DiLabio, 2008). An alternative method to the DFT-D approach grounded on system-dependent dispersion coefficients calculated from the first principles was originally suggested by Tkatchenko and Scheffler (vdW-TS method) (2009). The vdW-TS method was later improved by the addition of the self-consistently screened approach and Hirshfeld partitioning for getting the atomic polarizabilities (Ruiz et al., 2012) and atomic charges (Bučko et al., 2014), respectively. The problem of dispersion corrections in DFT is extensively studied, and there are numerous benchmark calculations and reviews available in the literature (e.g., Johnson et al., 2009; Bučko et al., 2010; Marom et al., 2011; Goerigk, 2014). In 2014, the DFT theory celebrates 50 years and its history, development, and perspectives can be found in the paper by Becke (2014).

6.2.3 Molecular Modeling Techniques: Summary

This section briefly summarizes three basic molecular modeling techniques: energy minimization, MD, and MC. There are many textbooks providing details on the molecular modeling techniques (some of them are cited in this chapter, e.g., Leach, 2001; Young, 2001; Frenkel and Smit, 2002; Cramer, 2004). Chapters 1 and 2 of this book also give much more details on the available sophisticated algorithms.

6.2.3.1 Energy Minimization

PES is a complex function of atomic position; for a system with N atoms, it is a function of $3N–6$ internal or $3N$ Cartesian coordinates. The aim of the energy minimization is to obtain the most stable configuration (the state with the lowest energy) for the system under study. In other words, the minimum of the PES is found by varying geometrical parameters. This method is also referred as a geometry optimization or a structural relaxation. It is expected that in the energy minimum the system has the lowest potential energy and the forces on all atoms are zero and the system is in equilibrium state. Note that "thermodynamic" condition for the optimization is $T = 0$ K. The global minimum is required for the calculation of molecular properties, for example, vibrational states. There are a variety of optimization procedures. Algorithms employing energy derivatives belong to the most frequently used methods (e.g., steepest descent, complex conjugate gradient, or Newton–Raphson). Such minimization procedures involve the calculation of the potential energy of the initial configuration and the potential energy derivatives with respect to atomic coordinates. New coordinates of the system are adjusted according to computed energy derivatives and are expected that the new configuration will have lower energy than the previous one. This procedure is repeated until defined tolerances for energies, gradients, forces, and deviations of atomic coordinates between successive steps are achieved. For complex energy hypersurfaces it is possible to use methods employing genetic algorithm or (MD) simulated annealing. During the optimization procedures some restrictions or constrains can be imposed. In this way, for example, symmetry can be preserved or positions of selected atoms can be fixed. For systems with translational periodicity, lattice parameters of the computational cell can be optimized as well.

6.2.3.2 Molecular Dynamics

MD is a deterministic method, which describes the movement of the system in the phase space defined by the coordinates and momenta of the particles in the system. The purpose of the MD is finding a trajectory of the system in the phase space by solving Newton's equations of motion. For many-body system (ensemble of particles), analytical solution is not possible and a numerical solution has to be used indeed. The core of MD is in the calculation of the forces acting on the particles in the system. The force for each particle can be determined as a derivative of the potential energy U with respect to the change in the particle's position. In the classical MD, the potential energy describes interactions among the particles in the system and is expressed in a particular type of the FF. Knowing positions, forces, and masses of the particles, it is possible to find new positions by performing a numerical integration of the equations of motion over the selected time interval. Usually, the MD has the following steps. At the beginning, for a given initial configuration of the system, initial velocities are assigned to the particles (e.g., using Boltzmann distribution of velocities). Then, forces are derived from the potential energy followed by the numerical integration of the equations of motion over a very small time step, Δt. From this integration new positions of the particles are obtained and new potential energy is calculated. The procedure is repeated over a desired time length. The time increment in the numerical solution has to be satisfactorily small. It is recommended to use a time step that is satisfactorily smaller than the period of the highest vibrational mode in the system. Thus, typical time steps in the MD simulations are few femtoseconds. The repeated step-by-step procedure results in a series of structural changes (configurations) over time. Several integration algorithms are used in MD, for example, Verlet, velocity Verlet, or leapfrog method (Verlet, 1967; Van Gunsteren and Berendsen, 1977; Ferrario and Ryckaert, 1985).

By calculating a partition function of the simulated system, it is possible to characterize its macroscopic behavior from the thermodynamic point of view. Various characteristics of the system can be calculated applying statistical ensemble averages of the various instantaneous values obtained during the MD run, for example, structural or energetic parameters. Moreover, time correlation functions (e.g., of velocities) can be used for the calculation of transport coefficients or power spectra (corresponding to vibrational spectra). MD simulations are also suitable for the prediction of transport properties such as diffusion and viscosity coefficients, but this type of the calculations usually requires a very long time.

Typical ensembles used in MD calculations are microcanonical (NVE—constant number of particles, volume, and energy), canonical (NVT—constant number of particles, volume, and temperature), or isothermal–isobaric (NpT—constant number of particles, pressure, and temperature). In NVT simulations, the system is embedded into a thermostat to keep a desired temperature using, for example, scaling of velocities or an external thermostat (e.g., Berendsen et al. (1984)).

The prediction of the thermodynamic quantities such as free energies and their changes is one of the most wished outputs from MD. Combination of the statistical averages with thermodynamic integration approach allows calculation of thermodynamic quantities such as free energies and entropy values. According to the ergodic principle, ensemble and time averages should be equal for the infinite time. However, in practice, MD evolves a finite-sized molecular configuration in limited time and possible errors have to be estimated carefully. Even though realistic classical MD simulations can approach milliseconds (Lindorff-Larsen et al., 2011), there are events that would not be sampled satisfactorily at such time scale. Such events are addressed as rare and are linked with barriers crossing on the energy hypersurface (e.g., phase transitions, chemical reactions, or protein folding). The simulations of these processes are challenging, and the corresponding theories and methods are under an intense development. For example, Hartmann et al. (2014) and Dellago and Hummer (2014) summarize MD methods and techniques for the calculations of the free energies and characterization of rare events.

Owing to a rapid progress in developing the DFT theory and increasing power of the computer sources, in 1985 appeared the pioneering work linking MD method with DFT theory (Car and Parrinello, 1985). This method, often referred as first-principles MD (FPMD) or *ab initio* MD (AIMD), is an accurate atomistic simulation and combines a QM description of electrons with a classical description of atomic nuclei. The Newton equations of motion are solved for all nuclei as in the classical MD. However, the forces acting on the nuclei are derived from a calculation of the electronic energy of the system at each discrete time step of the trajectory. Time and size scales in AIMD are smaller than in the classical MD because the electronic structure calculation is the most time-consuming part. Classical MD scales with $O(N \ln N)$ or $O(N^2)$ (depending on a type of FF) and the large-scale FF-MD simulations can be performed on models of approximately 1000 nm in size and milliseconds, while the AIMD scales with $O(N^3)$ having the time length approaching to microseconds and the systems consisting of thousands atoms. Indeed, the AIMD has at least two main advantages. Firstly, this method is not parameter dependent and can be used for any system, and the accuracy is determined by the accuracy of the DFT approach used. Secondly, the AIMD can be taken for the systems, where electronic properties and atomic dynamics are significantly dependent. It means that the AIMD can be used also for the simulation of real chemical reactions where chemical bonds are changed. Such simulations cannot be performed in the framework of the classical MD except when using a specific reactive FF (see Chapter 2).

6.2.3.3 Monte Carlo Method

MC is a simulation method widely used in molecular modeling of chemical systems. Unlike MD methods, MC methods are stochastic techniques based on a statistical probability. They are frequently used not only in chemistry but, for example, also in economics. The use of MC methods to model physical problems allows us to examine complex systems that cannot easily be solved by integral calculus or other numerical methods. With MC methods, a large system can be sampled in a number of random configurations, and those data can be used to describe the system as a whole. The most popular MC scheme is the Metropolis MC method (1953). In this method, the PES of the system, which is defined by a FF approach, is scanned over randomly generated configurations. The initial potential energy is calculated for the first configuration, and then, particles in the system (e.g., atoms) are randomly displaced to a new configuration for which the potential energy is again calculated. If this energy is smaller than the previous one, the system is in the more stable configuration and this state is accepted to the statistical ensemble. However, if the new energy is higher than the previous one, the difference between both energies is compared to a random number via a Boltzmann distribution. If the value is smaller than the random number, the new configuration is accepted and used for the generation of other new one. In the opposite, if the value is bigger than the random number, the configuration is rejected and the original configuration is used for the generation of the next configuration. In this way, the PES can be mapped for millions of configurations. Applying statistical ensemble averages various properties such as thermodynamic quantities can be calculated.

MC simulations are preferable if there are large intramolecular energy barriers, which can result into molecules being trapped in a few low-energy conformations in MD simulation. Randomly generated moves in the MC simulation can more readily lead to a barrier crossing. In the opposite case, the simulation of liquids by MD can be more effective than by MC. There is a large probability of selecting random moves in the MC simulation of liquids, in which two or more molecules can overlap leading to a large number of rejected configurations. This results in dropping of the efficiency of the MC sampling. However, there are improvements of the MC methods leading to a better performance, for example, configurational bias MC (Siepmann and Frenkel, 1992) or hybrid MC (Duane et al., 1987). Grand canonical MC version (constant chemical potential (μ), V, and T) is very suitable for the study of the interfacial phenomena, for example, for modeling of adsorption isotherm (Snurr et al., 1993; Liu and Monson, 2005).

6.2.4 Models: Clusters, Periodic Systems, and Environmental Effects

Molecular modeling of geochemical processes at surfaces and interfacial phenomena requires a special attention in building proper structural model covering both solid surface and interacting phase (molecule, liquid phase).

Clusters are molecular models prepared by cutting of a fragment, which should reflect typical features of the interacting site plus representative large part of the solid phase and surrounding solvent molecules. If there are dangling bonds, they should be saturated (e.g., by addition of hydrogen atoms or pseudo atoms). Cluster method has several drawbacks such as missing long-range interactions, potentially imbalanced stoichiometry, or unwanted impact of the terminal atoms. Additionally, computed physical properties usually depend on the cluster size (Sauer, 1989; Pacchioni et al., 1992). Moreover, the computed properties obviously do not converge continuously with the increasing size of the clusters but often oscillate, which brings difficulties in their extrapolation. For example, Sukrat et al. (2012) showed that the cluster size and cavity structure are very important for predicting energetic barriers for the proton exchange reactions of C2–C4 alkanes in ZSM-5 zeolite. A decrement up to 20 kcal/mol was observed when employing the periodic model instead of using the small cluster model.

The problem of the missing environment in the cluster model calculations is overcome in embedding methods. Continuum solvation models can also be assigned to this group. They are developed for the calculation of solvent effects and are based on the self-consistent reaction field (SCRF) approach (Rinaldi and Rivail, 1973). The polarizable continuum model (PCM) and its variants (Miertuš et al., 1981; Cances et al., 1997; Tomasi et al., 2005) and/or the conductor-like screening model (COSMO) (Klamt and Schürmann, 1993; Klamt et al., 1998) belong to the most widely used SCRF methods in quantum chemistry. Generally, these methods use a polarizable continuum representing solvent, in which a cavity for a solute molecule is created. The molecular free energy of solvation is computed as the sum of electrostatic, dispersion–repulsion, charge transfer, and cavitation energy terms. For more details, see the review on the QM continuum solvation models (Tomasi et al., 2005).

The simplest embedding scheme is a model cluster inserted in a background of classical point charges, which found numerous applications, for example, in the description of surfaces of ionic solids. More sophisticated schemes include embedding of the cluster treated at the QM level in the backbone of atoms described by MM potential. These hybrid schemes are denoted as QM/MM methods. They link the main advantages of QM (accuracy) and MM (speed) methods. Moreover, QM/MM schemes can be used in the study of chemical reactions in "realistic" environments that is not possible through the classical MM methods. Further, the QM/MM method is highly appropriate for large macromolecules or associates/aggregates of various smaller organic species (e.g., NOM) including their environments. There are numerous variants of the QM/MM schemes that can be used in different applications in various fields. Details on the methodology, development, advantages, problems, and applications can be found in several reviews (e.g., Gao and Thompson, 1998; Friesner and Guallar, 2005; Vreven and Morokuma, 2006; Senn and Thiel, 2009; Canuto and Sabin, 2010).

Bulk properties of the solid-state structures with the translational periodicity are simulated by using a computational cell corresponding to a crystallographic unit cell or its multiplications (supercell approach). The later approach is also carried out in the simulations of liquids and aperiodic solids. A different level of complexity for calculations with the periodic boundary conditions is the modeling of the solid-state surfaces. A periodic supercell slab model is another alternative to the cluster and embedding schemes. The slab is cut from the bulk structure along specific directions usually dictated by the desired surface termination that is determined by the crystallographic planes (in the case of periodic solids). The slab is periodic in two dimensions, whereas a vacuum is imposed in the third dimension. This vacuum should be adequately thick to avoid artificial interactions between slabs in the neighboring cells if the codes with only possibility of 3D computational cells are used. There are many additional factors that must be taken into account in the slab construction. For example, the slab has two surfaces and both should be the same if possible. This should minimize an artificial polarization effect in the perpendicular direction with respect to the surface. The thickness of the slab is another factor. The electronic structure in the bulk and at the surface is significantly different due to the broken bonds, not fully coordinated atoms at the surface, and the relaxed uppermost atomic planes. Thus, the surface properties planned to be studied have to be checked on a convergence with respect to the slab thickness. This convergence is usually different for systems with localized and delocalized bonding states. In the case of solids composed from two or more types of atoms arises a question about what atomic plane is the topmost. As the surfaces are formed from the unsaturated atoms with respect to the bulk having dangling bonds, these atoms are chemically reactive. Unlike vacuum termination in "realistic" environment (with air gases or liquid phase) they react and a final chemically stable termination can be formed (e.g., from hydroxyl groups). These forms of the surfaces are usually a subject of investigations of the interactions of the molecular species and/or solid–liquid interfaces, respectively. In such investigations it is important to use a slab also having two periodic dimensions large enough to minimize lateral interactions between the interacting molecule and its replicas in the neighboring cells and also to avoid an artificial periodicity for amorphous liquid phase imposed on the surface. Technically, solid–liquid interfaces are modeled in two ways: a water slab of defined thickness is deposited on the surface and then vacuum is inserted above to separate the slab in the neighboring cells or the water slab is confined between two surfaces of the solid slab. In the latter case the dimension of the computational cell in the perpendicular direction with respect to the surface has to be tuned with respect to some physical parameter, for example, pressure. There are many works reporting on molecular simulations of properties and interactions of the solid-state surfaces and interfaces and also Chapter 8 in this book offers examples.

6.3 Applications

This section summarizes primarily the results of our research group achieved approximately in the last decade. The examples represent mainly molecular modeling studies of interactions and mechanisms of binding several polar and nonpolar organic contaminants with typical minerals and natural organic matter (NOM). It has to be emphasized again that a characterization of structure, composition, and properties of solid phases is very complex subject for both experiment and molecular modeling. Mineral surface termination, surface morphology, hydration, hydroxylation, redox processes, charging, exchange processes, distribution, and characterization of surface sites and defects represent a complex set of problems that are intensively studied as it is evidenced by numerous papers in this field (see, e.g., other chapters in this book especially Chapter 8).

An even more complex situation we face is in the case of NOM. NOM, as the largest pool of the organic carbon on the Earth, plays an important role in many biogeochemical processes including binding, stabilization, and biodegradation of the organic contaminants (Senesi et al., 2009). NOM has a heterogeneous composition consisting of various groups of organic molecules possessing an extremely complex 3D macromolecular structure that is not completely known. Therefore, only approximate models are suggested that are built on knowledge from an elemental analysis and types and distribution of various functional groups achieved from experiments. Several models have been discussed how the NOM components are held together. The most frequently used models are the polymer and supramolecular models. In the polymer model, NOM is considered as a macromolecule containing amorphous and crystalline domains (Ghosh and Schnitzer, 1980; Leboeuf and Weber, 1997; Xing and Pignatello, 1997). The supramolecular model (Cozzolino et al., 2001; Piccolo, 2002) defines NOM as a physicochemical body dominated by nonbonding interactions such as hydrogen bonds, van der Waals, and weak hydrophobic interactions, which keep individual molecules of primary structure together. It is important to note that a micellar or membrane-like model taking into account the amphiphilic character of HSs was also suggested (Wershaw, 1993). Even though there is no definitive consensus on the structure of NOM, the functional groups of the organic molecules present in their composition have been well described. These include carboxyl, hydroxyl, phenolic, alcohol, carbonyl, and methoxy groups. Among these, carboxyl and phenolic groups are regarded the key functional groups responsible for the adsorption of NOM by soil minerals in forming stable organomineral aggregates (Sutton and Sposito, 2006; von Lützow et al., 2006).

6.3.1 Modeling of Surface Complexes of Polar Phenoxyacetic Acid-Based Herbicides with Iron Oxyhydroxides and Clay Minerals

The contamination of soils and water resources increases progressively with growing production and application of chemicals for agricultural activities that represents a serious problem throughout the world (Kurtz, 1990; Mathys, 1994; Larson et al., 1997). Nearly two million tons of pesticides are applied to agricultural land worldwide each year (Fenner et al., 2013). From a rich group of pesticides, derivatives of phenoxyacetic acid represent a broad spectrum of herbicides extensively used in agriculture. Their behavior in soil environments (solubility, transport, adsorption–desorption, chemical resistance, and biodegradation) is governed by their chemical structure (Scheme 6.1). Molecular structures of these modern herbicides contain polar carboxylic group. This group is responsible for the relatively high chemical activity of these herbicides in interactions with soil components and also contributes to a reduction of their persistence in the environment allowing a better control (Kästner et al., 2014).

For example, MCPA is currently used to reduce the spread of annual and perennial broad-leaved weeds on land under grasses and in vineyards. The agronomic MCPA dosage is typically as high as 1–2.5 kg/ha, and it has been classified by the US Environmental Protection Agency as a potential groundwater contaminant (Walker and Lawrence, 1992). In MCPA, one methyl group of the phenoxy moiety is replaced by a chlorine atom, whereas in 2,4-dichlorophenoxyacetic acid (2,4-D) both methyl groups are replaced by Cl atoms (Scheme 6.1). This substitution increases their polarity and, consequently, acidity and solubility (Haberhauer et al., 2000, 2001). Typical half-life of MCPA in soils is about 24 days (Thorstensen and Lode, 2001), and also a combination of processes such as dilution, sorption, and degradation contributes to disappearance of herbicides from the environment (Hiller et al., 2012).

Many experiments studied sorption/desorption processes of phenoxyacetic acid herbicides in soils and/or soil components, respectively (Bolan and Baskaran, 1996; Pignatello and Xing, 1996; DePaolis and Kukkonen, 1997; Sannino et al., 1997; Susarla et al., 1997; Benoit et al., 1998; Celis and Koskinen, 1999; Celis et al., 1999; Cox et al., 2000; Haberhauer et al., 2000, 2001; Clausen et al., 2001;

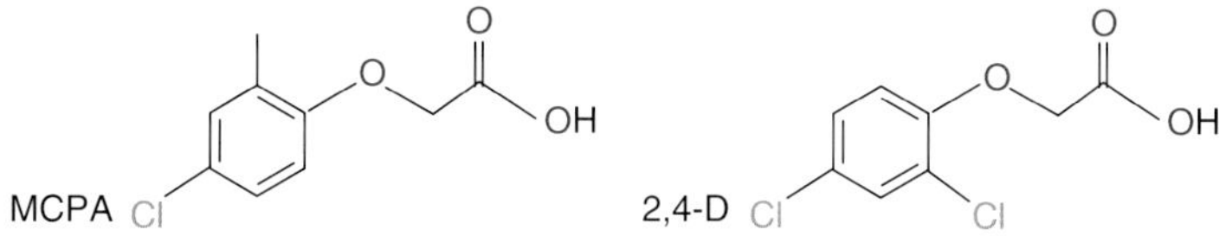

Scheme 6.1 *Skeletal formula of MCPA and 2,4-D molecules.*

Vasudevan et al., 2002; Spadotto and Hornsby, 2003). These experiments describe sorption from aqueous solutions to various solid matrices and provide mainly the soil–water distribution coefficient K_d and very rarely the adsorption energy or enthalpy calculated from fits to experimental data. It was observed that the adsorbed amount depends strongly on pH and a presence of electrolyte (e.g., $CaCl_2$) in solution (Cox et al., 2000; Vasudevan et al., 2002; Spadotto and Hornsby, 2003). The experiments demonstrated lower sorption capacity of the clay minerals at pH > 4 (what is above a pK_a value of 2,4-D (2.8) and MCPA (3.1)). A review on the distribution coefficients for sorbed phenoxyacetic acid-based herbicides in soils and soil components showed a mean $\log K_d$ of 0.16 ± 0.55 l/kg (2,4-D) and -0.10 ± 0.56 l/kg (MCPA), respectively (Werner et al., 2013).

6.3.1.1 Structure and Properties of Kaolinite, Montmorillonite, and Goethite

Clay minerals such as kaolinite and smectites and natural iron oxyhydroxides are common soil minerals. They are potential geosorbents for polar agrochemicals in soil having significant impact on the soil sorption capacity. Phenoxyacetic acids exist in the anionic form at pH >3–4 and may be physisorbed through deprotonated anionic carboxyl groups to positively charged (i.e., protonated) surface sites on minerals below their point of zero charge (PZC) (pH_{PZC}). Therefore, the low pH_{PZC} values of soil clays such as kaolinite or montmorillonite (Table 6.1) limit their sorption capacity with respect to anionic herbicides at ambient soil pH values (6–8), and direct sorption from the solution to neutral or negatively charged surfaces is not strongly preferred. Sorption of 2,4-D on clays was enhanced in the presence of ionic solution and was supposed a formation of a cation-bridged complex between herbicide anion and a negatively charged surface site of clay (Clausen et al., 2001). A ligand exchange mechanism was assumed in the sorption of 2,4-D to the synthetic chlorite-like complexes ($Al(OH)_x$ coated montmorillonite surfaces) in the presence of acetate or phosphate buffers at pH 5–6 (Sannino et al., 1997). The neutral form of phenoxyacetic acids dominates at low pH

Table 6.1 Typical physical and chemical characteristics of kaolinite, montmorillonite, and goethite.[a]

Geosorbent	Kaolinite	Montmorillonite	Goethite
Chemical composition	$Al_2Si_2O_5(OH)_4$	$Na_{x+y}(Si_{8-x}Al_x)(Al_{4-y}Mg_y)$ $O_{20}(OH)_4 \cdot nH_2O$	α-FeOOH
pH_{PZC}	~4	4–6	7.5–9.5
SSA^b (m^2/g)	10–30	300–600	50–200
CEC^c (cmol$_c$/kg)	3–15	80–150	Up to 100
Dominant/ typical surface	001	001	110
Permanent surface charge ($|e|$)	~0	0.1–0.5	~0
Surface termination	Surface OH (octahedral) basal surface oxygen atoms (tetrahedral)	Basal surface oxygen atoms + compensating cations (e.g., Na^+)	–OH μ-OH μ_3-OH
Typical particle shape	Plates, discs	Irregular flakes	Idles, rods
Particle size in soils (μm)	<2–3	<1–2	<1

[a]Specific references are not provided as the literature offers a variety of results.
[b]Specific surface area.
[c]Cation exchange capacity.

and can be sorbed onto clays or oils with a high clay content (Clausen et al., 2001; Vasudevan et al., 2002). In this form, neutral molecules are physisorbed through hydrogen bonds with the clay surfaces that can be terminated by oxygen atoms or hydroxyl groups. It has to be noticed that in the sorption of either neutral or ionic form of phenoxyacetic acid herbicides from solution, the competitive sorption of water molecules plays an important role. Celis et al. (1999) showed that dried clays apparently adsorbed water in preference to solved herbicides, leading to an increase in the pesticide concentration in the solution (negative sorption). A comprehensive overview of the mechanisms of sorption of 2,4-D on soils and soil components can be found in the work by Hyun and Lee (2005).

Apart from clays, significant sorption of phenoxyacetic acids to oxides and oxyhydroxides (e.g., γ-alumina, goethite, hematite, ferrihydrite) was observed (see review by Werner et al. (2013)). These minerals have more complicated surface morphology and structure with more defects than clay minerals. The enhanced sorption capacity can be assigned mainly to the presence of positively charged (protonated) surface sites.

Iron oxyhydroxides (FeOOH) contribute to large specific surface areas of soils, even though they are present in relatively low amounts in bulk soils (just a few mass percent) (Pronk et al., 2011). The Fe-oxyhydroxides are very good sorbents for phenoxyacetic acid herbicides because they have a high pH_{PZC} values and large specific surface areas (Table 6.1). The herbicide partition coefficient (K_d) values are three orders of magnitude higher for Fe-oxyhydroxides than for other soil minerals and one order of magnitude higher even than for SOM (Werner et al., 2013). Sorption of phenoxy herbicides to only Fe-oxyhydroxides was found sufficient to fully explain their K_d values taking into account the strong soil pH dependence (Clausen and Fabricius, 2001). Anionic MCPA is also barely adsorbed by HSs because of their low pH_{PZC} values (Iglesias et al., 2009). Iron oxyhydroxides in soil are often coated with a humic matter, but this was not found to decrease significantly their sorption capacities for phenoxy herbicides at ambient soil pH values (Iglesias et al., 2010).

The interaction mechanism of adsorption of phenoxyacetic acids to Fe-oxyhydroxide surfaces is still not fully clear. There are interpretations of adsorption isotherms by means of the formation of monodentate or bidentate inner sphere surface complexes but without further evidence (Pronk et al., 2011). The FTIR spectra of adsorbed phenoxyacetic acid herbicides also did not provide unambiguous confirmation about the mechanism of binding because measured spectra were complex and difficult to interpret with overlapping low-intensity bands in the spectroscopic regions of interest (Boily et al., 2000; Iglesias et al., 2009). Thus, both outer sphere and inner sphere complexes of MCPA with the surface sites of goethite were suggested as possible.

Complex surface chemistry of soil minerals can be also related to their particle size and shape. In soils, minerals exist in a form of small particles from nano- up to a few microns in size.

It means a large surface/volume ratio and also many defected sites due to broken bonds at edges and their termination. The properties such as cation exchange capacity (CEC) and specific surface area (SSA) depend strongly on the particle size, morphology, and isomorphic substitutions in natural samples (e.g., Al^{3+}/Si^{4+} in kaolinite (Ma and Eggleton, 1999)). The CEC and SSA values for kaolinite, montmorillonite, and goethite minerals are collected in Table 6.1.

Many microscopic techniques such as SEM, AFM, and TEM showed that typical particle shapes for *kaolinite* and *montmorillonite* minerals are thin plates or flakes, more irregular in the case of montmorillonite (Bauer and Berger, 1998; Zbik and Smart, 1998). These minerals have a layered structure with a strong anisotropic character (Figure 6.1a and b) where the bonding within the layers is much stronger than between layers.

Kaolinite is the dioctahedral phyllosilicate consisting of one octahedral and one tetrahedral sheet. The surface termination on the octahedral side is formed by a basal surface hydroxyl groups, $-OH_b$, whereas the tetrahedral sheet is terminated by basal surface oxygen atoms, O_b. The linking between

(a)   (001) Surface (–OH groups)

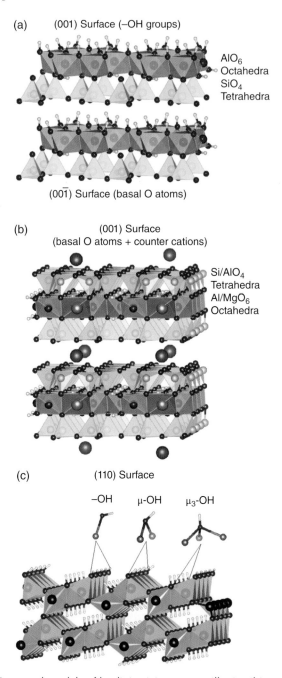

AlO$_6$
Octahedra
SiO$_4$
Tetrahedra

(00$\overline{1}$) Surface (basal O atoms)

(b)   (001) Surface
(basal O atoms + counter cations)

Si/AlO$_4$
Tetrahedra
Al/MgO$_6$
Octahedra

(c)   (110) Surface

–OH   μ-OH   μ$_3$-OH

Figure 6.1 *Structural models of kaolinite (a), montmorillonite (b), and goethite (c).*

layers is formed by hydrogen bonds between the surface hydroxyl groups of one layer and the basal oxygen atoms of the adjacent layer. These surfaces are parallel with (001) crystallographic plane and also represent the dominant surfaces of the kaolinite particles as they form about 70–90% of the total particle surface. The rest of the surface is formed by the broken edges where dangling bonds are saturated with chemically adsorbed water (Liu et al., 2012b). It is supposed that the hydroxyl groups formed at the broken surfaces are more acidic than the $-OH_b$ groups and contribute to the low PZC of kaolinite (Liu et al., 2012b). However, their concentration is much lower than the concentration of the $-OH_b$ groups. DFT calculations on the interactions of water and acetic acid molecules with the (001) kaolinite surfaces showed that the octahedral surface of the kaolinite layer is chemically more active than the tetrahedral surface because the hydrogen bonds formed by the basal surface hydroxyl groups are stronger than the weak hydrogen bonds to the basal oxygen atoms (Tunega et al., 2002a, 2004a). The calculated interaction energies for polar species were about two to three times larger for the octahedral surface than for the tetrahedral one.

Montmorillonite is a member of the rich smectite family and has more complicated chemistry comparing to kaolinite. It is dioctahedral 2 : 1 phyllosilicate having one layer composed from two tetrahedral sheets sandwiched over an octahedral sheet (Figure 6.1b). Therefore, a layer has only one type of the (001) surface formed from the basal surface oxygen atoms. Moreover, a low concentration of isomorphic substitutions in sheets is responsible for a permanent negative charge of the layers (Table 6.1) that are compensated by mono- or divalent cations such as Na^+ or Ca^{2+}. These cations are located in the interlayer space keeping the layers together by electrostatic interactions and can be found also on the top of the particle surfaces (Figure 6.1b). Cations are usually hydrated and, as their concentration is relatively low, interlayer forces are not very strong. Smectite minerals can easily swell by uploading a lot of water into the interlayer space. Generally, montmorillonite is considered as a better geosorbent than kaolinite. It is reflected in physical characteristics such as SSA and CEC (Table 6.1).

Goethite, α-FeO(OH), belongs to a widely distributed hydrous iron oxides in terrestrial soils, sediments, and ore deposits (Cornell and Schwertmann, 2003). Its bulk structure (*Pbnm* space group) is characterized by a slightly distorted hexagonal close-packed O-atom arrangement with Fe atoms occupying one-half of the octahedral interstices (Forsyth et al., 1968; Szytula et al., 1968; Hazemann et al., 1991). The FeO_6 octahedra share edges to form double chains, which are further linked to form a 3D structure by sharing vertices. The bulk structure contains H atoms bound to one O atom from the structural net, and the OH groups are involved in the intrastructural hydrogen bonds.

The surface structure, particle size, crystallinity, and morphology are important factors for the sorption affinity of goethite. In soils, goethite exits in a form of rodlike or lath-like particles with size up to a few μm, and its SSA is relatively high varying in a broad range of 50–200 m^2/g (Schwertmann and Cornell, 1991). Recently, the first-principles-based simulations predicted rhombohedral prisms as the thermodynamically most stable nanoparticles of goethite (Guo and Barnard, 2011). One of the most populated goethite surfaces is parallel to the (110) crystallographic plane (if *Pbnm* notation is used), and its model is displayed in Figure 6.1c. The surface is formed from three different hydroxyl sites at full bond saturation. According to the standard International Union of Pure and Applied Chemistry (IUPAC) nomenclature, these sites are labeled as (i) hydroxyl groups bound to one iron atom (–OH in Figure 6.1c) refer to hydroxo sites, (ii) hydroxyl groups connected to two iron atoms (μ-OH in Figure 6.1c) refer to μ-hydroxo sites, and (iii) sites, where oxygen atom is bound to three iron atoms (μ_3-OH in Figure 6.1c), refer to μ_3-hydroxo sites.

The acid–base properties of the goethite surface hydroxyl groups are of special interest as they are a key to understand a binding behavior of contaminants. It is practically impossible to obtain experimentally proton affinity constants, pK_a, for individual types of sites. Usually, acid–base potentiometric titrations provide PZC, an overall pH value, at which the mineral surface is charge balanced. For goethite, pH_{PZC} values are in a range of 7.5–9.5 (see, e.g., Venema et al. (1998) or Gaboriaud and Ehrhardt (2003)), meaning that a base character of goethite surface prevails. In the geochemical community a multisite complexation model (MUSIC) is frequently used to explain titration curves and to predict pK_a values for the individual sites (Hiemstra et al., 1989a, b). This model is empirical and, depending on the initial parameters, can provide different pK_a of the same site. For example, using different empirical parameters, protonation constants of 3.7, 7.7, and 11.7 were obtained for the –OH groups of the goethite surfaces (Hiemstra et al., 1989b). In this respect molecular simulations have shown up as an alternative tool to predict various physical properties including proton affinity constants. A cluster model approach (clusters with six Fe atoms) combined with explicit (up to six water molecules) and global solvation (COSMO model (Klamt and Schürmann, 1993; Klamt et al., 1998)) was used in a prediction of the pK_a values for the surface OH groups of goethite (Aquino et al., 2008a). In the study the spin-unrestricted B3LYP approach was used in combination with the SVP+sp basis set. SVP+sp is the split valence polarization basis set (Schäfer et al., 1992) augmented with diffuse s and p functions on heavy atoms in order to reduce the basis set superposition error (BSSE). It was found that the SVP+sp basis set reduces substantially the BSSE for hydrogen-bonded systems (up to 1–3 kcal/mol) (Aquino et al., 2002). The study (Aquino et al., 2008a) predicted a pK_a value of 12.1 for the –OH site (singly bound to Fe). Later, this value was revised to 7.0 by means of a periodic slab model of the goethite surface considering explicit solvation by a water slab and AIMD simulations and PMF calculations (Leung and Criscenti, 2012). This site is of a particular interest because at low pH it can easily bound an additional proton forming positively charged sites $-OH_2^+$ on the goethite surfaces, which can be more attractive for binding of polar or negatively charged species.

This section presents a comprehensive overview of our molecular modeling simulations on interactions of phenoxyacetic acid herbicides both in neutral and ionized forms with three typical soil minerals: kaolinite, montmorillonite, and goethite (Tunega et al., 2004b, 2007; Aquino et al., 2007a; Kersten et al., 2014). The studies were conducted as an effort to look at a basic sorption mechanism, the role of surface termination, pH, and solvent effect. The structural models of the most populated typical surfaces of all three minerals were investigated (Table 6.1 and Figure 6.1).

Methods. Except Aquino et al. (2007a) all calculations were performed on periodic slab models and details on their constructions can be found in the corresponding references. In the case of kaolinite and montmorillonite a single layer was selected to represent (001) surfaces, whereas for the goethite (110) surface a slab with two layers of the FeO_6 octahedra (Figure 6.1c) was constructed from the bulk structure. In all models a vacuum of some thickness (>15 Å) was added above the surface to avoid artificial interactions between periodic images. Prior to any type of calculations, lateral dimensions were optimized together with a relaxation of the atoms in the slab model.

In the work Aquino et al. (2007a) the cluster models of the goethite (110) surface with four, six, and eight Fe atoms were constructed to describe possible interaction sites of all types of the surface OH groups with 2,4-D herbicide (molecular and anionic). The valences of broken bonds were completed with hydrogen atoms, and the overall charge of the clusters was held zero. In the cluster calculations only the surface OH groups plus interacting moiety were optimized, and the remaining atoms in the clusters were fixed to preserve a surface geometry. As the clusters contained Fe atoms, the calculations were performed in a spin-unrestricted DFT formalism applying the B3LYP/SVP+sp approach. The cluster calculations were performed using the computer program TURBOMOLE (Ahlrichs et al., 1989).

The electronic structure calculations on the periodic slab models (Tunega et al., 2004b, 2007; Kersten et al., 2014) were performed using the VASP program (Kresse and Hafner, 1993; Kresse and Furthmüller, 1996a, b). In the program, the Kohn–Sham equations are solved variationally in a plane wave basis set. The electron–ion interactions were described using the projector augmented wave (PAW) method (Blöchl, 1994), and the electron exchange–correlation interactions were described by the standard functionals of the GGA theory. In all of our studies, we used the popular DFT functionals either PBE (Perdew et al., 1996) or PW91 (Perdew and Wang, 1992), respectively. Both functionals are generally considered as essentially equivalent, producing similar results, and for hydrogen bonding they are relatively precise. The core electrons and nuclei of the atoms were represented by the PAW pseudopotentials (Kresse and Joubert, 1999).

The Brillouin zone sampling was restricted to the Γ-point since the computational unit cells in all cases were sufficiently large. The plane wave cutoff energy, at least, of 400 eV was used in all cases of static structural relaxations which correspond to great precision in the calculation. The optimization of atomic positions was performed without any restrictions by means of a conjugate-gradient algorithm with a stopping criterion of 10^{-5} eV for the total energy change and 0.01 eV/Å for the maximal allowed forces acting on each atom.

The models of the kaolinite and montmorillonite slabs are electronically closed shell systems, whereas the goethite slab contains transition metal iron in the structure. Therefore, the calculations on the periodic models had to be performed in spin-polarized (open-shell) DFT formalism. Moreover, Fe d-electrons are strongly correlated and standard DFT is not very suitable for their correct electronic description. Instead of that we used the so-called DFT+U (Dudarev et al., 1998) approach developed for a better description of the strong Coulomb repulsion between the localized d-electrons. The effective on-site Coulomb and exchange interaction parameters for each iron atom were set to 4 and 1 eV, respectively, as used in a previous study of the bulk structure of goethite (Tunega, 2012).

It was shown experimentally (Forsyth et al., 1968) and later confirmed by the DFT calculations (Kubicki et al., 2008; Otte et al., 2009; Tunega, 2012) that the high-spin antiferromagnetic configuration is the most stable state of the bulk structure. However, it is not still clear what is the real spin configuration of iron atoms nearby the crystal surfaces. Therefore, in the calculations on the slab model, the spin configuration from the bulk was preserved.

As the studied surface complexes of phenoxyacetic acid herbicides are relatively weakly bound systems, in some cases explicit solvation was used by adding water molecules and extensive MD calculations were performed to achieve the dynamic features. Then, from statistical analysis of the MD data, some characteristics were calculated (e.g., radial distribution functions (RDFs)). The AIMD calculations were performed at finite temperature ($T = 300$ K) in the canonical (NVT) ensemble applying a Nosé–Hoover thermostat (Nosé, 1984). Newton's equations of motion were integrated using the velocity Verlet algorithm (Ferrario and Ryckaert, 1985) with a time step of 1 fs. The first phase of the MD was the equilibration in a duration of 10–15 ps (if necessary longer) followed by a production phase of at least 10 ps, from which typical characteristics (e.g., structural) were evaluated. The equilibration phase was controlled by a temporal evolution of the internal energy of the system.

6.3.1.2   *Kaolinite· · ·Phenoxy Herbicide Complexes*

Static relaxation and MD simulations were performed on models of three derivatives of the phenoxy-acetic acid, 2,4-D, MCPA, and 2,4-M, interacting with the active octahedral surface of the kaolinite layer terminated by the surface hydroxyl groups (Tunega et al., 2004b). The 2,4-M molecule has two methyl groups instead of two Cl atoms in 2,4-D in the positions 2 and 4. The structures of the optimized models are shown in Figure 6.2. The polar carboxyl group forms multiple hydrogen bonds with the surface OH groups. The carbonyl oxygen atom is involved in the interactions with three surface hydroxyl groups acting as proton donors in the hydrogen bond of a moderate strength (according the classification of hydrogen bonds (Scheiner, 1997)) with O· · ·H distances between 1.8 and 2.5 Å. The OH group of the carboxyl group forms a very strong bonding to one flexible surface hydroxyl group that acts as a proton acceptor. The carboxyl proton is almost in the center between the carboxyl oxygen atom and the surface OH group with O—H lengths of about 1.2 Å. A similar situation was also observed in the study of the interactions of acetic acid molecule with the same kaolinite surface (Tunega et al., 2002b). It can be concluded that the kaolinite surface

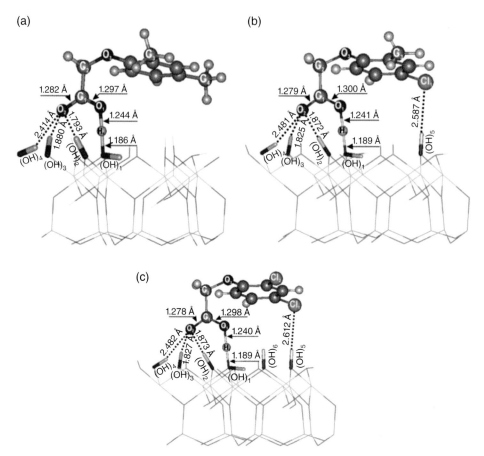

Figure 6.2   *Optimized structures of hydrogen-bonded complexes of 2,4-M (a), MCPA (b), and 2,4-D (c) on the (001) surface of kaolinite.*

hydroxyl groups are very flexible and able to act as proton donor and/or acceptor depending on the respective partner.

Calculated interaction energies or, in other words, complexation energies, ΔE_c, were obtained as a difference between total energies of the optimized surface complex and a sum of the total energies of the optimized individual partners in the complex. They are relatively large reflecting the number and a type of the hydrogen bonds formed (Table 6.2). Moreover, the binding energy increases with the increasing polarity of the phenoxy molecules (from 2,4-M to 2,4-D). Further inspection of the geometries showed that the aromatic parts of all molecules investigated are nearly parallel to the layer surface due to the flexibility of the torsional C—C—O—C angle and the formation of specific interactions between chlorine atoms and some surface hydroxyl groups (Figure 6.2). Additionally, attractive dispersion forces between the bonds of the benzene ring and the surface hydroxyl groups are expected to contribute to the interaction energy as well. Therefore, the presence of the phenoxylic group and Cl atoms enhances the adsorption energies of all three molecules in comparison to the corresponding value computed for acetic acid by about 3–8 kcal/mol (Table 6.2). However, it is expected that in the presence of a strong polar solvent such as water, the weak bonding will be disrupted because of the high affinity of the water molecules to the hydroxylated kaolinite surface.

Table 6.2 *Calculated interaction energies for complexes formed on octahedral (001) surface of kaolinite (K) and (110) surface of goethite (G).*

Model system	ΔE_c(kcal/mol)	Method/basis set/model	Figures
K$\cdots$H$_2$O	−14.7	PW91/PW/periodic slab	
K$\cdots$HAc	−20.8		
K$\cdots$2,4-M	−23.7		6.2a
K$\cdots$MCPA	−26.9		6.2b
K$\cdots$2,4-D	−28.0		6.2c
GFe$_6\cdots$H$_2$O (μ-OH−μ_3-OH)	−13.2	B3LYP/SVP+sp/cluster[a]	6.4a
GFe$_6\cdots$2,4-D (-μ-OH−μ_3-OH)	−25.9		6.4b
GFe$_6\cdots$2,4-D (-μ-OH−μ_3-OH)	−31.3		6.4c
G$\cdots$H$_2$O	−16.4/−17.6/−18.0	PW91/PW/periodic slab	
G$\cdots$MCPA[b]		PBE/PW/periodic slab	
nos-ng-OH−μ_3-OH	−32.0 PT[c]		6.5
nos-ng-OH−μ-OH	−25.7 PT		−
nos-ng-OH−OH → *nos-ng*-OH−μ_3-OH	−32.1 PT		6.5
nos-ng-μ-OH−μ_3-OH	−17.3		−
nos-ng-μ-OH−μ-OH	−16.7		−
nos-ng-μ-OH−OH	−11.1		−

[a]Without BSSE correction.
[b]*nos-ng, neutral outer sphere* complex on *neutral goethite* surface.
[c]PT, proton transfer.

Water molecule can also form multiple hydrogen bonds with the octahedral surface (Tunega et al., 2002a, 2004a) with a corresponding large interaction energy (Table 6.2) and will block available surface sites. Moreover, another destabilization effect in the adsorption from the solvent is relatively high solvation energy of the polar molecules (e.g., −14.2 kcal/mol for the MCPA molecule or −72.7 kcal/mol for the MCPA anion, our PBE/TZVP/COSMO results). The considerably large solvation energy of the MCPA anion can be a reason why at pH 6–8 phenoxyacetic acid herbicides sorb to kaolinite only minimally. Detectable sorption amount on kaolinite was observed at low pH values (Cox et al., 2000; Vasudevan et al., 2002), at which neutral form of herbicides dominates in water solution (p$K_a \sim 3$).

The interaction energies in Table 6.2 are relatively large (in absolute value). However, it is important to note that these energies are only differences of electronic energies for the association type of reaction (A + B → A$\cdots$B) not including zero-point vibrational energy, thermal corrections, and standard state correction energies for water. For the association reactions a large entropy lost effect is typical that can account for 10–12 kcal/mol (Aquino et al., 2002). Thus, for the estimation of the Gibbs free energies, the presented interaction energies in Table 6.2 should be reduced by about 12–15 kcal/mol. If we consider an exchange reaction mechanism instead of the association mechanism, then reaction energies are smaller. For example, for the exchange reaction on the hydroxylated kaolinite surface (Eq. 6.21), where one water molecule is replaced by one MCPA molecule, the reaction energy accounts for −10.4 kcal/mol:

$$K\cdots H_2O + MCPA \rightarrow K\cdots MCPA + H_2O \qquad (6.21)$$

The work Tunega et al. (2004b) also included short AIMD simulations that showed a dynamic picture of the strong bonding of the phenoxyacetic acids to the octahedral kaolinite surface. Frequent proton jumps of the carboxyl proton were observed during the simulations. The AIMD simulation also showed a coupling between the creation and breaking of the hydrogen bonds of the carbonyl

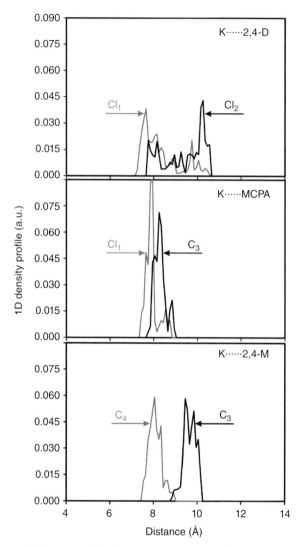

Figure 6.3 *One-dimensional density profiles for C (−CH₃ groups) and Cl atoms. Basal oxygen atoms from tetrahedral sheet represent a reference plane. Density profiles of kaolinite layer are not shown.*

oxygen atom on the one side and the chlorine atoms of the 2,4-D molecule on the other side. Figure 6.3 shows one-dimensional (1D) density profiles for C (CH₃ groups) and Cl and atoms in a perpendicular distance with respect to the plane of the basal oxygen atoms from the tetrahedral sheet (density profile of the kaolinite layer is not shown). Relatively broad peaks illustrate the torsional flexibility of the aromatic ring and alternating weak bonding of the Cl atoms in the 2,4-D molecule.

6.3.1.3 *Goethite···Phenoxy Herbicide Complexes*

Cluster models and B3LYP/SVP+sp approach were used in the study of the complexes formed on the (110) surface of goethite (Aquino et al., 2007a). Particularly, the interactions of water molecule,

(a)

(b)

(c)

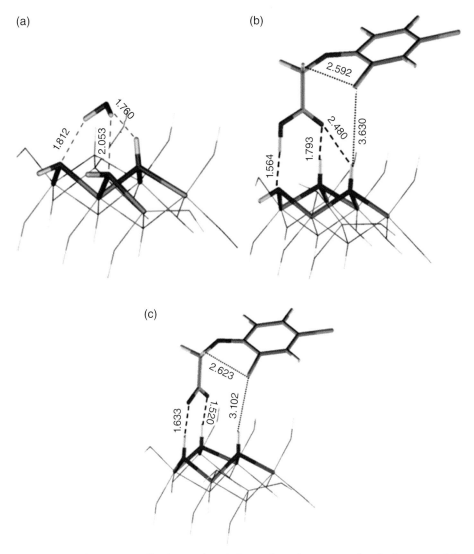

Figure 6.4 *Optimized structures of hydrogen-bonded complexes between molecular fragment of the goethite (110) surface and water (a), 2,4-D molecule (b), and 2,4-D⁻ anion (c). Distances are in Angstrom.*

acetic acid, 2,4-D molecule, and corresponding anion (2, 4-D⁻) were investigated. Figure 6.4b and c displays the optimized geometries for the cluster containing six Fe atoms (GFe_6). This cluster mimics the interaction sites formed by μ-OH and μ_3-OH hydroxyl groups, respectively. The size and number of the clusters in this study were limited as the QC calculations were extremely time-consuming because of a very slow SCF convergence. In spite of that it was found that the topology of the surface hydroxyl groups offers a variety of possibilities for hydrogen bonding with polar adsorbents. Hydroxo and μ-hydroxo groups have a sufficient flexibility allowing them to act as proton donors and/or acceptors. The third type, μ_3-OH, acts only as a proton donor because of its rigidity. Calculated interaction energies on different sites are about −13 to −15 kcal/mol for the water molecule. These values are in line with the number, type, and strength of hydrogen bonds formed. Table 6.2 shows the result for the μ-OH–μ_3-OH sites. Stronger interactions were observed for the neutral molecules of acetic acid and 2,4-D in comparison to the goethite/water complexes, for example, interaction energy of -25.9 kcal/mol was obtained for the 2,4-D (B3LYP/SVP+sp). The deprotonated, anionic form of acetic acid and 2,4-D showed even stronger interactions with energies between -30 and -50 kcal/mol, respectively. The hydrogen-bonding topology between neutral molecules and the surface hydroxyl groups is similar to that observed for the kaolinite octahedral surface. This is supported also by similar interaction energies collected in Table 6.2.

The complexes of the 2, 4-D$^-$ anion differ from those of the neutral molecule. The carboxylate oxygen atoms are the proton acceptors to the surface hydroxyl groups forming relatively strong hydrogen bonds with distances of 1.5–1.7 Å and the interaction energies greater than 30 kcal/mol (in absolute value). The particular example showed in Figure 6.4c displays hydrogen bonding with μ-OH and μ_3-OH sites. Chlorine atoms also contribute to the additional stabilization similarly to the neutral 2,4-D molecule.

However, the cluster model approach offered a limited number of possibilities to study the interactions of the phenoxyacetic acid molecules with the complex structure of the (110) surface of goethite. In ongoing molecular simulation work on interactions of the MCPA with the (110) goethite surface, we used a periodic slab model (Figure 6.1c) with the parameters of the orthorhombic computational cell ($a = 30.0$ Å, $b = 11.0$ Å, and $c = 9.1$ Å) and a vacuum spacing of 20 Å imposed along the a direction. The study also includes the simulation of different pH conditions achieved by adding proton(s) to the surface and/or removing acidic proton from the MCPA molecule. The periodic surface model has three types of the surface hydroxyl groups (Figure 6.1c), but according to their predicted pK_a values (Aquino et al., 2008a), only –OH groups are the candidates for a protonation below pH_{PZC}. However, it was not possible to protonate all –OH groups of the slab model. The test calculations led to a structural instability during the static geometry optimization because of extreme polarization and electrostatic repulsion effects. Therefore, only a partial protonation of the –OH groups was considered in a vicinity of the adsorbed MCPA moiety.

There are two basic mechanisms suggested for the binding of the MCPA molecule to the goethite surface: (i) outer sphere complex or (ii) inner sphere complex (Iglesias et al., 2010). In the former case the hydrogen bonds are formed between carboxyl or carboxylate and surface hydroxyl groups. In the latter case a direct chemical binding is formed and a firm Fe—O—C$_{MCPA}$ bond bridging is created. In other words, one surface hydroxyl group is replaced by carbonyl oxygen atom from the MCPA and a monodentate inner sphere complex is formed. In the case of the MCPA anion, bidentate binding can also be considered. However, from steric reasons these complexes are less probable.

Taking into account typical characteristics such as complex surface topology of the surface OH groups, protonation states (pH effect), and a form of binding (outer or inner sphere complexes), many surface complexes have to be considered. Therefore, we have created 26 models in total and carried out a full relaxation. The relative energetic stabilities of the particular configurations groups were then compared. Here we are presenting two aspects of these studies. On the example of the outer sphere complexes between the neutral MCPA molecule and the neutral goethite surface, a complexity involved in the formation of the complexes (denoted as *nos-ng* models in Table 6.2) is demonstrated. The second aspect is related to the fact that in the experiment the amount of adsorbed MCPA from water solution increased if the pH decreases reaching the adsorption edge at pH ~4 (Kersten et al., 2014). At such pH conditions the models of the outer and inner sphere complexes of the neutral MCPA on a partially protonated goethite (*pg*) surface were used in the calculations (denoted as *nos-pg* and *nis-pg* models, respectively). Furthermore, in order to explain the interaction mechanism of the MCPA from solvent, in the calculations an explicit solvation phase was included through a slab of water molecules with a thickness of approximately 18 Å. However, these complex models required more sophisticated approach than a simple static relaxation procedure used for the bare surface complexes to find an energetic minimum. Thus, extensive AIMD calculations

(T = 300 K and 25 ps MD) were performed on these two models in order to achieve the dynamic features of the surface complexes.

Nos-ng *Complexes.* Table 6.2 collects the calculated interaction energies for different configurations of the neutral MCPA molecule adsorbed on the neutral goethite surface. In total, six different initial configurations have been optimized. The MCPA molecule has two sites for hydrogen bonding (proton-donating –OH group and proton-accepting carbonyl oxygen). The labels of the six complexes in Table 6.2 contain initial positions of these two sites with respect to the surface hydroxyl groups. From the calculated interaction energies, it is evident that the –OH group is dominant in the surface complexation. The first three outer sphere complexes in Table 6.2 are bound much stronger than the remaining three. During their optimization process a proton transfer from the –COOH to the surface –OH group was observed. Moreover, during the optimization the complex *nos-ng*-OH–OH changed to the *nos-ng*-OH-μ₃OH configuration. The final most stable structure with interaction energy of about −30 kcal/mol is displayed in Figure 6.5. Hydrogen bonds formed between MCPA and the surface are depicted directly in Figure 6.5 together with their O··· H distances. Two strong hydrogen bonds are formed between the protonated surface –OH group and μ₃-OH group on one side and two carbonyl oxygen atoms of the –COO⁻ group on opposite side. The hydrogen bond formed between another –OH group and carbonyl oxygen atom is of a moderate strength with a length of 2.163 Å. The position of the MCPA anion allows also the formation of a

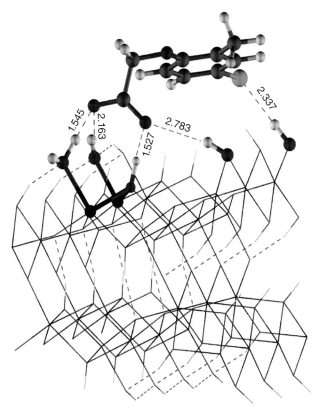

Figure 6.5 *The most stable optimized structure of outer sphere complex of MCPA⁻ anion formed on the (110) surface of goethite. Distances are in Angstrom.*

weak hydrogen bond with one μ-OH group. The stabilization of the MCPA on the surface is also enhanced by a hydrogen bond registered between Cl atom and one hydroxyl group from the μ-OH groups (Figure 6.5). The hydrogen bonding of the MCPA molecule to the goethite surface is similar to the hydrogen bonding observed between MCPA and the octahedral kaolinite surface (Figure 6.2b). The only difference is in the stabilization energies (−32.1 cf. −26.9 kcal/mol; Table 6.2).

After the proton transfer the MCPA–goethite complex can be regarded as the outer sphere complex formed between MCPA anion and partially protonated goethite surface. In this case the association energy is −42.0 kcal/mol. However, also here the same factors affecting the stability of the surface complex as discussed in the section about the adsorption of phenoxyacetic acid derivatives on kaolinite have to be considered. The calculated energies are without thermal corrections and solvent effects are also missing. The calculated interaction energies of several configurations of the isolated water molecule (Table 6.2) indicate that the surface solvation can be strong. It has been also shown that the solvation energies of the neutral and deprotonated MCPA are large. These effects lead to the strong destabilization of the surface complexes. However, their inclusion to the models makes the calculations complicated and much more time-consuming.

Nos-pg *and* **nis-pg** *Complexes.*   These two candidates were preselected as the most probable complexes for the outer and inner sphere binding of the neutral MCPA molecule to the partially protonated goethite surface at the adsorption edge of pH ~4 (Kersten et al., 2014). Inclusion of an explicit solvation by adding a water slab to the optimized bare complex required the change of the computational strategy from simple geometry optimization to the AIMD. Structural and dynamic features presented here were collected from 10 ps AIMD performed after the equilibration phase in a duration of 15 ps.

Figure 6.6 (upper part) displays the snapshot from the AIMD simulation of the outer sphere complex of the neutral MCPA molecule bound to the partially protonated goethite surface ($-OH_2^+$ site) and solvated by a water slab. Prior to the solvation different structural models were optimized and one distinct structure was found to be most stable. The carbonyl oxygen atom forms a moderate hydrogen bond with the protonated site, whereas the carboxylic proton is shared with a singly coordinated neighboring surface hydroxyl group. This model was explicitly solvated and then the MD simulation was performed. During the equilibration phase it was already observed that the water molecules strongly perturbed the outer sphere complex. The hydrogen bond of the carboxyl group was destabilized and proton from the carboxyl group migrated and was shared by neighboring water molecules. The water molecules also partially perturbed the second hydrogen bond between the positively charged surface site and the carbonyl oxygen atom as it is shown in Figure 6.6. The analysis of the production phase gave averaged distances of 1.0 and 1.9 Å in the O–H⋯O bridge. The bottom picture in Figure 6.6 displays 1D density profiles (across a perpendicular distance to the goethite surface) for selected atom types. Dashed arrows show the assignment of the several important peaks. The position of the MCPA molecule with respect to the surface is colored by cyan, and three peaks enhanced with the black lines display the positions of the oxygen atoms of the –COOH group from the MCPA molecule and the shared proton of the protonated surface site. The fourth black curve corresponds to the position of the Cl atom. It is evident that Cl is not in contact with the goethite surface as it was observed, for example, in the optimization of the bare complexes of the MCPA molecule (Figure 6.5). The hydrogen bond network is very dynamic and the distribution of the water molecules is interesting (blue line in Figure 6.6). Intensive peaks nearby the surface demonstrate a strong hydration of the surface and of the polar –COOH group that destabilizes the outer sphere complex. Then, the density of the water molecules decreases in the middle part because the space

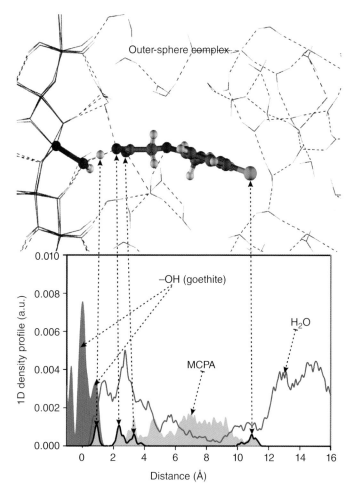

Figure 6.6   *One-dimensional density profiles for hydrated outer sphere complex of neutral MCPA molecule on the partially protonated (110) surface of goethite obtained from AIMD simulation.*

is filled by a hydrophobic part of the MCPA molecule (see clear cavity around aromatic ring in Figure 6.6). Chlorine atom forms weak hydrogen bonds with the water molecules distributed around and the remaining space is again filled out by water molecules.

A similar AIMD scenario was also applied for the inner sphere monodentate complex with the chemically bound MCPA to the goethite surface via the Fe—O—C bridge. The reaction can be formally expressed for the protonated –OH site as

$$[Fe-OH_2]^+ + MCPA \rightarrow [Fe-MCPA]^+ + H_2O \qquad (6.22)$$

The bare complex was stabilized in the geometry optimization process by the proton transfer to the neighboring –OH site accompanied by a formation of the strong hydrogen bond with an O···H distance of 1.5 Å. Following hydration by a water slab and performed AIMD simulations provided similar results to those observed for the hydrated outer sphere complex. Figure 6.7 displays a snapshot from the MD simulation and 1D density profiles for the same types of atoms as in Figure 6.6. The first peak enhanced with the black line corresponds to the oxygen atom in the Fe—O—C bridge.

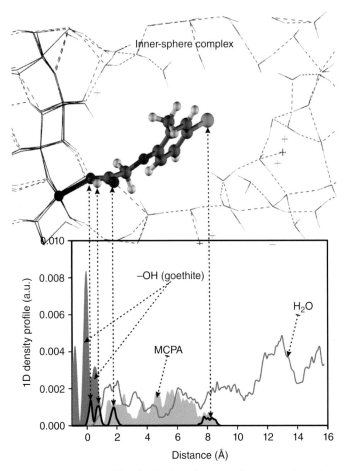

Figure 6.7 *One-dimensional density profiles for hydrated inner sphere complex of neutral MCPA molecule on the partially protonated (110) surface of goethite obtained from AIMD simulation.*

The averaged Fe—O bond is about 2.2 Å being more than typical Fe—O bonds in the bulk goethite structure (1.9–2.1 Å). The second black peak is assigned to the shared proton between the oxygen atom of the neighboring –OH group and the second oxygen atom of the carbonyl group (O_C, third peak in Figure 6.7). This oxygen atom is also open for the hydrogen bonding with surrounding water molecules destabilizing the binding to the goethite surface site. The position of the Cl atom and the distribution of the water molecules are very similar to those observed for the outer sphere complex.

The geometries (particularly C—O distances) obtained in the MD simulations on the two outer and inner sphere complexes were then successfully used in the CD-MUSIC modeling approach in order to explain experimentally observed dependency of the adsorption of MCPA on pH (Kersten et al., 2014).

6.3.1.4 *Montmorillonite···Phenoxy Herbicide Complexes*

In the work (Tunega et al., 2007), sorption of the anionic 2,4-D herbicide on the surface of the clay mineral montmorillonite was investigated using AIMD simulation at room temperature. Montmorillonite is common clay mineral widely distributed in various types of soils and is considered as a highly active geosorbent for organic contaminants. The role of the solute cations in the formation of complexes between negatively charged mineral surfaces and phenoxyacetic acid molecules (and their anionic forms) is still debated. It was supposed that cations could form surface complexes acting as cation bridges. However, Celis et al. (1999) and Cox et al. (2000) reported that pure natural mont-morillonites have only a minimal ability to sorb 2,4-D, thus questioning the relevance of cation bridges. In contrast, Pfeiffer observed significant sorption of 2,4-D and MCPA herbicides on mont-morillonite in the presence of $CaCl_2$ buffer, in other words, in the presence of the high Ca^{2+} con-centration (Pfeiffer, 1999). Therefore, this work was intended to test the hypothesis of the formation of the cation bridge between 2,4-D anion and a negatively charged clay surface, to reveal the role of the polar solvent molecules (H_2O) on the stability of this bridge, and to quantify the impact of the type of isomorphic substitution on the stability of the complexes.

Smectites possess a permanent, relatively low, negative layer charge (0.2–0.6 |e| per formula unit; Table 6.1), which is a consequence of isomorphic substitutions in the layers. This charge is localized on the regular layer surface formed from basal oxygen atoms (Figure 6.1b) representing about 75% of the total surface area of the montmorillonite particles (Tournassat et al., 2003). The isomorphic substitutions appear either in the octahedral or tetrahedral sheets (or in both), and a charge balance in natural smectites is achieved by hydrated cations located in the interlayer space or on the surface. The models of the layer used in this work were constructed on the base of the montmorillonite struc-ture determined by Tsipursky and Drits (1984). Two models of layers differing by the isomorphic substitution were constructed. In the first model one magnesium atom replaced octahedral alumi-num, whereas one Al atom replaced tetrahedral silicon in the second model. Both substitutions pro-duced a total excess layer charge of -1 |e| per computational cell. Since the anionic form of 2,4-D was investigated, the Ca^{2+} cation compensated for both negative charges leading to a neutral simulation box. In the simulations of the surface interactions of $2,4-D^-$, the first set of models represented a limited situation with a "dry" surface where no water molecules are present. It was found that in these models the Ca^{2+} cation has a stable position above the ditrigonal hole of the montmorillonite layer with average distances of 2.5 Å for the shorter and 3.1 Å for the longer distances between Ca^{2+} cation and the basal oxygen atoms of the surface. $2,4-D^-$ forms a very stable bidentate complex with the surface via the Ca^{2+} bridge with an average distance of 2.3 Å between carboxylate oxygen atoms and the cation.

In the second set, a partial hydration was added by distributing 12 water molecules in a random manner around the interaction site of the "dry" bidentate complex. The number of water molecules was estimated on the base of the molecular volume of a $2,4-D^- \cdots Ca^{2+}$ complex, the volume of the free space above the montmorillonite layer where the complex was located, and the density of the bulk water. In this way, two models with the direct cation-bridged binding to the surface were created (see MD snapshots A and C in Figure 6.8). In addition, other two models with "indirect" cation bridging were constructed as well (see MD snapshots B and D in Figure 6.8). They represent a surface complexation via hydrated Ca^{2+} cation where the contact of the cation with the ditrigonal hole is screened by the water molecules (outer sphere complex).

It was observed in the beginning of the MD equilibration that originally bidentate complexation of the carboxylate group quickly changed to a monodentate configuration where water molecule replaced one carboxylate oxygen atom. The AIMD simulations showed that the monodentate con-figuration is more stable than bidentate under the given thermodynamic conditions. Similar conclu-sions were presented for the hydrated Al^{3+} complexes with the acetate anion (Tunega et al., 2000). An important phenomenon observed in the AIMD process was hydrogen bond dynamics, which included formation and breaking of several types of hydrogen bonds. The water molecules were involved in mutual hydrogen bonds as well as in the hydrogen bonds with the basal surface oxygen

(a) (b)

(c) (d)

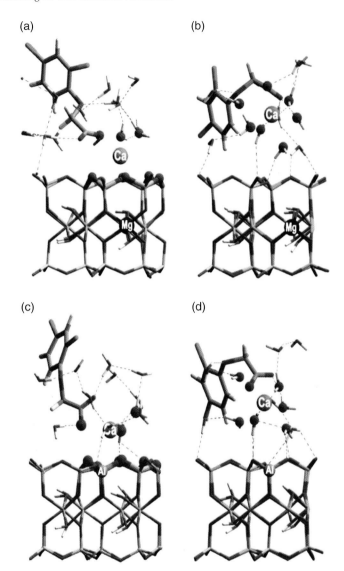

Figure 6.8 *Snapshots from AIMD simulations on four models of partially hydrated cation-bridged complexes of 2,4-D$^-$ anion on the (001) surface of montmorillonite.*

atoms. They behaved either as proton donors or proton acceptors. The most important hydrogen bonds were those formed between the carboxylate group of the 2,4-D$^-$ and water molecules. These hydrogen bonds contributed to the formation of the monodentate coordination of the Ca^{2+} cation by the 2,4-D$^-$.

The structural analysis was performed by calculating RDFs between Ca^{2+} cations and oxygen atoms and 1D density profiles in a perpendicular direction to the plane of the basal oxygen atoms (for models A and B in Figures 6.9 and 6.10, respectively). Corresponding curves for the models C and D are similar and not displayed.

Both the RDFs and the integrated RDFs (to obtain coordination number (*CN*)) show a more regular coordination of the Ca^{2+} cation in the models with hydrated cation bridge (B and D) than

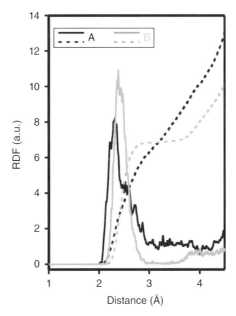

Figure 6.9 *Ca–O radial distribution functions for models A and B (Figure 6.8) calculated from AIMD simulations.*

in the models with direct cation bridging (A and C). The RDF peaks of the models B and D are relatively narrow and symmetric, having a maximum at about 2.4 Å, whereas the maximum of the RDF peaks of the A and C models is shifted to a smaller value. The *CN* is estimated to be 6.5–7 for all models, which is smaller than $CN = 8$ obtained for the Ca^{2+} cation in the pure water solvent using the DFT calculation (Schwenk et al., 2001). In our systems, the coordination space around the Ca^{2+} cation is more heterogeneous than in the pure water solvent. The presence either of the 2,4-D$^-$ or the montmorillonite surface in the first coordination sphere does not allow more water molecules to get closer to the cation.

Differences between two types of the surface complexation are also observed in the 1D density profiles presented in Figure 6.10. The maximum of the Ca^{2+} peak of the model A is at a distance of approximately 2.2 Å from the surface demonstrating its stable position above to the ditrigonal hole. The Ca^{2+} peak of the model B is located between 4 and 6 Å above the surface. This means that the complex formed between the Ca^{2+} cation, water, and the 2,4-D$^-$ has some freedom to move above the mineral surface. The curves representing a distribution of water (blue lines in Figure 6.10) show that the molecules are mostly concentrated around Ca^{2+} and the carboxylate group of 2,4-D$^-$. For the indirect complex (model B) the first peak of water is located between the surface of the basal oxygen atoms and the position of the calcium cation. The peaks representing the positions of the carboxylate oxygen atoms reflect the monodentate coordination of the $-COO^-$ group to the Ca^{2+} cation.

The relative stability of two different mechanisms of the surface complexation for the two types of the montmorillonite layer was estimated on the base of the averaged total MD energies of the models A–D. They are presented (together with standard deviations) in Table 6.3. For both pairs (A and B, and C and D), configurations with the direct cation bridge to the surface (A and C in Figure 6.10) are less stable than their corresponding partners (B and D in Figure 6.10). The differences between the averaged energies of both pairs indicate that indirect complexes are more stable than the

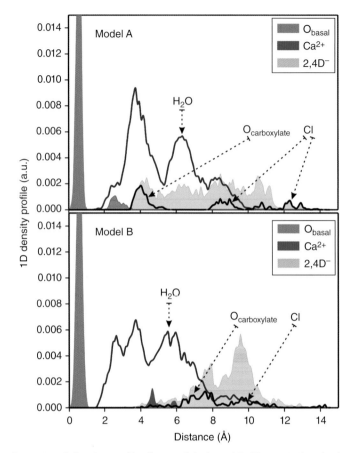

Figure 6.10 *One-dimensional density profiles for models A and B (Figure 6.8) calculated from AIMD simulations.*

corresponding direct complexes. Moreover, for the models with the tetrahedral substitution (C and D), the energy difference is only half of that for the pair A and B. Thus, the Al/Si defect localized nearby the surface has a larger effect on the stabilization of the Ca^{2+} cation close to the surface than in the case of the octahedral substitution. The reason is that the Coulomb interaction between the Ca^{2+} cation and the surface is stronger than in the models with the octahedral defect because of a greater localization of the excess negative layer charge on the basal oxygen atoms near the Si/Al substitution. However, this effect is still not strong enough to firmly attach the Ca^{2+} cation directly to the layer surface and to form a stable direct cation bridge between the montmorillonite surface and the 2,4-D$^-$, even though temporary surface complexes of this type can occur. The Ca^{2+} is better stabilized by hydration of more regular, monodentate complex with the 2,4-D$^-$. However, these observations have to be taken with a certain caution because the energy differences are nearly equal or less than the calculated standard deviations of the total energies (see Table 6.3).

It can be concluded that the MD simulations did not confirm that the direct cation bridge is a main binding mechanism in the formation of the surface complexes of the anionic form of 2,4-D and probably also MCPA to the montmorillonite basal surface. Calculations showed that hydrated 2,4-D$\cdots$Ca^{2+} complexes are thermodynamically more stable than complexes with the direct cation bridge to the surface. On the other hand, it is possible that phyllosilicates with a greater

Table 6.3   *Calculated averaged total energies, Ū, and their standard deviations for models A–D and differences for each corresponding pair (last column).*

Model	$\bar{U}$(kcal/mol)	$\Delta\bar{U}$(kcal/mol)
A (Mg/Al)	-20508.9 ± 9.4	9.7
B (Mg/Al)	-20518.6 ± 9.7	
C (Al/Si)	-20574.1 ± 9.4	4.9
D (Al/Si)	-20579.0 ± 9.7	

concentration of isomorphic substitutions (e.g., mica) could be able to form stable surface complexes with a direct cation-bridged mechanism.

In this section we presented results from the DFT-based modeling of the structure and properties of the surface complexes formed between polar molecules and anions of the phenoxyacetic acid derivatives and the hydroxyl-terminated surfaces of kaolinite (001) and goethite (110), and the (001) surface of montmorillonite. Multiple hydrogen bonding was found as a major mechanism for the outer sphere complexes. For goethite, the inner sphere complex was evaluated as another possible mechanism for the binding of the phenoxyacetic acid herbicides. It was shown that hydrogen-bonding network is destabilized by a presence of polar solvent molecules. For charged surface of montmorillonite, effective surface complexation is through cation bridges. However, they are also perturbed by a presence of polar solvent molecules. It was also shown that MD simulations represent an ultimate approach for better understanding and characterization of the surface complexation.

6.3.2   Modeling of Adsorption Processes of Polycyclic Aromatic Hydrocarbons on Iron Oxyhydroxides

PAHs are a class of nonpolar, hydrophobic, and persistent organic compounds belonging to the most widespread pollutants representing, therefore, a serious risk for the ecosystem and the human health (Harvey, 1997) with some of them even considered as potential carcinogens. Chemically, they are composed of at least two condensed aromatic rings. Naphthalene, anthracene, and phenanthrene represent the simplest PAHs (Fetzer, 2000). They can enter the environment during incomplete anthropogenic or natural combustion processes or can be also produced through coal gasification, petroleum cracking, crude oil refinement, organic biosynthesis, or volcanic eruptions. In the atmosphere, PAHs exist in the form of aggregates and aerosols. Owing to their low vapor pressure and low aqueous solubility, PAHs are predominantly found in sediments and soils. Especially soils, being a highly complex part of ecosystems, are capable to accumulate PAHs in rather large amounts (Means et al., 1980; Allen-King et al., 2002).

There are many studies on sorption conducted with natural geosorbents (e.g., Means et al., 1980; Chiou et al., 1998; Kleineidam et al., 1999; Rügner et al., 1999; Xia and Ball, 1999; Weigand et al., 2001a, b; Allen-King et al., 2002; Weigand et al., 2002; Zhou and Zhu, 2005; Ping et al., 2006; Jonker, 2008; Owabor et al., 2010; Yang et al., 2010; Chen and Yuan, 2011; Wang and Grathwohl, 2013; Chi, 2014; Zhang et al., 2014). However, a mechanism of PAH sorption is not completely clear, especially if natural sorbents are of complex character and composition such as NOM. It is supposed that mineral phase of soils can play a significant role in the retention of PAHs. Some sorption studies were also performed with pure or chemically modified mineral sorbents (e.g., Stauffer and McIntyre, 1986; Noll, 1987; Mader et al., 1997; Angove et al., 2002; Müller et al., 2007; Costa et al., 2012; Pei et al., 2012; Kaya et al., 2013; Jia et al., 2014; Joseph-Ezra et al., 2014; Wang et al., 2014).

Several competing mechanisms are proposed to explain sorption of PAHs to the mineral surfaces. A charge-induced dipole–dipole interaction mechanism is proposed for the interaction of nonpolar hydrophobic species such as PAHs with positively charged domains on mineral surfaces (Mader et al., 1997). Another suggestion is based on possible interactions between cationic sites and π-electrons of aromatic rings (Ma and Dougherty, 1997; Zaric, 2003). Entropy-driven partitioning is considered as a dominant mechanism in the sorption from polar solvents (Schwarzenbach et al., 2003). However, molecular-level studies of the nature of PAH interactions with mineral surfaces by QC methods are rare (Austen et al., 2008; Rimola et al., 2010).

In our work we performed the first systematic theoretical investigation on the interactions of benzene and four PAHs, naphthalene, anthracene, phenanthrene, and pyrene, with the (110) surface of the mineral goethite using DFT method and a slab model (Tunega et al., 2009). The work was later extended with further studies on mineral lepidocrocite (γ-FeOOH), a goethite polymorph (Wyckoff, 1963). Lepidocrocite is high-spin antiferromagnet, similarly to goethite but structurally is simpler. It crystallizes in the orthorhombic bipyramidal system. Distorted FeO_6 octahedra are linked in a form of two-dimensional layers that are terminated on both sides by hydroxyl groups similar to the μ-OH type in goethite (oxygen atom is linked to the two iron atoms). The layers are connected together by hydrogen bonds. The surface (010) formed from the μ-OH groups is dominant for lepidocrocite crystal particles, and its topology is simpler than more complex (110) goethite surface. Its structure can be seen from pictures of interacting PAHs with lepidocrocite in Figure 6.12.

The aim of our work (Tunega et al., 2009) was to find the most stable arrangements of different PAHs on the dominant surfaces of goethite and lepidocrocite and to elucidate the nature of these interactions. In the earlier stage of this study, it was clear that the standard DFT method would not be able to provide interaction energies in a satisfied accuracy. The dominant forces in the interactions of hydrophobic PAHs and polar surface hydroxyl groups are mainly of a dispersion type. It makes the calculations on these systems a real challenge. In spite of that, it was possible to show trends and mechanism of adsorption of PAHs to the goethite (110) surface already at the standard DFT level (Tunega et al., 2009). The optimized geometry showed PAHs in a practically parallel configuration to the surface plane (Figure 6.11). A specific phenomenon observed was a shape selectivity for PAH adsorption. The DFT calculated interaction energies regularly increased with increasing size of the linear chains of aromatic rings (Table 6.4). The strongest affinity to the goethite surface was found for anthracene. Two other PAHs with a nonlinear shape, phenanthrene and pyrene, were found to be less strongly bound to the surface although they have a similar (phenanthrene) or even larger size (pyrene) than anthracene (see calculated interaction energies in Table 6.4). This difference was explained by the specific configuration of the surface hydroxyl groups of goethite. The three types of the hydroxyl groups, μ-OH, μ_3-OH, and –OH, form a "valley"—the width of it fits very well to the molecular shape of the linear PAHs. It was also calculated that the linear PAHs can easily slide along the valley of OH groups with practically no barrier.

The calculated averaged perpendicular distances between planes of the PAH molecules and a plane formed from oxygen atoms of μ-OH and μ_3-OH groups (Figure 6.11) correspond very well to the calculated interaction energies. The distances are shorter for the linear PAHs then for phenanthrene and pyrene. Later calculations with the inclusion of the dispersion corrections (D2 type (Grimme, 2006)) to the PBE functional confirmed the trends observed already in the standard DFT calculations. The PBE-D2 interaction energies are much larger (in absolute values) and the distances are shorter than in the case of PW91 calculations.

The PBE-D2 calculations on the PAH···lepidocrocite models provided some difference in the results. The linear PAHs interact with the lepidocrocite surface weaker than in the case of goethite (e.g., for anthracene a difference is >6 kcal/mol). However, two nonlinear PAHs have the interaction energy comparable (phenanthrene) or larger (pyrene) than anthracene. These energies are also very

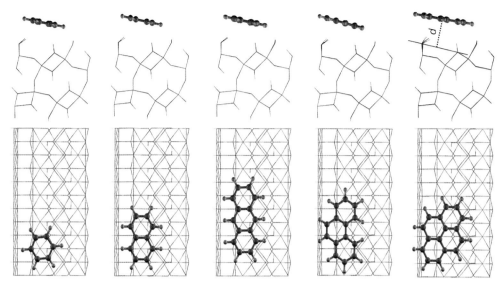

Figure 6.11 *DFT optimized geometries for selected PAH molecules adsorbed on the (110) surface of goethite.*

Table 6.4 *Calculated interaction energies and averaged distances for selected PAHs adsorbed on goethite (110) surface and lepidocrocite (010) surface.*

PAH	Goethite/PW91 (Tunega et al., 2009)		Goethite/PBE-D2		Lepidocrocite/PBE-D2	
	ΔE_c (kcal/mol)	d^a (Å)	ΔE_c (kcal/mol)	d^a (Å)	ΔE_c (kcal/mol)	d^a (Å)
Benzene	−4.8	3.30	−7.9	3.05	−6.4	3.05
Naphthalene	−7.6	3.31	−9.3	3.05	−8.6	2.95
Anthracene	−10.3	3.31	−21.2	3.01	−15.0	2.90
Phenanthrene	−5.9	3.32	−15.8	3.02	−15.2	2.95
Pyrene	−7.5	3.46	−16.9	3.22	−17.8	2.85

^{a} For definition of d see text and Figures 6.11 and 6.12.

similar with the interaction energies for goethite (Table 6.4). As the PAH molecules are not perfectly parallel with the plane of the surface OH groups of the lepidocrocite surface, the distances presented in Table 6.4 are averaged perpendicular distances of the molecular plane to the closest μ-OH groups (Figure 6.12). These distances do not vary so much as in the case of goethite. It can be concluded that for the linear shapes of the PAH molecules, goethite is better sorbent than lepidocrocite, while for larger and nonlinearly shaped PAHs, both FeOOH polymorphs are equivalent sorbents. The difference observed for the interactions of PAHs with goethite or lepidocrocite can be explained by the difference of the surface topology and the types of the surface hydroxyl groups.

Although it seems that the calculated interaction energies presented in Table 6.4 are relatively large, two important aspects, similarly as in the previous cases on the interactions of phenoxyacetic acid herbicides, have to be noted also here. First, the energy differences are calculated for an association type of reaction (adsorption from gas phase), for which large entropy lost effect is typical (Aquino et al., 2002). Thus, for the estimation of the Gibbs free energies, the interaction energies presented in Table 6.4 should be reduced by about 12–15 kcal/mol. Therefore, it seems that only

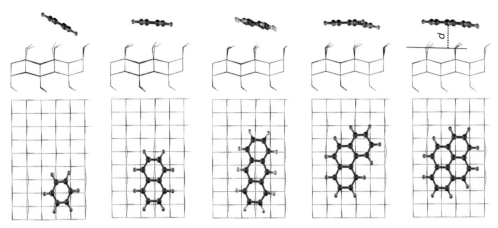

Figure 6.12 *DFT optimized geometries for selected PAH molecules adsorbed on the (010) surface of lepidocrocite.*

the complexes with larger PAHs can be thermodynamically stable. Second, also a certain solvent effect has to be included in the calculations if the adsorption from solution is intended to be simulated. This aspect has already been discussed in the previous section.

The nature of the interactions can be mainly characterized by the polarization of the π-system of PAHs by polar surface hydroxyl groups and in the formation of weak hydrogen bonds where the π-system acts as a proton acceptor, similar to the interactions in the water–benzene system (Ran and Hobza, 2009).

6.3.3 Modeling of Interactions of Polar and Nonpolar Contaminants in Organic Geochemical Environment

SOM is an important natural organic geosorbent representing a substantial part of soils. Owing to its porous, heterogeneous, and flexible structure, it can bind, trap, and stabilize various chemicals (polar, nonpolar, ions) in different SOM domains (hydrophobic, hydrophilic, charged) in large amounts. Moreover, the presence of water and cations in the SOM structures can have a significant impact on the stability of the complexes formed. For example, Borisover et al. observed an enhancement effect of water associated with SOM on the adsorption of carbamazepine (2011). A structurally complex, not completely determined, system such as SOM requires using a different strategy in modeling its properties and interactions with contaminants such as pesticides in comparison to minerals. In our earlier studies on the interactions of SOM, the goal was to elucidate basic mechanisms for binding of widely used herbicides (MCPA, 2,4-D) by means of reliable QC methods. Thus, taking a complex SOM model consisting from hundreds of atoms was impossible. Instead of that the focus was laid on specific representative fragment models reflecting knowledge about typical functional groups appearing in the SOM composition. The properties and structure of polar phenoxyacetic acid derivatives have already been discussed in a previous section of this chapter, and it was noticed that they are relatively strong organic acids with $pK_a \sim 3$. Thus, it implies that 2,4-D and MCPA exist mostly in their anionic form at the natural pH conditions (6–7) in most of the soils except highly acidic ones. However, due to the extreme structural heterogeneity of SOM, local domains with a high concentration of protons can be found where phenoxyacetic acid herbicides could be bound also as neutral molecules.

Table 6.5 Formation enthalpies and Gibbs free energies for complexes of 2,4-D/2,4-D⁻ with selected functional groups.

Association reactions[a]	Neutral 2,4-D				2,4-D⁻ anion			
	ΔH_g	ΔG_g	ΔH_{gs}	ΔG_{gs}	ΔH_g	ΔG_g	ΔH_{gs}	ΔG_{gs}
Me-CHO + X ⟶ Me-CHO···X	−8.6	1.2	0.8	10.0	−2.2	−0.3	0.8	2.8
Me-OH + X ⟶ Me-OH···X	−9.4	0.2	−4.6	4.1	−3.1	−1.1	0.2	2.2
Me-NH₂ + X ⟶ Me-NH₂···X	−11.1	−8.4	−2.0	4.6	−1.5	0.4	1.0	2.8
Me-NH₃⁺ + X ⟶ Me-NH₃⁺···X	−29.6	−79.1	−1.3	9.4	−5.7	−3.0	−0.3	2.5
Me-COOH + X ⟶ Me-COOH···X	−15.1	−16.7	1.3	12.4	−4.7	−2.1	−0.2	2.4

Subscript "g" denotes the gas phase calculations. Subscript "gs" denotes the results obtained with global solvation approach (COSMO model). All calculations were performed at the B3LYP/SVP+sp level of theory. Energies are BSSE corrected (For the global solvation, BSSE corrections were taken from the gas phase.) and given in kilocalories per mole.
[a] Me = CH₃, X = 2,4-D or 2,4-D⁻.

Table 6.6 Enthalpies and Gibbs free energies of exchange reactions between hydrated 2, 4-D/2, 4-D⁻ moieties and selected hydrated functional groups using micro+global solvation approach (subscript "mgs").

Exchange reactions[a]	Neutral 2,4-D		2,4-D⁻ anion	
	ΔH_{mgs}	ΔG_{mgs}	ΔH_{mgs}	ΔG_{mgs}
Me-CHO···2H₂O + X···2H₂O → Me-CHO···X + (H₂O)₄	−3.7	−3.4	−0.5	−1.7
Me-OH···2H₂O + X···2H₂O → Me-OH···X + (H₂O)₄	−8.6	−9.3	−7.8	−8.5
Me-NH₂···2H₂O + X···2H₂O → Me-NH₂···X + (H₂O)₄	−4.7	−5.4	3.8	2.8
Me-NH₃⁺···2H₂O + X···2H₂O → Me-NH₃⁺···X + (H₂O)₄	−0.1	3.0	2.2	4.9
Me-COOH···2H₂O + X···2H₂O → Me-COOH···X + (H₂O)₄	−0.2	−0.9	−3.7	−3.4

All calculations were performed at the B3LYP/SVP+sp level of theory. Energies are given in kilocalories per mole.
[a] Me = CH₃, X = 2,4-D or 2,4-D⁻.

6.3.3.1 Functional Groups Model

In the work by Aquino et al. (2007b), the adsorption of the 2,4-D herbicide and its anion with SOM moieties was investigated at the DFT level of theory using the B3LYP/SVP+sp approach. The SOM moieties were represented by acetaldehyde, methanol, methylamine, protonated methylamine, acetic acid, and water molecules. The pH effect was simulated by using neutral and deprotonated forms of the 2,4-D molecule. The environmental effect was considered by a combined micro- and global solvation approach. In the microsolvation case, solvent molecules are included in the calculations explicitly in order to describe local interactions such as hydrogen bonds. But this is a size-limited option and, additionally, the long-range effect of a polar solvent is not included. In the global solvation the solvent is represented by a macroscopic polarizable continuum surrounding a solute. In this way, long-range effects of a polar solvent are considered in the calculation. Therefore, a good compromise is to use the combination of both models. The solvent effect calculations were performed using the COSMO model (Klamt and Schürmann, 1993) available in the TURBOMOLE program package (Ahlrichs et al., 1989).

The results are shown in Tables 6.5 and 6.6 for a set of reactions between 2,4-D and different molecular systems mentioned previously. Table 6.5 shows that for the addition reactions the complexes are generally more stable in gas phase than in solution. The results look more realistic when the microsolvation approach is taken into account. In this case two water molecules are added to

2,4-D for each molecular system (Table 6.6) leading to an exchange reaction between water molecules and the 2,4-D complex. Now it is observed that the inclusion of the environmental effect stabilizes the system. The stronger hydrogen bond in each complex becomes even stronger due to the solvent effect as compared to the gas phase. Only in the complex with acetic acid the distances of both hydrogen bonds are almost the same as in the gas phase. In the complexes with nonequivalent hydrogen bonds, the weaker hydrogen bond breaks in solution and the originally cyclic shape of the interaction site opens. This means that the polar solvent environment stabilizes the strong hydrogen bonds on the cost of the weak ones which even can break under the effect of the solvent polarity. This pronounced effect of the solvent polarity is to be expected, as electrostatic and charge transfer effects are the major components of the interaction energy in hydrogen bonds (Scheiner, 1997). Similar effects were observed by Aquino et al. (2002) for acetic acid interactions with acetic acid, methanol, ammonia, phenol acetamide, and acetaldehyde via hydrogen bond formation.

Analogous calculations we also performed for the $2,4\text{-}D^-$ anion and the results are also collected in Tables 6.5 and 6.6. The $2,4\text{-}D^-$ loses the capability to make a hydrogen bond as a proton donor. Thus, the second hydrogen bond observed for complexes of 2,4-D molecule is absent in most of the complexes. The calculated complex formation energies for charged complexes do not differ too much from the neutral ones. However, a different mechanism driving these interactions is observed. Dipole–dipole interaction is the main contribution to the attractive energy in the neutral system, whereas in the charged ones the charge–dipole interaction prevails. In general, according the exchange reactions, the anionic form of 2,4-D is found to establish thermodynamically stable complexes in a polar solvent with aldehydic, hydroxyl, and carboxyl functional groups.

6.3.3.2 Oligomeric Model

As already mentioned elsewhere in this book chapter, HSs are major components of the NOM in soil and HA and FA with relatively high amount of the carboxyl groups are typical representatives of HS. Therefore, in the following study, polyacrylic acid (PAA) oligomers (OA2–OA5) were taken as a model of HSs to describe hydrogen-bonding and cation-bridging mechanisms through the adsorption process of polar species (Aquino et al., 2008b). Adsorption capacities of PAA oligomers with respect to the polar phenoxyacetic acid herbicides were investigated by means of the complexes formed by PAA subunits (up to five units) interacting with acetic acid, MCPA, and water. Water was chosen as solvation agent since it plays an important role in adsorption processes in soil. The structure and stability of different complexes were studied regarding to external conditions such as hydration effects and pH of the solution and the effect of next neighboring carboxyl groups in the complexes.

All calculations were performed at the DFT level using the B3LYP/SVP+sp approach. BSSE corrections were not computed in this work because SVP+sp basis set reduced BSSE substantially. The environment effect was considered by means of a combination of microsolvation and global solvation models as explained before.

It was found that the formed complexes are established through hydrogen bonds being stable structures as confirmed by harmonic frequency calculations. In Table 6.7 thermodynamic quantities of the exchange reactions are collected for the complexes formed between MCPA and four oligomers (plus HAc for comparison). Figure 6.13 shows the optimized structure of the complex formed by the oligomer consisting from five units (OA5) with MCPA. The results for the association reactions are not shown.

A neutral MCPA molecule, similarly as 2,4-D discussed previously, forms with the –COOH group a cyclic hydrogen-bonded complex having two nearly identical strong hydrogen bonds. Reference calculation with the acetic acid showed that MCPA forms weaker complex than 2,4-D (compare last

Table 6.7 *Enthalpies and Gibbs free energies of exchange reactions between hydrated MCPA and hydrated functional groups using micro+global solvation approach (subscript "mgs").*

	Micro		Micro+global	
Exchange reactions	ΔH_{ms}	ΔG_{ms}	ΔH_{mgs}	ΔG_{mgs}
$HAc\cdots(H_2O)_2 + MCPA\cdots(H_2O)_2 \rightarrow HAc\cdots MCPA + (H_2O)_4$	−0.3	−0.8	−0.7	−1.2
$OA2\cdots(H_2O)_2 + MCPA\cdots(H_2O)_2 \rightarrow OA2\cdots MCPA + (H_2O)_4$	−0.5	−3.4	−2.3	−5.2
$OA3\cdots(H_2O)_2 + MCPA\cdots(H_2O)_2 \rightarrow OA3\cdots MCPA + (H_2O)_4$	−0.5	−2.7	−2.3	−4.5
$OA4\cdots(H_2O)_2 + MCPA\cdots(H_2O)_2 \rightarrow OA4\cdots MCPA + (H_2O)_4$	−1.0	−3.6	−2.5	−5.7
$OA5\cdots(H_2O)_2 + MCPA\cdots(H_2O)_2 \rightarrow OA5\cdots MCPA + (H_2O)_4$	−0.7	−3.7	−2.9	−5.9

All calculations were performed at the B3LYP/SVP+sp level of theory. Energies are given in kilocalories per mole.

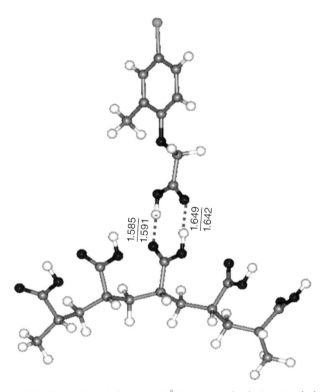

Figure 6.13 *Structure and hydrogen bond distances (Å) in gas and solution (underlined values) of the OA5···MCPA complex. Distances are in Angstrom. Underlined values correspond to COSMO calculations.*

row in Table 6.6 and first row in Table 6.7). This agrees with the fact that 2,4-D is stronger organic acid than MCPA. Further, comparing calculated reaction energies, the PAA oligomers form with MCPA significantly stronger complexes than HAc molecule. The length of the oligomer chains has a minor influence on the calculated interaction energies. The structure and the interaction energies of the cyclic hydrogen-bonded complexes showed little influence in dependence of the presence the neighboring –COOH. However, only static calculations were performed and only one of the possible stable configurations of the oligomeric chain was scrutinized.

6.3.3.3 Cation Bridges

Another important binding mechanism for adsorption of polar and ionic species in SOM is based on "cation bridges." Cation bridges constitute a powerful adsorption process for binding anionic species from solution not only to organic matter but also to mineral surfaces (Petrovic et al., 1999; Kang and Xing, 2007). The formation of cation bridges in SOM also plays a crucial role for the increasing stability and rigidity of the SOM itself by forming cross-links between charged sites (mostly carboxylate groups) in the supramolecular structure of SOM (Schaumann et al., 2006; Kalinichev and Kirkpatrick, 2007; Iskrenova-Tchoukova et al., 2010; Aquino et al., 2011; Kunhi Mouvenchery et al., 2013). It was also shown that polyvalent cations establish more effective cation bridges than monovalent cations (Kunhi Mouvenchery et al., 2013). Therefore, two complexes formed by the Ca^{2+} cation bridge were also studied—between 2,4-D⁻ anion and acetate (Aquino et al., 2007b) and between MCPA⁻ anion and monoanion of the OA3 oligomer (Aquino et al., 2008b) (displayed in Figure 6.14). The Ca^{2+} cation was selected because it naturally occurs in soil solutions and SOM matrices. Both micro- and global solvation approaches were used and the energies were computed for the following reaction:

$$Ca^{2+}(H_2O)_6 + 2,4-D^- \cdots 2H_2O + X \cdots 2H_2O \rightarrow 2,4-D^- \cdots Ca^{2+}(H_2O)_2 \cdots X + 2(H_2O)_4 \quad (6.23)$$

where $X = Ac^-$ or $OA3^-$. It is evident from the calculated reaction energies (Table 6.8) that the cation-bridged complexes of the anionic forms of the phenoxyacetic acid herbicides are thermodynamically very stable. Different phenoxyacetic acid structures and also the complexation partners can be an explanation for a relatively large difference in the reaction energies. Probably, the presence of the oligomeric chain with more acidic carboxyl groups than in the case of HAc enhanced the strength of the formation of the cation-bridged complex.

The results achieved with the static DFT calculations on hydrogen-bonded and cation-bridged complexes of the neutral and anionic forms of the phenoxyacetic acid herbicides demonstrated that most of the complexes were stable structures in both isolated and polar environments. The consideration of the combined solvation model is crucial for the evaluation of chemical reaction energies under either unsaturated or fully hydrated conditions. In the hydrogen-bonded complexes,

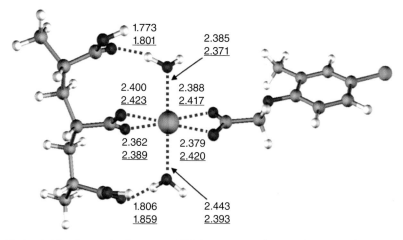

Figure 6.14 *Cation-bridged complex between MCPA⁻ anion and OA3 oligomeric fragment of polyacrylic acid. Distances are in Angstrom. Underlined values correspond to COSMO calculations.*

Table 6.8 *Reaction enthalpies and Gibbs free energies for the formation of the cation-bridged complexes of MCPA⁻/2,4-D⁻ (Eq. 6.23) using micro+global solvation approach ("mgs" subscript).*

Cation-bridged complex	ΔH_{mgs}	ΔG_{mgs}
$[2,4\text{-}D^- \cdots Ca^{2+}(H_2O)_2 \cdots Ac^-]$	−11.6	−6.6
$[MCPA^- \cdots Ca^{2+}(H_2O)_2 \cdots OA3^-]$	−43.7	−21.8

Energies are given in kilocalories per mole.

aldehydic, hydroxyl, and carboxyl groups are the most active and will play an important role in binding and stabilizing of polar but neutral molecules in the SOM matrices.

The negative ΔG values calculated for the formation reactions of the cation-bridged complexes with the MCPA and 2,4-D anions indicate that cations such as Ca^{2+} provide a promising binding mechanism at pH conditions where ionized carboxyl groups exist. In this way, anions can be quite strongly fixed in the SOM structure. The results obtained also signalize why under specific conditions HAs are able to form cross-linking cation bridges, which contribute to the increasing rigidity of SOM or, in other words, a transformation to a "glassy" form.

However, the approach used in previous examples has certain limitations. For example, complex formation energies may depend on the number of water molecules used in the microsolvation model. Therefore, much more calculations are needed to be performed to test the convergence with the increasing number of water molecules making the calculations very demanding. Moreover, with many water molecules explicitly inserted to the model, the probability to locate the global minimum in the optimization procedure decreases significantly. Further, it is difficult to compare directly this kind of calculated results with experiment.

6.3.3.4 Nanopore Model

The next step toward a more realistic model of HSs was to create a complex 3D structure featuring polar hydrophilic and nonpolar hydrophobic domains and also a pore structure. The purpose was to have the same model for the study of interactions of polar as well as nonpolar species. The model consists from aliphatic chains, each terminated by carboxyl group. Two chains are anchored to another aliphatic chain in nearly perpendicular configuration with a distance between them of about 12 Å having carboxyl groups oriented face to face. Three such constructions are arranged in a parallel set that forms a nanopore. The hydrophilic domain is represented by interacting carboxyl groups and aliphatic chains form the hydrophobic domain of the nanopore. The structure of the nanopore model is visible in Figures 6.15 and 6.16.

Recently, this model was taken to simulate trapping of MCPA (polar) and naphthalene (nonpolar) species inside of the nanopore. In unsaturated conditions, hydrophilic domains of the SOM structures are often microhydrated (varying amount of water depends on humidity). Therefore, the model was completed by adding water molecules atop of the hydrophilic domain of the nanopore. The goal was to estimate a change of free energy after trapping MCPA/naphthalene in the HS nanopore by performing constrained MD simulations. Of course, this was only possible using classical FF-MD approach due to the size of the models and a necessity to run long MD simulations. TIP3P FF (Jorgensen et al., 1983) was selected for water, and the rest atoms were described by OPLS-AA (Jorgensen et al., 1996). The calculations were performed with the program TINKER (Ponder and Richards, 1987) using NVT ensemble (Andersen thermostat (1980), $T = 300$ K). Newton's equations of motion were integrated using the velocity Verlet algorithm (Ferrario and

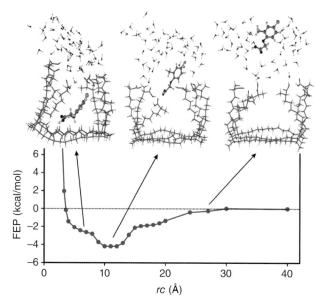

Figure 6.15 *Free energy profile for MCPA trapping in the nanopore model of HSs with partially hydrated hydrophilic domain.*

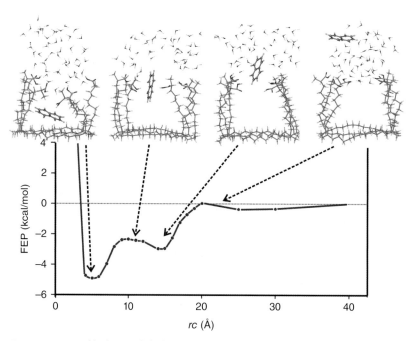

Figure 6.16 *Free energy profile for naphthalene trapping in the nanopore model of HSs with partially hydrated hydrophilic domain.*

Ryckaert, 1985) with a time step of 1 fs. In the first phase of each MD run, an equilibration was performed in a duration of approximately 1 ns followed by a production phase of additional 1 ns. The free energy change was mapped by using method of the potential of mean force (PMF) (Roux, 1995). To evaluate the PMF, the Blue Moon method was used (Sprik and Ciccotti, 1998) using constraints imposed on the positions of one carbon atom from the middle of the bottom aliphatic chains and the center of mass of the adsorbed molecule (see pink balls in Figure 6.16). The distance between these two points was selected as the reaction coordinate, rc, in the PMF calculations. Furthermore, during the MD calculations also terminal carbon atoms from the bottom aliphatic chains were fixed to keep a pore shape (yellow balls in Figure 6.16).

Figure 6.15 displays three snapshots from the MD simulations and free energy profile (FEP) with respect to the reaction coordinate. The reference state is the MCPA molecule out of the pore without contact with water molecules at a distance of 40 Å from the reference point of the pore. The minimum of the FEP is relatively broad at a distance of approximately 11 Å. Middle snapshot shows that at this distance the MCPA molecule interacts effectively with carboxyl groups of the aliphatic chains through hydrogen bonds. Moreover, also water molecules surrounding the hydrophilic part contribute to the stabilization through solvation of the rest part of the MCPA molecule. The depth of the FEP curve at minimum is about −4.2 kcal/mol. This number corresponds to DFT calculated free energies of the exchange reactions of the MCPA molecule interacting with PAA fragments (OA2-5) presented in Table 6.7.

The FEP profile for the naphthalene has a different shape (Figure 6.16). Two minima are observed with a small barrier between them. The first minimum is deeper (about −5.0 kcal/mol) at a distance of approximately 5 Å. In this position the naphthalene molecule is trapped nearby bottom aliphatic chains due to van der Waals interactions. Moreover, side aliphatic chains are flexible enough and they also contribute to the stabilization of the naphthalene molecule inside of the pore. Following barrier at a distance of approximately 10 Å is related to the situation when hydrogen bonds formed among terminal carboxyl groups are interrupted by the naphthalene molecule releasing the pore space. Then, the molecule penetrates to the water cluster where it is slightly stabilized by weak solvent effect (second minimum). The energetic depths for both molecules are similar evidencing that hydrophobic domains in the HSs can relatively strongly bind nonpolar species such as PAHs.

6.4 Perspectives and Future Challenges

The scale of problems in organic and contaminant geochemistry is broad and complex requiring integrative and interdisciplinary approaches where molecular simulations play undoubtedly an important role. This review has been aimed at introducing the basics of molecular simulation methods and techniques available for modeling interactions of organic chemicals and contaminants in geochemical environments. The literature review presented and examples of our own case studies have demonstrated the usability and importance of the molecular modeling methods in elucidating basic formation mechanisms of complexes between organic contaminants and typical geosorbents such as minerals or NOM. Perspectives and future challenges for the simulations in the organic geochemistry can be summarized as follows.

- A permanent challenge remains the construction of suitable representative models of complex structures of natural geosorbents. These models should reflect various factors such as structural and compositional heterogeneity of geomaterials and their interfaces. A good example is the modeling of the formation of organomineral associates and microaggregates that are typical for soils. Recent developments in computational methods at the nanoscale applicable for the simulation of interactions of NOM and layered minerals were reviewed by Greathause et al. (2014). Further, the models should include also environmental (e.g., solvation, partial solvation, drying/wetting cycles, pH, ionic strength) and external effects (e.g., radiation). Special attention has to be paid to modeling of the chemically bound organic contaminants, their degradation processes, and interactions of the metabolite intermediates. In these complex chemical processes, various physical and chemical phenomena can occur such as charge transfer, proton and electron transfers, redox reactions, electronic excitations, mineral surface dissolution, or transportation and diffusion in interfacial zones. Some of these problems are addressed directly in this book (see Chapters 9 and 11).

• Sustainable increase of the computational resources, massive parallelization, and a continuous development of algorithms and codes in the molecular modeling are key factors for performing large- and multiscale simulations in various fields of material science (Harding et al., 2008; Lyubartsev et al., 2009; Suter et al., 2009) that are applicable also in the field of computational geochemistry.

Improvements achieved in the development of the DFT functionals and dispersion corrections lead to an increase of the accuracy in predicting interaction energies for the organic contaminants where hydrogen-bonding and nonboding interactions dominate. Owing to effective computer codes and parallelization, it is possible to perform AIMD simulations on the systems approaching a size of thousands of atoms and time scale of hundreds of nanoseconds and providing structural and dynamic characteristics of the modeled systems.

There is a great potential in the application of advanced sampling (e.g., metadynamics, PMF) to AIMD techniques to compute thermochemical quantities of the adsorbed organic contaminants and their eventual chemical degradation processes. This approach is especially suitable for mineral surfaces. Its application for the natural organic geosorbents can be limited due to the size of the NOM models. However, this limitation can be overcome by the application of the embedding methods (e.g., QM/MM) linked with MD. An increasing number of FF-based molecular simulations can be expected for the computation of the interactions of organic substances and contaminants with mineral surfaces due to a systematic development of the FF parameters for organic–inorganic interfaces such as INTERFACE-FF (Heinz et al., 2013). The progress in the development of the reactive FFs (see, e.g., review by Liang et al. (2013)) gives a chance for broader applications of the FF methods in the simulation of the kinetics and dynamics of the chemical processes of the geochemical interfaces in the future. However, currently the application of such FFs is rather limited because the parameters are available for a limited set of atoms for certain classes of compounds. Moreover, the development of this type of FF requires an intensive testing and validation and also the transferability of these FF parameters is questionable.

An increasing number of the multiscale simulations are expected where the coarse-grained force fields (CG-FF) can be used. CG-FF-based simulations are applicable for the complex systems attacking mesoscale (μm) dimension and time scale of seconds. The CG-FF models comprise much fewer degrees of freedom compared to fully atomistic modes. In these models a group of atoms (e.g., functional groups but "graining" can go to a deeper level including also hundreds of atoms) is replaced by a pseudoparticle (bead). The potential energy is usually expressed as a sum of pairwise interactions between beads. The form of these pairwise interactions is similar or even simpler as in the atomistic FF models. In the CG-FF development two general approaches are mostly used, namely, the renormalization and reference potential approaches, which allow one to move back and forth between the coarse-grained and full atomistic models (Kamerlin et al., 2011). The CG-FF models are developed mainly for soft organic matter such as macromolecules in biochemistry (peptides, proteins, DNA) or polymers. The CG-FF simulations are frequently used in the simulations of processes such as self-assembling and folding of macromolecules. Thus, the applicability of the CG methods in the organic contaminant geochemistry is limited to the organic geosorbents. For example, there are possible applications in the simulation of the formation and stability of the associates and aggregates of the SOM using CG models for some smaller chemical entities such as HA and FA. It is also expected that the accuracy of the CG-FF method in the prediction of the interaction energies and binding mechanisms will be low because a chemical detail of the underlying atomistic system is lost in the CG model.

It is evident from the previous discussion that a tendency in the molecular simulation methods leads to a reduction of the spatial and temporal gaps between modeling and experiments by using "realistic" models (with reduced number of simplifications) and large and multiscale simulations. This trend is fully usable also in the simulations of structures and process of the organic and contaminant geochemistry. Thus, the combination of the advanced experimental and spectroscopic methods with the state-of-the-art molecular simulations is the right way in the geochemical research to improve our understanding of the behavior of organic contaminants in the environment and for evaluating critical geochemical processes.

Glossary

2,4-D	2,4-Dichlorophenoxyacetic acid
AFM	Atomic force microscopy
AIMD	*Ab initio* molecular dynamics
AMBER	Assisted Model Building with Energy Refinement
B3LYP	Becke, three-parameter, Lee–Yang–Parr
BLYP	Becke, Lee, Yang, and Parr
BSSE	Basis set superposition error
CC	Coupled clusters
CCSD(T)	Coupled cluster singles doubles + noniterative triples
CD-MUSIC	Charge distribution multisite complexation
CEC	Cation exchange capacity
CG	Coarse grained
CHARMM	Chemistry at HARvard Macromolecular Mechanics
CHELP	Charges from Electrostatic Potentials
CHELPG	Charges from Electrostatic Potentials using a Grid-based method
CI	Configuration interaction
CLAYFF	Clay force field
CN	Coordination number
COSMO	Conductor-like screening model
CVFF	Consistent valence force field
DCP	Dispersion-correcting pseudopotentials
DD	Dibenzo-*p*-dioxin
DDT	Dichlorodiphenyltrichloroethane
DEP	Diethyl phthalate
DFT	Density functional theory
DFTB	Density functional tight binding
DNA	Deoxyribonucleic acid
DNAN	2,4-Dinitroanisole
DNOC	Dinitro-ortho-cresol
DNT	2,4-Dinitrotoluene

EAM-FF	Embedded atom model force field
ESEM	Environmental scanning electron microscopy
ESP	Electrostatic potential
EXAFS	X-Ray Extended X-Ray Absorption Fine Structure
FA	Fulvic acid
FEP	Free energy profile
FF	Force field
FPMD	First-principles MD
FTIR	Fourier transform infrared
GGA	Generalized gradient approximation
GROMACS	GROningen MAchine for Chemical Simulations
HA	Humic acid
HF	Hartree–Fock
HMPA	Hexamethylphosphoramide
HS	Humic substance
IR	Infrared
LDA	Local density approximation
LJ	Lennard–Jones
LSDA	Local spin-density approximation
M05-2X, M06-2X	Minnesota global hybrid functionals
MC	Monte Carlo
MCPA	2-Methyl-4-chlorophenoxyacetic acid
MD	Molecular dynamics
MK	Merz–Singh–Kollman
MM	Molecular mechanics
MMFF	Merck molecular force field
MP	Møller–Plesset
NMR	Nuclear magnetic resonance
NOM	Natural organic matter
NTO	3-Nitro-1,2,4-triazole-5-one
OPLS-AA	Optimized Potentials for Liquid Simulations – All Atoms
PAA	Polyacrylic acid
PAHs	Polycyclic aromatic hydrocarbons
PAW	Projector augmented wave
PBC	Polychlorinated biphenyls
PBE	Perdew, Burke, and Ernzerhof
PCM	Polarizable continuum model
PES	Potential energy surface
PMF	Potential of mean force
PW91	Perdew and Wang's 1991 functional
PZC	Point of zero charge
QC	Quantum chemical
QM	Quantum mechanics
QSAR	Quantitative structure-activity relationship
RDF	Radial distribution function
Reactive FF	Reactive force field
RESP	Restrained electrostatic potential

SCF   Self-consistent field
SCRF   Self-consistent reaction field
SIMS   Secondary ion mass spectrometry
SOM   Soil organic matter
SPC   Simple point charge
SPC/E   Extended SPC
SSA   Specific surface area
SSC-DFTB   Self-consistent charge extension of DFTB
STM   Scanning tunneling microscopy
SVP   Split valence polarization
TIP3P   Transferable intermolecular potential with three interaction sites
TIP4P   Transferable intermolecular potential with four interaction sites
TNT   2,4,6-Trinitrotoluene
TZVP   Triple zeta valence polarization
UFF   Universal force field
VASP   Vienna Ab initio Simulation Package
vdW-DF   van der Waals density functional
vdW-TS   van der Waals Tkatchenko–Scheffler
XPS   X-ray photoelectron spectroscopy

References

V. Aggarwal, Y.Y. Chien, and B.J. Teppen, Molecular simulations to estimate thermodynamics for adsorption of polar organic solutes to montmorillonite, *European Journal of Soil Science*, **58**(4), 945–957 (2007).

R. Ahlrichs, M. Bär, M. Häser, H. Horn, and C. Kölmel, Electronic structure calculations on workstation computers: The program system Turbomole, *Chemical Physics Letters*, **162**(3), 165–169 (1989).

H.A. Al-Abadleh and V.H. Grassian, Oxide surfaces as environmental interfaces, *Surface Science Reports*, **52**(3–4), 63–161 (2003).

R.M. Allen-King, P. Grathwohl, and W.P. Ball, New modeling paradigms for the sorption of hydrophobic organic chemicals to heterogeneous carbonaceous matter in soils, sediments, and rocks, *Advances in Water Resources*, **25**(8), 985–1016 (2002).

H.C. Andersen, Molecular dynamics simulations at constant pressure and/or temperature, *Journal of Chemical Physics*, **72**(4), 2384–2393 (1980).

M.J. Angove, M.B. Fernandes, and J. Ikhsan, The sorption of anthracene onto goethite and kaolinite in the presence of some benzene carboxylic acids, *Journal of Colloid and Interface Science*, **247**(2), 282–289 (2002).

C.A.J. Appelo and D. Postma, (2005) *Geochemistry, Groundwater and Pollution*, 2nd ed., CRC Press, Boca Raton.

A.J.A. Aquino, D. Tunega, G. Haberhauer, M.H. Gerzabek, and H. Lischka, Solvent effects on hydrogen bonds: A theoretical study, *Journal of Physical Chemistry A*, **106**(9), 1862–1871 (2002).

A.J.A. Aquino, D. Tunega, G. Haberhauer, M.H. Gerzabek, and H. Lischka, Quantum chemical adsorption studies on the (110) surface of the mineral goethite, *Journal of Physical Chemistry C*, **111**(2), 877–885 (2007a).

A.J.A. Aquino, D. Tunega, G. Haberhauer, M.H. Gerzabek, and H. Lischka, Interaction of the 2,4-dichlorophenoxyacetic acid herbicide with soil organic matter moieties: A theoretical study, *European Journal of Soil Science*, **58**(4), 889–899 (2007b).

A.J.A. Aquino, D. Tunega, G. Haberhauer, M.G. Gerzabek, and H. Lischka, Acid-base properties of a goethite surface model: A theoretical view, *Geochimica et Cosmochimica Acta*, **72**(15), 3587–3602 (2008a).

A.J.A. Aquino, D. Tunega, H. Pasalic, G. Haberhauer, M.H. Gerzabek, and H. Lischka, The thermodynamic stability of hydrogen bonded and cation bridged complexes of humic acid models: A theoretical study, *Chemical Physics*, **349**(1–3), 69–76 (2008b).

A.J.A. Aquino, D. Tunega, G.E. Schaumann, G. Haberhauer, M.H. Gerzabek, and H. Lischka, The functionality of cation bridges for binding polar groups in soil aggregates, *International Journal of Quantum Chemistry*, **111**(7–8), 1531–1542 (2011).

L. Aristilde and G. Sposito, Binding of ciprofloxacin by humic substances: A molecular dynamics study, *Environmental Toxicology and Chemistry*, **29**(1), 90–98 (2010).

L. Aristilde, C. Marichal, J. Miehe-Brendle, B. Lanson, and L. Charlet, Interactions of oxytetracycline with a smectite clay: A spectroscopic study with molecular simulations, *Environmental Science & Technology*, **44**(20), 7839–7845 (2010).

P.W.M. Augustijn-Beckers, A.G. Hornsby, and R.D. Wauchope, SCS/ARS/CES Pesticide Properties Database for Environmental Decisionmaking II. Additional Properties, in G.W. Ware (Ed.), *Reviews of Environmental Contamination and Toxicology*, Springer-Verlag, New York, Vol. 137, pp. 1–82 (1994).

K.F. Austen, T.O.H. White, A. Marmier, S.C. Parker, E. Artacho, and M.T. Dove, Electrostatic versus polarization effects in the adsorption of aromatic molecules of varied polarity on an insulating hydrophobic surface, *Journal of Physics. Condensed Matter*, **20**(3), 035215 (2008).

A. Bauer and G. Berger, Kaolinite and smectite dissolution rate in high molar KOH solutions at 35 degrees and 80 degrees C, *Applied Geochemistry*, **13**(7), 905–916 (1998).

C.I. Bayly, P. Cieplak, W.D. Cornell, and P.A. Kollman, A well-behaved electrostatic potential based method using charge restraints for determining atom-centered charges: The RESP model, *Journal of Physical Chemistry*, **97**(40),10269–10280 (1993).

A.D. Becke, Density-functional exchange-energy approximation with correct asymptotic behavior, *Physical Review A*, **38**(6), 3098–3100 (1988).

A.D. Becke, Density-functional thermochemistry. III. The role of exact exchange, *Journal of Chemical Physics*, **98**(7), 5648–5652 (1993).

A.D. Becke, Perspective: Fifty years of density-functional theory in chemical physics, *Journal of Chemical Physics*, **140**(18), 18A301 (2014).

P. Benoit, E. Barriuso, and R. Calvet, Biosorption characterization of herbicides, 2,4-D and Atrazine, and two chlorophenols on fungal mycelium, *Chemosphere*, **37**(7), 1271–1280 (1998).

H.J.C. Berendsen, J.P.M. Postma, W.F. van Gunsteren, and J. Hermans, Interaction Models for Water in Relation to Protein Hydration, in B. Pullman (Ed.), *Intermolecular Forces*, Reidel Publishing, Dordrecht, pp. 331–342 (1981).

H.J.C. Berendsen, J.P.M. Postma, W.F. van Gunsteren, A. Dinola, and J.R. Haak, Molecular-dynamics with coupling to an external bath, *Journal of Chemical Physics*, **81**(8), 3684–3690 (1984).

H.J.C Berendsen, J.R. Grigera, and T.P. Straatsma, The missing term in effective pair potentials, *Journal of Physical Chemistry*, **91**(24), 6269–6271 (1987).

B. Berkowitz, I. Dror, and B. Yaron, (2008) *Contaminant Geochemistry: Interactions and Transport in the Subsurface Environment*, Springer, Berlin.

P.E. Blöchl, Projector augmented-wave method, *Physical Review B*, **50**(24), 17953–17979 (1994).

J. Blotevogel, T. Borch, Y. Desyaterik, A.N. Mayeno, and T.C. Sale, Quantum chemical prediction of redox reactivity and degradation pathways for aqueous phase contaminants: An example with HMPA, *Environmental Science and Technology*, **44**(15), 5868–5874 (2010).

J. Blotevogel, A.N. Mayeno, T.C. Sale, and T. Borch, Prediction of contaminant persistence in aqueous phase: A quantum chemical approach, *Environmental Science & Technology*, **45**(6), 2236–2242 (2011).

J.-F. Boily, P. Persson, and S. Sjöberg, Benzenecarboxylate surface complexation at the goethite (α-FeOOH)/water interface: II. Linking IR spectroscopic observations to mechanistic surface complexation models for phthalate, trimellitate, and pyromellitate, *Geochimica et Cosmochimica Acta*, **64**(20), 3453–3470 (2000).

N.S. Bolan and S. Baskaran, Biodegradation of 2,4-D herbicide as affected by its adsorption–desorption behaviour and microbial activity of soils, *Australian Journal of Soil Research*, **34**(6), 1041–1053 (1996).

M. Borisover, M. Sela, and B. Chefetz, Enhancement effect of water associated with natural organic matter (NOM) on organic compound-NOM interactions: A case study with carbamazepine, *Chemosphere*, **82**(10), 1454–1460 (2011).

C.M. Breneman and K.B. Wiberg, Determining atom-centered monopoles from molecular electrostatic potentials. The need for high sampling density in formamide conformational analysis, *Journal of Computational Chemistry*, **11**(3), 361–373 (1990).

T.D. Bucheli and O. Gustafsson, Quantification of the soot-water distribution coefficient of PAHs provides mechanistic basis for enhanced sorption observations, *Environmental Science & Technology*, **34**(24), 5144–5151 (2000).

T.D. Bucheli and O. Gustafsson, Soot sorption of non-ortho- and ortho-substituted PCBs, *Chemosphere*, **53**(5), 515–522 (2003).

T. Bučko, S. Lebegue, J.G. Angyan, and J. Hafner, Improved description of the structure of molecular and layered crystals: *Ab initio* DFT calculations with van der Waals corrections, *Journal of Physical Chemistry A*, **114**(43), 11814–11824 (2010).

T. Bučko, S. Lebegue, J.G. Angyan, and J. Hafner, Extending the applicability of the Tkatchenko-Scheffler dispersion correction via iterative Hirshfeld partitioning, *Journal of Chemical Physics*, **141**(3), 034114 (2014).

J. Buffle, F.L. Greter, and W. Haerdi, Measurement of complexation properties of humic and fulvic acids in natural waters with lead and copper ion-selective electrodes, *Analytical Chemistry*, **49**(2), 216–222 (1977).

U. Burkert and N.L. Allinger, (1982) *Molecular Mechanics*, ACS Monograph **177**, ACS, Washington, DC.

E. Cances, B. Mennucci, and J. Tomasi, A new integral equation formalism for the polarizable continuum model: Theoretical background and applications to isotropic and anisotropic dielectrics, *Journal of Chemical Physics*, **107**(8), 3032–3041 (1997).

S. Canuto and J.R. Sabin (Eds.), *Combining Quantum Mechanics and Molecular Mechanics. Some Recent Progresses in QM/MM Methods*, Advances in Quantum Chemistry, Vol. **59**, Academic Press, Amsterdam, (2010).

R. Car and M. Parrinello, Unified approach for molecular dynamics and density-functional theory, *Physical Review Letters*, **55**(22), 2471–2474 (1985).

R. Celis and W.C. Koskinen, An isotropic exchange method for the characterization of the irreversibility of pesticide sorption-desorption in soil, *Journal of Agricultural and Food Chemistry*, **47**(2), 782–790 (1999).

R. Celis M.C. Hermosin, L. Cox, and J. Cornejo, Sorption of 2,4-dichlorophenoxyacetic acid by model particles simulating naturally occurring soil colloids, *Environmental Science & Technology*, **33**(8), 1200–1206 (1999).

F.F. Charlotte, General Hartree-Fock program, *Computer Physics Communications*, **43**(3), 355–365 (1987).

B.L. Chen and M.X. Yuan, Enhanced sorption of polycyclic aromatic hydrocarbons by soil amended with biochar, *Journal of Soils and Sediments*, **11**(1), 62–71 (2011).

F.-H. Chi, The influence of black carbon on the sorption and desorption of two model PAHs in natural soils, *Bulletin of Environmental Contamination and Toxicology*, **92**(1), 44–49 (2014).

C.T. Chiou, S.E. McGroddy, and D.E. Kile, Partition characteristics of polycyclic aromatic hydrocarbons on soils and sediments, *Environmental Science & Technology*, **32**(2), 264–269 (1998).

L.E. Chirlian, and M.M. Francl, Atomic charges derived from electrostatic potentials: A detailed study, *Journal of Computational Chemistry*, **8**(6), 894–905 (1987).

J. Čížek, Origins of coupled cluster technique for atoms and molecules, *Theoretica Chimica Acta*, **80**(2–3), 91–94 (1991).

L. Clausen and I. Fabricius, Atrazine, isoproturon, mecoprop, 2,4-D, and bentazone adsorption onto iron oxides, *Journal of Environmental Quality*, **30**(3), 858–869 (2001).

L. Clausen, I. Fabricius, and L. Madsen, Adsorption of pesticides onto quartz, calcite, kaolinite and α-alumina, *Journal of Environmental Quality*, **30**(3), 846–857 (2001).

R.M. Cornell and U. Schwertmann, (2003) *The Iron Oxides*, VCH Verlag, Weinheim.

P. Cosoli, M. Fermeglia, and M. Ferrone, Molecular simulation of atrazine adhesion and diffusion in a saturated sand model, *Soil and Sediment Contamination*, **19**(1), 72–87 (2010).

A.A. Costa, W.B. Wilson, H.Y. Wang, A.D. Campiglia, J.A. Dias, and S.C.L. Dias, Comparison of BEA, USY and ZSM-5 for the quantitative extraction of polycyclic aromatic hydrocarbons from water samples, *Microporous and Mesoporous Materials*, **149**(1), 186–192 (2012).

L. Cox, R. Celis, M.C. Hermosin, and J. Cornejo, Natural soil colloids to retard simazine and 2,4-D leaching in soil, *Journal of Agricultural and Food Chemistry*, **48**(1), 93–99 (2000).

A. Cozzolino, P. Conte, and A. Piccolo, Conformational changes of humic substances induced by some hydroxy-, keto-, and sulfonic acids, *Soil Biology and Biochemistry*, **33**(4–5), 563–571 (2001).

C.J. Cramer, (2004) *Essentials of Computational Chemistry: Theories and Models*, John Wiley & Sons, Ltd, Chichester.

R.T. Cygan, J.J. Liang, and A.G. Kalinichev, Molecular models of hydroxide, oxyhydroxide, and clay phases and the development of a general force field, *Journal of Physical Chemistry B*, **108**(4) 1255–1266 (2004).

T. Darden, D. York, and L. Pedersen, Particle mesh Ewald: An $N \cdot \log(N)$ method for Ewald sums in large systems, *Journal of Chemical Physics*, **98**(12), 10089–10092 (1993).

M.S. Daw, S.M. Foiles, and M.I. Baskes, The embedded-atom method: A review of theory and applications, *Materials Science Reports*, **9**(7–8), 251–310 (1993).

C. Dellago and G. Hummer, Computing equilibrium free energies using non-equilibrium molecular dynamics, *Entropy*, **16**(1), 41–61 (2014).

A. Delle Site, Factors affecting sorption of organic compounds in natural sorbent/water systems and sorption coefficients for selected pollutants. A review, *Journal of Physical and Chemical Reference Data*, **30**(1), 187–439 (2001).

F. DePaolis and J. Kukkonen, Binding of organic pollutants to humic and fulvic acids: Influence of pH and the structure of humic material, *Chemosphere*, **34**(8) 1693–1704 (1997).

G.A. DiLabio, Accurate treatment of van der Waals interactions using standard density functional theory methods with effective core-type potentials: Application to carbon-containing dimers, *Chemical Physics Letters*, **455**(4–6), 348–353 (2008).

M. Dion, H. Rydberg, E. Schroder, D.C. Langreth, and B.I. Lundqvist, Van der Waals density functional for general geometries, *Physical Review Letter*, **92**(24), 246401 (2004).

S. Duane, A.D. Kennedy, B.J. Pendleton, and D. Roweth, Hybrid Monte Carlo, *Physics Letters B*, **195**(2), 216–222 (1987).

S.L. Dudarev, G.A. Botton, S.Y. Savrasov, C.J. Humphreys, and A.P. Sutton, Electron energy loss spectra and the structural stability of nickel oxide: An LSDA+U study, *Physical Review B*, **57**(3), 1505–1509 (1998).

A.C.T. van Duin, S. Dasgupta, Siddharth, F. Lorant, and W.A. Goddard, ReaxFF: A reactive force field for hydrocarbons, *Journal of Physical Chemistry A*, **105**(41), 9396–9409 (2001).

M. Elstner, D. Porezag, G. Jungnickel, J. Elsner, M. Haugk, T. Frauenheim, S. Suhai, and G. Seifert, Self-consistent-charge density-functional tight-binding method for simulations of complex materials properties, *Physical Review B*, **58**(11), 7260–7268 (1998).

C.M. Fang, S.C. Parker, and G. de With, Atomistic simulation of the surface energy of spinel $MgAl_2O_3$, *Journal of American Ceramic Society*, **83**(8), 2082–2084 (2000).

K. Fenner, S. Canonica, L.P. Wackett, and M. Elsner, Evaluating pesticide degradation in the environment: Blind spots and emerging opportunities, *Science*, **341**(6147), 752–758 (2013).

P.A. Fenter, M.L. Rivers, N.C. Sturchio, and S.R. Sutton (Eds.), *Applications of Synchrotron Radiation in Low-Temperature Geochemistry and Environmental Sciences*, Reviews in Mineralogy and Geochemistry, Vol. **49**, Mineralogical Society of America, Washington, DC, (2002).

M. Ferrario and J.P. Ryckaert, Constant pressure–constant temperature molecular dynamics for rigid and partially rigid molecular systems, *Molecular Physics*, **54**(3), 587–603 (1985).

J.C. Fetzer, (2000) *The Chemistry and Analysis of the Large Polycyclic Aromatic Hydrocarbons*, John Wiley & Sons, Inc., New York.

J.B. Forsyth, I.G. Hedley, and C.E. Johnson, The magnetic structure and hyperfine field of goethite (α-FeOOH), *Journal of Physics Part C Solid State Physics*, **1**(1), 179–188 (1968).

D. Frenkel and B. Smit, (2002) *Understanding Molecular Simulation*, 2nd ed., Academic Press, San Diego.

R.A. Friesner and V. Guallar, *Ab initio* quantum chemical and mixed quantum mechanics/molecular mechanics (QM/MM) method for studying enzymatic catalysis, *Annual Review of Physical Chemistry*, **56**, 389–427 (2005).

F. Gaboriaud and J.J. Ehrhardt, Effects of different crystal faces on the surface charge of colloidal goethite (α-FeOOH) particles: An experimental and modeling study, *Geochimica et Cosmochimica Acta*, **67**(5), 967–983 (2003).

J. Gao and M.A. Thompson (Eds.), *Combined Quantum Mechanical and Molecular Mechanical Methods*, ACS Symposium Series, **712**, American Chemical Society, Washington, DC (1998).

K. Ghosh and M. Schnitzer, Macromolecular structure of humic substances, *Soil Science*, **129**(5), 266–276 (1980).

L. Goerigk, How do DFT-DCP, DFT-NL, and DFT-D3 compare for the description of London-dispersion effects in conformers and general thermochemistry? *Journal of Chemical Theory and Computation*, **10**(3), 968–980 (2014).

J.A. Greathouse, K.L. Johnson, and H.C. Greenwell, Interaction of natural organic matter with layered minerals: Recent developments in computational methods at the nanoscale, *Minerals*, **4**(2), 519–540 (2014).

M.L. Greenfield, Molecular modelling and simulation of asphaltenes and bituminous materials, *International Journal of Pavement Engineering*, **12**(4), 325–341 (2011).

S. Grimme, Accurate description of van der Waals complexes by density functional theory including empirical corrections, *Journal of Computational Chemistry*, **25**(12), 1463–1473 (2004).

S. Grimme, Semiempirical GGA-type density functional constructed with a long-range dispersion correction, *Journal of Computational Chemistry*, **27**(15), 1787–1799 (2006).

S. Grimme, Density functional theory with London dispersion corrections, *WIREs: Computational Molecular Science*, **1**(2), 211–228 (2011).

S. Grimme, J. Antony, S. Ehrlich, and H. Krieg, A consistent and accurate *ab initio* parameterization of density functional dispersion correction (DFT-D) for the 94 elements H-Pu, *Journal of Chemical Physics*, **132**(15), 154104 (2010).

H. Guo and A.S. Barnard, Thermodynamic modelling of nanomorphologies of hematite and goethite, *Journal of Materials Chemistry*, **21**(31), 11566–11577 (2011).

G. Haberhauer, L. Pfeiffer, and M.H. Gerzabek, Influence of molecular structure on sorption of phenoxyalkanoic herbicides on soil and its particle size fractions, *Journal of Agricultural and Food Chemistry*, **48**(8), 3722–3727 (2000).

G. Haberhauer, L. Pfeiffer, M.H. Gerzabek, H. Kirchmann, A.J.A. Aquino, D. Tunega, and H. Lischka, Response of sorption processes of MCPA to the amount and origin of organic matter in a long-term field experiment, *European Journal of Soil Science*, **52**(2), 279–286 (2001).

T.A. Halgren, Merck molecular force field. I. Basis, form, scope, parameterization, and performance of MMFF94, *Journal of Computational Chemistry*, **17**(5–6), 490–519 (1996).

J.H. Harding, D.M. Duffy, M.L. Sushko, P.M. Rodger, D. Quigley, and J.A. Elliott, Computational techniques at the organic-inorganic interface in biomineralization, *Chemical Reviews*, **108**(11), 4823–4854 (2008).

C. Hartmann, R. Banisch, M. Sarich, T. Badowski, and C. Schutte, Characterization of rare events in molecular dynamics, *Entropy*, **16**(1), 350–376 (2014).

R.G. Harvey, (1997) *Polycyclic Aromatic Hydrocarbons*, John Wiley & Sons, Inc., New York.

J.L. Hazemann, J.F. Bérar, and A. Manceau, Rietveld studies of the aluminum iron substitution in synthetic goethite, *Materials Science Forum*, **79**, 821–826 (1991).

W.J. Hehre, (1998) *A Brief Guide to Molecular Mechanics and Quantum Chemical Calculations*, Wavefunction, Inc., Irvine.

H. Heinz, R.A. Vaia, B.L. Farmer, and R.R. Naik, Accurate simulation of surfaces and interfaces of face-centered cubic metals using 12-6 and 9-6 Lennard-Jones potentials, *Journal of Physical Chemistry C*, **112**(44), 17281–17290 (2008).

H. Heinz, T.J. Lin, R.K. Mishra, and F.S. Emami, Thermodynamically consistent force fields for the assembly of inorganic, organic, and biological nanostructures: The INTERFACE force field, *Langmuir*, **29**(6), 1754–1765 (2013).

G.S. Henderson, D.R. Neuville, and R.T. Downs (Eds.), *Spectroscopic Methods in Mineralogy and Material Sciences*, Reviews in Mineralogy and Geochemistry, Vol. **78**, Mineralogical Society of America, Washington, DC, (2014).

B. Hess, C. Kutzner, D. van der Spoel, and E. Lindahl, GROMACS 4: Algorithms for highly efficient, load-balanced, and scalable molecular simulation, *Journal of Chemical Theory and Computation*, **4**(3), 435–447 (2008).

T. Hiemstra, J.M.C. Dewit, and W.H. Van Riemsdijk, Multisite proton adsorption modeling at the solid-solution interface of (hydr)oxides: A new approach. 2. Application to various important (hydr)oxides, *Journal of Colloid and Interface Science*, **133**(1), 105–117. (1989a).

T. Hiemstra, W.H. Van Riemsdijk, and G.H. Bolt, Multisite proton adsorption modeling at the solid-solution interface of (hydr)oxides: A new approach. 1. Model description and evaluation of intrinsic reaction constants, *Journal of Colloid and Interface Science*, **133**(1), 91–104 (1989b).

E. Hiller, V. Tatarková, S. Šimonovičová, and M. Bartaľ, Sorption, desorption, and degradation of (4-chloro-2-methylphenoxy) acetic acid in representative soils of the Danubian lowland, Slovakia, *Chemosphere*, **87**(5), 437–444 (2012).

P. Hohenberg and W. Kohn, Inhomogeneous electron gas, *Physical Review B*, **136**(3B), 864–871 (1964).

A.G. Hornsby, R.D. Wauchope, and A.E. Herner, (1996) *Pesticide Properties in the Environment*, Springer-Verlag, New York.

S. Hyun and L.S. Lee, Quantifying the contribution of different sorption mechanisms for 2,4–dichlorophenoxyacetic acid sorption by several variable-charge soils, *Environmental Science & Technology*, **39**(8), 2522–2528 (2005).

A. Iglesias, R. Lopez, D. Gondar, J. Antelo, S. Fiol, and F. Arce, Effect of pH and ionic strength on the binding of paraquat and MCPA by soil fulvic and humic acids, *Chemosphere*, **76**(1), 107–113 (2009).

A. Iglesias, R. Lopez, D. Gondar, J. Antelo, S. Fiol, and F. Arce, Adsorption of MCPA on goethite and humic acid-coated goethite, *Chemosphere*, **78**(11), 1403–1408 (2010).

E. Iskrenova-Tchoukova, A.G. Kalinichev and R.J. Kirkpatrick, Metal cation complexation with natural organic matter in aqueous solutions: Molecular dynamics simulations and potentials of mean force, *Langmuir*, **26**(20), 15909–15919 (2010).

H.Z. Jia, J.C. Zhao, L. Li, X.Y. Li, and C.Y. Wang, Transformation of polycyclic aromatic hydrocarbons (PAHs) on Fe(III)-modified clay minerals: Role of molecular chemistry and clay surface properties, *Applied Catalysis B: Environmental*, **154**, 238–245 (2014).

E.R. Johnson and A. Otero-de-la-Roza, Adsorption of organic molecules on kaolinite from the exchange-hole dipole moment dispersion model, *Journal of Chemical Theory and Computation*, **8**(12), 5124–5131 (2012).

E.R. Johnson, I.D. Mackie, and G.A. DiLabio, Dispersion interactions in density-functional theory, *Journal of Physical Organic Chemistry*, **22**(12), 1127–1135 (2009).

M.T.O. Jonker, Absorption of polycyclic aromatic hydrocarbons to cellulose, *Chemosphere*, **70**(5), 778–782 (2008).

W.L. Jorgensen, J. Chandrasekhar, J.D. Madura, R.W. Impey, and M.L. Klein, Comparison of simple potential functions for simulating liquid water, *Journal of Chemical Physics*, **79**(2), 926–935 (1983).

W.L. Jorgensen, D.S. Maxwell, and J. Tirado-Rives, Development and testing of the OPLS all-atom force field on conformational energetics and properties of organic liquids, *Journal of the American Chemical Society*, **118**(45), 11225–11236 (1996).

H. Joseph-Ezra, A. Nasser, and U. Mingelgrin, Surface interactions of pyrene and phenanthrene on Cu-montmorillonite, *Applied Clay Science*, **95**, 348–356 (2014).

A.G. Kalinichev and R.J. Kirkpatrick, Molecular dynamics simulation of cationic complexation with natural organic matter, *European Journal of Soil Science*, **58**(4), 909–917 (2007).

S.C.L. Kamerlin, S. Vicatos, A. Dryga, and A. Warshel, Coarse-grained (multiscale) simulations in studies of biophysical and chemical systems, *Annual Review of Physical Chemistry*, **62**, 41–64 (2011).

S. Kang and B.S. Xing, Adsorption of dicarboxylic acids by clay minerals as examined by in situ ATR-FTIR and ex situ DRIFT, *Langmuir*, **23**(13), 7024–7031 (2007).

M. Kästner, K.M. Nowak, A. Miltner, S. Trapp, and A. Schaffer, Classification and modelling of non-extractable residue (NER) formation of xenobiotics in soil: A synthesis, *Critical Reviews in Environmental Science and Technology*, **44**(19), 2107–2171 (2014).

E.M.O. Kaya, A.S. Ozcan, O. Gok, and A. Ozcan, Adsorption kinetics and isotherm parameters of naphthalene onto natural- and chemically modified bentonite from aqueous solutions, *Adsorption: Journal of the International Adsorption Society*, **19**(2–4), 879–888 (2013).

M. Kersten, D. Tunega, I Georgieva, N. Vlasova, and R. Branscheid, Adsorption of the herbicide 4-chloro-2-methylphenoxyacetic acid (MCPA) by goethite, *Environmental Science & Technology*, **48**(20), 11803–11810 (2014).

A. Klamt and G. Schürmann, COSMO: A new approach to dielectric screening in solvents with explicit expressions for the screening energy and its gradient, *Journal of the Chemical Society, Perkin Transactions*, **2**(5), 799–805 (1993).

A. Klamt, V. Jonas, T. Burger, and J.W.C. Lohrenz, Refinement and parametrization of COSMO-RS, *Journal of Physical Chemistry A*, **102**(26), 5074–5085 (1998).

A. Klamt, F. Eckert, M. Hornig, M.E. Beck, and T. Burger, Prediction of aqueous solubility of drugs and pesticides with COSMO-RS, *Journal of Computational Chemistry*, **23**(2), 275–281 (2002).

A. Klamt, F. Eckert, and M. Diedenhofen, Prediction of the free energy of hydration of a challenging set of pesticide-like compounds, *Journal of Physical Chemistry B*, **113**(14), 4508–4510 (2009).

S. Kleineidam, H. Rügner, B. Ligouis, and P. Grathwohl, Organic matter facies and equilibrium sorption of phenanthrene, *Environmental Science & Technology*, **33**(10), 1637–1644 (1999).

W. Kohn and L.J. Sham, Self-consistent equations including exchange and correlation effects, *Physical Review A*, **140**(4A), 1133–1140 (1965).

L.N. Konda, I. Czinkota, G. Füleky, and G. Morovján, Modeling of single-step and multistep adsorption isotherms of organic pesticides on soil, *Journal of Agricultural Food and Chemistry*, **50**(25), 7326–7331 (2002).

G. Kresse and J. Furthmüller, Efficiency of *ab initio* total energy calculations for metals and semiconductors using a plane-wave basis set, *Computational Materials Science*, **6**(1), 15–50 (1996a).

G. Kresse and J. Furthmüller, Efficient iterative schemes for *ab initio* total-energy calculations using a plane-wave basis set, *Physical Review B*, **54**(16), 11169–11186 (1996b).

G. Kresse and J. Hafner, *Ab initio* molecular dynamics for open-shell transition metals, *Physical Review B*, **48**(17), 13115–13118 (1993).

G. Kresse and D. Joubert, From ultrasoft pseudopotentials to the projector augmented-wave method, *Physical Review B*, **59**(3), 1758–1775 (1999).

J.D. Kubicki, Molecular mechanics and quantum mechanical modeling of hexane soot structure and interactions with pyrene, *Geochemical Transactions*, **1**(1), 41–46 (2000).

J.D. Kubicki, Computational chemistry applied to studies of organic contaminants in the environment: Examples based on benzo[a]pyrene, *American Journal of Science*, **305**(6–8), 621–644 (2005).

J.D. Kubicki, Molecular simulations of benzene and PAH interactions with soot, *Environmental Science & Technology*, **40**(7), 2298–2303 (2006).

J.D. Kubicki and S.E. Apitz, Models of natural organic matter and interactions with organic contaminants, *Organic Geochemistry*, **30**(8B), 911–927 (1999).

J.D. Kubicki, K.W. Paul, and D.L. Sparks, Periodic density functional theory calculations of bulk and the (010) surface of goethite, *Geochemical Transactions*, **9**(1), 4 (2008).

Y. Kunhi Mouvenchery, A. Jäger, A.J.A. Aquino, D. Tunega, M. Diehl, M. Bertmer, and G.E. Schaumann, Restructuring of a peat in interaction with multivalent cations: Effect of cation type and aging time, *Plos One*, **8**(6), e65359 (2013).

D.A. Kurtz (Ed.), *Long-Range Transport of Pesticides*, Lewis, Chelsea, (1990).

D.C. Langreth and M.J. Mehl, Beyond the local-density approximation in calculations of ground-state electronic properties, *Physical Review B*, **28**(4), 1809–1834 (1983).

S.J. Larson, D. Capel, and M.S. Majewski, (1997) *Pesticides in Surface Waters: Distribution, Trends, and Governing Factors*, Pesticides in the Hydrologic System, Vol. 3, Ann Arbor Press, Inc., Chelsea.

A. Leach, (2001) *Molecular Modelling: Principles & Applications*, 2nd ed., Pearson Education Limited, London.

E.J. Leboeuf and W.J. Weber Jr., A distributed reactivity model for sorption by soils and sediments. 8. Sorbent organic domains: Discovery of a humic acid glass transition and an argument for a polymer-based model, *Environmental Science & Technology*, **31**(6), 1697–1702 (1997).

C. Lee W. Yang, and R.G. Parr, Development of the Colle-Salvetti correlation-energy formula into a functional of the electron density, *Physical Review B*, **37**(2), 785–789 (1988).

K. Lee, E.D. Murray, L.Z. Kong, B.I. Lundqvist, and D.C. Langreth, Higher-accuracy van der Waals density functional, *Physical Review B*, **82**(8), 081101 (2010).

K. Leung and L.J. Criscenti, Predicting the acidity constant of a goethite hydroxyl group from first principles, *Journal of Physics. Condensed Matter*, **24**(12), 124105 (2012).

T. Liang, Y.K. Shin, Y.T. Cheng, D.E. Yilmaz, K.G. Vishnu, O. Verners, C.Y. Zou, S.R. Phillpot, S.B. Sinnott, and A.C.T. van Duin, Reactive potentials for advanced atomistic simulations, in D.R. Clarke (Ed.), *Annual Review of Materials Research*, **43**, 109–129 (2013).

O.A. von Lilienfeld, I. Tavernelli, U. Rothlisberger, and D. Sebastiani, Optimization of effective atom centered potentials for London dispersion forces in density functional theory, *Physical Review Letters*, **93**(15), 153004 (2004).

K. Lindorff-Larsen, S. Piana, R.O. Dror, and D.E. Shaw, How fast-folding proteins fold, *Science*, **334**(6055), 517–520 (2011).

J.C. Liu and P.A. Monson, Molecular modeling of adsorption in activated carbon: Comparison of Monte Carlo simulations with experiment, *Adsorption: Journal of the International Adsorption Society*, **11**(1), 5–13 (2005).

C. Liu, H. Li, B.J. Teppen, C.T. Johnston, and S.A. Boyd, Mechanisms associated with the high adsorption of dibenzo-p-dioxin from water by smectite clays, *Environmental Science and Technology*, **43**(8), 2777–2783 (2009).

C. Liu, H. Li, C.T. Johnston, S.A. Boyd, and B.J. Teppen, Relating clay structural factors to dioxin adsorption by smectites: Molecular dynamics simulations, *Soil Science Society of America Journal*, **76**(1), 110–120 (2012a).

X.D. Liu, X.C. Lu, R.C. Wang, E.J. Meijer, H.Q. Zhou, and H.P. He, Atomic scale structures of interfaces between kaolinite edges and water, *Geochimica et Cosmochimica Acta*, **92**, 233–242 (2012b).

R. Lohmann, J.K. MacFarlane, and P.M. Gschwend, Importance of black carbon to sorption of native PAHs, PCBs, and PCDDs in Boston and New York harbor sediments, *Environmental Science & Technology*, **39**(1), 141–148 (2005).

J.P. Lowe and K.L. Peterson, (2005) *Quantum Chemistry*, 3rd ed., Academic Press, New York.

M. von Lützow, I. Kogel-Knabner, K. Ekschmitt, E. Matzner, G. Guggenberger, R. Marschner, and H. Flessa, Stabilization of organic matter in temperate soils: Mechanisms and their relevance under different soil conditions. A review, *European Journal of Soil Science*, **57**(4), 426–445 (2006).

A. Lyubartsev, Y.Q. Tu, and A. Laaksonen, Hierarchical multiscale modelling scheme from first principles to mesoscale, *Journal of Computational and Theoretical Nanoscience*, **6**(5), 951–959 (2009).

J.C. Ma and D.A. Dougherty, The cation-π interaction, *Chemical Reviews*, **97**(5), 1303–1324 (1997).

C. Ma and R.A. Eggleton, Cation exchange capacity of kaolinite, *Clays and Clay Minerals*, **47**(2), 174–180 (1999).

A.D. MacKerell Jr., D. Bashford, M. Bellott, R.L. Dunbrack, J.D. Evanseck, M.J. Field, S. Fischer, J. Gao, H. Guo, S. Ha, D. Joseph-McCarthy, L. Kuchnir, K. Kuczera, F.T.K. Lau, C. Mattos, S. Michnick, T. Ngo, D.T. Nguyen, B. Prodhom, W.E. Reiher, B. Roux, M. Schlenkrich, J.C. Smith, R. Stote, J. Straub, M.

Watanabe, J. Wiorkiewicz-Kuczera, D. Yin, and M. Karplus, All-atom empirical potential for molecular modeling and dynamics studies of proteins, *Journal of Physical Chemistry B*, **102**(18), 3586–3616 (1998).

B.T. Mader, K.U. Goss, and S.J. Eisereich, Sorption of nonionic, hydrophobic organic chemicals to mineral surfaces, *Environmental Science & Technology*, **31**(4), 1079–1086 (1997).

N. Marom, A. Tkatchenko, M. Rossi, V. Gobre, O. Hod, and M. Scheffler, Dispersion interactions with density-functional theory: Benchmarking semiempirical and interatomic pairwise corrected density functionals, *Journal of Chemical Theory and Computation*, **7**(12), 3944–3951 (2011).

W. Mathys, Pesticide pollution of groundwater and drinking water by the processes of artificial groundwater enrichment or coastal filtration: Underrated sources of contamination, *Zentralblatt für Hygiene und Umweltmedizin*, **196**(4), 338–359 (1994).

J.C. Means, S.G. Wood, J.J. Hassett, and W.L. Banwart, Sorption of polynuclear aromatic hydrocarbons by sediments and soils, *Environmental Science & Technology*, **14**(12), 1524–1528 (1980).

N. Metropolis, A.W. Rosenbluth, M.N. Rosenbluth, A.H. Teller, and E. Teller, Equations of state calculations by fast computing machines, *Journal of Chemical Physics*, **21**(6), 1087–1092 (1953).

A. Michalková, L. Gorb, M. Ilchenko, O.A. Zhikol, O.V. Shishkin, and J. Leszczynski, Adsorption of sarin and soman on dickite: An *ab initio* ONIOM study, *Journal of Physical Chemistry B*, **108**(6), 1918–1930 (2004a).

A. Michalková, M. Ilchenko, L. Gorb, and J. Leszczynski, Theoretical study of the adsorption and decomposition of sarin on magnesium oxide, *Journal of Physical Chemistry B*, **108**(17), 5294–5303 (2004b).

A. Michalková, J. Martinez, O.A. Zhikol, L. Gorb, O.V. Shishkin, D. Leszczynska, and J. Leszczynski, Theoretical study of adsorption of Sarin and Soman on tetrahedral edge clay mineral fragments, *Journal of Physical Chemistry B*, **110**(42), 21175–21183 (2006).

B. Miehlich, A. Savin, H. Stoll, and H. Preuss, Results obtained with the correlation energy density functionals of Becke and Lee, Yang, and Parr, *Chemical Physics Letters*, **157**(3) 200–206 (1989).

S. Miertuš, E. Scrocco, and J. Tomasi, Electrostatic interaction of a solute with a continuum. A direct utilization of ab initio molecular potentials for the prevision of solvent effects, *Chemical Physics*, **55**(1), 117–129 (1981).

C. Møller and M.S. Plesset, Note on an approximation treatment for many-electron systems, *Physical Review*, **46**(7), 618–622 (1934).

S. Müller, K.U. Totsche, and I. Kögel-Knaber, Sorption of polycyclic aromatic hydrocarbons to mineral surfaces, *European Journal of Soil Science*, **58**(4), 918–938 (2007).

J. Murgich, J. Rodriguez, A. Izquierdo, L. Carbognani, and E. Rogel, Interatomic interactions in the adsorption of asphaltenes and resins on kaolinite calculated by molecular dynamics, *Energy and Fuels*, **12**(2), 339–343 (1998).

J. Murgich, J.A. Abanero, and O.P. Strausz, Molecular recognition in aggregates formed by asphaltene and resin molecules from the Athabasca oil sand, *Energy & Fuels*, **13**(2), 278–286 (1999).

J. Murgich, E. Rogel, O. Leon, and R. Isea, A molecular mechanics-density functional study of the adsorption of fragments of asphaltenes and resins on the (001) surface of Fe_2O_3, *Petroleum Science and Technology*, **19**(3–4), 437–455 (2001).

M. Négre, H.R. Schulten, M. Gennari, and D. Vindrola, Interaction of imidazolinone herbicides with soil humic acids. Experimental results and molecular modeling, *Journal of Environmental Science and Health B*, **36**(2), 107–125 (2001).

L.A. Noll, Adsorption of aqueous benzene onto hydrophobic and hydrophilic surfaces, *Colloids and Surfaces*, **28**(2–4), 327–329 (1987).

S. Nosé, A unified formulation of the constant temperature molecular dynamics methods, *Journal of Chemical Physics*, **81**(1), 511–519 (1984).

P.A. O'Day, Molecular environmental geochemistry, *Reviews of Geophysics*, **37**(2), 249–274 (1999).

K. Otte, R. Pentcheva, W.W. Schmahl, and J.R. Rustad, Pressure-induced structural and electronic transitions in FeOOH from first principles, *Physical Review B*, **80**(20), 205116 (2009).

C.N. Owabor, S.E. Ogbeide, and A.A. Susu, Adsorption and desorption kinetics of naphthalene, anthracene, and pyrene in soil matrix, *Petroleum Science and Technology*, **28**(5), 504–514 (2010).

G. Pacchioni, P.S. Bagus, and F. Parmigiani (Eds.), *Cluster Models for Surface and Bulk Phenomena*, NATO ASI, Series B, Vol. **283**, Physics Plenum, New York (1992).

J. Paldus, The Beginnings of Coupled-Cluster Theory: An Eyewitness Account, in C. Dykstra G. Frenking, K. Kim, and G. Scuseria (Eds.), *Theory and Applications of Computational Chemistry: The First Forty Years*, Elsevier, Amsterdam, pp. 115–148 (2005).

R.G. Parr and W.T. Yang, (1989) *Density-Functional Theory of Atoms and Molecules*, Oxford University Press, New York.

D.A. Pearlman, D.A. Case, J.W. Caldwell, W.S. Ross, T.E. Cheatham, S. Debolt, D. Ferguson, G. Seibel, and P.A. Kollman, AMBER, a package of computer programs for applying molecular mechanics, normal mode analysis, molecular dynamics and free energy calculations to simulate the structural and energetic properties of molecules, *Computer Physics Communications*, **91**(1–3), 1–41 (1995).

Z.G. Pei, J.J. Kong, Y.Q. Shan, and B. Wen, Sorption of aromatic hydrocarbons onto montmorillonite as affected by norfloxacin, *Journal of Hazardous Materials*, **203**, 137–144 (2012).

J.P. Perdew and Y. Wang, Accurate and simple analytic representation of the electron-gas correlation energy, *Physical Review B*, **45**(23), 13244–13249 (1992).

J.P. Perdew, J.A. Chevary, S.H. Vosko, K.A. Jackson, M.R. Pederson, D.J. Singh, and C. Fiolhais, Atoms, molecules, solids, and surfaces: Applications of the generalized gradient approximation for exchange and correlation, *Physical Review B*, **46**(11), 6671–6687 (1992).

J.P. Perdew, K. Burke, and M. Ernzerhof, Generalized gradient approximation made simple, *Physical Review Letters*, **77**(18), 3865–3868 (1996).

M. Petrovic, M. Kastelan-Macan, and A.J.M. Horvat, Interactive sorption of metal ions and humic acids onto mineral particles, *Water Air Soil Pollution*, **111**(1–4), 41–56 (1999).

L. Pfeiffer, Adsorption und Desorption von Phenoxysäure-Herbiziden und Chlorphenolen an natürlichen Böden, deren Korngrössenfraktionen, Huminsäuren und Montmorillonit, PhD thesis, Department für Lebenswissenschaften, Universität Wie, pp. 93–97 (1999).

A. Piccolo, The supramolecular structure of humic substances: A novel understanding of humus chemistry and implications in soil science, *Advances in Agronomy*, **75**, 57–134 (2002).

J.J. Pignatello and B. Xing, Mechanism of slow sorption of organic chemicals to natural particles, *Environmental Science & Technology*, **30**(1), 1–11 (1996).

L.F. Ping, Y.M. Luo, L.H. Wu, W. Qian, J. Song, and P. Christie, Phenanthrene adsorption by soils treated with humic substances under different pH and temperature conditions, *Environmental Geochemistry and Health*, **28**(1–2), 189–195 (2006).

J.W. Ponder and F.M. Richards, An efficient Newton-like method for molecular mechanics energy minimization of large molecules, *Journal of Computational Chemistry*, **8**(7), 1016–1024 (1987).

G.J. Pronk, K. Heister, and I. Kögel-Knabner, Iron oxides as major available interface component in loamy arable topsoils, *Soil Science Society of America Journal*, **75**(6), 2158–2168 (2011).

J.P.P. Ramalho, A.V. Dordio, and A.J.P. Carvalho, Adsorption of two phenoxyacid compounds on a clay surface: A theoretical study, *Adsorption: Journal of the International Adsorption Society*, **19**(5), 937–944 (2013).

J. Ran and P. Hobza, On the nature of bonding in lone pair π-electron complexes: CCSD(T)/complete basis set limit calculations, *Journal of Chemical Theory and Computation*, **5**(4), 1180–1185 (2009).

A.K. Rappé and C.J. Casewit, (1997) *Molecular Mechanics across Chemistry*, University Science Books, Sausalito.

A.K. Rappé, C.J. Casewit, K.S. Colwell, W.A. Goddard, and W.M. Skiff, UFF, a full periodic table force field for molecular mechanics and molecular dynamics simulations, *Journal of the American Chemical Society*, **114**(25), 10024–10035 (1992).

A. Rimola, B. Civalleri, and P. Ugliengo, Physisorption of aromatic organic contaminants at the surface of hydrophobic/hydrophilic silica geosorbents: A B3LYP-D modeling study, *Physical Chemistry Chemical Physics*, **12**(24), 6357–6366 (2010).

D. Rinaldi and J.-L. Rivail, Molecular polarizability and dielectric effect of medium in liquid-phase: Theoretical study of water molecule and its dimers, *Theoretica Chimica Acta*, **32**(1), 57–70 (1973).

J. Rombke, C. Bauer, and A. Marschner, Hazard assessment of chemicals in soil. Proposed ecotoxicological test strategy. *Environmental Science and Pollution Research*, **3**(2), 78–82 (1996).

B. Roux, The calculation of the potential of mean force using computer simulations, *Computer Physics Communications*, **91**(1–3), 275–282 (1995).

H. Rügner, S. Kleineidam, and P. Grathwohl, Long term sorption kinetics of phenanthrene in aquifer materials, *Environmental Science & Technology*, **33**(10), 1645–1651 (1999).

V.G. Ruiz, W. Liu, E. Zojer, M. Scheffler, and A. Tkatchenko, Density-functional theory with screened van der Waals interactions for the modeling of hybrid inorganic-organic systems, *Physical Review Letters*, **108**(14), 146103 (2012).

F. Sannino A. Violante, and L. Gianfreda, Adsorption-desorption of 2,4-D by hydroxy aluminium montmorillonite complexes, *Pesticide Science*, **51**(4), 429–435 (1997).

P. Saparpakorn, J.H. Kim, and S. Hannongbua, Investigation on the binding of polycyclic aromatic hydrocarbons with soil organic matter: A theoretical approach, *Molecules*, **12**(4), 703–715 (2007).

J. Sauer, Molecular-models in ab-initio studies of solids and surfaces- from ionic crystals and semiconductors to catalysts, *Chemical Reviews*, **89**(1), 199–255 (1989).

A. Schäfer, H. Horn, and R. Ahlrichs, Fully optimized contracted Gaussian basis sets for atoms Li to Kr, *Journal of Chemical Physics*, **97**(4), 2571–2577 (1992).

G.E. Schaumann, Matrix relaxation and change of water state during hydration of peat, *Colloids and Surfaces A: Physicochemical and Engineering Aspects*, **265**(1–3), 163–170 (2005).

G.E. Schaumann and M. Bertmer, Do water molecules bridge soil organic matter molecule segments?, *European Journal of Soil Science*, **59**(3), 423–429 (2008).

G.E. Schaumann and E.J. LeBoeuf, Glass transitions in peat: Their relevance and the impact of water, *Environmental Science & Technology*, **39**(3), 800–806 (2005).

G.E. Schaumann, F. Lang, and J. Frank, Do Multivalent Cations Induce Crosslinks in DOM Precipitates? in *Humic Substances: Linking Structure to Functions*, F.H. Frimmel and G. Abbt-Braun (Eds.), *Proceedings of the 13th Meeting of the International Humic Substances Society in Karlsruhe*, Universität, Karlsruhe, pp. 941–944 (2006).

S. Scheiner, (1997) *Hydrogen Bonding. A Theoretical Perspective*, Oxford University Press, New York.

T. Schneckenburger, G.E. Schaumann, S.K. Woche, and S. Thiele-Bruhn, Short-term evolution of hydration effects on soil organic matter properties and resulting implications for sorption of naphthalene-2-ol, *Journal of Soils and Sediments*, **12**(8), 1269–1279 (2012).

H.R. Schulten and M. Schnitzer, A state-of-the-art structural concept for humic substances, *Naturwissenschaften*, **80**(1), 29–30 (1993).

H.R. Schulten, M. Thomsen, and L. Carlsen, Humic complexes of diethyl phthalate: Molecular modelling of the sorption process, *Chemosphere*, **45**(3), 357–369 (2001).

R.P. Schwarzenbach, P.M. Gschwend, and D.M. Imboden, (2003) *Environmental Organic Chemistry*, John Wiley & Sons, Inc., New York.

C.F. Schwenk, H.H. Loeffler, and B.M. Rode, Molecular dynamics simulations of Ca^{2+} in water: Comparison of a classical simulation including three-body corrections and Born–Oppenheimer *ab initio* and density functional theory quantum mechanical/molecular mechanics simulations, *Journal of Chemical Physics*, **115**(23), 10808–10813 (2001).

U. Schwertmann and R.M. Cornell, (1991) *Iron Oxides in the Laboratory*, VCH, Weinheim.

A.M. Scott, L. Gorb, E.A. Mobley, F.C. Hill, and J. Leszczynski, Predictions of Gibbs free energies for the adsorption of polyaromatic and nitroaromatic environmental contaminants on carbonaceous materials: Efficient computational approach, *Langmuir*, **28**(37), 13307– 13317 (2012).

A.M. Scott, L. Gorb, E.A. Burns, S.N. Yashkin, F.C. Hill, and J. Leszczynski, Toward accurate and efficient predictions of entropy and Gibbs free energy of adsorption of high nitrogen compounds on carbonaceous materials, *Journal of Physical Chemistry C*, **118**(9), 4774–4783 (2014).

G. Seifert, and J.O. Joswig, Density-functional tight binding: An approximate density- functional theory method, *Wiley Interdisciplinary Reviews: Computational Molecular Science*, **2**(3), 456–465 (2012).

G. Seifert, D. Porezag, and T. Frauenheim, Calculations of molecules, clusters, and solids with a simplified LCAO-DFT-LDA scheme, *International Journal of Quantum Chemistry*, **58**(2), 185–192 (1996).

N. Senesi, Binding mechanisms of pesticides to soil humic substances, *Science of the Total Environment*, **123**, 63–76 (1992).

N. Senesi, B. Xing, and P.M. Huang, (2009) *Biophysico-Chemical Processes Involving Natural Nonliving Organic Matter in Environmental Systems*, John Wiley & Sons, Inc., Hoboken.

H.M. Senn and W. Thiel, QM/MM methods for biomolecular systems, *Angewandte Chemie*, **48**(7), 1198–1229 (2009).

R. Shankar, (1994) *Principles of Quantum Mechanics*, 2nd ed., Plenum Press, New York.

J.I. Siepmann and D. Frenkel, Configurational bias Monte Carlo: A new sampling scheme for flexible chains, *Molecular Physics*, **75**(1), 59–70 (1992).

U.C. Singh and P.A. Kollman, An approach to computing electrostatic charges for molecules. *Journal of Computational Chemistry*, **5**(2), 129–145 (1984).

R.Q. Snurr, A.T. Bell, and D.N. Theodorou, Prediction of adsorption of aromatic- hydrocarbons in silicalite from grand-canonical Monte-Carlo simulations with biased insertions, *Journal of Physical Chemistry*, **97**(51), 13742–13752 (1993).

C.A. Spadotto and A.G. Hornsby, Soil sorption of acidic pesticides. Modeling pH effects, *Journal of Environmental Quality*, **32**(3), 745–750 (2003).

D.L. Sparks, (1995) *Environmental Soil Chemistry*, Academic Press, New York.

D.L. Sparks, Kinetics and Mechanisms of Chemical Reactions at the Soil Mineral/Water Interface, in D.L. Sparks (Ed.), *Soil Physical Chemistry*, 2nd ed., CRC Press, Boca Raton, pp. 135–191 (1999).

D.L. Sparks, A.M. Scheidegger, D.G. Stawn, and K.G. Schecker, Kinetics and Mechanisms of Metal Sorption at the Mineral-Water Interface, in D.L. Sparks and T.J. Grundl (Eds.), *Mineral/Water Interface in Mineral-Water Interfacial Reactions, Kinetics and Mechanisms*, ACS Symposium Series **715**, American Chemical Society, Washington, DC, pp. 135–191 (1999).

G. Sposito, (1984) *The Surface Chemistry of Soil*, Oxford University Press, New York.

M. Sprik and G. Ciccotti, Free energy from constrained molecular dynamics, *Journal of Chemical Physics*, **109**(18), 7737–7744 (1998).

J.F. Stanton and R.J. Bartlett, The equation of motion coupled-cluster method. A systematic biorthogonal approach to molecular excitation energies, transition probabilities, and excited state properties, *Journal of Chemical Physics*, **98**(9), 7029–7039 (1993).

T.B. Stauffer and W.G. McIntyre, Sorption of low-polarity organic compounds on oxide minerals and aquifer material, *Environmental Toxicology and Chemistry*, **5**(11), 949–955 (1986).

F.J. Stevenson, (1982) *Humus Chemistry Genesis, Composition, Reactions*, Wiley Interscience, New York.

K. Sukrat, D. Tunega, A.J.A. Aquino, H. Lischka, and V. Parasuk, Proton exchange reactions of C2-C4 alkanes sorbed in ZSM-5 zeolite, *Theoretical Chemistry Accounts*, **131**(6), 1–12 (2012).

H. Sun, S.J. Mumby, J.R. Maple, and A.T. Hagler, An *ab-initio* CFF93 all-atom force field for polycarbonates, *Journal of the American Chemical Society*, **116**(7), 2978–2987 (1994).

S. Susarla, G.V. Bhaskar, and R. Bhamidimarri, Competitive adsorption-desorption kinetics of phenoxyacetic acids and a chlorophenol in volcanic soil, *Environmental Technology*, **18**(9), 937–943 (1997).

J.L. Suter, R.L. Anderson, H.C. Greenwell, and P.V. Coveney, Recent advances in large-scale atomistic and coarse-grained molecular dynamics simulation of clay minerals, *Journal of Materials Chemistry*, **19**(17), 2482–2493 (2009).

R. Sutton and G. Sposito, Molecular simulation of humic substance-Ca-montmorillonite complexes, *Geochimica et Cosmochimica Acta*, **70**(14), 3566–3581 (2006).

R. Sutton, G. Sposito, M.S. Diallo, and H.R. Schulten, Molecular simulation of a model of dissolved organic matter, *Environmental Toxicology and Chemistry*, **24**(8), 1902–1911 (2005).

A. Szabo and N.S. Ostlund, (1989) *Modern Quantum Chemistry: Introduction to Advanced Electronic Structure Theory*, Dover, New York.

P.G. Szalay, T. Müller, G. Gidofalvi, H. Lischka, and R. Shepard, Multiconfiguration self- consistent field and multireference configuration interaction methods and applications, *Chemical Reviews*, **112**(1), 108–181 (2012).

Szytula, A., A. Burewicz, Z. Dimitrij, S. Krasnick, H. Rzany, J. Todorovi, A. Wanic, and W. Wolski, Neutron diffraction studies of α-FeOOH, *Physica Status Solidi*, **26**(2), 429–434 (1968).

T. Thonhauser, V.R. Cooper, S. Li, A. Puzder, P. Hyldgaard, and D.C. Langreth, Van der Waals density functional: Self-consistent potential and the nature of the van der Waals bond, *Physical Review B*, **76**(12), 125112 (2007).

C.W. Thorstensen and O. Lode, Laboratory degradation studies of bentazone, dichlorprop, MCPA and propiconazole in Norwegian soils, *Journal of Environmental Quality*, **30**(3), 947–953 (2001).

A. Tkatchenko and M. Scheffler, Accurate molecular van der Waals interactions from ground-state electron density and free-atom reference data, *Physical Review Letters*, **102**(7), 073005 (2009).

J. Tomasi, M. Benedetta, and R. Cammi, Quantum mechanical continuum solvation models, *Chemical Reviews*, **105**(8), 2999–3094 (2005).

C. Tournassat, A. Neaman, F. Villieras, D. Bosbach, and L. Charlet, Nanomorphology of montmorillonite particles: Estimation of the clay edge sorption site density by low-pressure gas adsorption and AFM observations, *American Mineralogist*, **88**(11–12), 1989–1995 (2003).

O. Tsendra, A.M. Scott, L. Gorb, A.D. Boese, F.C. Hill, M.M. Ilchenko, D. Leszczynska, and J. Leszczynski, Adsorption of nitrogen-containing compounds on the (100) α-quartz surface: *Ab initio* cluster approach, *Journal of Physical Chemistry C*, **118**(6), 3023–3034 (2014).

S.I. Tsipursky and V.A. Drits, The distribution of octahedral cations in the 2:1 layers of dioctahedral smectites studied by oblique-texture electron diffraction, *Clay Minerals*, **19**(2), 177–193 (1984).

D. Tunega, Theoretical study of properties of goethite (α-FeOOH) at ambient and high-pressure conditions, *Journal of Physical Chemistry C*, **116**(11), 6703–6713 (2012).

D. Tunega, G. Haberhauer, M.H. Gerzabek, and H. Lischka, Interaction of acetate anion with hydrated Al^{3+} cation: A theoretical study, *Journal of Physical Chemistry B*, **104**(29), 6824–6833 (2000).

D. Tunega, G. Haberhauer, M.H. Gerzabek, and H. Lischka, Theoretical study of adsorption sites on the (001) surfaces of 1:1 clay minerals, *Langmuir*, **18**(1), 139–147 (2002a).

D. Tunega, L. Benco, G. Haberhauer, M.H. Gerzabek, and H. Lischka, *Ab initio* molecular dynamics study of adsorption sites on the (001) surfaces of 1:1 dioctahedral clay minerals, *Journal of Physical Chemistry B*, **106**(44), 11515–11525 (2002b).

D. Tunega, M.H. Gerzabek, and H. Lischka, Ab initio molecular dynamics study of a monomolecular water layer on octahedral and tetrahedral kaolinite surfaces, *Journal of Physical Chemistry B*, **108**(19), 5930–5936 (2004a).

D. Tunega, G. Haberhauer, M.H. Gerzabek, and H. Lischka, Sorption of phenoxyacetic acid herbicides on the kaolinite mineral surface: An *ab initio* molecular dynamics simulation, *Soil Science*, **169**(1), 44–54 (2004b).

D. Tunega, G. Haberhauer, M.H. Gerzabek, and H. Lischka, Formation of 2,4-D complexes on montmorillonites: An *ab initio* molecular dynamics study, *European Journal of Soil Science*, **58**(3), 680–691 (2007).

D. Tunega, M.H. Gerzabek, G. Haberhauer, K.U. Totsche, and H. Lischka, Model study on sorption of polycyclic aromatic hydrocarbons to goethite, *Journal of Colloid and Interface Science*, **330**(1), 244–249 (2009).

US Environmental Protection Agency, (1989) Environmental Fate and Effects Division, Pesticide Environmental Fate One Line Summary: DDT (p, p'), USA USEPA, Washington, DC.

W.F. Van Gunsteren and H.J.C. Berendsen, Algorithms for macromolecular dynamics and constraint dynamics, *Molecular Physics*, **34**(5), 1311–1327 (1977).

D. Vasudevan, E.M. Cooper, and O.L. van Exem, Sorption-desorption of ionogenic compounds at the mineral-water interface: Study of metal oxide-rich soils and pure-phase minerals, *Environmental Science & Technology*, **36**(3), 501–511 (2002).

P. Venema, T. Hiemstra, P.G. Weidler, and W.H. van Riemsdijk, Intrinsic proton affinity of reactive surface groups of metal (hydr)oxides: Application to iron (hydr)oxides, *Journal of Colloid and Interface Science*, **198**(2), 282–295 (1998).

L. Verlet, Computer "experiments" on classical fluids. I. Thermodynamical properties of Lennard–Jones molecules, *Physical Review*, **159**(1), 98–103 (1967).

S.J. Vosko, L. Wilk, and M. Nusair, Accurate spin-dependent electron liquid correlation energies for local spin density calculations: A critical analysis, *Canadian Journal of Physics*, **58**(8), 1200–1211 (1980).

T. Vreven and K. Morokuma, Chapter 3 Hybrid methods: ONIOM(QM:MM) and QM/MM, *Annual Reports in Computational Chemistry*, **2**, 35–51 (2006).

M. Walker and H. Lawrence, (1992) *EPA's Pesticide Fact Sheet Database*, Lewis Publishers, Chelsea.

G. Wang and P. Grathwohl, Isosteric heats of sorption and desorption of phenanthrene in soils and carbonaceous materials, *Environmental Pollution*, **175**, 110–116 (2013).

T. Wang, X. Jiang, C. Wang, F. Wang, Y. Bian, and G. Yu, Adsorption of phenanthrene on Al (oxy) hydroxides formed under the influence of tannic acid, *Environmental Earth Sciences*, **71**(2), 773–782 (2014).

W.J. Weber, P.M. McGinley, and L.E. Katz, Sorption phenomena in subsurface systems: Concepts, models and effects on contaminant fate and transport, *Water Resources*, **25** (5), 499–528 (1991).

W.J. Weber Jr., E.J. Leboeuf, T.M. Young, and W. Huang, Contaminant interactions with geosorbent organic matter: Insights drawn from polymer sciences, *Water Resources*, **35**(4), 853–868 (2001).

H. Weigand, K.U. Totsche, B. Huwe, and I. Kogel-Knabner, PAH mobility in contaminated industrial soils: A Markov chain approach to the spatial variability of soil properties and PAH levels, *Geoderma*, **102**(3–4), 371–389 (2001a).

H. Weigand, K.U. Totsche, T. Mansfeldt, and I. Kogel-Knabner, Release and mobility of polycyclic aromatic hydrocarbons and iron-cyanide complexes in contaminated soil, *Journal of Plant Nutrition and Soil Science*, **164**(4), 643–649 (2001b).

H. Weigand, K.U. Totsche, I. Kogel-Knabner, E. Annweiler, H.H. Richnow, and W. Michaelis, Fate of anthracene in contaminated soil: Transport and biodegradation under unsaturated flow conditions, *European Journal of Soil Science*, **53**(1), 71–81 (2002).

D. Werner, J.A. Garratt, and G. Pigott, Sorption of 2,4-D and other phenoxy herbicides to soil, organic matter, and minerals, *Journal of Soils and Sediments*, **13**(1), 129–139 (2013).

R.L. Wershaw, Model for humus in soils and sediments, *Environmental Science & Technology*, **27**(5), 814–816 (1993).

M.F. Wolfe and I.N. Seiber, Environmental activation of pesticides, *Occupational Medicine*, **8**(3), 561–573 (1993).

G.Z. Wu, L. He, and D.Y. Chen, Sorption and distribution of asphaltene, resin, aromatic and saturate fractions of heavy crude oil on quartz surface: Molecular dynamic simulation, *Chemosphere*, **92**(11), 1465–1471 (2013).

R.W.G. Wyckoff, (1963) *Crystal Structures 1*, 2nd ed., Interscience Publishers, New York, pp. 290–295.

G. Xia and W.P. Ball, Adsorption-partitioning uptake of nine low-polarity organic chemicals on a natural sorbent, *Environmental Science & Technology*, **33**(2), 262–269 (1999).

B. Xing and J.J. Pignatello, Dual-mode sorption of low-polarity compounds in glassy poly (vinyl chloride) and soil organic matter, *Environmental Science & Technology*, **31**(3), 792–799 (1997).

Y. Yang, S. Tao, N. Zhang, D.Y. Zhang, and X.Q. Li, The effect of soil organic matter on fate of polycyclic aromatic hydrocarbons in soil: A microcosm study, *Environmental Pollution*, **158**(5), 1768–1774 (2010).

B. Yaron, R. Calvet, and R. Prost, (1996) *Soil Pollution: Processes and Dynamics*, Springer-Verlag, Berlin.

D. Young, (2001) *Computational Chemistry: A Practical Guide for Applying Techniques to Real World Problems*, Wiley-Interscience, New York.

C.-H. Yu, S.Q. Newton, M.A. Norman, L. Schafer, and D.M. Miller, Molecular dynamics simulations of adsorption of organic compounds at the clay mineral/aqueous solution interface, *Structural Chemistry*, **14**(2), 175–185 (2003).

S.D. Zaric, Metal ligand aromatic cation-π interactions, *European Journal of Inorganic Chemistry*, **12**, 2197–2209 (2003).

M. Zbik and R.St.C. Smart, Nanomorphology of kaolinites: Comparative SEM and AFM studies, *Clays and Clay Minerals*, **46**(2), 153–160 (1998).

X.Y. Zhang, Y.G. Wu, S.H. Hu, C. Lu, and H.R. Yao, Responses of kinetics and capacity of phenanthrene sorption on sediments to soil organic matter releasing, *Environmental Science and Pollution Research*, **21**(13), 8271–8283 (2014).

Y. Zhao and D.G. Truhlar, Density functionals with broad applicability in chemistry, *Accounts of Chemical Research*, **41**(2), 157–167 (2008a).

Y. Zhao and D.G. Truhlar, The M06 suite of density functionals for main group thermochemistry, thermochemical kinetics, noncovalent interactions, excited states, and transition elements: Two new functionals and systematic testing of four M06-class functionals and 12 other functionals, *Theoretical Chemistry Accounts*, **120**(1–3), 215–241 (2008b).

Y. Zhao, N.E. Schultz, and D.G. Truhlar, Exchange-correlation functional with broad accuracy for metallic and nonmetallic compounds, kinetics, and noncovalent interactions, *Journal of Chemical Physics*, **123**(16), 161103 (2005).

J. Zhong, P. Wang, Y. Zhang, Y.G. Yan, S.Q. Hu, and J. Zhang, Adsorption mechanism of oil components on water-wet mineral surface: A molecular dynamics simulation study, *Energy*, **59**, 295–300 (2013).

W.J. Zhou and L.Z. Zhu, Distribution of polycyclic aromatic hydrocarbons in soil-water system containing a nonionic surfactant, *Chemosphere*, **60**(9), 1237–1245 (2005).

7

Petroleum Geochemistry

Qisheng Ma[1] and Yongchun Tang[2]

[1]*Department of Computational and Molecular Simulation, GeoIsoChem Corporation, Covina, CA, USA*
[2]*Geochemistry Division, Power Environmental Energy Research Institute, Covina, CA, USA*

7.1   Introduction: Petroleum Geochemistry and Basin Modeling

Petroleum geochemistry is a mature science that has been revitalized by recent advances in computational modeling and simulation techniques. Prior to the successful application of quantum chemistry calculations, conventional wisdom built upon extensive data correlations and empirical formulations is predominant in the early development of our knowledge about geologic systems and organic chemistry. Extensive progress has been made in many aspects of petroleum geochemistry from theoretical, laboratory, and field studies. In order to further exert the power of the molecular modeling in petroleum geochemistry, it is essential to better understand the developed geochemistry modeling methods, empirically or semiempirically, as well as key experimental and field testing protocols.

The term *geochemistry* is believed to be first introduced in 1838 by a German–Swiss chemist Christian Friedrich Schönbein, who had also discovered ozone and the principle of fuel cells (Reinhardt, 2008):

> In a word, a comparative geochemistry out to be launched, before geochemistry can become geology, and before the mystery of the genesis of our planets and their inorganic matter may be revealed.

The birth of the *organic geochemistry* was attributed to the pioneering works of Parker Trask (Trask, 1932; Trask and Pathnode, 1942) in the United States and Alfred E. Treibs (1936) in Germany in early 1930s, which linked the organic matter in sediments and sedimentary rocks to the sources of the crude oils (Dow, 2014). Alfred Treibs, also known as the "father of the organic

Molecular Modeling of Geochemical Reactions: An Introduction, First Edition. Edited by James D. Kubicki.
© 2016 John Wiley & Sons, Ltd. Published 2016 by John Wiley & Sons, Ltd.

geochemistry" (Kvenvolden, 2006), first discovered the metalloporphyrins (an organic molecule) in crude oils and demonstrated its biogenic origin. This discovery established that there is a linkage between the biochemical compounds in living matter and in geochemical organic matters (Kvenvolden, 2002).

Petroleum geochemistry has been considered as one aspect of organic geochemistry, with a focus on applying fundamental principles of chemistry to explain geologic systems that involve evolution of organic matter (hydrocarbons and/or nonhydrocarbons) in the sedimentary basin (Hunt, 1979). For example, in 1956, Phillippi (1956) and Hunt and Jamieson (1956) pioneered applications of chemical methods to study the petroleum systems, followed by a group of geochemists (Bray and Evans, 1961; Hedberg, 1965; Weeks, 1958). In 1959, an international conference entitled "General Petroleum Geochemistry Symposium" was held at Fordham University in New York. Petroleum geochemistry has gained official recognition as a scientific discipline. A number of reviews on the early developments of the petroleum geochemistry have also been conducted (Huc, 2003; Hunt et al., 2002; Katz et al., 2008; Martinelli, 2009; Peter and Fowler, 2002).

The concept of a petroleum system (originally proposed as the oil system) which correlates the genetic differences among crude oils to specific source rocks was proposed in Williams and Dow in 1970s (Dow, 1974; William, 1974). Such a concept was explored later in an AAPG Memoir, entitled "The Petroleum System: From Source to Trap" (Margoon and Dow, 1994). Typically, a petroleum system consists of key elements, namely, (i) the *source rock* that can generate hydrocarbons (oil/gas), (ii) the *reservoir* with sufficient porosity and permeability for hydrocarbons to migrate and transport, (iii) the *seal* that is impermeable for hydrocarbons to trap and accumulate hydrocarbons, etc. Fundamental aspects of formation, evolution, and assessment of the petroleum systems are covered in the *Basin Modeling* or *Basin Analysis*, which deals with the burial history, thermal history, maturity history, and the formation history of a petroleum system (Hantschel and Kauerauf, 2010; Lerche, 1990).

Computer simulation techniques have been extensively applied in the basin modeling for illustration, interpretation, and modeling of the petroleum system. A number of commercial software are developed and widely used in petroleum exploration and production companies:

- **BasinMod**: Basin Modeling Software, by Platte River Associates, Inc.
- **PetroMod**: Petroleum System Modeling Software, by Schlumberger
- **Genesis and Trinity**: Interactive Petroleum System Tools, by ZetaWare
- **BMT**[TM]: Basin Modeling Toolbox, by Tectonor
- **Migrino**: Migration Simulator, by Migris AS
- **Permedia**[TM]: Petroleum Systems Modeling, by Halliburton
- **Novva**: Geologic modeling, by Sirius Softwares, Pte, Ltd
- **GOR-Isotopes**: PC-based hydrocarbon generation simulator, by GeoIsoChem Corp.
- **Kinetics05**: Reaction Kinetics Analysis, by GeoIsoChem Corp.

7.2 Technology Development of the Petroleum Geochemistry

7.2.1 Thermal Maturity and Vitrinite Reflectance

Thermal maturity is one of the most important parameters used in petroleum geochemistry, and vitrinite reflectance (VRo) is a commonly used thermal maturity indicator. Vitrinite is a type of organic component of coal, which has a shiny appearance. Through measuring the percentage of light reflected off the vitrinite maceral at 500× magnification in oil immersion, VRo can be ranked. According to the classification of the International Committee for Coal and Organic Petrology

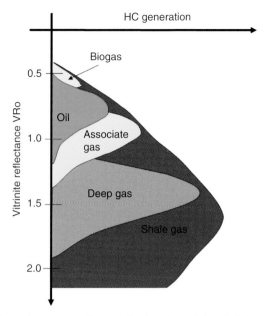

Figure 7.1 *Hydrocarbon generations as the function of the vitrinite reflectance (VRo).*

(ICCP) (1998), the VRo parameter is based on the change of reflection properties of sedimentary organic matter of terrestrial origin (type III) and can be the result of structural and molecular alteration (ordering) as a function of mainly temperature and subordinately time and pressure (Fedor and Hámor-Vidó, 2003). In the beginning, the VRo of coals has been ranked to calibrate their degree of maturation (Davis, 1978; MrCartney and Teichmüller, 1972). Later, it has been widely used for source rocks, that is, kerogen and shale, of the hydrocarbons (Dow, 1977; Tissot and Welte, 1984). Advanced measurement techniques have been developed to improve accuracy (Mukhopadhyay and Dow, 1994; Philippi, 1965). A number of standard test methods have also been established (American Society for Testing and Materials (ASTM), 2011).

The most significant feature of VRo is its sensitivity to the temperature ranges that correspond to the thermal chemical reactions of hydrocarbon compounds (Tissot et al., 1974) (Figure 7.1). Hydrocarbon generation involves thermal transformation processes in sedimentary basins at low temperature (100–200°C) over long geological times (typically a few million years). In order to simulate those geochemical processes under laboratory conditions, thermal/chemical reactions are carried out at higher temperature (250–600°C) either isothermally or nonisothermally, in opened or closed systems (Behar et al., 1997a; Brunham and Braun, 1990; Lewan, 1993a; Ungerer, 1990). VRo has become an important linkage to extrapolate laboratory-derived reaction kinetics to the geological setting. A simplified version of a vitrinite maturation model, based on changes in vitrinite composition with time and temperature, was developed by Sweeney and Burnham (1990). The simplified model, called EASY%Ro, uses an Arrhenius first-order parallel reaction approach with a distribution of activation energies.

7.2.2 Rock-Eval Pyrolysis

For certain types of source rocks, such as marine shale, in where the amount of vitrinite maceral is small, alternative methods can be used to determine the thermal maturity. Rock-Eval pyrolysis measurement is conducted by programmed heating a small piece of rock in an inert atmosphere (helium or nitrogen) to measure the amount of hydrocarbon and nonhydrocarbon (CO_2) gases at different temperature ranges (Behar and Pelet, 1985). Specifically, four basic data can be obtained from the Rock-Eval pyrolysis experiments:

1. S_1 = the amount of free hydrocarbons in the sample
2. S_2 = the amount of hydrocarbons generated from thermal cracking of nonvolatile organic matters
3. S_3 = the amount of CO_2 produced during pyrolysis
4. T_{max} = the temperature associated with the maximum of the S_2 peak

The latest developed Rock-Eval 6 product line (Behar et al., 2001; Lafargue et al., 1998), which can simultaneously measure the total organic content (TOC) of the source, enables the determination of several important parameters that can be derived as follows:

- HI = hydrogen index $(HI = 100^* S_2/TOC)$
- OI = oxygen index $(OI = 100^* S_3/TOC)$
- PI = production index $(PI = S_1/S_1 + S_2)$
- PC = pyrolyzable carbon $(PC = 0.083^*[S_1 + S_2])$

7.2.3   Kerogen Pyrolysis and Gas Chromatography Analysis

Oil and gas generations result from biologic (biogenic) or chemical/thermal (abiogenic) processes of the mixture of organic chemical compounds (often referred to as *kerogen*) that are deposited into fine-grained source rocks. Kerogen is the most abundant component of the sedimentary organic matter when source rocks are formed. As sediments are buried, kerogen undergoes thermal cracking with leads to petroleum generation. Due to the limitations of current technologies, accurate knowledge of both the nature and the absolute amount of all chemical bonds in the kerogen network is not possible. The classification of kerogen is mainly based on their sources and potentials to generation hydrocarbons (Vandenbroucke and Largeau, 2007), that is, the hydrogen to carbon (H/C) ratio and the oxygen to carbon ratio (O/H) are two of the most important parameters for kerogen types. Specifically, there are currently five basic types of kerogen, namely, type I, type II, type IIS, type III, and type IV (Table 7.1). Van Krevelen diagram is a cross-plot of H/C as a function of the O/C atomic ratio (Van Krevelen, 1950). Chemical and isotopic compositions of individual petroleum hydrocarbons derived from different kerogen types are significantly different (Behar et al., 1997b, 2008a, 2010; Tang and Behar, 1995; Tomió et al., 1995). Variations in thermal maturity can significantly affect the compound-specific carbon isotopic composition of petroleum hydrocarbons (Berner and Faber, 1996; Clayton, 1991a, 1991b; Clayton and Bjorøy, 1994).

Table 7.1   *Classification of kerogen types and their basic properties.*

Kerogen type	Kerogen form	H/C	O/C	HC generation
I	Sapropelic (algal), rich in lipids (most saturated)	>1.25	<0.15	Primarily oils
II	Planktonic (liptinic), algae + zooplankton	<1.25	0.03–0.08	Both oil and gas
IIS	Sulfurous, similar to type II, but high in sulfur content	<1.25	0.03–0.08	Both oil and gas
III	Humic, rich in aromatic compounds	<1.0	0.03–0.3	Primarily gas, some oils
IV	Residue	<0.5	—	None

Laboratory pyrolysis experiments have been extensively used for decades to investigate the thermal degradation (artificial maturation) of large-molecular organic compounds into smaller volatiles species, and pyrolysis–gas chromatography (Py-GC) is the most important method to determine the composition or structure of the original samples (Sobeih et al., 2008). Three types of pyrolysis systems have been applied: open system, anhydrous closed system, and hydrous closed system.

Hydrous closed-system pyrolysis (Lewan, 1993b, 1997) heats the source rock samples in the presence of water under subcritical water temperatures (<374°C) in a stainless steel reactor for days. The gas components are collected after reactions and analyzed by gas chromatography (GC) techniques.

Anhydrous closed-system pyrolysis experiments (Behar et al., 1991; Hill et al., 1996) are typically carried out in sealed gold tubes. Gold is chosen as the material of the pyrolysis reaction because of its chemical inertness and physical flexibility which allows volume expansion and contraction *via* external hydraulic pressure controlled by the large box furnace and the efficient contact between the gas-phase natural gas inside the organic porosity and solid surface. The loaded gold tubes were flushed and welded sealed under argon. The stainless steel autoclaves with temperature control to ±1°C were used to hold the sealed gold tube in the box furnace.

Multiple cold-trap pyrolysis–gas chromatography (Liu and Tang, 1997; Tang and Stauffer, 1991, 1994) (MCTP-GC) instrument permits pyrolysis in an open system under isothermal condition. Hydrocarbons generated during certain time period (6 min) at given temperature were vented through a valve and then captured in the liquid-nitrogen-filled cold traps. Totally 22 ports of cold traps are automated in order to contain a "slice" of the pyrolysis products, which can be quantitatively analyzed by GC.

Upon artificial maturation through pyrolysis in laboratory, chemical components in gas phase, liquid phase, and solid phase can be collected and analyzed. The changes of these chemicals (hydrocarbons and nonhydrocarbons) with time, temperature, pressure, heating rate, etc. can be used to derive and calibrate the kinetic models of the kerogen pyrolysis.

7.2.4 Kinetic Modeling of Kerogen Pyrolysis

The temperature dependence of the rate coefficient (k) of hydrocarbon cracking is commonly expressed by the semiempirical Arrhenius equation (Arrhenius, 1889; Laidler, 1987):

$$k = A_f \exp\left(-\frac{E_a}{RT}\right) \tag{7.1}$$

Here A_f is the preexponential factor (s^{-1}), also named as the "frequency factor"; E_a is the activation energy (in units of kJ/mol, or kcal/mol); T is the temperature (in K); and R is the gas constant (in J/mol/K or kcal/mol/K). The average rate of decomposition of a collection of chemical compounds can be expressed in terms of a set of independent parallel reaction kinetics:

$$X_i \begin{cases} \xrightarrow{P_{i,1}, k_{i,1}} a_{i,1,1}X_1 + \cdots + a_{i,1,h\neq i}X_{h\neq i} + \cdots + a_{i,1,m}X_m \\[2ex] \xrightarrow{P_{i,2}, k_{i,2}} a_{i,2,1}X_1 + \cdots + a_{i,2,h\neq i}X_{h\neq i} + \cdots + a_{i,2,m}X_m \\[2ex] \qquad\qquad\qquad \vdots \\[2ex] \xrightarrow{P_{i,n}, k_{i,n}} a_{i,n,1}X_1 + \cdots + a_{i,n,h\neq i}X_{h\neq i} + \cdots + a_{i,n,m}X_m \end{cases} \tag{7.2}$$

where X_i represents a group of chemical compounds and $X_{j \neq i}$ are other groups of chemical compounds that can be generated from the decompositions of X_i. $k_{i,j}$ is the rate constant of the jth decomposition reaction of X_i to generate $X_{j \neq I}$, and $P_{i,j}$ is the weight fraction of X_i that decompose through the jth pathway. $a_{i,j,h}$ is the stoichiometric coefficient for the conversion of X_i into $X_{h \neq I}$ in the jth reaction pathway. m is the total number of chemical groups in the modeling system, and n is the number of total possible decomposition pathways of the X_i group. Therefore, the kinetics for a group of compound $X_{h \neq I}$ from X_i can be determined from

$$X_i \sum_j^n \xrightarrow{P_{i,j}, k_{i,j}} a_{i,j} X_{h \neq i} \tag{7.3}$$

Within this scheme, all hydrocarbon and nonhydrocarbon compounds, including those in solid, liquid, and gas phases during thermal cracking reactions under laboratory pyrolysis, can be grouped into a set of X_i. For instance, in the kerogen pyrolysis experiments, a class of chemical compounds can be grouped as follows:

• C_{15}^+-asphaltene (ASPH)	Asphaltene compounds in heavier oils (C_{15}^+)
• C_{15}^+-NSO-resin (NSO)	N, S, and O compounds in heavier oils (C_{15}^+)
• C_{15}^+-saturates (SAT)	Saturated compounds in heavier oils (C_{15}^+)
• C_{15}^+-aromatics (ARO)	Aromatic compounds in heavier oils (C_{15}^+)
• C_6–C_{14}-saturates (SAT6)	Saturated compounds in light oils (C_6–C_{14})
• C_6–C_{14}-aromatics (ARO6)	Aromatic compounds in light oils (C_6–C_{14})
• C_3–C_5 (C3–5)	Condensates
• C_2 (C2)	Ethane
• C_1 (C1)	Methane
• CO_2 (CO2)	Carbon dioxide
• H_2S (H2S)	Hydrogen sulfite
• N_2 (N2)	Nitrogen
• Coke (COKE)	Unreactive solid residues

A complete reaction network scheme is illustrated in Figure 7.2 where a dashed arrow is used to represent the potential generation reaction from a parent compound (arrow out) to a child compound (arrow in). Therefore, a special group such as C_6–C_{14}-saturates can be generated from a number of compounds including kerogen and heavier oil compounds; and it can also further decompose to produce lower molecular weight compounds. These dashed arrow lines also represent a collection of all possible chemical reactions from the parent compound to one child compound. Also, because chemicals within one special group often have similar molecular structures, we can assume that they will follow similar decomposition mechanism.

If we use the Arrhenius equation to calculate the reaction rate, the differences among A_f and E_a will be relatively small. Therefore, we can use

$$X_i \xrightarrow{A_f \sum_j^n P_{i,j} * \exp\left(\frac{-E_{a,i,j}}{RT}\right)} a_{i,h} X_{h \neq i} \tag{7.4}$$

That is, the thermal decompositions of X_i to generate $X_{h \neq I}$ follow a set of parallel first-order reaction with the distributed activation energies $E_{a,i,j}$ but have the same frequency factor $A_{f,i}$. Such a discrete distribution model was incorporated into the PC-based kinetic modeling software (KINETICS) by Burnham (Braun and Burnham, 1987; Burnham et al., 1987), which used the

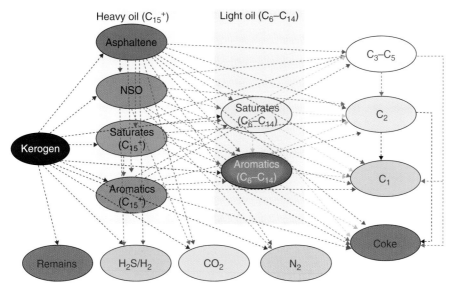

Figure 7.2 *A comprehensive reaction network of the kerogen pyrolysis to generate hydrocarbon and nonhydrocarbon compounds.*

Gaussian distribution activation energies, and then been further explored to various schemes (Burnham et al., 1987; Peters et al., 2006; Sundararaman et al., 1992). Another frequently used distributed activation energy scheme is the uniformly discrete energies ranging from 160 to 320 kJ/mol, with an increment of 8 kJ/mol (Behar et al., 2008b). The asymmetric Weibull distribution energy scheme has also been used (Burnham and Braun, 1999; Lakshmanan and White, 1994; Weibull, 1951).

In principle, the reaction kinetics from X_i to a special $X_{h \neq i}$ group will not be the same. However, if we assume the thermal cracking energy of a group of compound is independent to the child compounds that it can generate, we can use the same set of kinetics for the thermal decomposition of X_i; therefore we have

$$X_i \xrightarrow{A_f \sum_j^n P_{i,j}*\exp\left(\frac{-E_{a,i,j}}{RT}\right)} \sum_{h=1,h\neq i}^{m} a_{i,h}X_h \tag{7.5}$$

or simply

$$X_i \xrightarrow{k_i} \sum_{h=1,h\neq i}^{m} a_{i,h}X_h \tag{7.6}$$

and

$$k_i = A_f \sum_j^n P_{i,j}*\exp\left(\frac{-E_{a,i,j}}{RT}\right) \tag{7.7}$$

Therefore, a complete reaction network as shown in Figure 7.2 will consist of the following reactions:

- $$Kerogen \rightarrow a_{1,1}ASPH + a_{1,2}NSO + a_{1,3}SAT + a_{1,4}ARO + a_{1,5}SAT6 + a_{1,6}ARO6$$
$$+ a_{1,7}C3-5 + a_{1,8}C2 + a_{1,9}C1 + a_{1,10}CO2 + a_{1,11}H2S + a_{1,12}N_2 + a_{1,13}COKE \quad (7.8)$$

- $$ASPH \rightarrow a_{2,1}NSO + a_{2,2}SAT + a_{2,3}ARO + a_{2,4}SAT6 + a_{2,5}ARO6$$
$$+ a_{2,6}C3-5 + a_{2,7}C2 + a_{2,8}C1 + a_{2,9}CO2 + a_{2,10}H2S + a_{2,11}N2 + a_{2,12}COKE \quad (7.9)$$

- $$NSO \rightarrow a_{3,1}SAT + a_{3,2}ARO + a_{3,3}SAT6 + a_{3,4}ARO6 + a_{3,5}C3-5$$
$$+ a_{3,6}C2 + a_{3,7}C1 + a_{3,8}CO2 + a_{3,9}H2S + a_{3,10}N2 + a_{3,11}COKE \quad (7.10)$$

- $$SAT \rightarrow a_{4,1}ARO + a_{4,2}SAT6 + a_{4,3}ARO6 + a_{4,4}C3-5$$
$$+ a_{4,5}C2 + a_{4,6}C1 + a_{4,7}CO2 + a_{4,8}H2S + a_{4,9}N2 + a_{4,10}COKE \quad (7.11)$$

- $$ARO \rightarrow a_{5,1}SAT6 + a_{5,2}ARO6 + a_{5,3}C3-5$$
$$+ a_{5,4}C2 + a_{5,5}C1 + a_{5,6}CO2 + a_{5,7}H2S + a_{5,8}N2 + a_{5,9}COKE \quad (7.12)$$

- $$SAT6 \rightarrow a_{6,1}ARO6 + a_{6,2}C3-5 + a_{6,3}C2 + a_{6,4}C1 + a_{6,5}COKE \quad (7.13)$$

- $$ARO6 \rightarrow a_{7,1}C2 + a_{7,2}C1 + a_{7,3}COKE \quad (7.14)$$

- $$C3-5 \rightarrow a_{8,1}C2 + a_{8,2}C1 + a_{8,3}COKE \quad (7.15)$$

- $$C2 \rightarrow a_{9,1}C1 + a_{9,2}COKE \quad (7.16)$$

Here we assume that the nonhydrocarbon gases CO_2, H_2S, and N_2 are mainly generated from higher molecular weight compounds (i.e., kerogen and heavier oils). We also notice that the aromatic C_6–C_{14} compounds consisting of one aromatic ring molecular structure gain extra stability due to their aromaticity, such that they could be even more stable than the condensates C_3–C_5. So we eliminate the possible generation reactions of C_3–C_5 from ARO6. The first reaction is considered as the primary generation reaction, while the other reaction is secondary cracking.

For the primary generation reaction, kinetic energy differences of the kerogen pyrolysis to generate different compounds could be more significant, such that the assumption of Equation 7.8 might yield greater errors. A precursor scheme could be used. For instance, if we are particularly interested in the special reaction kinetics from X_i to generate X_g, we can use three reaction kinetic equations:

$$X_i \xrightarrow{k_0} \sum_{h=1, h\neq i, h\neq g}^{m} a_{i,h}(\text{pre-X})_h + a_{i,g}(\text{pre-X})_g \quad (7.17)$$

$$(\text{pre-X})_g \xrightarrow{k_{i,g}} X_g \quad (7.18)$$

$$(\text{pre-X})_h \xrightarrow{k_{i,h}} X_h \quad (7.19)$$

where $(\text{pre-X})_h$ and $(\text{pre-X})_g$ are the precursors of the group compounds X_h and X_g, and we use the activation energy for the kinetics k_o sufficiently small such that the decompositions of X_i to X_h and

X_g are instantaneous. Therefore, the first equation allows us to focus on the optimized stoichiometric coefficients $a_{i,h}$ and $a_{i,g}$ and the following equations to optimize the reaction kinetics separately.

7.2.5 Natural Gases and C/H Isotopes

Natural gas often suffers from a limited transportation infrastructure and relatively limited demand (Katz, 2011). Natural gases produced with the crude oil have been regarded as the "associated gas" that often is flared directly without utilization, and natural gas discovered in remote locations is considered as "stranded" gas and could go undeveloped for decades. However, because natural gases are dominated by a few simple, low molecular weight hydrocarbons, important genetic information can be commonly obtained from analysis of the compositional and isotopic ratios. Especially, the stable isotope measurement on C_1–C_5 hydrocarbons provides a "fingerprint" that can be used to assess the natural and thermal maturity of potential source beds, the pathways by which gas migration occurred, the presence of mixed-source gases, and, more controversially, reservoir accumulation and loss histories.

Basin models built on the compositional and isotopic information of natural gases initially were based primarily on field data collected and correlations. These include shallow, low-temperature bacterial gases (Claypool and Kaplan, 1974; Coleman et al., 1988; Jenden and Kaplan, 1986; Martini et al., 1996; Rice, 1992), high-temperature "thermogenic" gases that often associated with oil (Galimov, 1974; Rooney et al., 1995; Schoell, 1980; Stahl and Carey, 1975) and coalbed and shale-hosted gases (Colombo et al., 1970; Rice, 1993). Theoretical interpretations (Chung et al., 1988; Schoell, 1983; Stahl, 1977) were primarily empirical and were often valid only under certain geological conditions.

Early mathematical models (Galimov, 1988; Gaveau et al., 1987; Sundberg and Bennett, 1983; Waples and Tornheim, 1978) involved sophisticated algorithm and complicated parameters and assumptions that are difficult to be applied. Nevertheless, there are studies establishing linkages between observable isotopic fractionations and hydrocarbon generation kinetic models. As a result, they have paved the pathways of fundamental understandings on the relationships of the isotopic fractionation changes with the natural gas generation rate, yield, and reservoir accumulation histories. Specifically, the mathematical models (Berner et al., 1995; Clayton, 1991a; Rooney et al., 1995; Tang and Jenden, 1988) based on the Rayleigh distillation theory have been applied to explicitly address variations in the isotopic composition of the source organic matter and the accumulation history (instantaneous vs. cumulative). It is now accepted that kinetic model for gas generation and isotope fractionation can be modeled by a multicomponent reaction network scheme (Lorant et al., 1988), which consists of a series of *n*-parallel first-order reaction kinetics (Cramer et al., 2001; Ungerer and Pelet, 1987). The underlying idea is that the complex process of formation of a hydrocarbon entity from the kerogen macromolecule is controlled by one rate-determining reaction step and that the corresponding reaction rate is directly proportional to the residual generation potential (Cramer et al., 1998).

7.3 Computational Simulations in Petroleum Geochemistry

7.3.1 *Ab Initio* Calculations of the Unimolecular C–C Bond Rapture

Previously, kinetic parameters have been primarily derived from laboratory pyrolysis experiments and/or from calibrations with the field data. The models developed were basin dependent, in that the model derived from one basin often does not work for others, and some schemes are contradictory (Jenden and Kaplan, 1989; Lorant et al., 1998). Furthermore, a given data set may give rise to very different interpretations, particularly when postgenerative processes such as diffusive fractionation are invoked (Jenden et al., 1993; Prinzhofer and Huc, 1995; Prinzhofer and Pernaton, 1997). Introduction of quantum chemistry density functional theory (DFT) calculations into the kinetic and isotopic modeling has significantly strengthened the quantitative basin models in terms of providing reaction mechanism on the C–C bond rupture and formation of the alkyl radicals (Goldstein et al., 1998; Van der Hart, 1999).

Oil and gas generations involve hydrocarbon cracking by which higher carbon number hydrocarbons (including kerogen) are converted to lower molecular weight hydrocarbons (Olah and Molnar, 1995; Poutsma, 2000). There are basically three generation mechanisms that had been used to describe the hydrocarbon cracking: thermal cracking through the free-radical interactions has been the predominant one, although the catalytic cracking by the transition metals (Mango et al., 1994) and hydrocracking involving clays and water catalytic processes have also been proposed (Goldstein, 1983; Helgeson et al., 1993; Huc et al., 1986; Johns, 1979; Seewald, 1994). The free-radical chain reactions have been extensively studied (Corma and Wojciechowski, 1985; Kossiakoff and Rice, 1943; Rice, 1933; Rice and Herzfeld, 1934).

With most commercially or academically distributed software packages that allow the user to predict absolute rate constant using transition state theory (TST), numerous *quantum mechanical* computational calculations of the bond rupture energy and isotopic changes have been conducted (Fernández-Ramos, 2006; Jursic, 1996; Liu et al., 1996; Xiao, 2001; Xiao et al., 1997). The conventional TST based on the Rice–Ramsperger–Kassel–Marcus (RRKM) theory, developed in 1952 by Marcus (1952) who combined the TST with the theory of chemical reactivity developed by Rice and Ramsperger (RR) (1927) in 1927 and Kassel (1928) in 1928, enabled the computation of simple estimates of the unimolecular reaction rates from the potential energy surface (PES).

The classical RR theory is based on the idea of Hinshelwood (1956) that the molecule is a collection of classical oscillators and that their density of states is related to their vibrational frequencies. Kassel proposed to replace the classical model of vibrations with the quantum harmonic oscillators. Chemical processes such as the simple bond dissociation or radical recombination generally have no saddle point on the PES for the initial association step (Garrett and Truhlar, 1979; Gilber and Smith, 1990; Wardlaw and Marcus, 1986). Computational evaluations of such a barrierless reaction resulted in considerable variability in the location of the transition state. Significant efforts have been conducted to provide fundamental basis for identifying the transition state of the bond rupture. The basic assumption in the RRKM theory is that the reactant of each elementary step is in microcanonical equilibrium; therefore, the same reaction rates result no matter how energy is deposited in a molecule (or complex) (Bunker and Hase, 1973; Hase, 1979). It is therefore possible to define universally applicable elementary rate coefficients and to bypass the problem of computing the intramolecular dynamics of the complexes entirely. Phase space theory (PST) provides a useful, and easily implemented, reference theory (Light, 1964; Truhlar, 1969). The basic assumption in PST is that the interaction between the two reaction fragments is isotropic and does not affect the internal fragment motions. Consequently, in order to obtain meaningful results for the transition state, numerous calculations should be run to obtain the trajectories such that a PES could be depicted in the phase space (coordinates and momenta). Unfortunately, this level of theory is currently limited to processes that involve small molecules and radicals.

Simplified methodologies are available. Lorant et al. (2001) used a generalized TST, where transition states are defined along rate constant profiles and not along potential energy curves. The advantage of using such a method is that it does not require specific programming, because it makes use of current quantum chemistry tools. However, to limit the number of quantum mechanical calculations, only one path across the PES will be considered. The Gibbs free energy $\Delta G^{\ddagger}$ and the contributing energies E_{SCFE}, E_{ZPE}, E_H, and E_S at different temperatures along the reaction pathway

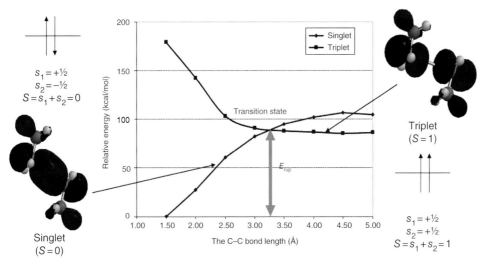

Figure 7.3 *A computational approach to locate the transition state of the C–C bond rupture based on the electron distribution. The C–C bond length has been fixed from 1.5 to 5.0 Å with an increment of 0.5 Å, while the rest of the atoms of the butane molecule is allowed to fully be optimized. Both the relative Gibbs free energy differences (referred to the optimized stable configuration with C–C bond length of 1.53 Å) of both the singlet and triplet configurations at room temperature T = 298.13 K were determined. All calculations were conducted at DFT/B3LYP/6-31G∗(d,p) level.*

were calculated, where E_{SCFE}, E_{ZPE}, E_H, and E_S denote the energy terms of the electron, zero-point energy (ZPE), enthalpy, and entropy changes. The reaction rate constant k as functions of the distance along the reaction pathway and temperature can then be determined.

Another approach is from the examination of the electronic spin interaction (Figure 7.3) (Ni et al., 2011). When a pair of electrons is shared by two C atoms, a covalent C–C bond is formed. According to Pauli's exclusion principle, no two electrons in a multielectron atom can have identical values of four quantum numbers, namely, the principal quantum number n, the azimuthal quantum number l, the magnetic quantum number m_l, and the spin quantum number m_s. Because the electrons in the covalent bond share the same n, l, and m_l numbers, their m_s must be different, and two electrons will have opposite spins (one for $s_1 = + \frac{1}{2}$ and another one for $s_2 = - \frac{1}{2}$). Therefore, the net spin $s = s_1 + s_2 = 0$ (the spin state $S = 2s + 1 = 1$ is the so-called singlet). This does not exclude the existence of a triplet state ($s_1 = s_2 = + \frac{1}{2}$, so $s = 1$ and the spin state $S = 2s + 1 = 3$). It only indicates that under such a typical atomic configuration (i.e., C–C bond distance), the triplet state has much higher energy than the singlet state, such that the probability for us to observe electrons at such a state is very small. Computationally, we can determine energies of both singlet and triplet states at any giving C–C bond length.

With the continuous increase of the C–C bond length (the C–C bond rupture), the paired electrons gradually return to the C atoms; and the energy difference between the singlet and triplet states gradually diminishes. Importantly, when two carbonyl radicals are far apart, the energy of their triplet state becomes lower than that of the singlet state (Hund's multiplicity rule) (Hund, 1925). The transition state of the C–C bond rupture can be located at the energy crossover point of the singlet/triplet curves. At the transition state, one of the paired electrons will flip its spin state. The exact mechanism of this observed phenomenon is still under debate (Rioux, 2007). One plausible explanation is that these two electrons from a pair interact with the same external field to compete with the expulsion force from Pauli's exclusion. When the C–C bond length is smaller than that of the transition state, Pauli's exclusion is predominant, so the singlet state has the lower energy, and vice versa. The molecular orbital of a singlet state is the elliptical S-like, whereas the triplet state is a dumbbell-shaped P-like. The transition state is located in the immediate transformation from the S-like to the P-like orbital.

7.3.2 Quantum Mechanical Calculations on Natural Gas ^{13}C Isotopic Fractionation

The C-isotope composition ($^{13}\delta$C) (in unit per mil, ‰) of each hydrocarbon component in natural gas is defined as

$$\delta^{13}C = \left[\frac{\left(^{13}C/^{12}C \right)}{\left(^{13}C/^{12}C \right)_r} - 1 \right] * 1000 \tag{7.20}$$

where the subscript r refers to the carbon isotopic ratio of the standard reference, which is taken from the international Pee Dee Belemnite standard with a generally accepted absolute $\left(^{13}C/^{12}C \right)_r$ ratio of 0.0112372 (Werner and Brand, 2001). When there is a certain amount of heavier C isotopic component (^{13}C) in hydrocarbons, the energy barrier and the kinetic rate constant of the bond rupture between ^{12}C–^{12}C and ^{12}C–^{13}C will have the substantial difference. Typically, it will be more difficult to break a ^{12}C–^{13}C bond than a ^{12}C–^{12}C bond, resulting in isotopically lighter, less ^{13}C isotopic component in the gaseous products. The ratio of the kinetic rate constants to break the ^{12}C–^{12}C than to break the ^{12}C–^{13}C bond is called kinetic isotope effect (KIE), which is be calculated from first principle (see Chapters 6 and 13).

Using ab initio DFT calculation, Tang and Jenden (1995) in 1995 estimated ZPE differences for different bond positions in *n*-octane, toluene, and dimethyl ether and determined the energy different values for ^{13}C substitution ranged from 80 to 240 J/mol with the ZPE differences. They also found that the ZPE differences of isotopic fractionation associated with C–O and C–S bonds were slightly smaller than C–C bonds. On the other hand, the values of the D-substitution (1300–1800 J/mol) were many times larger than those for ^{13}C substitution. In 2000, Tang et al. (2000) continued the studies to include higher levels of computational (DFT/B3LYP/6-31G*, DFT/B3LYP/6-31G**, DFT/B3LYP/6-311G** and MP2/6-31G* and even larger cc-pVTZ(-F) + basis set) to obtain more accurate bond dissociation energy of methyl, ethyl, and propyl radicals from more than 60+ different hydrocarbon molecules. From these studies, they concluded that within the parallel first-order hydrocarbon gas generation kinetic model, the thermal cracking rates of isotopically substituted ($k*$) and unsubstituted (k), which can be represented by the equation

$$\frac{k^*}{k} = \frac{A_f^*}{A_f} \exp\left(\frac{-\Delta E_a}{RT} \right), \tag{7.21}$$

have an average frequency factor ratio $A_f^*/A_f \sim 1.021$ and the average activation energy difference $\Delta E_a = \left(E_a^* - E_a \right) \approx 176\,J/mol$. Specifically, the energies for cleavage of the methyl, ethyl, and propyl radicals fall into the ranges

- C*H$_3$: $A_f^*/A_f = 1.021$ $\Delta E_a = 175.44\,J/mol$
- C*H$_2$CH$_3$: $A_f^*/A_f = 1.017$ $\Delta E_a = 148.64\,J/mol$
- C*H$_3$CH$_2$CH$_3$: $A_f^*/A_f = 1.017$ $\Delta E_a = 151.61\,J/mol$

The degree of fractionation for isotopic substitution away from the ruptured bond is comparatively small:

- CH_2C*H_3: $A_f^*/A_f = 1.003$ $\Delta E_a = 19.60\,J/mol$
- $CH_2C*H_2CH_3$: $A_f^*/A_f = 1.003$ $\Delta E_a = 22.32\,J/mol$
- $CH_2CH_2C*H_3$: $A_f^*/A_f = 1.003$ $\Delta E_a = 24.04\,J/mol$

From these relationships, the isotopic fractionation between total carbon in the cleaved radical and that in the precursor molecule can be estimated at any temperature. Moreover, it has been found that the entropy term $\left(A_f^*/A_f\right)$ correlated very well with the enthalpy term (ΔE_a) in the Arrhenius expression for isotopic fractionation. A straight line with slope of 6.69×10^{-4} and an intercept of 0.999 $\left(R^2 = 0.724\right)$ for the correlation of A_f^*/A_f and ΔE_a can provide an important simplifying constraint for isotope modeling based on the Arrhenius equation approach.

Another important relationship between ΔE_a and E_a had also been derived based on the sigmoid model:

$$\Delta E_a = \beta_L + (\beta_H - \beta_L)f(E_a) \tag{7.22}$$

where β_H and β_L are two constants standing for the minimum and maximum probable value of ΔE_a, respectively, and $f(E_a)$ is the cumulative Gaussian distribution as a function of E_a:

$$f(E_a) = \int_E \frac{1}{\sigma\sqrt{2\pi}} e^{-(E_a - \mu)^2/2\sigma^2} dE_a \tag{7.23}$$

Here, σ is the variance, μ is the mean distribution, and E_a is the activation energy for gas generation.

Although DFT calculations can provide important constraints for modeling isotopic fractionations, the results cannot be extrapolated to geologic conditions without additional information. The biggest obstacle is the complex nature of hydrocarbon generation, a process that is simply not amenable to treatment from first principles. Different types of kerogen may have different initial C-isotope compositions and gas generation kinetics, which influence the relationship between isotope composition and formation temperature of natural gas. In addition to the conventional mechanisms of C–C rupture gas generation from kerogen pyrolysis and oil cracking, there are other mechanisms. For example, early methane generation originated from bacterial degradation of organic matter often results in lighter isotope compositions (Burruss and Laughrey, 2010). In addition, C-isotope fractionations could also been affected, if there are exchanging reactions occurring (Prinzhofer and Huc, 1995).

Gas migration processes might also affect gas isotopic fractionations, especially when there are leakages from a reservoir (Felipe et al., 2005). However, it has also been found that the change of isotopic compositions of natural gas due to primary diffusive fractionation is generally very small (<1‰ for methane) (Fuex, 1980). The adsorption and desorption of natural gas could have more significant effects on the isotopic fractionation (Xia and Tang, 2012). Stronger adsorption affinity and larger adsorption capacity result in a remarkably negative isotopic composition of early gas, but this is not the case of the weak adsorption of methane on organic matter. Large isotopic fractionation due to gas transport may happen under laboratory conditions with sufficient degassing, but the fractionation is limited under geological conditions.

Gas accumulation history affects isotopic changes. Gas generation is considered a continuous process, and the instantaneous C-isotope ratios of the generated gas change continuously. We consider two extreme cases for gas accumulations: (i) all generated gases are eventually trapped in a single reservoir, which we call "cumulative" such that the gas isotope composition is a weighted average over all maturity ranges, and (ii) gases generated at different maturity stages are trapped in different reservoirs, that is, the "instantaneous" model in which the gas isotope ratio will reflect the maturity of a short period of time. In actual petroleum systems, the gas isotope changes are in between the two extreme cases.

By taking the aforementioned factors into consideration, Tang et al. (2000) developed a general procedure to predict the ^{13}C isotope fractionation of methane generation from kerogen pyrolysis and oil cracking. The theoretical model was first applied to fit the laboratory data of the closed-system isothermal cracking of *n*-octadecane (*n*-C$_{18}$) at 400–500°C (Jeffery, 1981; Sackett et al., 1970) and has been widely applied to many different petroleum systems, as well as laboratory kerogen pyrolysis studies (Etiope et al., 2011; Ni et al., 2012; Strápoć et al., 2010; Zhang and Drooss, 2001).

7.3.3 Deuterium Isotope Fractionations of Natural Gas

Because deuterium (D) and hydrogen (H) have the largest relative mass difference of any stable isotope pair, hydrogen isotopes exhibit the greatest fractionation in natural systems (Bigeleisen, 1965). The deuterium isotope (δD) changes of natural gases due to the C–C rupture is about six times greater than that of the carbon isotope changes. That is, the stable δD isotopic fractionation related to the gas generation is in range of 1250–1500 J/mol, which is much greater than that of ^{13}C ($\approx$200 per mol). This feature makes compound-specific H-isotope analysis a valuable complement to ^{13}C values (Li et al., 2001).

δD values in hydrocarbons in a petroleum system could be affected by the type of the source organic matter (Hoefs, 1987), hydrogen-exchange processes with water (Sessions et al., 2004) and/or with clay minerals (Alexander et al., 1982), and thermal maturation processes (Schimmelmann et al., 1999, 2001). Additionally, water washing, biodegradation, and migration all have the potential to alter δD values; however, the magnitude of these effects is not currently well known (Schimmelmann et al., 2005). Quantum mechanical calculations have been applied to establish the equilibrium H-isotope fractionation in various organic molecules (Wang et al., 2009a, 2009b, 2013).

Compound-specific isotope analysis (CSIA) is an advanced GC analytic process. For C isotopic analyses, separated compounds were converted to CO_2 by passing the eluting analyte stream through a ceramic oxidation reaction at a temperature of 940°C. The standard CO_2 gas sample, of which the known δ^{13}C value is known, was injected before and after and species of interest to permit the calculation of δ^{13}C values. In the case of H isotopic analyses, individual compounds separated by gas chromatograph were converted to H_2 at 1440°C in a pyrolysis reactor (Burgoyne and Hayes, 1998; Hayes, 1983; Hilkert et al., 1999). The early-stage approach was offline, involving the steps of alkane separation and oxidation, water collection and reduction, and H_2 injection (Dumke et al., 1989). The modern online pyrolytic method invented by Burgoyne and Hayes (1998) simplified the operation when combined with gas chromatographic alkane separation (Henning et al., 2007), but pyrolysis needs a much higher temperature (over 1400°C) compared with oxidation (1000°C), and fractionation during pyrolysis may cause analytical error if a total conversion is not achieved. In addition to the pretreatment, the small *z/e* ratio of H_2 also makes the H isotopic composition measurement more sensitive to the stability of the electromagnet field compared with C.

Reference materials are critical to standardize measurements along internationally accepted isotopic scale, in order to harmonize data sets from different laboratories technique and sound calibration system. The compound-specific C and hydrocarbon isotopic compositions of three natural gas round robins were liberated by ten laboratories carrying more than 800 measurements including both online and offline methods (Dai et al., 2012). The three natural gas samples are NG1 (coal-related gas), NG2 (biogas), and NG3 (oil-related gas). Two-point calibrations were performed with international measurement standards for hydrogen isotope ration with Vienna Standard Mean Ocean Water (VSMOW) and Standard Light Antarctic Precipitation (SLAP); and the C-isotopes (in unit of Vienna Pee Dee Belemnite (VPDB)) were calibrated with the NBS 19 (also referred to as TS-Limestone) (Coplen et al., 2006; Friedman et al., 1982; Hut, 1987) and LSVEC, lithium carbonate methods (Coplen et al., 2006; Flesch et al., 1973; Stichler, 1995). The measured values and uncertainties derived from the maximum likelihood estimation (MLE) based on the offline measurements are summarized as follows:

NG1 (coal-related gas)

Methane:	$\delta^{13}C_{VPDB} = -34.18‰ \pm 0.10‰$	$\delta^2H_{VSMOW} = -185.1‰ \pm 1.2‰$
Ethane:	$\delta^{13}C_{VPDB} = -24.66‰ \pm 0.11‰$	$\delta^2H_{VSMOW} = -156.3‰ \pm 1.8‰$
Propane:	$\delta^{13}C_{VPDB} = -34.18‰ \pm 0.10‰$	$\delta^2H_{VSMOW} = -185.1‰ \pm 1.2‰$
i-Butane:	$\delta^{13}C_{VPDB} = -21.62‰ \pm 0.12‰$	
n-Butane:	$\delta^{13}C_{VPDB} = -21.74‰ \pm 0.13‰$	
CO_2:	$\delta^{13}C_{VPDB} = -5.00‰ \pm 0.12‰$	

NG2 (biogas)

Methane:	$\delta^{13}C_{VPDB} = -68.89‰ \pm 0.12‰$	$\delta^2H_{VSMOW} = -237.0‰ \pm 1.2‰$

NG3 (oil-related gas)

Methane:	$\delta^{13}C_{VPDB} = -43.61‰ \pm 0.09‰$	$\delta^2H_{VSMOW} = -167.6‰ \pm 1.0‰$
Ethane:	$\delta^{13}C_{VPDB} = -40.24‰ \pm 0.10‰$	$\delta^2H_{VSMOW} = -164.1‰ \pm 2.4‰$
Propane:	$\delta^{13}C_{VPDB} = -33.79‰ \pm 0.09‰$	$\delta^2H_{VSMOW} = -138.4‰ \pm 3.0‰$

The kinetic model of the D- and ^{13}C-isotope enrichments has been calibrated with the laboratory data from the artificial thermal maturation of a North Sea crude oil under anhydrous closed-system conditions that have been studied (Tang et al., 2005) and from the geologically observed data of the natural gas isotope fractionation (Ni et al., 2011). The kinetic model fits very well with the experimental data with a constant frequency factor ratio of $A*/A = 1.02$, and the activation energy difference of $\Delta E_a = 961.4\,J/mol$ (Figure 7.4) is consistent with the theoretical modeling results (Tang et al., 2000).

Following the same molecular modeling approach with the assumption that the hydrocarbon cracking is the barrierless C–C bond rupture process, such that the transition state can be represented by two well-separated radical species, we calculated the frequency factor ratio for the hydrogen isotopic fractionation of 1.20 (Figure 7.5). However, the poor fitting results suggested that there is substantial difference between the true transition state and the separated radical species that could affect the D-isotope fractionation results. Following the molecular modeling procedure as outlined in Figure 7.3, we can determine the transition state along the C–C bond rupture pathway and, from that, determine the frequency factor ratio of $A*/A = 1.07$. Using this value and a ZPE difference value of 1340 J/mol (320 cal/mol) and the activation energy distribution determined from the kinetic fitting of the pyrolysis data, we were able to obtain a predicted δD curve with improved fitting.

Kinetic isotope fractionations of the D-isotope fractionations of hydrocarbon gases have also been studied. Using the $A*/A = 1.07$ and the activation energy difference of 1332.6, 1177.5, and 1171.2 J/mol for methane, ethane, and propane, respectively, we can apply the established kinetic model to quantify natural gas maturity from the measured D-isotope values. Theoretical predictions

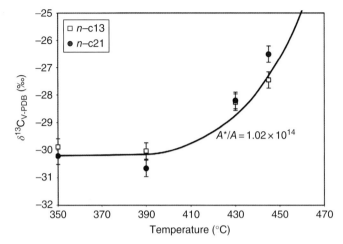

Figure 7.4 $\delta^{13}C$ isotope fractionation of C_{13} to C_{21} n-alkane as a function of final pyrolysis temperature (points) and the theoretical kinetic model using a $^{13}C/^{12}C$ isotope substitution $\Delta\Delta H^{\ddagger}$ value of 230 J/mol (55 cal/mol), a frequency factory ratio ($A*/A$) of 1.02, and an initial $\delta^{13}C$ value of -30.35‰. From Tang et al. (2005). Reproduced with permission from Elsevier.

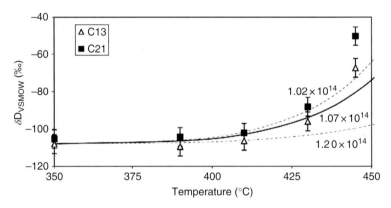

Figure 7.5 δD isotope fractionation of C_{13} to C_{22} (filled) and C_{13}–C_{15} n-alkanes (nonfilled) as a function of final pyrolysis temperature and the theoretical kinetic model using a D/H isotope substitution $\Delta\Delta H^{\ddagger}$ value of 1340 J/mol (320 cal/mol), an initial δD value of -107.8‰, and three different frequency factor ratios $A*/A =$ 1.02, 1.07, and 1.20. From Tang et al. (2005). Reproduced with permission from Elsevier.

offered a good correlation with the field observed data from the United States, China, Thailand, Australia, Angola, and Canada.

7.3.4 Molecular Modeling of the ^{13}C and D Doubly Substituted Methane Isotope

Recent studies based on the multiple isotope signatures of CO_2 show great promise toward the possibility to determine carbonate paleothermometry by measuring $^{13}C_{18}O^{16}O$ concentration (Affek and Eiler, 2006; Eiler and Schauble, 2004; Ghosh et al., 2006a, 2006b; Rahn and Eiler, 2001; Wang et al., 2004). Methane (CH_4), as the primary chemical component of natural gases, also has a potential of forming the doubly substituted isotopologues ($^{13}CH_3D$). The total abundance of $^{13}CH_3D$ in the natural gas is controlled by both the temperature-independent randomly populated process and the isotopic exchange due to the thermal equilibrium that is dependent on the surrounding temperature. Even though at the current stage it is still uncertain if the natural gas formation would reach fully thermal equilibrium, a close examination of their thermal equilibrium behavior will provide important insights to the measurement requirements and the temperature dependence of the thermodynamic equilibration. Assuming that the thermal equilibrium condition is achieved, we should have

$$^{13}CH_4 + CH_3D \xrightleftharpoons{K_{eq}} CH_4 + {}^{13}CH_3D \tag{7.24}$$

where the thermal equilibrium constant K_{eq} as a function of temperature (T) can be determined from first principles calculations using the Urey model (Urey, 1947) or Bigeleisen and Mayer equation (Bigeleisen and Mayer, 1947) or directly computed from the Gibb free energy difference (the so-called delta-G method) (Ma et al., 2008). Thus the abundance of the doubly substituted methane isotopologues $N(^{13}CH_3D)$ as a function of temperature can be determined:

$$N\left(^{13}CH_3D\right) = K_{eq} * N_C * N_D / N_0 \tag{7.25}$$

where N_C, N_D, and N_0 are the concentrations of isotopic ^{13}C and D contents and the total methane concentration. One of the direct applications of this modeling study is to offer the reasonable range of the detection limit in order to measure the $^{13}CH_3D$ concentration change within the given temperature range (say, 50°C) in the natural gas system (Figure 7.6).

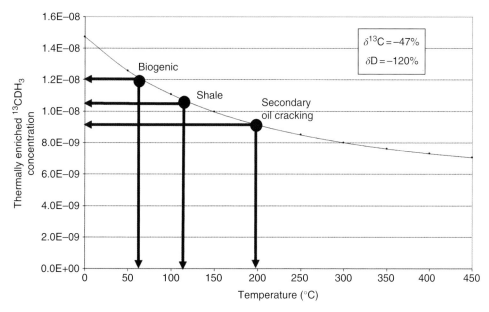

Figure 7.6 *Molecular modeling of the computed doubly substituted methane isotopologue concentrations related to the paleo gas formation conditions. In order to distinguish different methane sources, a high-resolution determination of the $^{13}CH_3D$ at the level of 10^{-9} will be required.*

7.4 Summary

Petroleum geochemistry and basin modeling are special scientific disciplines that involve many interactive efforts from various disciplines, such as geology, organic chemistry, analytic chemistry, physical chemistry, instrumentation, and theoretical modeling. Computational simulations based on modern quantum mechanical methods have gained substantial successes in providing information on the details of the relevant hydrocarbon and nonhydrocarbon transformation under both laboratory and geologic conditions. First principles theoretical calculations offer potentials to investigate the geological and chemical processes related to the generation and cracking of the petroleum and natural gas systems. Especially the quantum mechanical methods are uniquely suited to the studies of the natural of the transition state and the kinetic processes, which are essential to the studies of the compositional and isotopic changes of the petroleum systems as functions of time, temperature, and pressure. Nevertheless, it is worth noting that molecular modeling technique needs to be combined and integrated with the vast amount of the knowledge that has been accumulated through almost a century efforts from all scientific aspects, as well as the advanced laboratory and geological studies, in order to truly become a powerful tool that will play an important role in the quantification of oil and gas generations.

References

Affek, H. P. and Eiler, J. M. (2006) Abundance of mass $^{47}CO_2$ in urban air, car exhaust, and human breath. *Geochimica et Cosmochimica Acta* **70**, 1–12.

Alexander, R., Kagi, R. I. and Larcher, A. V. (1982) Clay catalysis of aromatic hydrogen-exchange reactions. *Geochimica et Cosmochimica Acta* **46**, 219–222.

American Society for Testing and Materials (ASTM) (2011) Standard test method for microscopical determination of the reflectance of vitrinite dispersed in sedimentary rocks: West Conshohocken, PA, ASTM International, Annual book of ASTM standards: Petroleum products, lubricants, and fossil fuels; Gaseous fuels; coal and coke, sec. 5, v. 5.06, D7708-11, p. 823–830, doi: 10.1520/D7708-11, http://www.astm.org/Standards/D7708.htm.

Arrhenius, S. A. (1889) Über die Dissociationswärme und den Einfluß der Temperatur auf der Dissociationsgrad der Elektrolyte. *Zeitschrift für Physikalische Chemie* **4**, 96–116.

Behar, F. and Pelet, R. (1985) Pyrolysis-gas chromatography applied to organic geochemistry, structural similarities between kerogens and asphaltenes from related rock extracts and oils. *Journal of Analytical and Applied Pyrolysis* **8**, 173–187.

Behar, F., Kressmann, S., Rudkiewicz, J. L. and Vendenbroucke, M. (1991) Experimental simulation in a confined system and kinetic modeling of kerogen and oil cracking. *Organic Geochemistry* **19**, 173–189.

Behar, F., Vandenbroucke, M., Tang, Y., Marquis, F. and Espitalié, J. (1997a) Thermal cracking of kerogen in open and closed systems: determination of kinetic parameters and stoichiometric coefficients for oil and gas generation. *Organic Geochemistry* **26**, 321–339.

Behar, F., Tang, Y. and Liu, J. (1997b) Comparison of rate constants for some molecular tracers generated during artificial maturation of kerogens: influence of kerogen type. *Organic Geochemistry* **26**(3–4), 281–287.

Behar, F., Beaumont, V. and De B. Penteado, H. L. (2001) Rock-Eval 6 technology: performances and developments. *Oil & Gas Science and Technology* **56**(2), 111–134.

Behar, F., Lorant, F. and Lewan M. (2008a) Role of NSO compounds during primary cracking of a Type II kerogen and a Type III lignite. *Organic Geochemistry* **39**, 1–22.

Behar, F., Lorant, F. and Mazeas L. (2008b) Elaboration of a new compositional kinetic schema for oil cracking. *Organic Geochemistry* **39**, 764–782.

Behar, F., Roy, S. and Jarvie D. (2010) Artificial maturation of a Type I kerogen in closed system: mass balance and kinetics modeling. *Organic Geochemistry* **41**, 1235–1247.

Berner, U. and Faber, E. (1996) Empirical carbon isotope/maturity relationship for gases from algal kerogens and terrigenous organic matter, based on dry, open-system pyrolysis. *Organic Geochemistry* **24**, 947–955.

Berner, U., Faber, E., Scheeder, G. and Panten, D. (1995) Primary cracking of algal and land plant kerogens: kinetic modeling of kerogen and oil cracking. *Organic Geochemistry* **126**, 233–245.

Bigeleisen, J. (1965) Chemistry of isotopes. *Science* **147**, 463–471.

Bigeleisen, J. and Mayer, M. G. (1947) Calculation of equilibrium constants for isotopic exchange reactions. *The Journal of Chemical Physics* **15**(5), 261–267.

Braun R. L. and Burnham A. K. (1987) Analysis of chemical reaction kinetics using a distribution of activation energies and simpler models. *Energy & Fuels* **1**, 153–161.

Bray, E. E. and Evans, E. D. (1961) Distribution of *n*-paraffins as a clue to the recognition of source beds. *Geochimica et Cosmochimica Acta* **22**, 2–15.

Brunham, A. K. and Braun R. L. (1990) Development of a detailed model of petroleum formation, destruction, and expulsion from lacustrine and marine source rocks. *Organic Geochemistry* **16**, 27–39.

Bunker, D. L. and Hase, W. L. (1973) On non-RRKM unimolecular kinetics: molecules in general, and CH3NC in particular. *The Journal of Chemical Physics* **59**, 4621–4632.

Burgoyne, T. W. and Hayes, J. M. (1998) Quantitative production of H_2 by pyrolysis of gas chromatographic effluents. *Analytical Chemistry* **70**, 5136–5141.

Burnham, A. K. and Braun, R. L. (1999) Global kinetic analysis of complex materials. *Energy and Fuels* **13**, 1–22.

Burnham, A. K., Braun, R. L., Gregg, H. R. and Samoun A. M. (1987) Comparison of methods for measuring kerogen pyrolysis rates and fitting kinetic parameters. *Energy and Fuels* **1**(6), 452–458.

Burruss, R. C. and Laughrey, C. D. (2010) Carbon and hydrogen isotopic reversals in deep basin gas: evidence for limits to the stability of hydrocarbons. *Organic Geochemistry* **41**, 1285–1296.

Chung, H. M. Gormly, J. R. and Squires, R. M. (1988) Origin of gaseous hydrocarbons in subsurface environments: theoretical consideration of carbon isotope distribution. *Chemical Geology* **71**, 97–103.

Claypool, G. E. and Kaplan, I. R. (1974) The origin and distribution of methane in marine sediments. In Kaplan, I. R. (Ed.) *Natural Gases in Marine Sediments*. Plenum Publishing, New York, p. 99–139.

Clayton, C. J. (1991a) Carbon isotope fractionation during natural-gas generation from kerogen. *Marine and Petroleum Geology* **8**, 232–240.

Clayton, C. J. (1991b) Effect of maturity on carbon isotope ratios or oils and condensates. *Organic Geochemistry* **17**, 887–899.

Clayton, C. J. and Bjorøy, M. (1994) Effect of maturity on $^{13}C/^{12}C$ ratios of individual compounds in North-Sea oils. *Organic Geochemistry* **21**, 737–750.

Coleman, D. D., Liu, C. L. and Riley, K. M. (1988) Microbial methane in the shallow Paleozoic sediments and glacial deposits of Illinois, U.S.A. *Chemical Geology* **71**, 23–40.

Colombo, U., Gazzarini, F., Gonfiantini, R., Kneuper, G., Teichmüller, M. and Teichmüller R. (1970) Carbon isotope study on methane from German coal deposits. In Hobson, G. D. and Speers, G. C. (Eds.) *Advances in Organic Geochemistry 1966*. Pergamon Press, Oxford, p. 1–25.

Coplen, T. B., Brand, W. A., Gehre, M., Gröning, M., Meijer, H. A. J., Toman, B. and Verkouteren, R. M. (2006) New guidelines for 13C measurements. *Analytical Chemistry* **78**(7), 2439–2441.

Corma, A. and Wojciechowski, B. W. (1985) The chemistry of catalytic cracking. *Catalysis Reviews: Science and Engineering* **27**, 29–150.

Cramer, B., Krooss, B. M. and Littke, R. (1998) Modeling isotope fractionation during primary cracking of natural gas: a reaction kinetic approach. *Chemical Geology* **149**, 235–250.

Cramer B., Faber, E., Gerling, P. and Krooss, B. M. (2001) Reaction kinetics of stable isotopes in natural gas – insights from dry, open system pyrolysis experiments. *Energy and Fuels* **15**, 517–532.

Dai, J., Xia, X., Li, Z., Coleman, D. D., Dias, R. F., Gao, L., Li, J., Deev, A., Li, J., Dessort, D., Duclerc, D., Li, L., Liu, J., Schloemer, S., Zhang, W., Ni, Y., Hu, G., Wang, X. and Tang, Y. (2012) Inter-laboratory calibration of natural gas round robins for $\delta^2 H$ and $\delta^{13}C$ using off-line and on-line techniques. *Chemical Geology* **310–311**, 49–55.

Davis, A. (1978) The reflectance of coal. In Karr, C. Jr. (Eds.) *Analytic Methods for Coal and Coal Products*, vol. **1**. Academic Press, New York, p. 27–81.

Dow, W. G. (1974) Application of oil-correlation and source rock data to exploration in Williston basin. *American Association of Petroleum Geologists Bulletin* **58**, 1253–1262.

Dow, W. G. (1977) Kerogen studies and geological interpretations. *Journal of Geochemical Exploration* **7**(2), 79–99.

Dow, W. (2014) Musings on the History of Petroleum Geochemistry – From My Perch. Article # 80375, AAPG, Houston, TX, April 6–9.

Dumke, I., Faber, E. and Poggenburg, R. (1989) Determination of stable carbon and hydrogen isotopes of light hydrocarbons. *Analytical Chemistry* **61**, 2149–2154.

Eiler, J. M. and Schauble, E. A. (2004) $^{18}O^{13}C^{16}O$ in Earth's atmosphere. *Geochimica et Cosmochimica Acta* **68**(23), 767–4777.

Etiope, G., Baciu, C. L. and Schoell, M. (2011) Extreme methane deuterium, nitrogen and helium enrichment in natural gas from the Homorod seep (Romania). *Chemical Geology* **280**(1–2), 89–96.

Fedor, F. and Hámor-Vidó, M. (2003) Statistical analysis of vitrinite reflectance data – a new approach. *International Journal of Coal Geology* **56**, 277–294.

Felipe, M. A., Kubicki, J. D. and Freeman, K. H. (2005) A mechanism for carbon isotope exchange between aqueous acetic acid and CO_2/HCO_3^-: an ab initio study. *Organic Geochemistry* **36**, 835–850.

Fernández-Ramos, A. (2006) Modeling the kinetics of bimolecular reactions. *Chemical Reviews* **106**, 4518–4584.

Flesch, G. D., Anderson Jr., A. R. and Svec, H. J. (1973) A secondary isotopic standard for $^6Li/^7Li$ determinations. *International Journal of Mass Spectrometry and Ion Physics* **12**(3), 265–272.

Friedman, I., O'neil, J. and Cebula, G. (1982) Two new carbonate stable isotope standards. *Geostandards Newsletter* **6**(1), 11–12.

Fuex, A. A. (1980) Experimental evidence against an appreciable isotopic fractionation of methane during migration. *Physics and Chemistry of the Earth* **12**, 725–732.

Galimov, E. (1974) Characteristics of the kinetic isotope effect in the degradation of organic macromolecules. *Russian Journal of Physical Chemistry* **48**, 811–814.

Galimov, E. (1988) Sources and mechanisms of formation of gaseous hydrocarbons in sedimentary rocks. *Chemical Geology* **71**, 77–95.

Garrett, B. C. and Truhlar, D. G. (1979) Accuracy of tunneling corrections to transition state theory for thermal rate constants of atom transfer reactions. *Journal of Physical Chemistry* **83**, 200–203.

Gaveau, B., Letolle, R. and Monthioux, M. (1987) Evaluation of kinetic parameters from ^{13}C isotopic effect during coal pyrolysis. *Fuel* **66**, 228–231.

Ghosh, P., Garzione, C. N. and Eiler, J. M. (2006a) Rapid uplift of the Altiplano revealed through $^{13}C–^{18}O$ bonds in paleosol carbonates. *Science* **311**(5760), 511–515.

Ghosh, P., Adkins, J., Affek, H., Balta, B., Guo, W., Schauble, E. A., Schrag, D. and Eiler, J. M. (2006b) $^{13}C–^{18}O$ bonds in carbonate minerals: a new kind of paleothermometer. *Geochimica et Cosmochimica Acta* **70**, 1439–1456.

Gilber, R. G. and Smith, S. C. (1990) *Theory of Unimolecular and Recombination Reactions*. Blackwell Scientific Publications, Cambridge, MA.

Goldstein, T. P. (1983) Geocatalytic reactions in formation and maturation of petroleum. *American Association of Petroleum Geologists Bulletin* **67**, 152–159.

Goldstein, E., Haught, M. and Tang, Y. (1998) Evaluation of density functional theory in the bond rupture of octane. *Journal of Computational Chemistry* **19**(2), 154–167.

Hantschel, T. and Kauerauf, A. I. (2010) *Fundamentals of Basin and Petroleum Systems Modeling*. Springer, Berlin, Heidelberg.

Hase, W. L. (1979) Overview of unimolecular dynamics. In Truhlar, D. G. (Ed.) *Potential Energy Surfaces and Dynamics Calculations*. Plenum Press, New York, p. 1–35.

Hayes, J. M. (1983) Practice and principles of isotopic measurements in organic geochemistry. In: Meinschein, W. G. (Ed.) *Organic Geochemistry of Contemporaneous and Ancient Sediments*. Society of Economic Paleontologists and Mineralogists, Bloomington, IN, pp. 5-1–5-31.

Hedberg, H. D. (1965) Significance of high-wax oils with respect to genesis of petroleum. *AAPG Bulletin* **52**, 736–750.

Helgeson, H. C., Knox, A. M., Owens, C. E. and Shock, E. L. (1993) Petroleum, oil field waters, and authigenic mineral assemblages: are them in metastable equilibrium in hydrocarbon reservoirs? *Geochimica et Cosmochimica Acta* **57**, 3295–3339.

Henning, M., Strąpoć, D., Lis, G. P., Sauer, P., Fong, J., Schimmelmann, A. and Pratt, L. M. (2007) Versatile inlet system for on-line compound-specific δD and δ13C GC-ox/red-IRMS analysis of gaseous mixtures. *Rapid Communications in Mass Spectrometry* **21**, 2269–2272.

Hilkert, A. W., Douthitt, C. B., Schluter, H. J. and Brand, W. A. (1999) Isotope ratio monitoring gas chromatography mass spectrometry of D/H by high temperature conversion isotope ratio mass spectrometry. *Rapid Communications in Mass Spectrometry* **13**, 1226–1230.

Hill, R. J., Tang, Y., Kaplan, I. R. and Jenden, P. D. (1996) Pressure effect on oil cracking. *Energy & Fuels* **10**, 873–882.

Hinshelwood, C. N. (1956) Chemical kinetics in the past few decades, Nobel Lecture. http://www.nobelprize.org/nobel_prizes/chemistry/laureates/1956/hinshelwood-lecture.pdf (accessed January 8, 2016).

Hoefs, J. (1987) *Stable Isotope Geochemistry*. Springer-Verlag, Berlin.

Huc, A. Y. (2003) Petroleum geochemistry at the dawn of the 21st century. *Oil & Gas Science and Technology, Rev. IFP* **58**(2), 233–241.

Huc, A. Y., Durand, B., Roucache, J., Vandenbroucke, M. and Pittion, J. L. (1986) Comparison of three series of organic matter of continental origin. *Organic Geochemistry* **10**, 65–72.

Hund, F. (1925) Zur Deutung verwickelter Spektren, insbesondere der Elemente Scandium bis Nickel. *Zeitschrift für Physik* **33**, 345–371.

Hunt, J. M. (1979) *Petroleum Geochemistry and Geology*. Freeman, San Francisco, CA.

Hunt, J. M. and Jamieson, G. W. (1956) Oil and organic matter in source rocks of petroleum. *AAPG Bulletin* **40**, 477–488.

Hunt, J. M., Philp, R. P. and Kvenvolden, K. A. (2002) Early developments in petroleum geochemistry. *Organic Geochemistry* **33**, 1025–1052.

Hut, G. (1987) Consultants' group meeting on stable isotope reference samples for geochemical and hydrological investigations, Report to the Director General, International Atomic Energy Agency, Vienna.

ICCP (1998) The new vitrinite classification (ICCP System 1998). *Fuel* **77**, 349–358.

Jeffery, A. W. A. (1981) Thermal and clay catalyzed cracking in the formation of natural gas, Ph.D. dissertation, Texas A&M University, pp. 100–107.

Jenden, P. D. and Kaplan, I. R. (1986) Comparison of microbial gases from the Middle America Trench and Scripps Submarines Canyon: implications for the origin of natural gas. *Applied Geochemistry* **1**, 631–646.

Jenden, P. D. and Kaplan, I. R. (1989) Origin of natural gas in Sacramento basin, California. *American Association of Petroleum Geologists Bulletin* **73**, 431–453.

Jenden P. D., Drazan D. J. and Kaplan I. R. (1993) Mixing of thermogenic natural gases in northern Appalachian basin. *American Association of Petroleum Geologists Bulletin* **77**, 980–998.

Johns W. D. (1979) Clay mineral catalysis and petroleum generation. *Annual Review of Earth and Planetary Sciences* **7**, 183–198.

Jursic, B. S. (1996) The evaluation of nitrogen containing bond dissociation energies using the *ab initio* and density functional methods. *Journal of Molecular Structure (THEOCHEM)* **366**, 103–108.

Kassel, L. S. (1928) Studies in homogeneous gas reaction. I. *The Journal of Physical Chemistry* **32**(2), 225–242.

Katz, B. J. (2011) Microbial processes and natural gas accumulations. *The Open Geology Journal* **5**, 75–83.

Katz, B. J., Mancini, E. A. and Kitchka, A. A. (2008) A review and technical summary of the AAPG Hedberg Research Conference on "Origin of petroleum-biogenic and/or abiogenic and its significance in hydrocarbon exploration and production". *AAPG Bulletin* **92**, 549–556.

Kossiakoff, A. and Rice, F. O. (1943) Thermal decomposition of hydrocarbons, resonance stabilization and isomerization of free radicals. *Journal of the American Chemical Society* **65**, 590–594.

Kvenvolden, K. A. (2002) History of the recognition of organic geochemistry in geoscience. *Organic Geochemistry* **33**, 517–521.

Kvenvolden, K. A. (2006) Organic geochemistry – a retrospective of its first years. *Organic Geochemistry* **37**, 1–11.

Lafargue, E., Marquis, F. and Pillot, D. (1998) Rock-Eval 6 applications in hydrocarbon, exploration, production and soil contamination studies. *Oil & Gas Science and Technology* **53**(4), 421–437.

Laidler, K. J. (1987) *Chemical Kinetics*, 3rd Edition. Harper & Row, New York, p. 42.

Lakshmanan, C. C. and White, N. (1994) A distribution activation energy model using Weibull distribution for the representation of complex kinetics. *Energy and Fuels* **8**(6), 1158–1167.

Lerche, I. (1990) *Basin Analysis, Vol. 2: Quantitative Methods*. Academic Press, San Diego.

Lewan, M. (1993a) Laboratory simulation of petroleum formation: hydrous pyrolysis. In Engel, M. H. and Macko, S. (Eds.) *Organic Geochemistry Principles and Applications*. Plenum Press, New York, p. 419–442.

Lewan, M. D. (1993b) Laboratory simulation of petroleum formation – hydrous pyrolysis. In Engle, M. and Macko, S. (Eds.) *Organic Geochemistry*. Plenum, New York, p. 419–442.

Lewan, M. D. (1997) Experiments on the role of water in petroleum formation. *Geochimica et Cosmochimica Acta* **61**, 3692–3723.

Li, M. W., Huang, Y. S., Obermajer, M., Jiang, C. Q., Snowdon, L. R. and Fowler, M. G. (2001) Hydrogen isotopic compositions of individual alkanes as a new approach to petroleum correlation: case studies from the Western Canada Sedimentary Basin. *Organic Geochemistry* **32**, 1387–1399.

Light, J. C. (1964) Phase-space theory of chemical kinetics. *The Journal of Chemical Physics* **40**, 3221–3229.

Liu, J. and Tang, Y. (1997) Kinetics of petroleum generation determined from multiple cold trap pyrolysis gas chromatography. *Chinese Science Bulletin* **42**, 254–258.

Liu, R., Morokuma, K., Mebel, A. M. and Lin, M. C. (1996) *Ab initio* study of the mechanism for the thermal decomposition of the phenoxyl radical. *The Journal of Physical Chemistry* **100**, 9314–9322.

Lorant, F., Prinzhofer, A., Behar, F. and Huc, A. Y. (1988) Carbon isotope and molecular constraints on the formation and the expulsion of thermogenic hydrocarbon gases. *Chemical Geology* **147**, 249–264.

Lorant, F., Prinzhofer, A., Behar, F. and Huc, A.-Y. (1998) Carbon isotopic and molecular constraints on the formation and the expulsion of thermogenic hydrocarbon gases. *Chemical Geology* **147**, 240–264.

Lorant, F., Behar, F., Goddard, W. A. III and Tang, Y. (2001) *Ab initio* investigation of ethane dissociation using generalized transition state theory. *The Journal of Physical Chemistry. A* **105**, 7896–7904.

Ma, Q., Wu, S. and Tang, Y. (2008) Formation and abundance of doubly-substituted methane isotopologues ($^{13}CH_3D$) in natural gas systems. *Geochimica et Cosmochimica Acta* **72**, 5446–5456.

Mango, F. D., Hightower, J. W. and James, A. T. (1994) Role of transition-metal catalysis in the formation of natural gas. *Nature* **368**, 536–538.

Marcus, R. A. (1952) Unimolecular dissociations and free radical recombination reactions. *The Journal of Chemical Physics* **20**, 359–364.

Margoon, L. B. and Dow, W. G. (1994) *The Petroleum System: From Source to Trap*, AAPG Memoirs. American Association of Petroleum Geologists, Tulsa, OK.

Martinelli, G. (2009) Petroleum geochemistry. In Mesini, E. and Macini, P. (Eds.) *Petroleum Engineering – Upstream. Encyclopedia of Life Support Systems (EOLSS)*. Developed under the Auspices of the UNESCO, Eolss Publishers, Oxford, Chapter 4. p. 193–216. http://www.eolss.net (accessed January 29, 2015).

Martini, A., Budai, A., Walters, L. and Schoell, M. (1996) Economic accumulation of biogenic methane. *Nature* **383**, 153–158.

MrCartney, J. T. and Teichmüller, M. (1972) Classification of coals according to degree of coalification by reflectance of the vitrinite component. *Fuel* **61**, 64–68.

Mukhopadhyay, P. K. and Dow, W. G. (Eds.) (1994) *Vitrinite Reflectance as a Maturity Parameter – Applications and Limitations*. ACS Symposium Series, vol. **570**. American Chemical Society, Washington, DC, 294pp.

Ni, Y., Ma, Q., Ellis, G. S., Dai, J., Katz, B., Zhang, S. and Tang, Y. (2011) Fundamental studies on kinetic isotope effect (KIE) of hydrogen isotope fractionation in natural gas systems. *Geochimica et Cosmochimica Acta* **75**, 2696–2707.

Ni, Y., Liao, F., Dai, J., Zou, C., Zhu, G., Zhang, B. and Liu, Q (2012) Using carbon and hydrocarbon isotope to quantify gas maturity, formation temperature, and formation age – specific applications for gas fields from the Tarim Basin, China. *Energy Exploration & Exploitation* **30**(2), 273–294.

Olah, G. G. and Molnar, A. (1995) *Hydrocarbon Chemistry*. John Wiley & Sons, Inc., New York.

Peter, K. E. and Fowler M. G. (2002) Applications of petroleum geochemistry to exploration and reservoir management. *Organic Geochemistry* **33**, 5–36.

Peters, K. E., Walters, C. C. and Mankiewicz, P. J. (2006) Evaluation of kinetic uncertainty in numerical models of petroleum generation. *American Association of Petroleum Geologists Bulletin* **90**(3), 387–403.

Philippi, G. T. (1956) Identification of oil source beds by chemical means. 20th International Geological Congress. Mexico City, Mexico Section III, Geology of Petroleum, September, 1956. Reprint, p. 25–28.

Philippi, G. T. (1965) On the depth, time, and mechanism of petroleum generation. *Geochimica et Cosmochimica Acta* **29**, 1021–1049.

Poutsma, M. L. (2000) Fundamental reactions of free radicals relevant to pyrolysis reactions. *Journal of Analytical and Applied Pyrolysis* **54**, 109–126.

Prinzhofer, A. A. and Huc, A. Y. (1995) Genetic and post-genetic molecular and isotopic fractionations in natural gases. *Chemical Geology* **126**, 281–290.

Prinzhofer, A. and Pernaton, E. (1997) Isotopically light methane in natural gas: bacterial imprint or diffusive fractionation? *Chemical Geology* **142**, 193–200.

Rahn, T. and Eiler, J. M. (2001) Experimental constraints on the fractionation of $^{13}C/^{12}C$ and $^{18}O/^{16}O$ ratios due to desorption of CO_2 on mineral substrates at conditions relevant to the surface of Mars. *Geochimica et Cosmochimica Acta* **65**, 839–846.

Reinhardt, C. (2008) *Chemical Sciences in the 20th Century: Bridging Boundaries*. John Willy & Sons, Weinheim, p. 161.

Rice, F. O. (1933) The thermal decomposition of organic compounds from the standpoint of free radicals. III. The calculation of the products formed from paraffin hydrocarbons. *Journal of the American Chemical Society* **55**, 3035–3040.

Rice, D. D. (1992) Controls, habitat and resource potential of ancient bacterial gas. In Vially, R. (Ed.) *Bacterial Gas*. Editions Technip, Paris, p. 91–118.

Rice, D. D. (1993) Composition and origins of coalbed gas. In Law, B. E. and Rice, D. D. (Eds.) *Hydrocarbons from Coal*. Editions Technip, Paris, p. 91–118.

Rice, F. O. and Herzfeld, K. F. (1934) The thermal decomposition of organic compounds from the standpoint of free radicals. VI. The mechanism of some chain reactions. *Journal of the American Chemical Society* **56**, 284–289.

Rice, O. K. and Ramsperger, H. C. (1927) Theories of unimolecular gas reaction at low pressures. *Journal of the American Chemical Society* **49**(7), 1617–1629.

Rioux, F. (2007) Hund's multiplicity rule revisited. *Journal of Chemical Education* **84**, 358–360.

Rooney, M. A. Claypool, G. E. and Chung, H. M. (1995) Modeling thermogenic gas generation using carbon isotope ratios of natural gas hydrocarbon. *Chemical Geology* **126**, 219–232.

Sackett, W. M., Nakaparksin, S. and Dalrymple, D. (1970) Carbon isotope effects in methane production by thermal cracking. In Hobson, G. D. and Speers, G. C. (Eds.) *Advances in Organic Geochemistry*. Pergamon Press, Oxford, p. 37–53.

Schimmelmann, A., Lewan, M. D. and Wintsch, R. P. (1999) D/H isotope ratios of kerogen, bitumen, oil, and water in hydrous pyrolysis of source rocks containing kerogen types I, II, IIs and III. *Geochimica et Cosmochimica Acta* **63**, 3751–3766.

Schimmelmann, A., Boudou, J. P., Lewan M. D. and Wintsch, R. P. (2001) Experimental controls on D/H and $^{13}C/^{12}C$ ratios of kerogen, bitumen and oil during hydrous pyrolysis. *Organic Geochemistry* **32**, 1009–1018.

Schimmelmann, A., Sessions, A. L., Boreham, C. J., Edwards, D. S., Logan, G. A. and Summons, R. (2005) D/H ratios in petroleum systems with terrestrial sources. *Organic Geochemistry* **35**(10), 1169–1195.

Schoell, M. (1980) The hydrogen and carbon isotopic composition of methane from natural gases of various origins. *Geochimica et Cosmochimica Acta* **44**, 649–661.

Schoell, M. (1983) Genetic characterization of natural gases. *American Association of Petroleum Geologists Bulletin* **67**, 2225–2238.

Seewald, J. W. (1994) Evidence for metastable equilibrium between hydrocarbon under hydrothermal conditions. *Nature* **370**, 285–287.

Sessions, A. L., Sylva, S. P., Summons, R. E. and Hayes, J. M. (2004) Isotopic exchange of carbon-bound hydrogen over geologic timescales. *Geochimica et Cosmochimica Acta* **68**, 1545–1559.

Sobeih, K. L., Baron, M. and Gonzalez-Rodriguez, J. (2008) Recent trends and developments in pyrolysis-gas chromatography. *Journal of Chromatography A* **1186**, 51–66.

Stahl, W. (1977) Carbon and nitrogen isotopes in hydrocarbon research and exploration. *Chemical Geology* **20**, 121–149.

Stahl, W. and Carey, B. D. (1975) Source-rock identification by isotope analyses of natural gases from fields in the Val Verde and Delaware basins, West Texas. *Chemical Geology* **16**, 257–267.

Stichler, W. (1995) Interlaboratory comparison of new materials for carbon and oxygen ratio measurements. Proceedings of a consultants meeting held in Vienna, December 1–3, 1993. IAEA-TECDOC-825, p. 67–74.

Strápoć, D., Mastalerz, M., Schimmelmann, A., Drobniak, A. and Hasenmueller, N. R. (2010) Geochemical constraints on the origin and volume of gas in the New Albany Shale (Devonian-Mississippian), eastern Illinois Basin. *American Association of Petroleum Geologists Bulletin* **94**(11), 1713–1740.

Sundararaman, P., Merz, P. H. and Mann R. G. (1992) Determination of kerogen activation energy distribution. *Energy and Fuels* **6**(6), 793–803.

Sundberg, K. R. and Bennett, C. R. (1983) Carbon isotope paleothermometry of natural gas. In Bjoroy, M. (Ed.) *Advances in Organic Geochemistry 1981*. John Wiley & Sons, New York, p. 769–774.

Sweeney, J. J. and Burnham, A. K. (1990) Evaluation of a simple model of vitrinite reflectance based on chemical kinetics. *American Association of Petroleum Geologists Bulletin* **74**(10), 1559–1570.

Tang, Y. and Behar, F. (1995) Rate constants of *n*-alkanes generation from type II kerogen in open and closed pyrolysis systems. *Energy & Fuels* **9**, 507–512.

Tang, Y. and Jenden, P. D. (1988) Modeling early methane generation in coal. *Energy and Fuels* **10**, 659–671.

Tang, Y. and Jenden, P. D. (1995) Theoretical modeling of carbon and hydrogen isotope fractionations in natural gas. In Grimalt, J. O. and Dorronsoro, C. (Eds.) *Organic Geochemistry, Developments and Applications to Energy, Climate, Environment and Human History*. AIGOA, Donostia-San Sebastian, p. 1067–1096.

Tang, Y. and Stauffer, M. (1991) Development of multiple cold trap pyrolysis. *Journal of Analytical and Applied Pyrolysis* **28**(2), 167–274.

Tang, Y. and Stauffer, M. (1994) Multiple cold trap pyrolysis gas chromatography: a new technique for modeling hydrocarbon generation. *Organic Geochemistry* **22**, 863–872.

Tang, Y., Perry, J. K., Jenden, P. D. and Schoell M. (2000) Mathematical modeling of stable carbon isotope ratios in natural gases. *Geochimica et Cosmochimica Acta* **64**(15), 2673–2687.

Tang, Y., Huang, Y., Ellis, G. S., Wang, Y., Kralert, P. G., Gillaizeau, B., Ma, Q. and Hwang, R. (2005) A kinetic model for thermally induced hydrocarbon and carbon isotope fractionation of individual *n*-alkanes in crude oil. *Geochimica et Cosmochimica Acta* **69**(18), 4505–4520.

Tissot, B. P. and Welte, D. H. (1984) *Petroleum Formation and Occurrence*, second revised and enlarged edition. Springer-Verlag, New York, 699p.

Tissot, B., Durand, B., Espitalié, J. and Combaz, A. (1974) Influence of nature and diagenesis of organic matter in formation of petroleum. *AAPG Bulletin* **58**(3), 499–506.

Tomió, J., Behar, F., Vandenbroucke, M. and Tang, Y. (1995) Artificial maturation of Monterey kerogen (Type II-S) in a closed system and comparison with Type II kerogen: implications on the fate of sulfur. *Organic Geochemistry* **23**(7), 647–660.

Trask, P. D. (1932) *Origin and Environment of Source Sediments*. The Gulf Publishing Company, Houston, TX, 323p.

Trask, P. D. and Pathnode, H. W. (1942) *Source Beds of Petroleum*. AAPG, Tulsa, OK, 566p.

Treibs, A. (1936) Chlorophyll- und Häminderivate in organischen Mineralstoffen. *Angewandte Chemie* **49**, 682–688.

Truhlar, D. G. (1969) Statistical phase-space theory of the reaction $C^+ + D_2$ including threshold behavior. *The Journal of Chemical Physics* **51**, 4617–4623.

Ungerer, P. (1990) State of the art research in kinetic modeling of oil formation and expulsion. *Organic Geochemistry* **16**, 1–25.

Ungerer, P. and Pelet, R. (1987) Extrapolation of oil and gas formation kinetics from laboratory experiments to sedimentary basins. *Nature* **327** (6117), 52–54.

Urey, H. C. (1947) The thermodynamic properties of isotopic substances. *Journal of the Chemical Society* (Resume), 562–581.

Van der Hart, W. J. (1999) *Ab initio* calculations on the isomerization of alkene radical cations. *Journal of the American Society for Mass Spectrometry* **10**, 575–586.

Van Krevelen, D. W. (1950) Graphical-statistical method for the study of structure and reaction processes of coal. *Fuel* **29**, 269–284.

Vandenbroucke, M. and Largeau, C. (2007) Kerogen origin, evolution and structure. *Organic Geochemistry* **38**, 719–833.

Wang, Z., Schauble, E. A. and Eiler, J. M. (2004) Equilibrium thermodynamics of multiply substituted isotopologues of molecular gases. *Geochimica et Cosmochimica Acta* **68**(23), 4779–4797.

Wang, Y., Sessions, A. L., Nielsen, R. J. and Goddard, W. A. (2009a) Equilibrium H-2/H-1 fractionation in organic molecules: I. Experimental calibration of ab initio calculations. *Geochimica et Cosmochimica Acta* **73**, 7060–7075.

Wang, Y., Sessions, A. L., Nielsen, R. J. and Goddard, W. A. (2009b) Equilibrium H-2/H-1 fractionation in organic molecules: II. Linear alkanes, alkenes, ketones, carboxylic acids, esters, alcohols and ethers. *Geochimica et Cosmochimica Acta* **73**, 7076–7086.

Wang, Y., Sessions, A. L., Nielsen, R. J. and Goddard, W. A. (2013) Equilibrium H-2/H-1 fractionation in organic molecules: III. Cyclic ketones and hydrocarbons. *Geochimica et Cosmochimica Acta* **107**, 82–95.

Waples, D. W. and Tornheim, D. W. (1978) Mathematical model for petroleum-forming processes: carbon isotope fractionation. *Geochimica et Cosmochimica Acta* **42**, 467–472.

Wardlaw, D. M. and Marcus, R. A. (1986) Unimolecular reaction rate theory for transition states of any looseness. 3. Application to methyl radical recombination. *The Journal of Physical Chemistry* **90**, 5383–5393.

Weeks, L. G. (1958) Habitat of oil and some factors that control it. In Week, L. G. (Ed.) *Habitat of Oil*. AAPG, Tulsa, p. 58–59.

Weibull, W. (1951) A statistical distribution function of wide applicability. *Journal of Applied Mechanics* **18**, 293–296.

Werner, R. A. and Brand, W. A. (2001) Referencing strategies and techniques in stable isotope ratio analysis. *Rapid Communications in Mass Spectrometry* **15**, 501–519.

William, J. A. (1974) Characterization of oil types in Williston basin. *American Association of Petroleum Geologists Bulletin* **58**, 1243–1252.

Xia, X. and Tang, Y. (2012) Isotope fractionation of methane during natural gas flow with coupled diffusion and adsorption/desorption. *Geochimica et Cosmochimica Acta* **77**, 489–503.

Xiao, Y. (2001) Modeling the kinetics and mechanisms of petroleum and natural gas generation: a first principles approach. In Cygan, R. T. and Kubicki, J. D. (Eds.) *Molecular Modeling Theory: Applications in the Geosciences*, Reviews in Mineralogy 42. Mineralogical Society of America and the Geochemical Society, Washington DC, p. 383–436.

Xiao, Y., Longo, J. M., Hieshima, G. B. and Hill, R. J. (1997) Understanding the kinetics and mechanisms of hydrocarbon thermal cracking: an *ab initio* approach. *Industrial and Engineering Chemistry Research* **36**, 4033–4040.

Zhang, T. and Drooss, B. M. (2001) Experimental investigation on the carbon isotope fractionation of methane during gas migration by diffusion through sedimentary rocks at elevated temperature and pressure. *Geochimica et Cosmochimica Acta* **65**(16), 2723–2741.

8

Mineral–Water Interaction

Marie-Pierre Gaigeot[1,2] and Marialore Sulpizi[3]

[1]*LAMBE CNRS UMR 8587, Université d'Evry val d'Essonne, Evry, France*
[2]*Institut Universitaire de France, Paris, France*
[3]*Department of Physics, Johannes Gutenberg Universitat, Mainz, Germany*

8.1 Introduction

The atomic-level structure of water at mineral surfaces is an important factor in controlling interfacial reactions such as crystal growth and dissolution, ion adsorption and incorporation, and redox reactions. Recent advances in diverse experimental and computational techniques have now made possible a more detailed structural characterisation of mineral–water interfaces at the atomic scale. In particular, X-ray scattering techniques have permitted to probe water molecular structure at the surfaces of a range of mineral classes including oxides (Catalano et al. 2006, 2007, 2009; Eng et al. 2000; Tanwar et al. 2007; Trainor et al. 2004; Zhang et al. 2007), carbonates (Fenter et al. 2000a, 2007; Geissbuhler et al. 2004; Jun et al. 2007) and silicates (Cheng et al. 2001; Fenter et al. 2000b, 2003; Schlegel et al. 2002). Despite the tremendous progresses, X-ray experiments still suffer from some limitations amongst which hydrogen atoms that are not directly visible, thus creating a difficulty in probing the protonation state of surface groups. Interface selective spectroscopy, such as second harmonic generation (SHG) (Chen et al. 1981; Eisenthal 1996, 2006) and sum frequency generation (SFG) (Shen and Ostroverkhov 2006; Tian and Shen 2014; Zhu et al. 1987), has become in the last 20 years a powerful technique to address properties of interfaces and in particular of interfacial fluids. SFG and SHG have been widely employed for the characterisation of oxide–water interfaces (see, for instance, Dewand et al. 2013; Eisenthal 2006; Geiger 2009; Hayes et al. 2010; Ostroverkhov et al. 2005; Sung et al. 2011; Zhang et al. 2008). However, also interfacial spectroscopy suffers from some limitations, amongst which the depth of the probed interface or a missing spatial lateral information.

Molecular Modeling of Geochemical Reactions: An Introduction, First Edition. Edited by James D. Kubicki.
© 2016 John Wiley & Sons, Ltd. Published 2016 by John Wiley & Sons, Ltd.

Simulations have been established as a complementary approach to experiments, which can help to provide a detailed molecular interpretation of the experimental data on solid–liquid interfaces. In particular in connection with the described interface sensitive experiments, molecular dynamics (MD) simulations can help to assign protonation states, can test multiple models and can provide a detailed atomistic description of interfaces. Simulations can also provide knowledge on the thermodynamics of the interfaces, whose data are difficult to assess from experiments. Classical MD simulations can currently access system sizes in the order of nanometers. The main limitation of such an approach is the transferability of the force field parameters and its use in vibrational spectroscopy calculations. Ab initio (electronic structure-based) MD simulations can overcome such problems, providing an accurate description of the heterogeneous environment, including polarisation effects (also including the electronic polarisation). The drawback of ab initio molecular dynamics (AIMD) is the relatively high computational cost, which limits the model size to a few tenths of angstroms and also confine the accessible timescale to a few tenths of picoseconds (probably of the order of 100 ps, depending on the system size, level of representation and computational accessibility on high-performance machines (Khaliullin et al. 2013; VandeVondele and Hutter 2012)) when using the electronic density functional theory (DFT) representation. In the case of interfaces of geochemical relevance, the liquid in contact with the solid is mainly water. For such a fluid, the timescales accessible by AIMD are already enough to provide a detailed picture of the solvent structural and dynamical properties at the interface.

When addressing chemical (including geochemical) reactivity, the use of ab initio method is mandatory since we are addressing bond breaking and formation and we need to establish an accurate estimate of the free energy landscape associated with reactive processes. Note that classical reactive force fields (see, e.g. Lockwood and Garofalini 2014; van Duin et al. 2003; Zhang et al. 2004) are still demanding a large amount of work for parameterisation and still remain quantitatively less precise than ab initio representations. For example, ab initio simulations have contributed to understanding dissolution of silicates in water. Ab initio dissolution studies of silicates were pioneered by Lasaga and co-workers about two decades ago using the Hartree–Fock method and a small basis set for small silicate clusters (Kubicki et al. 1993; Lasaga and Gibbs 1990; Xiao and Lasaga 1994, 1996). Since then, several new density functional methods and larger basis set calculations have been used to calculate the barrier heights of dissolution reactions (Zhang et al. 2007). Pelmenschikov et al. (2000) used ab initio calculations to study dissolution reactions in β-cristobalite and proposed that dissolution occurs preferentially from less coordinated surface sites. Another study by Criscenti et al. used a protonated silicate cluster to show that dissolution at low pH range has lower barrier height than the neutral case (Criscenti et al. 2006). More recent studies have also addressed the role of ions in dissolution (DelloStritto et al. 2014).

Another class of reactions that requires electronic structure-based methods is the redox chemistry, for example, iron-containing minerals are widespread hosts of reduction–oxidation-active iron species with impact on a great variety of geochemical and environmental reaction chemistry. A full chapter by Kevin Rosso in this book (Chapter 12) is dedicated to redox processes at interfaces, and we refer the reader to this chapter for a detailed overview on methodology and applications.

One specific question we will address in this contribution is the acidity of surface groups. Surface acidity plays a central role in the chemical processes occurring at interfaces. Experimentally, surface acidity is manifested in a net surface charge changing over from positive to negative with increasing pH. Because of the inherent difficulty for titration measurements to distinguish between various functional groups, modelling of surface acidity has always played an important role in the understanding of relevant quantities, such as the pH at the point of zero charge (PZC).

An approach commonly encountered in the literature is to parameterise empirical models for surface acidity using the pK_a values of solution monomers, which are chemically better characterised and more easily measured. The most successful and popular of these methods, the modified MUSIC model (Hiemstra et al. 1996) uses bond valence to predict pK_a values, as does the method of Bickmore et al. (2003). The appeal of bond valence methods for pK_a prediction is the use of structural information, which is translated in the prediction of chemical behaviour. Bond lengths of relaxed structures determined by diffraction experiments or ab initio structure optimisations can be used to refine the bond valence methods (Bickmore et al. 2004). A similar approach underlies the single site solvation bond strength and electrostatic (SBE) model (Sverjensky 1994, 1996), which is based on a thermodynamic dielectric continuum approach, whereas the modified MUSIC model depends on explicit hydration. The advantages and disadvantages of both models have been discussed, for example, in Sahai (2000). The MUSIC model relies on the correlation between the pK_a values of surface functional groups with those of the analogous groups in solution monomers (Sahai 2002). However that is not always the case, for example, for silica, which appears to be an outlier in such correlation relations (Sahai 2002; Sverjensky 1994, 1996). Moreover, differentiating between specific surface site geometries can be difficult for empirical models, which normally use one or two generic sites with equilibrium constants and fitted density of sites. One of the key issues in the pK_a calculations is an adequate treatment of solvent effects.

State-of-the-art self-consistent solvent reaction field methods can achieve an accuracy of 0.5 of a pK_a unit in the prediction of pK_a of molecules in solution (Liptak and Shields 2001; Saracino et al. 2003). The situation is certainly more complicated in the case of strong and specific interactions of solvent molecules with the solute. In this respect DFT-based MD simulations present the advantage that both the solute (oxide surface in the present interfacial context) and the solvent (electrolyte solution) are treated at the same level of theory, which includes the full electronic structure in a consistent way.

One of the first work on the calculation of acidity constants for surface deprotonable groups is the work from K. Leung on silica model surfaces (Leung et al. 2009). The main conclusion of that work was to show that surface silanols can have relative acidic values, but strong acidity can only be reached on strained surfaces. Constraint dynamics simulations were used in Leung et al. (2009) and Leung and Criscenti (2012) to calculate deprotonation free energies. A limitation of this method resides in the difficulty to follow the proton dissociation beyond the first solvation shell of the surface group. The proton insertion method described in Cheng and Sprik (2010), Cheng et al. (2009), Costanzo et al. (2011), Gaigeot et al. (2012), Mangold et al. (2011), Sulpizi and Sprik (2008, 2010) and Sulpizi et al. (2012) allows to overcome such a problem since the proton is completely eliminated from the system. Details of this methodology will be presented in Section 8.3. The proton insertion method has been successfully applied to quite a few surfaces of interest for the geochemistry community (Cheng and Sprik 2010; Churakov et al. 2014; Gaigeot et al. 2012; Liu et al. 2014; Sulpizi et al. 2012; Tazi et al. 2012).

In particular in this chapter, we will be discussing a few selected examples chosen amongst the recent literature. These include the calculation of surface pK_a for silica, which is a particularly interesting case, since it has been a highly discussed example in the literature due to the difficulty to assign a microscopic origin to the bimodal behaviour observed in the experiments. We will also present results for alumina and compare the two oxides (Sulpizi et al. 2012). Finally we will also present the application to clays, where an experimental determination of the edge group reactivity is particularly challenging and the computational prediction of group acidity becomes particularly precious in order to model the charge distribution at the interface (Tazi et al. 2012).

We finish this introduction and the motivation to our works by addressing the relevance of ab initio techniques for the description of vibrational spectroscopy at interfaces and its use in the interfacial structural characterisation. Since the pioneering works of the groups of K. B. Eisenthal and Y. R. Shen (see their excellent reviews in Eisenthal (2006), Shen and Ostroverkhov (2006), and Tian and Shen (2014)), non-linear vibrational spectroscopy at interfaces has become an essential experimental tool for structural characterisation at interfaces. Beyond the knowledge of spectroscopic data, surface harmonic generation (SHG) experiments are also used to obtain interfacial titration curves. From there the protonation state of interfaces can be inferred at different pH values, which is of high importance, for instance, for chemical reactivity at interfaces, surface dissolution, etc. The groups of K. B. Eisenthal, F. M. Geiger and J. Gibbs-Davis are amongst the most representative of SHG experiments at solid–liquid interfaces. See, for instance, review articles and/or recent publications in Azam et al. (2014), Eisenthal (2006) and Geiger (2009). SFG is now part of several laboratory tools for the structural characterisation of solid–liquid and liquid–air interfaces. See typical reviews and publications from the groups of Y. R. Shen (Shen and Ostroverkhov 2006; Tian and Shen 2014), E. Borguet (Dewand et al. 2013), M. Bonn (Arnolds and Bonn 2010; Zhang et al. 2011), H. C. Allen (Allen et al. 2009; Gopalakrishnan et al. 2006; Jubb et al. 2012) and G. Richmond (Hopkins et al. 2005). The initial 'static' SFG spectroscopy has been recently extended to provide 'phase-sensitive SFG' spectroscopy (Shen 2013), thus directly probing the orientation of molecules at the interface. Two other very recent extensions of the SFG method concern time dependence with time-resolved SFG (Eftekhari-Bafrooei and Borguet 2010; Zhang et al. 2011) and multi-dimensions with 2D-SFG (Zhang et al. 2011), where direct couplings between modes and molecules can be directly investigated at interfaces (Lis et al. 2014).

AIMD simulations are an excellent and reliable theoretical tool for anharmonic vibrational spectroscopy calculations at finite temperature. See, for instance, reviews (Gaigeot 2010; Thomas et al. 2013) where DFT-MD has been shown to provide excellent agreements with gas phase and liquid-phase spectroscopy experiments. Beyond the excellent agreements between theoretical dynamical spectra and experimental spectra, theoretical spectra can be interpreted at the microscopic level in great detail, providing a precise assignment of vibrational bands in terms of molecular movements (mainly stretch and bending for water molecules in the frequency domain of interest to SFG spectroscopy) and a precise assignment in terms of molecules that give rise to the signal. Vibrational anharmonicities are naturally taken into account in theoretical dynamical spectroscopy, making this method an essential partner of the experiments. It is also important to note that classical force field-based MD simulations are essentially unable to provide theoretical vibrational spectra that can be of relevance. Indeed, force fields are essentially harmonic in essence, thus not providing the anharmonic aspects necessary to SFG spectroscopy, and the development of classical force fields for spectroscopy is an ample work that not many groups are willing to undertake. Morita's developments of such force fields have to be emphasised in the context of the water–air interface (see, for instance, Morita and Ishiyama (2008)). Transferability of such force fields to a different condensed domain (solid oxide–liquid interfaces in our case) and/or to different environments (adding, for instance, ions at the interface) is however very much questionable. AIMD simulations do not suffer such limitations, to the extent that size and timescales are limited to much smaller values than classical MD would allow.

Our chapter is organised with the following topics. We review AIMD simulations in the framework of DFT and its implementations in terms of Car–Parrinello and Born–Oppenheimer (BO) approaches. These two parts are related to the implementations in the two packages CPMD (Car and Parrinello 1985; The CPMD Consortium 2009) and CP2K (CP2K Developers Group n.d.; VandeVondele et al. 2005). This is presented in Section 8.2. As mentioned earlier, our theoretical investigations to mineral–water interfaces have in particular focused on the characterisation of surface site acidities and on vibrational spectroscopy at these interfaces. Theories behind these two issues are respectively presented in Sections 8.3, 8.4 and 8.5 in the context of AIMD simulations.

We then take examples from our recent works on silica–water interfaces (Section 8.6.1), alumina–water interfaces (Section 8.6.2) and clay–water interfaces (Section 8.6.5). Beyond the structural organisation of the surface–water interface, these works attempt to provide relationships between structure, acidity constants and vibrational spectroscopy.

8.2   Brief Review of AIMD Simulation Method

8.2.1   *Ab Initio* Molecular Dynamics and Density Functional Theory

In this section, we present a short review of the AIMD simulations techniques, which are currently used in material science and which have also applications for the description of mineral–water interfaces.

In AIMD the interatomic forces $F_I = -\nabla_{R_I}\Phi(R)$ are determined on the fly using first-principles electronic structure methods. This means that AIMD is not relying on any adjustable parameter, but only on the atomic coordinates of the model system R. This makes AIMD particularly suitable to address heterogeneous environments, such as interfaces where the transferability of force field parameters, which have usually been developed for the bulk phase, can be an issue. The drawback is the high cost required to solve the electronic structure problem, namely, to find the antisymmetric ground-state eigenfunctions $|\psi_0\rangle$ of the corresponding many-body Hamiltonian at each MD step.

The first important step in the AIMD is the BO approximation (Born and Oppenheimer 1927) (see Chapter 1), which allows for a *product ansatz* of the total wavefunction consisting of the nuclear and of the electronic wavefunctions. Due to the large separation of the nuclear and electronic masses, the electrons can be expected to be in instantaneous equilibrium with the much heavier nuclei, so that the electronic subsystem can be treated independently at constant nuclear coordinates R. Hence the Hamiltonian $\mathcal{H}_e(\{r_i\}; R)$ depends only parametrically on the nuclear coordinates.

Applying the BO approximation, the potential energy function for the system $\Phi(R)$ can be written as

$$\Phi(R) = \langle\psi_0|\mathcal{H}_e(\{r_i\}; R)|\psi_0\rangle + E_{II}(R), \tag{8.1}$$

where $\mathcal{H}_e(\{r_i\}; R)$ is the electronic many-body Hamiltonian that depends on the electronic coordinates $\{r_i\}$ and parametrically on the nuclear degrees of freedom R.

Nevertheless, we still need to solve the electronic, non-relativistic, time-independent, many-body Schrödinger equation:

$$\mathcal{H}_e(\{r_i\}; R)|\psi_0(\{r_i\})\rangle = \varepsilon_0(R)|\psi_0(\{r_i\})\rangle \tag{8.2}$$

which is a high-dimensional eigenvalue problem, with eigenfunctions $\psi_0(\{r_i\})$ and eigenvalues $\varepsilon_0(R)$, respectively.

One ingenious solution to the solution of the electronic structure problem is represented by DFT (Hohenberg and Kohn 1964; Kohn and Sham 1965). DFT has nowadays become the most used approach for the description of the electronic structure properties of materials thanks to a favourable compromise between computational cost and accuracy. DFT has its foundation in the two famous Hohenberg–Kohn (HK) theorems (see Chapter 1). The first HK theorem proves the existence of a one-to-one correspondence between the ground-state density $\rho_0(r)$ and an external potential $v(r)$.

The electronic density $\rho(r) = \int r_2 \cdots \int r_{N_e} |\psi(\{r_i\})|^2$, which depends on just three electronic degrees of freedom, becomes the central quantity in DFT in place of the more complex $3N_e$-dimensional many-body wavefunction.

The second HK theorem introduces a variational principle for the total energy in the electronic density space:

$$E^{\text{DFT}}[\rho_0] = \langle \psi_0 | \mathcal{H}_e | \psi_0 \rangle \leq \langle \psi' | \mathcal{H}_e | \psi' \rangle = E^{\text{DFT}}[\rho'], \tag{8.3}$$

for which equality holds if and only if $\rho_0 = \rho'$. Thanks to the second HK theorem, Equation 8.2 can not only be solved by iteratively diagonalising $\mathcal{H}_e[\rho]$ within a self-consistent field (SCF) procedure but also by minimising the quantum expectation value of $\mathcal{H}_e[\rho]$, that is,

$$E^{\text{DFT}}[\rho_0] = \min_{\psi} \langle \psi | \mathcal{H}_e | \psi \rangle = \min_{\rho} \langle \psi[\rho] \mathcal{H}_e[\rho] | \psi[\rho] \rangle$$

$$= \min_{\rho} E^{\text{DFT}}[\rho]. \tag{8.4}$$

In the following, we will assume atomic units and consider the following Hamiltonian of a system composed of N_e electrons and N nuclei (see also Chapter 1, for a similar introduction to DFT):

$$\mathcal{H}_e = \frac{1}{2} \sum_{i=1}^{N_e} \nabla_i^2 + \sum_{i<j}^{N_e} \frac{1}{|r_i - r_j|} + \sum_{I,i}^{N,N_e} \frac{Z_I}{|R_I - r_i|} \tag{8.5a}$$

$$= \hat{T} + \hat{U} + \hat{V}, \tag{8.5b}$$

where Z_I is the proton number, $\hat{T}$ the kinetic energy operator of the electrons, $\hat{U}$ is the electron–electron interaction and $\hat{V} = \sum_i v(r_i)$ the electron–ion operator. It is quite appealing that in DFT we can get the ground-state energy of a many-electron system as minimum of the energy functional:

$$E^{\text{DFT}}[\rho(r)] = T[\rho(r)] + U[\rho(r)] + V[\rho(r)]. \tag{8.6}$$

The problem is now to provide an explicit form for the three terms appearing in Equation 8.6. The solution is provided by the Kohn–Sham approach to DFT. Kohn and Sham introduce a non-interacting reference system for which the electron density is the same as the density of the full interacting system.

The kinetic energy functional for such reference system is

$$T_s[\rho(r)] = -\frac{1}{2} \sum_{i=1}^{N_e} \int dr \psi_i^*(r) \nabla^2 \psi_i(r)$$

$$= T_s[\{\psi_i[\rho(r)]\}]. \tag{8.7}$$

and the density can be written in terms of $\psi_i(r)$ as

$$\rho(r) = \sum_{i=1}^{N_{\text{occ}}} f_i \psi_i(r) \psi_i^*(r), \tag{8.8}$$

where N_{occ} is the number of occupied orbitals and f_i the occupation number of state i, so that

$$\sum_{i=1}^{N_{\text{occ}}} f_i = N_e. \tag{8.9}$$

The KS energy functional is then simply given by

$$E^{KS}[\rho(r)] = E^{KS}[\{\psi_i[\rho(r)]\}] = T_S[\{\psi_i[\rho(r)]\}]$$
$$+ U_H[\rho(r)] + V[\rho(r)] + E_{XC}[\rho(r)] \tag{8.10a}$$

$$= -\frac{1}{2}\sum_{i=1}^{N} f_i \int dr \psi_i^*(r)\nabla^2\psi_i(r)$$

$$+ \frac{1}{2}\int dr \int dr' \frac{\rho(r)\rho(r')}{|r-r'|} \tag{8.10b}$$

$$+ \int dr v_{ext}(r)\rho(r) + E_{XC}[\rho(r)],$$

where $E_{XC}[\rho(r)] = (T[\rho(r)] - T_s[\{\psi_i[\rho(r)]\}]) + (U[\rho(r)] - U_H[\rho(r)])$ is the exchange and correlation (XC) energy functional and $v_{ext}(r) = \delta V[\rho(r)]/\delta\rho(r)$ is the external potential.

The XC energy functional is the unknown part of the energy functional. Its definition shows that a significant part of $E_{XC}[\rho(r)]$ is due to correlation effects of the kinetic energy.

Using the variational principle, it is then possible to derive from Equation 8.10a the corresponding Euler–Lagrange equation of the non-interacting system within the potential v. The KS scheme permits to map the full interacting many-body problem with the electron–electron interaction $\hat{U}$, onto an equivalent fictitious single-body problem, with an effective potential operator $\hat{V}_{KS} = \hat{U}_S + \hat{V}_H + \hat{V}_{XC}$. The simplest expression for the $E_{XC}[n(\mathbf{r})]$ is provided by the local density approximation (LDA). In LDA the value of $E_{XC}[n(\mathbf{r})]$ is approximated by the exchange–correlation energy of an electron in a homogeneous electron gas of the same density $n(\mathbf{r})$, that is,

$$E_{XC}^{LDA}[n(\mathbf{r})] = \int \epsilon_{XC}(n(\mathbf{r}))n(\mathbf{r})d\mathbf{r}. \tag{8.11}$$

The most accurate data for $\epsilon_{XC}(n(\mathbf{r}))$ come from quantum Monte Carlo calculations (Ceperley and Alder 1980). The LDA is often surprisingly accurate and for systems with slowly varying charge densities, it generally gives very good results. The failures of the LDA representation are now well established: it has a tendency to favour more homogeneous systems and over-binds molecules and solids. In weakly bonded systems, these errors are exaggerated and bond lengths are too short. In good systems where the LDA works well, often those mostly consisting of sp bonds, geometries are good and bond lengths and angles are accurate to within a few percent. Quantities such as the dielectric and piezoelectric constant are approximately 10% too large.

Despite the remarkable success of the LDA, its limitations mean that care must be taken in its application. For example, in strongly correlated systems where an independent particle picture breaks down, the LDA is very inaccurate. An example is given by the transition metal oxides XO (X = Fe, Mn, Ni), which are all Mott insulators, but the LDA predicts that they are either semi-conductors or metals.

The success of the LDA has been shown to result from a real-space cancellation of errors in the LDA exchange and correlation energies. The LDA does not account for van der Waals bonding and gives a very poor description of hydrogen bonding. These phenomena are essential for a correct description of bulk water and interfaces with water.

An obvious approach to improving the LDA is to include gradient corrections, by making E_{XC} a functional of the density and its gradient:

$$E_{XC}^{GGA}[n(\mathbf{r})] = \int \in_{XC}(n(\mathbf{r}))n(\mathbf{r}) + \int F_{XC}[n(\mathbf{r}), |\nabla n(\mathbf{r})|]d\mathbf{r} \qquad (8.12)$$

where F_{XC} is a correction chosen to satisfy one or several known limits for E_{XC}. For solids, the most commonly used GGA functional is the one proposed by Perdew, Burke and Ernzherhof (PBE) (1936). Another popular GGA functional is the BLYP functional (Becke 1988; Lee et al. 1988).

However, it has been shown that dispersion forces are poorly described by GGA functionals. A series of empirical corrections have been proposed, which can improve the structural properties without raising the computational costs and which have become now a routine choice. In particular amongst such empirical approaches, we would like to mention the Grimme corrections (Grimme et al. 2010), which we have used in our applications to mineral surfaces. Other approaches include the dispersion-corrected atom-centred potentials (DCACPs) (von Lilienfeld et al. 2005). Corrections based on the RPA method are developed by the group of Furche (Burow et al. 2014) and many-body van der Waals interactions by Tkatchenko et al. (see, e.g. Santra et al. 2013; Tkatchenko et al. 2012).

We finally turn to the description of the MD approach. In computational material science, the two most popular AIMD approaches are the Born–Oppenheimer MD (BOMD) and Car–Parrinello MD (CPMD) approaches.

In BOMD the potential energy $E[\{\psi_i\}; R]$ is minimised at every MD step with respect to $\{\psi_i(r)\}$ under the holonomic orthonormality constraint $\langle \psi_i(r)|\psi_j(r)\rangle = \delta_{ij}$. This leads to the following Lagrangian:

$$\mathcal{L}_{BO}(\{\psi_i\}; R, \dot{R}) = \frac{1}{2}\sum_{I=1}^{N} M_I \dot{R}_I^2 - \min_{\{\psi_i\}} E[\{\psi_i\}; R]$$
$$+ \sum_{i,j} \Lambda_{ij}(\langle \psi_i|\psi_j\rangle - \delta_{ij}), \qquad (8.13)$$

where Λ is a Hermitian Lagrangian multiplier matrix. By solving the corresponding Euler–Lagrange equations, one obtains the associated equations of motion (EOM):

$$M_I \ddot{R}_I = -\nabla_{R_I}\left[\min_{\{\psi_i\}} E[\{\psi_i\}; R]\Big|_{\{\langle\psi_i|\psi_j\rangle = \delta_{ij}\}}\right]$$

$$= -\frac{\partial E}{\partial R_I} + \sum_{i,j} \Lambda_{ij}\frac{\partial}{\partial R_I}\langle\psi_i|\psi_j\rangle \qquad (8.14a)$$

$$-2\sum_i \frac{\partial\langle\psi_i|}{\partial R_I}\left[\frac{\delta E}{\delta\langle\psi_i|} - \sum_j \Lambda_{ij}|\psi_j\rangle\right]$$

$$0 \lesssim -\frac{\delta E}{\delta\langle\psi_i|} + \sum_j \Lambda_{ij}|\psi_j\rangle$$

$$= -\hat{H}_e\langle\psi_i| + \sum_j \Lambda_{ij}|\psi_j\rangle \qquad (8.14b)$$

Here the first term on the right-hand side of Equation 8.14a is the so-called Hellmann–Feynman force (Feynman 1939; Hellmann 1937). The second term is denoted 'Pulay' (1969) or wavefunction force F_{WF}, is a constraint force due to the holonomic orthonormality constraint and is non-vanishing if and only if the basis set functions ϕ_j explicitly depend on $\boldsymbol{R}$. The last term stems from the fact that, independently of the particular choice of the basis set, there is always an implicit dependence on the atomic positions through the expansion coefficient $c_{ij}(\boldsymbol{R})$ within the common linear combination of atomic orbitals ϕ_j:

$$\psi_i(\boldsymbol{R}) = \sum_j c_{ij}(\boldsymbol{R})\phi_j \tag{8.15}$$

The inequality in Equation 8.14b stems from numerical approximations, meaning that in practice the equality is rarely achieved.

In CPMD (Car and Parrinello 1985), a coupled electron–ion dynamics is instead performed. Additional electronic degrees of freedom are added to the Lagrangian as a classical term:

$$\mathcal{L}_{CP}(\{\psi_i\}; \boldsymbol{R}, \dot{\boldsymbol{R}}) = \frac{1}{2}\mu \sum_i \langle \dot{\psi}_i | \dot{\psi}_i \rangle + \frac{1}{2}\sum_{I=1}^{N} M_I \dot{\boldsymbol{R}}_I^2$$
$$- E[\{\psi_i\}; \boldsymbol{R}] \tag{8.16}$$
$$+ \sum_{i,j} \Lambda_{ij}(\langle \psi_i | \psi_j \rangle - \delta_{ij}),$$

where the electronic degrees of freedom carry a fictitious mass parameter μ and are also characterised by orbital velocities $\{\dot{\psi}_i\}$. As in the case of the BO dynamics, applying the Euler–Lagrange equations leads to the following EOM:

$$M_I \ddot{\boldsymbol{R}}_I = -\nabla_{\boldsymbol{R}_I}\left[E[\{\psi_i\};\boldsymbol{R}]\Big|_{\{\langle\psi_i|\psi_j\rangle = \delta_{ij}\}}\right]$$
$$= -\frac{\partial E}{\partial \boldsymbol{R}_I} + \sum_{i,j}\Lambda_{ij}\frac{\partial}{\partial \boldsymbol{R}_I}\langle\psi_i|\psi_j\rangle \tag{8.17a}$$

$$\mu\ddot{\psi}_i(r,t) = -\frac{\delta E}{\delta\langle\psi_i|} + \sum_j \Lambda_{ij}|\psi_j\rangle$$
$$= -\hat{H}_e\langle\psi_i| + \sum_j \Lambda_{ij}|\psi_j\rangle, \tag{8.17b}$$

where $-\delta E/\delta\langle\psi_i|$ are the electronic forces to propagate the electronic degrees of freedom in time within a fictitious Newtonian dynamics. At variance to BOMD, no SCF cycle is required to quench the electrons to the BO surface and to force them to adiabatically follow the nuclei.

In CPMD in order to ensure the adiabatic energy-scale separation of the nuclear and the electronic degrees of freedom and to prevent energy transfer between them, the highest ionic phonon frequency ω_I has to be much smaller than its lowest electronic analogue ω_e. This is ensured by a proper choice of the fictitious mass.

A question on which method, either BOMD or CPMD, is favoured turns out to be rather subtle (Tangney 2006) and depends largely on the definition of accuracy, as well as on the particular application.

In the applications presented in this chapter, we always perform BOMD as currently implemented within the CP2K package (CP2K Developers Group).

Mixed plane waves and gaussian basis sets are used in CP2K. Only the valence electrons are taken into account and pseudo-potentials of the Goedecker–Teter–Hutter (GTH) form are used (Goedecker et al. 1996; Hartwigsen et al. 1998). We use the Becke, Lee, Yang and Parr (BLYP) gradient-corrected functional (Becke 1988; Lee et al. 1988) for the exchange and correlation terms. Dispersion interactions have been included with the Grimme D2 and D3 corrections (Grimme et al. 2010). Calculations are restricted to the Γ point of the Brillouin zone. We employ plane-wave basis sets with a kinetic energy cutoff usually around 340 Ry and gaussian basis sets of double-ζ (DZVP) to triple-ζ (TZVP) qualities from the CP2K library. Periodic boundary conditions are applied on non-cubic cells. Our dynamics are strictly microcanonical (NVE ensemble), once thermalisation has been achieved (through NVE and velocity rescaling periods of time or through NVT dynamics). Once thermalisation has been achieved, trajectories are accumulated for typical length times of 10–20 ps for analyses in terms of interfacial structures, dynamics of water, acidity constant calculations and vibrational spectroscopy.

At this point, one should be aware of possible limitations of the DFT-MD methodology. One is the time length of the dynamics of a few tens of picoseconds, which is enough for the properties of interfacial structures, acidity constants and vibrational spectroscopy, but might not be long enough for diffusion of water molecules between different layers, typically diffusion from inner-sphere adsorption to the bulk liquid or vice versa. Hydrogen bonds between surface and water, and between water molecules, are well represented at the BLYP(+D2) level of representation mentioned earlier, employing small/intermediate basis sets such as DZVP and TZVP. See Gaigeot and Sprik (2003), Gaigeot et al. (2012) and Sulpizi et al. (2012) for typical comparisons of the bulk liquid water structure obtained with these levels of calculations to literature. One has also to keep in mind that GGA functionals such as BLYP are well-known for underestimating energy barriers for proton transfers.

8.3   Calculation of the Surface Acidity from Reversible Proton Insertion/Deletion

The acidity constants of surface groups are computed using the reversible proton insertion/deletion method. The method was initially developed and tested on a series of aqueous compounds (Cheng and Sprik 2010; Cheng et al. 2009; Costanzo et al. 2011; Mangold et al. 2011; Sulpizi and Sprik 2008, 2010) and then applied to the calculation of the acidity of surfaces groups of several oxides in contact with water (Gaigeot et al. 2012; Sulpizi et al. 2012). The advantage of this approach is that the solid (mineral) surface and the solvent (water) are treated at the same level of theory and therefore the approach is particularly suitable for heterogeneous environments such interfaces. The acidity constants are calculated starting from the free energy of transferring a proton from a group on the surface to a water molecule in the bulk solution. The following reaction is considered:

$$-MOH + H_2O(aq) \rightarrow -MO^- + H_3O^+(aq) \tag{8.18}$$

where M stands for a generic metal atom, for example, in the case of metal oxides. The transfer process is implemented using the thermodynamic integration. The charge of the acidic proton of a surface OH group is gradually switched off transforming the proton into a neutral 'dummy' particle. Simultaneously a similar dummy proton attached to a water molecule is charged up creating an hydronium in the eigenform. The fractional charges on the two groups always add up to unity. Following the approach that we have discussed in detail in Costanzo et al. (2011), the discharge integral is calculated according to

$$\Delta_{dp}A = \int_0^1 d\eta \langle \Delta_{dp}E \rangle_{r\eta} \tag{8.19}$$

where $\Delta_{dp}E$ is the vertical energy gap, defined as the potential energy difference between product P $\left(MO^- + H_3O^+(aq)\right)$ in Equation 8.18 and reactant R $\left(-MOH + H_2O(aq)\right)$ for instantaneous configurations of an MD trajectory. The subscript rη indicates that the averages are evaluated over the restrained mapping Hamiltonian

$$\mathcal{H}_\eta = (1-\eta)\mathcal{H}_R + \eta\mathcal{H}_P, \tag{8.20}$$

where η is a coupling parameter that is gradually increased from $0\left(-MOH + H_2O(aq)\right)$ to $1\left(-MO^- + H_3O^+(aq)\right)$. This Hamiltonian also contains an harmonic restraining potential V_{restr} keeping the dummy atom close to the equilibrium position of the H^+ nucleus in the protonated system. The Simpson rule (three-point approximation) is used to evaluate the integral in Equation 8.19:

$$\Delta A_{TP} = \frac{1}{6}\left(\langle \Delta E \rangle_0 + \langle \Delta E \rangle_1\right) + \frac{2}{3}\langle \Delta E \rangle_{0.5} \tag{8.21}$$

This requires the generation of three trajectories corresponding to values of $\eta = 0, 0.5, 1$. This formula is often a good compromise between computational cost and accuracy of the free energy change (Hummer and Szabo 1996; Sulpizi and Sprik 2008). However, if the curvature of $\langle \Delta E \rangle_\eta$ is large, more integration points may be required.

The pK_a value of a surface group is obtained from the proton transfer integral of Equation 8.19 by adding a thermochemical correction. The leading term in this correction adds in the translational entropy generated by the acid dissociation. This term is missing in reactions 8.18, which is formally a proton transfer reaction conserving the number of translational degrees of freedom. Indeed this translational entropy term is related to the definition of our solvated proton, which is assumed to be in the eigenform, namely, as $H_3O^+(aq)$, where instead the proton is actually free to diffuse around and should be instead represented as $H^+(aq)$. The simplification is motivated by the necessity to have a well-defined position for the proton insertion. A detailed explanation about this issue can be found in Costanzo et al. (2011). Further corrections are needed for the difference in zero-point motion of a proton in an acidic group and in H_3O^+ and a possible mismatch between the frequencies of dummy proton and real protons. These differences can be neglected at zero-order approximation (Costanzo et al. 2011). Assuming the low-temperature limit for vibrational partition functions, the overall formula used for the pK_a calculation becomes

$$2.30k_B T pK_a = \int_0^1 d\eta \langle \Delta_{dp}E \rangle_{r\eta} + k_B T \ln\left[c^\circ \Lambda_{H^+}^3\right] \tag{8.22}$$

where $c^\circ = 1\,mol\,dm^{-3}$ is the unit molar concentration and Λ_{H+} is the thermal wavelength of the proton. The logarithm of the product $c^\circ\Lambda_{H+}^3$ accounts for the liberation entropy of the proton and is responsible for a correction of -3.2 pK units to the thermodynamic integral.

Finite system size effects are of course a point of concern in these calculations (Costanzo et al. 2011). We have checked this issue in the case of quartz where the calculation of the silanol pK_a has been carried out in two different cells, increasing the water separation between two slabs *i* in period boundary conditions. In the case of quartz, we did not see major differences between the results in the two different cell sizes, neither for the water structure at the interface and neither for the values of the calculated acidity constants.

8.4 Theoretical Methodology for Vibrational Spectroscopy and Mode Assignments

Within statistical mechanics and linear response theory (see, for instance, Kubo et al. (1985) and McQuarrie (1976); see also Chapter 10), an infrared (IR) spectrum is calculated as the Fourier transform of the time correlation function of the fluctuating dipole moment vector of the absorbing molecular system as

$$I(\omega) = \frac{2\pi\beta\omega^2}{3cV} \int_{-\infty}^{\infty} dt \langle \delta\mathbf{M}(t) \cdot \delta\mathbf{M}(0) \rangle \exp(i\omega t) \tag{8.23}$$

where $\beta = 1/kT$, T is the temperature, c is the speed of light in vacuum and V is the volume. The angular brackets represent a statistical average of the correlation function, where $\delta\mathbf{M}(t) = \mathbf{M}(t) - <\mathbf{M}>$ with $<\mathbf{M}>$ the time average of $\mathbf{M}(t)$. The calculation in Equation 8.23 is done in the absence of an applied external field. For the prefactor in Equation 8.23, we have taken here into account an empirical quantum correction factor (multiplying the classical line shape) of the form $\beta\hbar\omega/(1 - \exp(-\beta\hbar\omega))$, which was shown by us and others to give accurate results on calculated IR intensities (Ahlborn et al. 2000; Gaigeot and Sprik 2003; Iftimie and Tuckerman 2005). For more detailed discussions on quantum corrections, see, for instance, Borysow et al. (1985), Kim and Rossky (2006), Lawrence and Skinner (2005) and Ramirez et al. (2004).

This is the standard way used in statistical mechanics for calculating the IR spectrum of molecular assemblies, that is, isolated molecules, liquids, solutes in the liquid phase and solids, taking into account vibrational anharmonicities and temperature. MD simulations are adapted to the calculation of the evolution in time of the dipole moment of the system and therefore its time correlation. The main advantages of the MD approach in Equation 8.23 for the calculation of IR spectra (also called dynamical spectra) can be listed as follows:

- There are no approximations made in Equation 8.23 apart from the hypothesis of linear response theory, that is, a small perturbation from the applied external electric field on the absorbing molecular system. Such condition is always fulfilled in vibrational spectroscopy of interest here.
- There are no harmonic approximations made, be they on the potential energy surface or on the dipole moment, contrary to static calculations performed in the gas phase community or to instantaneous normal mode (INM) analysis performed in the condensed phase community.
- The quality of the potential energy surface is entirely contained in the 'ab initio' force field used in the dynamics, calculated at the DFT/BLYP (+dispersion when needed) level in the works presented in this chapter (see details in Section 8.2). The good to excellent agreements of the absolute (and relative) positions of the different active bands obtained in our theoretical works (see, for instance, dynamical spectra in the gas phase (Beck et al. 2013; Cimas et al. 2009; Marinica et al. 2006; Sediki et al. 2011), in the liquid phase (Bovi et al. 2011; Gaigeot and Sprik 2003) and at solid–liquid and liquid–air interfaces, including the SFG spectrum (Gaigeot et al. 2012; Sulpizi et al. 2012, 2013)) are a demonstration (though *a posteriori*) that this level of theory is correct.
- Equation 8.23 gives the whole IR spectrum of a molecular system in *one single calculation*, that is, the band positions, the band intensities and the band shapes. There are no approximations applied, in particular the shape and broadening of the vibrational bands result from the underlying dynamics and mode couplings in the system at a given temperature.

As already emphasised, all investigations presented in the next sections have employed Born–Oppenheimer MD (BOMD). We apply no scaling factor of any kind to the vibrations extracted from the dynamics. The sampling of vibrational anharmonicities, that is, potential energy surface, dipole anharmonicities, mode couplings and anharmonic modes, being included in our simulations, *by construction*, application of a scaling factor to the band positions is therefore not required.

An accurate calculation of anharmonic dynamical IR spectra is one goal to achieve, and the assignment of the active bands into individual atomic displacements or vibrational modes is another one. This issue is essential to the understanding of the underlying molecular structural and dynamical properties. In MD simulations, interpretation of the vibrational bands into individual atomic displacements is traditionally and easily done using the vibrational density of states (VDOS) formalism. The VDOS is obtained through the Fourier transform of the atomic velocity autocorrelation function:

$$\text{VDOS}(\omega) = \sum_{i=1,N} \int_{-\infty}^{\infty} \langle \mathbf{v}_i(t) \cdot \mathbf{v}_i(0) \rangle \exp(i\omega t)dt \qquad (8.24)$$

where i runs over all atoms of the investigated system. $\mathbf{v}_i(t)$ is the velocity vector of atom i at time t. As in Equation 8.23, the angular brackets in Equation 8.24 represent a statistical average of the correlation function. The VDOS spectrum provides all vibrational modes of the molecular system. However, only some of these modes will be IR or Raman active, so VDOS spectra can by no means directly substitute for IR or Raman spectra.

The VDOS can further be decomposed according to atom types, or to groups of atoms, or to chemical groups of interest, in order to get a detailed assignment of the vibrational bands in terms of individual atomic motions. This is done by restraining the sum over i in Equation 8.24 to the atoms of interest only. Such individual signatures are straightforward to interpret in terms of movements for localised vibrational modes that involve only a few atomic groups, typically in the 3000–$4000\,\text{cm}^{-1}$ domain of bond stretchings, but the interpretation of the VDOS becomes more complicated when delocalised vibrations take place through several groups of atoms or when there are couplings between different atomic movements. Fourier transforms of intramolecular coordinate (IC) time correlation functions can also be used: $\int_{-\infty}^{\infty} \langle \text{IC}(t) \cdot \text{IC}(0) \rangle \exp(i\omega t)dt$. Such approach requires to have *a priori* knowledge of the relevant IC(s), as one cannot easily analyse all possible combinations.

More relevant theoretical methods have been developed in order to extract vibrational modes from the dynamics, especially bypassing the limitations on mode delocalisation in VDOS assignments; see, for instance, Gaigeot et al. (2007), Martinez et al. (2006), Mathias et al. (2012) and Nonella et al. (2003). Such methods usually provide 'effective normal modes', similar to the well-known normal modes obtained by a Hessian diagonalisation in static harmonic calculations, but maintain a certain degree of mode couplings and temperature from the dynamics in the final modes. We have no space here to describe these techniques.

Of special interest to solid–liquid interfaces is SFG vibrational spectroscopy, experimentally pioneered by Shen and co-workers (Chen et al. 1981; Shen and Ostroverkhov 2006; Zhu et al. 1987) and now applied by several groups worldwide as reviewed in the introduction of this chapter. Recent theory of the SFG signal by Morita (2004) and Shen (2012) has shown that the resonant electric dipole susceptibility ($\chi^{(2)}$) is the origin of the surface SFG signal, although higher-order quadrupolar terms arising from the bulk should also participate. However, no MD-derived SFG spectrum has yet been obtained including these quadrupolar terms. Our modelling of SFG from DFT-MD follows the methods introduced by Morita et al. (Ishiyama et al. 2012; Morita and Hynes 2002; Morita and Ishiyama 2008; Nihonyanagi et al. 2011).

The classical expression for resonant $\chi^{(2)}$ is

$$\chi_{\text{pqr}}^{(2)}(\omega) = \frac{i\omega}{k_{\text{B}}T} \int_0^\infty dt \exp(i\omega t) \langle A_{\text{pq}}(t) M_{\text{r}}(0) \rangle \tag{8.25}$$

where A_{pq} and M_{r} are, respectively, the components of the polarisability tensor and dipole moment of the whole modelled system, (p, q, r) are any direction amongst (x, y, z), k_{B} and T are the Boltzmann constant and temperature, respectively, and $\langle \ldots \rangle$ denotes an average over the trajectory.

See Sulpizi et al. (2013) for details of calculations of this signal (square and phase resolved) from DFT-MD simulations.

8.5   Property Calculations from AIMD: Dipoles and Polarisabilities

The knowledge of the evolution with time of the molecular dipole moments is mandatory for the calculation of IR spectra with MD simulations. In the modern theory of polarisation, the dipole moment of the (periodic) box cell is calculated with the Berry phase representation, as implemented in the CP2K package (Bernasconi et al. 1998). Briefly, in the limit where the Γ point approximation applies, the electronic contribution to the cell dipole moment $\mathbf{M}_\alpha^{\text{el}}$ (where $\alpha = x, y, z$) is given by Resta (1998):

$$\mathbf{M}_\alpha^{\text{el}} = \frac{e}{|\mathbf{G}_\alpha|} \Im \ln z_N \tag{8.26}$$

where $\Im \ln z_N$ is the imaginary part of the logarithm of the dimensionless complex number $z_N = \left\langle \Psi \left| e^{-i\mathbf{G}_\alpha \cdot \hat{\mathbf{R}}} \right| \Psi \right\rangle$ and $\mathbf{G}_\alpha$ is a reciprocal lattice vector of the supercell ($\mathbf{G}_1 = 2\pi/L_1(1,0,0)$, $\mathbf{G}_2 = 2\pi/L_2(0,1,0)$, $\mathbf{G}_3 = 2\pi/L_3(0,0,1)$), and $\hat{\mathbf{R}} = \sum_{i=1}^N \hat{\mathbf{r}}_i$ denote the collective position operators of the N electrons (or in other words the centre of the electronic charge distribution). ψ is the ground-state wave function. The quantity $\Im \ln z_N$ is the Berry phase, which in terms of a set of occupied Kohn–Sham orbitals $\psi_k(\mathbf{r})$ is computed as $\Im \ln z_N = 2\Im \ln \det \mathbf{S}$ with elements of the matrix $\mathbf{S}$ given by $S_{kl} = \langle \psi_k | e^{-i\mathbf{G}_\alpha \cdot \hat{\mathbf{r}}} | \psi_l \rangle$ Resta (1998).

When calculating IR absorption in the gas phase, $\mathbf{M}(t)$ in Equation 8.23 is simply the dipole moment of one molecule or one cluster. This dipole is well identified. In case of condensed phases, that is, liquids, solids and their interfaces, $\mathbf{M}(t)$ is the total dipole moment of the supercell modelling the extended system. The Fourier transform of this dipole cell vector will most of the time not be informative, as it provides the signatures of the whole system, either dominated by the solvent response or by the solid response. However, what we are usually interested in are the detailed signatures from a thin layer of water molecules, typically around an immersed solute or at an interface with a solid as of interest here.

A method based on maximally localised Wannier functions has been devised in order to extract these molecular information, and details can be found in Marzari and Vanderbilt (1997), Silvestrelli (1999), Silvestrelli and Parrinello (1999a, b) and Silvestrelli et al. (1998), with practical details in Gaigeot and Sprik (2003). A detailed description of this method is out of the scope of this chapter, and we present only the main ingredients.

Wannier functions $\omega(\vec{r})$ are formally related to the Kohn–Sham orbitals $\Psi_k(\vec{r})$ by a unitary transformation:

$$\omega_n(\vec{r}) = \sum_{k=1}^{N/2} U_{nk} \Psi_k(\vec{r}) \tag{8.27}$$

and are therefore orthogonal (the definition earlier takes into account a system of N perfectly paired electrons described by $N/2$ spin-restricted one-electron orbitals). Localisation can then be optimised by finding the unitary matrix U in Equation 8.27 minimising the total spread:

$$\Omega = \sum_{n=1}^{N/2} \left(\langle \omega_n | r^2 | \omega_n \rangle - \langle \omega_n | \vec{r} | \omega_n \rangle^2 \right) \tag{8.28}$$

This defines a set of maximally localised Wannier states. The expectation value of position $\vec{r}_n = \langle \omega_n | \vec{r} | \omega_n \rangle$ in Equation 8.28 can be interpreted as the location of the centre of the Wannier function ω_n and is often referred to as a Wannier function centre (WFC). The first calculation of this kind was carried out by Silvestrelli and co-workers (1998).

The electronic contribution to the polarisation can be directly computed from the position of the WFC:

$$\vec{M}^{\text{el}} = 2e \sum_{n=1}^{N/2} \vec{r}_n \tag{8.29}$$

$\vec{r}_n = (x_n, y_n, z_n)$, each component of $\vec{r}_n$ being

$$x_n = -\frac{L}{2\pi} \Im \left\langle \omega_n | e^{-i2\pi x/L} | \omega_n \right\rangle \tag{8.30}$$

As seen in the previous section, the theoretical signal for SFG requires the supplementary knowledge of individual molecular polarisability tensors $\alpha_{pq}^i(t)$ where i is the label for the molecule within the system and (p, q) are the choice of directions amongst (x, y, z). We also turn to the localised Wannier functions and WFC in order to extract these tensors. To that end, we resort to the calculation of the induced dipole moment of each molecule of the system, calculated through the application of an external field. Each induced dipole moment requires the knowledge of all the WFC of the whole interfacial system and the use of the electronic dipole moment discussed in Equation 8.29 assigned to each individual molecule of the system. From there, the polarisability tensor is known, as both quantities are directly related. Such calculations typically involve finite electric fields of 0.001 −0.0001a.u. intensity and are performed independently along the x, y and z directions at each time step of the dynamics to extract the individual molecular polarisabilities. Useless to say that this is a rather costly procedure for DFT-MD simulations. This method has been developed by Madden and Salanne (Heaton and Madden 2008; Heaton et al. 2006; Salanne et al. 2008) and has been applied in our proof-of-principle demonstration on the SFG spectroscopy of the water–liquid–air interface in Sulpizi et al. (2013).

8.6   Illustrations from Our Recent Works

8.6.1   Organisation of Water at Silica–Water Interfaces: (0001) α-Quartz Versus Amorphous Silica

Silicates are the most abundant metal oxides and the main component of the Earth's crust. Its behaviour in contact with water thus plays a critical role in a variety of geochemical and environmental processes. Despite its key role, the details of the aqueous silica interface at the microscopic molecular level are still elusive. Our reference system is the hydroxylated (0001) α-quartz–water interface, but certainly this represents a very specific system from the point of view of surface structure, and when addressing more realistic geochemistry conditions, a key question is how the picture of the water organisation changes when going from the crystalline quartz–liquid water interface to a more disordered silica interface. Characterising the aqueous interface of non-crystalline silica is therefore relevant from that point of view.

In this section we review some of our recent results on two silica–water interfaces, either in the crystalline phase (quartz) or in a model of an amorphous phase. The results presented later are built on our recent publications (Cimas et al. 2014; Gaigeot et al. 2012; Sulpizi et al. 2012) where we have characterised the aqueous interface of crystalline α-quartz (Gaigeot et al. 2012; Sulpizi et al. 2012) and of an amorphous silica (Cimas et al. 2014) with DFT-MD simulations. This latest publication presented the first characterisation of an amorphous silica–water interface using this level of theory.

For the sake of clarity in our following discussions, we define in the succeeding text our nomenclature for the silanol groups (SiOH) at silica surfaces. We schematically review in Figure 8.1 the main kinds of silanols encountered in the more general cases at silica surfaces. At the dry silica surface, one can roughly define three types of silanols, that is, isolated, geminal and vicinal silanols, following Rimola et al. (2013), that can also be defined as Q^3 sites (isolated and vicinal silanols) and Q^2 sites (geminal silanols), following Stebbins (1987) and Skelton et al. (2011). An isolated silanol SiOH is not involved in H-bond interactions with close-by surface SiOH groups, and roughly speaking

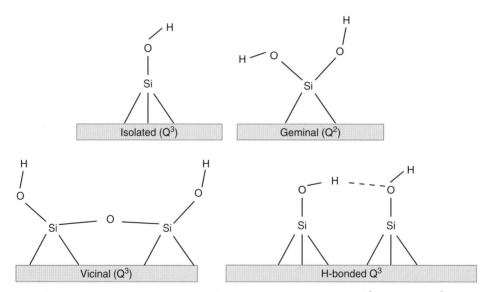

Figure 8.1   *Scheme of silanols encountered at the surface of silica: Isolated (Q^3), geminal (Q^2), vicinal (Q^3) and H-bonded with a neighbouring silanol. This scheme does not take into account the presence of an aqueous interface.*

silanol groups separated by more than approximately 3.3 Å can be considered as unable to establish mutual hydrogen bonds, as specified in Rimola et al. (2013). Such isolated silanols will consequently be free to possibly establish H-bond interactions with adsorbate molecules at the dry surface and act as H-bond donors and/or acceptors with water molecules at the interface with liquid water. Pairs of silanols belonging to tetrahedra that share a common oxygen vertex are called vicinals, the two hydroxyl groups being typically separated by less than 3 Å at the dry surface. These silanols could possibly establish mutual hydrogen bonds at the dry surface. Two OH groups linked to the same surface silicon atom to give the Si—(OH)$_2$ moiety are called geminals. Even though they are very close, the two geminal OH groups are typically oriented in such a way that they cannot be involved in mutual H bonds at the dry surface. On disordered silica surfaces, it is also possible that silanols that do not belong to directly connected tetrahedra (vicinals) but are closer than approximately 3.3 Å establish mutual H bonds at the dry surface. They are called interacting or H-bonded silanols.

Our theoretical investigations consist in establishing a detailed map of the interactions and associated hydrogen bond network that can be formed between the silica surface silanols and between the surface silanols and the water molecules, once the surface is wetted with liquid water. Our interest is to establish how the H-bonding network that possibly exists at the dry surface can be reorganised once liquid water is present at the interface. We compare in the succeeding text such H-bond maps for the ordered crystalline (0001) α-quartz–water interface and for the disordered amorphous silica–water interface. These investigations have been published in Cimas et al. (2014), Gaigeot et al. (2012) and Sulpizi et al. (2012), and we summarise our main findings here, discussing similarities and differences from the perspective of the interfacial structure (surface and interfacial water). The perspective from the point of view of vibrational spectroscopy will be presented in Section 8.6.4.

Before discussing the interfaces with water, one striking remark has to be made for the organisation of the silanols at these two dry surfaces. At the dry crystalline planar quartz surface, all geminal silanols are H bonded within the surface, thus forming a regular in-plane H-bond zigzagging network, as illustrated in Figure 8.2. At the non-planar dry amorphous surface, the non-crystalline local topology also dictates the structural arrangement between the three types of silanols (isolated,

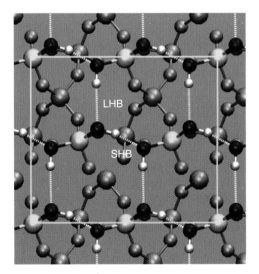

Figure 8.2 *Organisation of silanols, SiOH, at the dry quartz surface. The silanols are all located in plane and form a H-bonded network within the surface plane. Sulpizi et al. (2012). Reproduced with permission of the American Chemical Society.*

geminal, vicinal, Figure 8.1), resulting in a different silanol–silanol H-bond network from the α-quartz. In particular, geminals form H bonds with neighbouring vicinal silanols at the dry surface (see Cimas et al. (2014)). Due to the non-planar topology, and thus the existence of concave/convex zones at the surface, the silanols at the amorphous silica surface display varied orientations with respect to the surface, but they are mostly found pointing out of the surface, without being involved into specific surface H-bond network. This is at odds with the dry (0001) crystalline quartz surface where all silanols are located in the plane of the surface and forming in-plane H bonds.

Once liquid water is added to the (0001) crystalline α-quartz surface, the geminal silanols are distributed into two populations, that is, one that keeps surface in-plane orientations and one that adopts surface out-of-plane orientations, with average angle values with respect to the normal to the surface, respectively, of approximately 30° (in-plane) and 100° (out-of-plane). See the resulting bimodal distribution in Figure 8.3 (black line). This shows the breakage of the regular in-plane silanol H-bond network that was observed at the surface of dry quartz. These two silanol populations are therefore, on average, respectively, acting as H-bond donors to water molecules (out-of-plane silanols) and as H-bond acceptors to water molecules (in-plane silanols). See Sulpizi et al. (2012) for all details showing these results. One can consequently observe an interesting alternate and repeated motif of the reciprocal structural arrangements between silanols (Si)O—H and water O—H groups within the interfacial quartz–water layer (first adsorbed water layer), schematically illustrated in Figure 8.4, left. Note that no hydrogen bonds between these interfacial water molecules have been found. This schematic motif is repeated in the two directions of the surface plane.

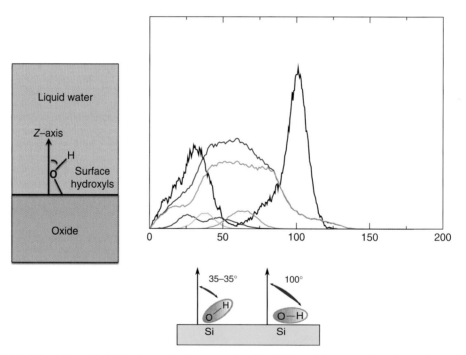

Figure 8.3   *Left: scheme defining the angle between the (Si—)OH and the normal to the surface (denoted Z-axis on the scheme). Right: distribution of (Si—)OH angles (in degrees) with respect to the surface normal for aqueous (0001) quartz interface (black) and aqueous amorphous interface (red). For this latest interface, decomposition of the angles is also presented in terms of Si—OH group types: geminals belonging to a concave zone (green), geminals belonging to a convex zone (blue), 'isolated' (light blue) and vicinals (brown). Bottom: scheme illustrating the bimodal character of Si—OH orientations at the crystalline α-quartz–water interface.*

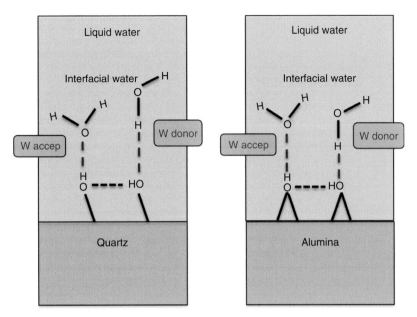

Figure 8.4   *Schematic organisation of the repeated motif of water molecules adsorbed within the first layer of aqueous water at the interface with crystalline (0001) α-quartz (left) and (0001) α-alumina (right). These motifs are repeated over the 2D surfaces.*

At the amorphous silica–water interface, all silanol groups donate H bonds to the water molecules in the first adsorbed interfacial layer. Please keep in mind that the silanols were mostly pointing out of the dry amorphous surface, which makes it easier to directly interact and form hydrogen bonds with interfacial water molecules at the aqueous surface. The strength of such H bonds is overall comparable to that of the H bonds donated by the geminal silanols of the (0001) crystalline quartz surface in contact with water. The H-bond strength is here measured from the relative intermolecular H-bond distance between donor and acceptor atoms, that is, the smaller, the stronger. Geminal silanols and (most of the) vicinal silanols are involved in hydrogen bonds with water molecules, similar to the quartz–water interface. Although all silanols from the amorphous surface generally point towards the adsorbed water layer, that is, with angles with respect to the normal to the amorphous surface in the 30–80° range (see Figure 8.3, red line), there is however no privileged directionalities of the silanol/water H bonds. Indeed there is one broad distribution of orientations of the silanols with respect to the normal amorphous surface (30–80°) contrary to the two well-separated orientations observed at the quartz–water interface. Note that the single broad distribution observed at the amorphous silica interface is already present at the dry surface and is thus maintained once the surface is wetted. This is the consequence of the topography of the amorphous surface, and this is nicely seen in the decomposition of silanol orientations in Figure 8.3. However, geminals display distinct orientations with respect to the surface that reflect their specific surface topographic environment, whether they belong to a convex or concave region of the surface. Vicinals on the other hand display a broad distribution of orientations that show that they are distributed over the surface without specific topographic environments.

As a consequence, the water molecules that belong to the first adsorbed interfacial layer at the amorphous silica surface can be roughly grouped into three categories: either pointing their dipole moment towards silanols and thus acting as H-bond donors to the surface or having their dipole moment roughly lying parallel to the surface and acting roughly as H-bond donors to the surface

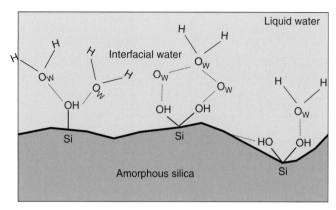

Figure 8.5 *Schematic organisation of the interfacial region between amorphous silica and water. Cimas et al. (2014). Reproduced with permission of Institute of Physics.*

or pointing their dipole moment out of the surface and acting as H-bond acceptors to the surface. See Cimas et al. (2014) for a detailed description.

The average amorphous silica–water H-bond network view thus obtained from our simulations is schematically illustrated in Figure 8.5. On average, the vicinal silanols form two simultaneous H bonds with two separate water molecules, that is, one H-bond donor and one H-bond acceptor. The same is obtained for the 'isolated silanol' (this silanol was defined as isolated at the *dry* surface) that becomes involved into two H bonds with two water molecules when the surface is wetted (this silanol thus simultaneously acts as H-bond donor and H-bond acceptor). Roughly the same average H-bond distances as for vicinals are obtained. The two silanols of the convex geminal simultaneously donate one H bond to one water molecule; strong H bonds are formed, according to the very short H-bond distances of 1.56 and 1.66 Å. These silanols are acting only as H-bond donors towards water and are simultaneously acting as H-bond acceptors to neighbouring vicinal silanols, forming H bonds of 1.65–1.75 Å distances on average. For the concave geminal, both silanols are H-bond donors to water, and one silanol of the pair is a H-bond acceptor from a neighbour vicinal. We have also observed striking local nests of water–water H bonds linking the silanol pairs of these geminals: 4- and 5-membered H-bond rings have been observed. This is strikingly different from our observations at the crystalline quartz–water interface, and this can only be obtained because of the non-planar surface of amorphous silica. The rugosity of this amorphous surface indeed provides in particular 'deep' concave zones that can be 'filled' by nests of H-bonded water molecules. All these results are schematically illustrated in Figure 8.5.

Beyond the specific first adsorbed water layer at the aqueous surfaces, water quickly recovers its bulk liquid structural organisation. This is shown by atomic density profiles of water oxygens and hydrogens through the water layers along the *z*-direction perpendicular to the quartz surface and by comparing radial distribution functions of the water molecules within these layers to the ones of bulk liquid water. This can be found in Cimas et al. (2014), Gaigeot et al. (2012) and Sulpizi et al. (2012) for the silica interfaces discussed here.

As will be discussed in Section 8.6.4 on the vibrational spectroscopy at oxide–liquid water interfaces, the three water layers beyond the interfacial adsorbed water layer have IR signatures that indeed correspond to liquid water. Similar conclusions were obtained by Mamontov et al. (2007) on titanium oxide from QENS experiments and classical MD simulations.

8.6.2 Organisation of Water at Alumina–Water Interface: (0001) α-Alumina Versus (101) Boehmite

A similar comparison on the organisation of liquid water at the crystalline (0001) α-alumina interface and at the defective (101) boehmite interface can be made. Our investigations have been published in Gaigeot et al. (2012) and Motta et al. (2012), and a similar investigation by Galli's group can be found in Huang et al. (2014) on (0001) α-alumina–water interface.

Aluminum oxides, hydroxides and oxi-hydroxides are present in soils, natural aquatic environments, and the Earth's crust, where they naturally interact with water. We consider here α-alumina, also known as corundum, which is the most common crystalline form of alumina and the gamma-AlOOH boehmite. Boehmite is a stable natural Al_2O_3 polymorph in water (Digne et al. 2002). It is more stable than corundum (α-Al_2O_3) and gibbsite ($Al(OH)_3$) at 298 K and has a water activity of unity (Peryea and Kittrick 1988). Besides its natural occurrence, the –AlOOH boehmite is an important object to study as nanoparticles of AlOOH boehmite are synthesised for applications in heterogeneous catalysis (Mercuri et al. 2009; Raybaud et al. 2001). Interestingly, boehmite rather than gibbsite precipitation has been proposed as a mechanism for the production of alumina (Konigsberger et al. 2011).

In the succeeding text is discussed the structural organisation of interfacial water at the surface of the hydroxylated crystalline (0001) α-alumina, and the summary scheme can be found in Figure 8.4. Similar to the hydroxylated crystalline (0001) α-silica interface described in Section 8.6.1, the interfacial water molecules at the (0001) α-alumina adopt a structural arrangement with a repeating motif based on alternate orientation of their dipole moment, which is associated with a motif of alternate orientation of the aluminols at the alumina surface. Hence, as illustrated in Figure 8.4, the basic structural motif that is repeated in the two directions of the surface plane is composed of one in-plane aluminol H bonded to a neighbouring out-of-plane aluminol, the H bond being roughly located within the plane of the alumina surface. Associated with these surface site orientations, the water molecules are organised in an alternating motif of H-bond donor towards one in-plane aluminol and H-bond acceptor towards one out-of-plane aluminol. The strength of the associated hydrogen bonds formed between the water molecules and the aluminols is also depicted by the two colours employed in Figure 8.4, that is, blue and red colours, respectively, for strong and weak water–aluminol H bonds. Our discussion on H-bond strength is based on the average value of the distances OW—HW···O(—H)Al and OW···H—O(Al) observed during the DFT-MD simulations. We find that the water molecules that act as H-bond donors to in-plane aluminols form the strongest H bonds with the surface, with an average OW—HW···O (short) distance of 1.70 Å, while the ones acting as H-bond acceptors to out-of-plane aluminols form weaker H bonds with an average OW···H—O (longer) distance of 2.00 Å. Conversely, the covalent OW—HW bond lengths follow the trend: (long) 1.004 Å for the water H-bond donors to the surface and (shorter) 0.994 Å for the H-bond acceptors to the surface.

On the other hand, (101) boehmite is an aluminum oxide with a defective surface that presents a step at the surface, thus delineating two separate terraces at the surface. See Figure 8.6 for a representative scheme. One can note that theoretical investigations of steps is not common in the community, in comparison with planar ideal surfaces.

The boehmite surface is composed of four aluminol species with the following average repartition: $0.48\mu_2$-OH + $0.26\mu_1$-OH_2 + $0.24\mu_1$-OH + $0.02\mu_2$-OH_2. All details can be found in Motta et al. (2012). Figure 8.6 illustrates how these sites are distributed along the surface in the box cell. Note that μ_1-OH_2 groups are located at the step and low edges.

At the boehmite–liquid water interface, we observe a network of hydrogen bonds simultaneously formed between adjacent aluminols and between aluminols and water within the first interfacial layer, which is illustrated in Figure 8.6. On average, we find one μ_1-OH_2···μ_1-O(H) H bond formed

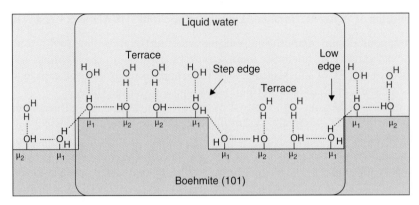

Figure 8.6 *Scheme of the (101) boehmite interface with liquid water. Water is either explicitly represented schematically within the first adsorbed layer at the surface or is implicitly represented by the blue colour for the rest of the bulk water. The scheme presents nomenclature of the boehmite surface sites, the two terraces, the step edge and the low edge. Motta et al. (2012). Reproduced with permission of American Chemical Society.*

at the step (both partners act equally well as donor or acceptor and interchange their respective role along the time, that is, with μ_1-$OH_2\cdots\mu_1$-$O(H)$ or μ_1-$OH\cdots\mu_1$-$O(H_2)$ H bonds formed), while μ_2-$OH\cdots\mu_1$-$O(H)$ and μ_2-$OH\cdots\mu_1$-$O(H_2)$ are, respectively, formed on the terraces and at the step edge (the μ_2-OH sites predominantly act as H-bond donors in these hydrogen bonds, showing that they have a more acidic character than the μ_1-OH sites). We never observe μ_2-$OH\cdots\mu_2$-$O(H)$ H bonds amongst vicinal groups at the terraces.

From the point of view of H bonds formed between the surface and interfacial water, we find that the boehmite surface acts on average more as H-bond acceptor (~60% of the surface O—H groups) than as H-bond donor (~40% of the surface O—H groups). We obtain the following average trends that illustrate the respective acid/base character of the boehmite surface sites: (i) μ_1-OH groups simultaneously act as H-bond acceptors from a neighbouring μ_2-OH (1 H bond on average) and as H-bond donors to one water molecule (0.66 H bond on average). If one uses these H-bond properties as reference, one can thus infer that μ_1-OH sites have a more basic than acidic character. (ii) μ_2-OH sites act as H-bond donors to μ_1-OH sites (1 H bond on average) and as H-bond acceptors from one water (0.66 H bond on average). They thus display a more acidic than basic character. (iii) Interfacial water molecules are on average donating 1 H bond to μ_2-OH sites and receive on average 0.7 H bond from μ_1-OH sites. Statistical analysis of the orientation of the interfacial water molecules with respect to the normal to the surface plane shows that 60% of these waters are oriented with one HW down while the remaining 40% are oriented with one HW up. These interfacial water molecules are not mobile (at least over the timescale of the DFT-MD simulations), and no preferential H bonding within these waters was observed.

At the defective step separating the two terraces, there is an interesting chain of H-bonded water molecules bridging the μ_1-OH_2 at the step edge of the upper terrace to the μ_1-OH at the lower terrace. A snapshot from the trajectory reported in Figure 8.7 illustrates the three-water chain hence formed. Again, these three water molecules are not mobile and remain H bonded as such over the entire trajectory. In this chain, μ_1-OH_2 acts as donor of H bond to the adjacent water molecule of the chain, and all subsequent water molecules from the chain also act as H-bond donor to the adjacent one, up to the last water being a H-bond donor to the μ_1-OH site on the lower terrace of the surface. This particular arrangement of H bonds within the chain enables a barrierless proton transfer from μ_1-OH_2 on the upper terrace to μ_1-OH on the lower terrace, through the water chain at the step. This has

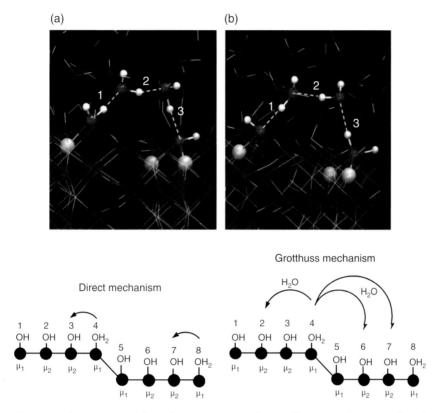

Figure 8.7 *Top: snapshots extracted from the DFT-MD simulations illustrating the water chain formed at the step edge of (101) aqueous boehmite and the proton transfer occurring along that chain (a: before proton transfer, b: after proton transfer). Bottom: schemes of proton transfers occurring at the boehmite aqueous surface observed along the DFT-MD. Motta et al. (2012). Reproduced with permission of American Chemical Society.*

been spontaneously observed in our trajectory. Once the proton transfer is achieved, the water molecules of the chain reorient but maintain successive H-bond donations with their partners along the chain. Other proton transfers have been observed at this surface, showing how chemically reactive a defective surface can be. They are schematically illustrated at the bottom of Figure 8.7. No such proton transfer events were observed on the crystalline alumina surface (over the same trajectory time length).

8.6.3 How Surface Acidities Dictate the Interfacial Water Structural Arrangement

We now look at our results on the structural arrangements of interfacial water molecules at the interface with the two crystalline (0001) α-quartz and (0001) α-alumina oxide surfaces from a more chemical point of view, that is, in terms of silanol/aluminol acidities and how these acidities might dictate the structure of interfacial water. Acidities of surface sites have been calculated via the methodology presented in Section 8.3 from DFT-MD simulations. Results presented later for the crystalline oxide–liquid water interfaces have been detailed in Gaigeot et al. (2012) and Sulpizi et al. (2012). Similar calculations are currently done for some of the non-crystalline surfaces in contact with liquid water described earlier.

For the hydroxylated (0001) α-quartz/liquid water interface, we have found a bimodal behaviour of the surface silanols pK_a acidities depending whether they are located in plane or out of plane (see Section 8.6.1 for the structural analyses). The out-of-plane silanols are the most acidic with a calculated pK_a value of 5.6, while the in-plane silanols are less acidic with a pK_a value of 8.5. We thus find the bimodal behaviour of the hydroxyls at the silica surface initially observed by Eisenthal and colleagues (Ong et al. 1992; Ostroverkhov et al. 2005) with SHG spectroscopy, where pK_a values of 4.5 and 8.5 were obtained. The difference we find between the two types of silanols is somewhat smaller than the experimental estimation (within error bars of both calculations and experiments), because our acidic site ($pK_a = 5.6$) is not as acidic as inferred from experiment ($pK_a = 4.5$). Errors on the theoretical pK_a values are of the order of 0.6–0.8 unit of pK_a; see details in our paper (Sulpizi et al. 2012). The lower pK_a of the out-of-plane silanols with respect to the in-plane silanols is also in agreement with the results of Leung et al. (2009) that intermolecular hydrogen bonds on the surface can lower acidity. Furthermore, even though our pK_a value for the out-of-plane silanols is not as low as experimentally inferred from SHG spectroscopy, this value is still very much in line with the observation in Eisenthal (1996) and Ong et al. (1992) that very acidic groups could only be found on highly strained surface groups.

One other quantity of high relevance is the PZC of the quartz surface, which is directly comparable to experimental measurements. It is calculated from the pK_a values of the $-SiOH$ and $-SiOH_2^+$ groups, which determine the possible protonation states of the surface for different values of pH. The pK_a of the $-SiOH_2^+$ groups is found at -5.0. Combining the calculated pK_a values of the in-plane/out-of-plane $-SiOH$ and $-SiOH_2^+$ groups at the quartz surface, we calculate a PZC of 1.0, which is compatible with the experimental value reported for silica and quartz (Kosmulski 2002) (PZC in the interval 2–4) and with the value of 1.9 calculated with the MUSIC model in Hiemstra et al. (1996).

Following the same theoretical strategy, we find that out-of-plane aluminols at the (0001) α-alumina/liquid water interface have a basic character with a pK_a of 16.6, while water in liquid water has a pK_a of 15.4 (see again Gaigeot et al. (2012) and Sulpizi et al. (2012) for more details on the 0.6–0.8 unit of pK_a error in our calculations).

With these theoretical pK_a values in mind for the hydroxylated sites at the quartz and alumina crystalline interfaces with liquid water, together with the pK_a of the O—H groups in pure liquid water, we can now comment more on the structural organisation of interfacial water at these interfaces and show that the local chemistry of interfacial water is modulated by the acidity of the surface groups.

Hence, out-of-plane O—H silanols at the surface of (0001) α-quartz are stronger acids than water O—H groups (respectively, pK_as of 5.6 and 15.4 from DFT-MD simulations), so that they are able to strongly donate and share their proton into H bonding with interfacial water molecules. As rather strong acids, out-of-plane silanols consequently form very short H bonds with water, with an average (Si)O—H$\cdots$OW distance of 1.64 Å, while the (Si)O—H bond length is very much elongated with an average value of 1.003 Å (thus highly sharing its proton with its water partner). Such surface-water H bond is shorter (and therefore energetically stronger) than the ones observed in pure liquid water, where an average OW—HW$\cdots$OW distance of 1.76 Å is obtained (DFT-MD simulations employing exactly the same set-up as the one for the interfaces discussed here; see Gaigeot et al. (2012) and Sulpizi et al. (2012)).

The in-plane O—H silanols are less acidic than the out-of-plane silanols and are not acidic enough in order to participate to the surface-water intermolecular H-bond network in terms of (Si)O—H donors to interfacial water molecules. Instead, they share their proton with neighbour out-of-plane silanols, forming (Si)O—H$\cdots$O(H—Si) hydrogen bonds of an average 1.65 Å distance. Once such intra-surface H bond is formed, the in-plane oxygen of the silanol is accessible to interfacial water molecules, thus forming OW—HW$\cdots$O(Si) H bonds of 1.82 Å on average. In such H bond, the water proton is not very much shared as the covalent OW—HW bond length is 0.988 Å on average, very close to what is observed in pure liquid water (0.990 Å).

On the contrary, out-of-plane aluminols at the (0001) α-alumina surface are very basic sites, therefore releasing their proton to the aqueous solution only at relatively high pH values. In our simulations where all surface sites are hydroxylated, the pH is estimated around 2. The aluminols thus retain their proton and form weak H bonds with interfacial water molecules. As a consequence, (Al)$_2$O—H$\cdots$OW H-bond distances of 2.00 Å are obtained on average at the interface with water, where the OW—HW water covalent bonds are 0.994 Å on average. Interfacial water molecules become strong H-bond donors to in-plane aluminols, forming short OW—HW$\cdots$O(H—Al$_2$) H bonds of 1.70 Å on average. Proton sharing in these H bonds is well illustrated by the average OW—HW covalent bond of 1.004 Å.

Can we therefore characterise the interfacial water molecules at the (0001) α-quartz/liquid water and (0001) α-alumina/liquid water interfaces in terms of 'liquid-like' and 'ice-like' molecules, and what definitions do we infer behind these terms? If we use the average OW—HW$\cdots$OW H-bond distances and average OW—HW covalent bond distances obtained from DFT-MD simulations of pure liquid water (using the same set-up as the one for the interfaces), that is, 0.990 and 0.999 Å for the covalent bonds of, respectively, liquid water and ice, as definitions and reference to the 'liquid-like' and 'ice-like' states, we find that the interfacial water molecules H-bond donors to in-plane silanols and the water H-bond acceptors from out-of-plane aluminols have H-bond and covalent bond length structural properties that are compatible with the ones observed in liquid water. They therefore could be qualified as 'liquid-like' interfacial molecules from that point of view. On the contrary, the interfacial water molecules H-bond acceptors from out-of-plane silanols and the ones H-bond donors to in-plane aluminols have H-bond and covalent bond length structural properties that are compatible with the ones observed in ice. They therefore could be qualified as 'ice-like' from that point of view. In Figure 8.4, we have therefore drawn 'liquid-like' and 'ice-like' water molecules with a different colour.

8.6.4 Vibrational Spectroscopy at Oxide–Liquid Water Interfaces

For our discussion on vibrational spectroscopy at oxide–liquid water interfaces, we take here again the comparison between the crystalline (0001) α-quartz/liquid water interface and (0001) α-alumina/liquid water interface. Here again, we will try to understand what spectral features can be associated with the 'liquid-like' and with the 'ice-like' terms used in the literature (Shen and Ostroverkhov 2006; Tian and Shen 2014) and see whether these terms have similar meaning as we have seen in the previous Section 8.6.3. Figure 8.8 reports the DFT-MD IR spectra of the first interfacial water layer at these two interfaces. By 'first interfacial water layer', we mean that only the water molecules that belong to the first adsorbed layer at the surface of the oxide are taken into account in the IR spectrum calculation. These water molecules are the ones that we previously characterised as H-bond donors or H-bond acceptors towards the surface hydroxylated sites (in-plane and out-of-plane); see Sections 8.6.1 and 8.6.2 and Figure 8.4 for associated schemes. For details of calculations on vibrational spectra from DFT-MD, please refer to Section 8.4. More details on the results presented later can be found in Gaigeot et al. (2012) and Sulpizi et al. (2012).

As Figure 8.8 shows, the interfacial IR spectra at the α-quartz–water and α-alumina–water interfaces have very similar general patterns, but as we will describe later, very interestingly the assignment of the vibrational bands is reversed between the two interfaces.

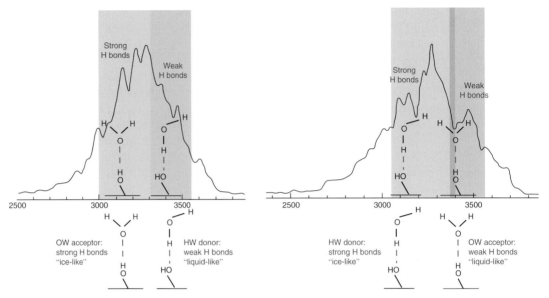

Figure 8.8 *Infrared (IR) spectra of the interfacial water at the (0001) quartz–water (left) and (0001) alumina–water (right) interfaces from DFT-MD simulations. The bands are colour coded according to 'weak' or 'strong' H bonds formed between water and hydroxyl surface sites; see text for details. The schemes below the bands correspond to the assignment of the vibrational bands in terms of interfacial water molecules being H-bond acceptors or donors to the surface sites. Sulpizi et al. (2012). Reproduced with permission of American Chemical Society.*

Hence, both IR spectra show activities in the $3000\text{–}3800\,\text{cm}^{-1}$ domain, with three separate domains coloured with three different colours in Figure 8.8. Going from 3800 to $3000\,\text{cm}^{-1}$, the three domains, respectively, roughly extend in the 3300–3500, 3200–3300 and $3000\text{–}3200\,\text{cm}^{-1}$. Grossly speaking, the two extreme domains are, respectively, associated with water molecules weakly H bonded to their environment ($3300\text{–}3500\,\text{cm}^{-1}$) and strongly H bonded to their environment ($3000\text{–}3200\,\text{cm}^{-1}$). In more details, as schematically represented in Figure 8.8, we find very striking assignments for each spectral domain. The active $3300\text{–}3500\,\text{cm}^{-1}$ band arises from the interfacial water H-bond *donors* to in-plane silanols at the α-quartz–water, while it arises from the interfacial water H-bond *acceptors* to out-of-plane silanols at the α-alumina–water. Similarly, the active $3000\text{–}3200\,\text{cm}^{-1}$ band is due to the interfacial waters H-bond *acceptors* to out-of-plane aluminols at the α-quartz–water, while it is due to the interfacial waters H-bond *donors* to in-plane aluminols at the α-alumina–water.

We have therefore found that there is a reversal of spectral assignments of the water molecules at the two crystalline oxide–water interfaces, although the IR bands are located at the same frequencies in the final IR spectra. The interfacial water molecules H-bond donors to in-plane silanols form the weakest H bonds at the quartz–water interface, and thus they give rise to the $3300\text{–}3500\,\text{cm}^{-1}$ vibrational band, while the interfacial water molecules H-bond acceptors to out-of-plane silanols form the strongest H bonds at the quartz–water interface, and thus they give rise to the more downshifted $3000\text{–}3200\,\text{cm}^{-1}$ vibrational band. On the contrary, the interfacial water molecules H-bond acceptors to out-of-plane aluminols form the weakest H bonds at the alumina–water interface, and thus they give rise to the $3300\text{–}3500\,\text{cm}^{-1}$ band, while the interfacial water molecules H-bond donors to in-plane aluminols form the strongest H bonds at this interface, and thus they give rise to the more downshifted $3000\text{–}3200\,\text{cm}^{-1}$ band.

As we have seen in Section 8.6.3, these weak/strong water-hydroxylated surface site H bonds are the result of the surface site acidities, which in turn modulate the H-bond strengths. What we see with the calculated IR spectra and their direct assignments from DFT-MD simulations is that such H-bond modulations have direct consequences on spectral features (band positions) and have very subtle consequences in terms of band assignments.

These interfacial water molecules have furthermore been qualified in Section 8.6.3 as 'liquid-like' and 'ice-like' based on their intrinsic structural properties of covalent OW—HW average distances and of H-bond intermolecular distances. As we have seen, the 'liquid-like' interfacial water molecules are the ones for which these two structural properties are similar to the ones obtained for water molecules in pure liquid water. Conversely, the 'ice-like' interfacial water molecules display covalent and H-bond structural properties similar to the ones of water molecules in ice. The terms 'liquid-like' and 'ice-like' bands are thus employed to qualify the structural covalent and H-bond distances of the interfacial water molecules in reference to the same properties obtained in pure liquid water and ice, respectively. These terms obviously do not mean that there is coexistence of liquid water and ice water at the oxide–water interfaces, but these terms refer to structural properties of the interfacial water molecules and their consequences on spectral signatures. Spectroscopically speaking, 'liquid-like' interfacial water molecules are the ones forming the 'weaker' H bonds with the surface hydroxyls, while 'ice-like' interfacial waters form the 'stronger' H bonds with the surface hydroxyls. Consequently, the former give rise to the spectroscopic signatures in the 3300–$3500 \, \mathrm{cm}^{-1}$ domain, whatever the oxide surface they interact with, while the latter give rise to the spectroscopic signatures in the 3000–$3200 \, \mathrm{cm}^{-1}$ domain, also without any regard to the oxide surface they interact with. Thus the 'liquid-like' and 'ice-like' vibrational bands are systematically observed for the interfacial water molecules at quartz or alumina crystalline surfaces. The nature of the oxide surface in terms of acidity of its hydroxyl sites enters in the fine details of these spectroscopic band assignments as typically shown here with the reversal of assignment of the 'liquid-like' (resp. 'ice-like') band from water H-bond *donors* to in-plane silanols at the quartz–water interface and from water H-bond *acceptors* to out-of-plane silanols at the alumina–water interface.

Figure 8.9 illustrates that subsequent layers of water molecules at the (0001) α-quartz–water interface have the signature of pure liquid water, although these waters are slightly compressed between the surface and its replica in the simulation box. We have compared in this figure the DFT-MD IR spectrum of liquid water (blue line), ice water (orange line), and the one of the layers of water beyond the interfacial layer at the quartz–water interface (black line). One can see that the spectra of bulk liquid water and the water beyond the interfacial adsorbed water molecules at the surface are identical, although the artificial compression of these later layers by the surface replica gives rise to an approximately $100 \, \mathrm{cm}^{-1}$ redshift of their signatures. Interestingly also is the DFT-MD IR spectrum of ice, which displays its band position at $3200 \, \mathrm{cm}^{-1}$, similar to the 'ice-like' band of the interfacial water layer.

Also of particular interest for interfacial vibrational signatures are the vibrational signatures of the hydroxylated surface sites. We have hence calculated the vibrational signatures of silanols and aluminols at the quartz–water and alumina–water interfaces. Please refer to Section 8.4 for the definition of VDOS and how VDOS spectra differ from IR spectra. Interestingly, we have shown (see Gaigeot et al. (2012) for details) that the in-plane and out-of-plane aluminols at the alumina–water interface have two distinct vibrational signatures: out-of-plane aluminols (forming the weakest H bonds with interfacial water molecules) have a vibrational signature at about $3700 \, \mathrm{cm}^{-1}$, while the in-plane aluminols (forming strong H bonds with interfacial water acceptors and simultaneously forming weak H bonds with neighbouring out-of-plane aluminols within the surface plane) have a vibrational signature downshifted from the previous ones by approximately $300 \, \mathrm{cm}^{-1}$. On the

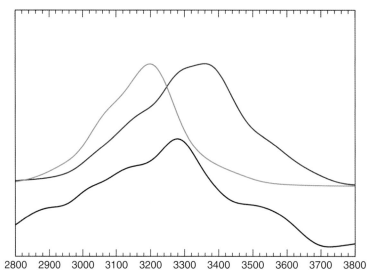

Figure 8.9 *DFT-MD infrared (IR) spectra of bulk liquid water (blue), ice water (orange) and the water layers beyond the interfacial layer at the quartz–water interface (black).*

contrary, in-plane and out-of-plane silanols at the quartz–water interface have similar vibrational signatures in the broad 3600–3000 cm^{-1} domain, independent of the strength of the H bonds they form with their environment. Their signatures thus overlap with interfacial water vibrational features.

What have we learned with these two interface examples, and what can we say in order to provide a detailed and microscopic interpretation of SFG experimental spectra? We remind the reader that we have calculated earlier the IR spectrum of the interfacial water layer at the oxide–water interface, but we have not calculated the direct SFG spectrum of that layer. We have not done this calculation at oxide–water interfaces yet, only at the liquid water–air interface; see Sulpizi et al. (2013). From the theoretical SFG signal expression in Equation 8.25, one can immediately see that for the SFG signal to be active, one needs a simultaneous IR and Raman activity. Calculating the interfacial IR signal is therefore one (mandatory) part of the SFG signal, and we can therefore draw some conclusions from the above spectra that are useful for the final understanding of the SFG features.

Before doing that, let us state a few more points. If we take again the two examples presented earlier on the (0001) quartz–water and (0001) alumina–water interfaces and look at the $|\chi^{(2)}(\omega)|^2$ spectra recorded by Shen and collaborators in particular (see, for instance, the review by Shen and Ostroverkhov (2006)), one can see that both spectra (in the same pH conditions, here we refer to a pH where the surfaces are fully hydroxylated) surprisingly show striking resemblance: two active bands, one located at approximately 3400 cm^{-1}, called 'liquid-like band', and one at approximately 3200 cm^{-1}, called 'ice-like band'. One more important fact is that these two bands are systematically recorded whatever the oxide surface, possibly with different relative intensities depending on the surface oxide. Also these two bands are enhanced if the interface is charged (protonation/deprotonation of surface, presence of ions at the interface; see, e.g. Shen and Ostroverkhov (2006)); again the enhancement possibly depends on the nature of the oxide surface. The first obvious and naive interpretation of these SFG spectra would be that the structural arrangement of the interfacial water molecules responsible for these spectral features is exactly the same, thus giving rise to the same spectral signatures.

We have just seen in the previous sections that the organisation of the water molecules within the first interfacial layer of (0001) quartz–water and (0001) alumina–water interfaces is systematically composed of alternating water molecules H-bond donors to in-plane hydroxyls and H-bond acceptors from out-of-plane hydroxyls (see again Figure 8.4). The interfacial water layer thus appears identical for the two (0001) quartz–water and (0001) alumina–water interfaces, which seems to indeed confirm the simple above spectral interpretation.

However, we have also shown that this structural arrangement is due to the surface sites acidities. We have shown that out-of-plane silanols are acidic while out-of-plane aluminols are basic. This in turn provokes a change in the strength of interfacial water–hydroxyl interactions: water H-bond acceptors to silanols are thus forming the stronger H bonds with the surface sites in (0001) quartz–water, while water H-bond donors to aluminols are forming the stronger H bonds with the surface sites in (0001) alumina–water. As shown in Figure 8.8, the direct consequence is that water H-bond acceptors to silanols are responsible for the ice-like band in the SFG spectrum of (0001) quartz–water, while water H-bond donors to aluminols are responsible for the ice-like band in the SFG spectrum of (0001) alumina–water (and the complementary assignments for the liquid-like band). As a consequence, although there is a spectral signature identically located around $3200 \, \mathrm{cm}^{-1}$ for both interfaces, the detailed microscopic assignment is reversed between the two interfaces – the same for the liquid-like band at $3400 \, \mathrm{cm}^{-1}$. In other words, although the general structural organisation of the water molecules at both interfaces is identical, that is, alternate H-bond donors/acceptors to surface sites, the surface site acidities have imposed different H-bond strengths, which in turn give rise to different detailed spectral assignments. In other words, the spectral microscopic interpretation is not only contained in the general structural organisation of water at the interface but is mostly contained in the detailed substructures and H-bond strengths in relation with the surface site acidities.

8.6.5 Clay–Water Interface: Pyrophyllite and Calcium Silicate

In this paragraph, we review our recent work on clay–water interfaces (Tazi et al. 2012) and provide a microscopic understanding of the solvation structure and reactivity of the edges of neutral clays. In particular we address the tendency to deprotonation of the different reactive groups on the (010) face of pyrophyllite. Such information cannot be inferred directly from titration experiments, which do not discriminate between different sites and whose interpretation resorts to macroscopic models.

The acid–basic properties of clay minerals play a key role in their surface chemistry. The increased acidity of water confined in smectite clay interlayers (Bergaya et al. 2006), for example, contributes to its catalytic properties. The protonation state of the edges of clay layers has an impact both on the stability of colloidal clay suspensions (Tombácz and Szekeres 2004) and on the sorption of ions (Baeyens and Bradbury 1997; Bradbury and Baeyens 1997). This is particularly important since the retention of contaminants by argillaceous rocks is pivotal in the current concepts for the geological disposal of long-lived high-level nuclear waste in several countries such as Switzerland, France and Belgium (ANDRA, 2005; Bradbury and Baeyens, 2003). Clays are lamellar aluminosilicate minerals. While the negative charge of their basal surface builds up because of isomorphic substitutions in the mineral lattice, that of edge sites (which arise from the finiteness of the sheets) is due to the deprotonation or protonation of surface groups. The protonation state is then controlled by the acidity of the various groups and the pH of the solution.

Several difficulties arise when trying to address the question of acidity from an experimental point of view. The most problematic one for the assignment of individual pK_as to the different surface groups is that macroscopic measurements such as titration experiments cannot distinguish between them. Moreover, Bourg et al. have shown recently that the analysis of such experiments must be performed with great care, in particular to correctly take into account the surface charge at the onset of titration (Bourg et al. 2007).

In the wait for the use of site-specific techniques (e.g. spectroscopic ones), the most promising approach has been to interpret experiments within the framework of a conceptual model, relying on assumptions (e.g. on the reactions to consider or how to decouple electrostatic from chemical contributions) and associated parameters (equilibrium constants for the various reactions) (Baeyens and Bradbury 1997; Bradbury and Baeyens 1997). The predictions of the model are then compared to the limited and global experimental data. In the best case, a favourable comparison could still occur by a compensation of errors between the different assumptions and parameters underlying the prediction (Bourg et al. 2007). In order to limit this risk, Tournassat et al. have proposed to constrain a large number of parameters from structure-based models (Tournassat et al. 2004). More precisely, the density of each site can be predicted from the structure of the mineral layer and the corresponding acidity constants estimated using semi-empirical bond valence-based models such as MUSIC (Bickmore et al. 2004, 2006; Brown 2009; Hiemstra et al. 1996; Tournassat et al. 2004). Nevertheless, there has been to date no direct way to probe the charge density on the edge and to measure the deprotonation free energies of the different sites. Another indirect strategy to determine the overall protonation state of clay edges has been proposed by Zhao et al. who found using atomic force microscopy that the force between a silica tip and a surface consisting of muscovite edges turned from repulsive at pH ~ 10 to attractive at pH ~ 6, from which they concluded that these edge surfaces had a PZC of pH 7–8 (Zhao et al. 2008).

In this respect atomistic simulations have emerged as a powerful tool to address the microscopic properties of clays. While force-field-based, classical molecular simulations have, for example, provided insights on bulk clays (including interlayer water and ion dynamics (Ferrage et al. 2011; Marry and Turq 2003; Marry et al. 2002; Rinnert et al. 2005), swelling (Boek et al. 1995; Tambach et al. 2004) and ion exchange (Rotenberg et al. 2009; Teppen and Miller 2006)) or on ion sorption (Greathouse and Cygan 2005) and water at basal surfaces (Marry et al. 2008; Rotenberg et al. 2011; Wang et al. 2006, 2009), their application to clay edges has been so far limited (Rotenberg et al. 2007, 2010). One of the main reasons is that the description of interfaces often requires to address polarisation effects and, in the case of clay edges, reactivity. Electronic structure calculations provide a direct way to include such effects describing both solid and liquid in a consistent way. In the last years, DFT-based MD simulations have provided a first-principles understanding of bulk clays such as the properties of interlayer ions (Boek and Sprik 2003; Suter et al. 2008) and their vibrational properties (Larentzos et al. 2007) or of water at basal surfaces (Bridgeman et al. 1996; Tunega et al. 2002, 2004). More recently, the acidity of interlayer water (Churakov and Kosakowski 2010; Liu et al. 2011) and the structure and reactivity of clay edges (Bickmore et al. 2003; Kremleva et al. 2011; Liu et al. 2008, 2012) have also been investigated. In particular, Churakov has studied water confined between edges of pyrophyllite. He evidenced transient proton exchange between surface groups mediated by surface water molecules (Churakov 2007) and proposed a first attempt to provide an acidity scale of the surface groups from their deprotonation energy in vacuum (Churakov 2006). In this context, it has also been suggested that ab initio simulations could provide refinements of the bond valence-based models of acidity by introducing the relaxation of bond lengths (Bickmore et al. 2003) (whose importance was already suggested by White and Zelazny (1988)) or the solvation structure (coordination and bond lengths) (Machesky et al. 2008) determined from DFT simulations.

In our recent work (Tazi et al. 2012), we used free energy calculation as described in Section 8.3 to provide the first direct calculation of the pK_a of the different groups, which are present on the (010) face of pyrophyllite.

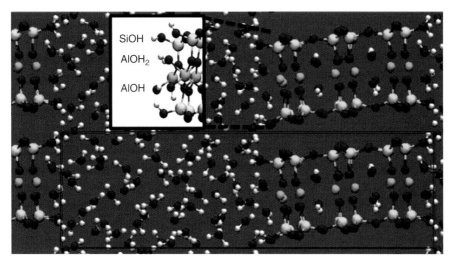

Figure 8.10 *A snapshot from the (010) pyrophyllite/water interface. Si atoms are represented in yellow, Al in green, O in red and H in white. The simulation box is also indicated (blue line), and periodic boundary conditions in all directions are used. The inset shows the deprotonable groups, namely, SiOH, AlOH$_2$ and AlOH.*

Three different deprotonable groups are present on the (010) edge, namely, SiOH, AlOH$_2$ and AlOH as can be seen on the inset of Figure 8.10.

The most acidic site is SiOH, with a pK_a of 6.8 ± 0.4, closely followed by AlOH$_2$ (p$K_a = 7.6 \pm 1.3$). Finally the AlOH group is the least likely to loose a proton, with a pK_a of 22.1 ± 1.0. As mentioned previously, direct comparison with experimental data is not possible, since such measurements for individual sites on a given clay edge are to date not available. However a comparison is possible with previous estimates obtained with empirical methods based on bond valence-based models. In particular in Tazi et al. (2012), we have compared our results for clays with the different pK_a estimates from the MUSIC models. Different MUSIC models are available for clays, using different values for the coordination numbers and valence units for the neighbouring H (attached to the O or received from H bonds) and metal. In particular the first version from Tournassat et al. uses the a priori recommendations of the original MUSIC model for coordination numbers and valences (2004). In the second model (model II), coordination numbers from the DFT simulations have been also used. Finally, in the revised version (model III), both coordination numbers and valences are determined from the DFT simulations (Machesky et al. 2008). The estimates of the original model are in remarkable agreement with the ab initio pK_a estimates (Tazi et al. 2012). The values of pK_a for AlOH in Machesky et al. (2008) and Tazi et al. (2012) are very close, out of the water domain; those for SiOH and AlOH$_2$ are on the alkaline side, with a pK_a for the latter approximately one unit larger than for the former, although both are larger than ours. Interestingly, for the modelling of titration curves, which was the main purpose of Tournassat et al., a smaller value than the one predicted by MUSIC was used for SiOH (8.2 instead of 9.1) to obtain a better agreement with the experiments. This suggests that our prediction in Tazi et al. (2012) of a more acidic SiOH group might reflect more accurately the experimental situation.

As it has been discussed in Tazi et al. (2012), the actual coordination numbers are not very different from the a priori estimates, so that the predictions of model II are still reasonable (although shifted in the wrong direction for SiOH and AlOH$_2$). Introducing the valences from DFT, computed as explained in Tazi et al. (2012), only worsens all predictions, yielding in particular very acidic SiOH and AlOH sites. From this comparison, it is possible to conclude that the original method provides the closest agreement with the prediction of thermodynamic integration, but this agreement is fortuitous, since including information on the correct structure and valence, which are the physical ingredients underlying this model, does not improve the result. This suggests that including such information on the deprotonated state only is not sufficient to capture the free energy of reaction, which accounts for the changes between the initial and final states.

It is interesting to compare the findings of Tazi et al. (2012) with previous calculations on simple oxides (Gaigeot et al. 2012; Sulpizi et al. 2012) (also described in Section 8.6.3). For example, for the (0001) surface of quartz, we found two types of silanols, 'out-of-plane' and 'in-plane', with different acidity constants, namely, 5.6 and 8.5 (Gaigeot et al. 2012; Sulpizi et al. 2012). In the case of the clay edge, the silanols are preferentially out of plane. They donate a H bond to water (nH$\cdots$Ow = 0.92) and accept H bonds mainly from water H (0.80 H bonds) and occasionally from nearby surface groups (0.26 H bond from AlOH$_2$). Such strong H bonds donated by silanols to water have also been observed on the (0001) quartz surface (Gaigeot et al. 2012; Sulpizi et al. 2012). The higher pK_a of the silanol groups on the (010) edges of pyrophyllite with respect to the out-of-plane silanols on the quartz (0001) surface could be due to the nearby surface groups and in particular the less strong H bond accepted from AlOH$_2$.

We would like to close this paragraph on clays mentioning some recent results of the application of the proton insertion method to the calculation of absolute pK_a values to the case of montmorillonite and kaolinite. In particular in Liu et al. (2013), the surface acidity of (010)-type edges of montmorillonite and kaolinite, which are representatives of 2:1 and 1:1-type clay minerals, respectively, has been addressed. The main conclusion is that Si—OH and AlOH$_2$—OH groups of kaolinite have pK_as of 6.9 and 5.7 and those of montmorillonite have pK_as of 7.0 and 8.3, respectively. For each mineral, the calculated pK_as are consistent with the experimental ranges derived from fittings of titration curves, indicating that Si—OH and AlOH$_2$—OH groups are the major acidic sites responsible to pH-dependent experimental observations. The effect of Mg substitution in montmorillonite was also investigated, and it was found that Mg substitution increases the pK_as of the neighbouring Si—OH and Si—OH$_2$ groups by 2 pK_a units. Furthermore, the pK_a of edge Mg(OH$_2$)$_2$ was found to be as high as 13.2, indicating that the protonated state dominates under common pH. These calculated acidity constants would suggest that Si- and Al(OH)$_2$-groups are the most probable edge sites for complexing heavy metal cations.

One last remark on the accuracy of DFT functionals in these works – pyrophyllite does not pose particular challenges to the gradient-corrected (GGA) functionals and can be very accurately described with PBE. In particular, the accuracy of the GGA functionals to reproduce the bulk structure of pyrophyllite has been discussed in the previous work from Churakov (2007). Lattice parameters could be reproduced within 1% of the experimental value. The small (positive) deviations of the lattice parameters are in line with typical performance of GGA exchange correlation functionals, which are known to systematically overestimate the molecular bond length and lattice constants of solids.

8.7   Some Perspectives for Future Works

In this contribution, we have shown that AIMD simulations can provide a precious tool to address properties at the mineral–water interfaces, which are not so easily accessible from experiments. In particular, we have discussed how ab initio simulations can be successfully employed to calculate pK_as of mineral surfaces (in other words dissociation constants of surface sites) and to understand the microscopic molecular origin of the vibrational properties at the mineral–water interfaces. We have also seen how these properties, that is, interfacial structure, pK_a and vibrational features, are intertwined. Several challenges remain to be tackled in order to move towards more realistic and predictive models of interest for geochemistry but also for industrial purposes. We would like here to highlight some of those that we think are important and that we aim to address in a near future. This is certainly far from being an exhaustive list of the open issues in the field, and what is actually presented later is more reflecting our personal and limited perspectives.

The first challenge we foresee is certainly to move from crystalline well-defined planar surfaces to more realistic amorphous mineral surfaces. So far most of the ab initio work has been limited to well-defined crystalline models or eventually to small clusters where structural 'flexibility' and 'inhomogeneity' can be introduced. In one of the examples we have presented in this chapter, we have started to address such issue with the structural investigation of an amorphous silica surface in contact with water. pK_a calculations on such more complex inhomogeneous interfaces are currently under way and the characterisation of different models representing an amorphous interface.

Another challenge is the microscopic characterisation of charged mineral–water interfaces. One has to unravel how the ions are structurally organised at these interfaces and how these microscopically charged systems compare to the debated electric double-layer or Langmuir models at mineral–water interfaces. Furthermore, one has to understand how this structural organisation possibly impacts electrostatic properties, interactions, dissolution and chemical reactivity. Directly or indirectly related is the investigation of inhomogeneous catalysis at mineral–water interfaces. This is especially of interest for industrial purposes where optimisations of chemical reactions are the focus.

As already stated, our theoretical investigations are very much centred on vibrational spectroscopy at interfaces, especially extracting SFG signals of mineral–water interfaces. The method described in this chapter and applied in an earlier publication of ours for the SFG spectrum of liquid water–air interface is computationally costly when using AIMD simulations. In order to apply the method to several mineral–water interfaces and make detailed interpretations of experiments under various thermodynamics and ionic and pH conditions, one has to reduce the computational cost of such an approach. This is one of our current focuses.

Combining the two preceeding points, one has to keep in mind that the structural and vibrational investigations presented in this chapter have focused on charge-neutral surfaces, where we have interpreted the 'liquid-like' and 'ice-like' spectroscopic bands in terms of surface-water hydrogen bond strengths. The next challenge is to investigate deprotonated or protonated surfaces and characterise the changes in the structural organisation of the interfacial water molecules and their consequences on the existence of the 'liquid-like'/'ice-like' spectroscopic bands and on their respective intensities, as has been done experimentally by Shen and co-workers by varying the bulk pH (Shen and Ostroverkhov 2006). One typical question is whether water molecules hydrogen bonded to Si–O$^-$ or Si–OH$_2^+$ surface groups all become 'liquid-like' because of the formation of stronger hydrogen bonds. This is currently investigated.

Another challenge is the use of ab initio methods in a multiscale approach that would allow the prediction of macroscopic properties. An example in this direction is our recent work on calcium silicate hydrates (C–S–H) (Churakov et al. 2014). There the ion sorption equilibrium was modelled applying ab initio calculated pK_as in titrating grand canonical Monte Carlo simulations using a coarse-grained model for C–S–H/solution interface in the framework of the primitive model for electrolytes. The modelling results provided direct comparison with available data from electrophoretic measurements. The model predictions were found to reproduce quite well the available experimental data.

References

Ahlborn H, Space B and Moore PB 2000. *J. Chem. Phys.* **112**, 8083.

Allen HC, Casillas-Ituarte NN, Sierra-Hernandez MR, Chen X and Tang CY 2009. Shedding light on water structure at air-aqueous interfaces: ions, lipids, and hydration. *Phys. Chem. Chem. Phys.* **11**, 5538.

ANDRA (2005) Evaluation de la faisabilité du stockage géologique en formation argileuse. Dossier 2005 Argile: Synthèse, Châtenay-Malabry, France.

Arnolds H and Bonn M 2010. *Surf. Sci. Rep.* **65**, 45.

Azam M, Darlington A and Gibbs-Davis J 2014. *J. Phys. Condens. Matter* **26**, 244107.

Baeyens B and Bradbury MH 1997 A mechanistic description of Ni and Zn sorption on Na-montmorillonite part I: titration and sorption measurements. *J. Contam. Hydrol.* **27**, 199–222.

Beck JP, Gaigeot MP and Lisy JM 2013. *Phys. Chem. Chem. Phys.* **15**, 16736.

Becke A 1988 Density-functional exchange-energy approximation with correct asymptotic behavior. *Phys. Rev. A* **38**(6), 3098–3100.

Bergaya F, Theng BGK and Ladaly D 2006 *Handbook of clay science*. Elsevier, Amsterdam.

Bernasconi M, Silvestrelli PL and Parrinello M 1998. *Phys. Rev. Lett.* **81**, 1235.

Bickmore B, Rosso K, Tadanier C, Bylaska E and Doud D 2006 Bond-valence methods for pK_a prediction. II. Bond-valence, electrostatic, molecular geometry, and solvation effects. *Geochim. Cosmochim. Acta* **70**, 4057–4071.

Bickmore BR, Rosso KM, Nagy KL, Cygan RT and Tadanier CJ 2003 Ab initio determination of edge surface structures for dioctahedral 2:1 phyllosilicates: implications for acid-base for reactivity. *Clays Clay Miner.* **51**, 359–371.

Bickmore BR, Tadanier CJ, Rosso KM, Monn WD and Eggett DL 2004 Bond-valence methods for pK_a prediction: critical reanalysis and a new approach. *Geochim. Cosmochim. Acta* **68**, 2025–2042.

Boek E, Coveney P and Skipper N 1995 Monte Carlo molecular modelling studies of hydrated Li-, Na- and K-smectites: understanding the role of potassium as a clay swelling inhibitor. *J. Am. Chem. Soc.* **117**(50), 12608–12617.

Boek ES and Sprik M 2003 Ab initio molecular dynamics study of the hydration of a sodium smectite clay. *J. Phys. Chem. B* **107**, 3251–3256.

Born M and Oppenheimer J 1927 Zur quantentheorie der molekeln. *Ann. Phys.* **84**, 457.

Borysow J, Moraldi M and Frommhold L 1985. *Mol. Phys.* **56**, 913.

Bourg IC, Sposito G and Bourg ACM 2007 Modeling the acid-base surface chemistry of montmorillonite. *J. Colloid Interface Sci.* **312**, 297–310.

Bovi D, Mezzetti A, Vuilleumier R, Gaigeot MP, Chazallon B, Spezia R and Guidoni L 2011. *Phys. Chem. Chem. Phys.* **13**, 20954.

Bradbury MH and Baeyens B 1997 A mechanistic description of Ni and Zn sorption on Na-montmorillonite part II: modelling. *J. Contam. Hydrol.* **27**, 223–248.

Bradbury M and Baeyens B (2003) Near Field Sorption Data Bases for Compacted MX-80 Bentonite for Performance Assessment of a High-Level Radioactive Waste Repository in Opalinus Clay Host Rock. PSI Bericht Nr. 03-07, Nagra NTB 02-18. Paul Scherrer Institut, Switzerland.

Bridgeman C, Buckingham A, Skipper N and Payne M 1996 Ab-initio total energy study of uncharged 2:1 clays and their interaction with water. *Mol. Phys.* **89**, 879–888.

Brown ID 2009 Recent developments in the methods and applications of the bond valence model. *Chem. Rev.* **109**, 6858–6919.

Burow A, Bates J, Furche F and Eshuis H 2014. *J. Chem. Theor. Comput.* **10**, 180.

Car R and Parrinello M 1985. *Phys. Rev. Lett.* **55**, 2471.

Catalano J, Fenter P and Park C 2007. *Geochim. Cosmochim. Acta* **71**, 5313.

Catalano J, Fenter P and Park C 2009. *Geochim. Cosmochim. Acta* **73**, 2242.

Catalano J, Park C, Zhang Z and Fenter P 2006. *Langmuir* **22**, 4668.

Ceperley DM and Alder BJ 1980. *Phys. Rev. Lett.* **45**, 566.

Chen C, Heinz T, Ricard D and Shen Y 1981. *Phys. Rev. Lett.* **46**, 1010.

Cheng J and Sprik M 2010 Acidity of the aqueous rutile TiO_2(110) surface from density functional theory based molecular dynamics. *J. Chem. Theory Comput.* **6**, 880.

Cheng J, Sulpizi M and Sprik M 2009 Redox potentials and pK_a for benzoquinone from density functional theory based molecular dynamics. *J. Chem. Phys.* **131**, 154504.

Cheng L, Fenter P, Nagy K, Schlegel M and Sturchio N 2001. *Phys. Rev. Lett.* **87**, 156103.

Churakov S, Labbez C, Pegado L and Sulpizi M 2014. *J. Phys. Chem. C* **118**, 11752.

Churakov SV 2006 Ab initio study of sorption on pyrophyllite: structure and acidity of the edge sites. *J. Phys. Chem. B* **110**, 4135–4146.

Churakov SV 2007 Structure and dynamics of the water films confined between edges of pyrophyllite: a first principle study. *Geochim. Cosmochim. Acta* **71**, 1130–1144.

Churakov SV and Kosakowski G 2010 An ab initio molecular dynamics study of hydronium complexation in Na-montmorillonite. *Philos. Mag.* **90**, 2459–2474.

Cimas A, Tielens F, Sulpizi M, Gaigeot M and Costa D 2014. *J. Phys. Condens. Matt.* **26**, 244106.

Cimas A, Vaden TD, de Boer TSJA, Snoek LC and Gaigeot MP 2009. *J. Chem. Theory Comput.* **5**, 1068.

The CPMD Consortium 2009. CPMD version 3.13.2, CPMD, http://www.cpmd.org/, Copyright IBM Corp 1990–2015, Copyright MPI für Festkörperforschung Stuttgart 1997–2001.

Costanzo F, Sulpizi M, Della Valle R.G. and Sprik M 2011 The oxidation of tyrosine and tryptophan studied by a molecular dynamics normal hydrogen electrode. *J. Chem. Phys.* **134**, 244508.

CP2K Developers Group n.d. http://cp2k.org (Accessed January 7, 2016).

Criscenti L, Kubicki J and Brantley S 2006. *J. Phys. Chem. A* **110**, 198.

DelloStritto MJ, Kubicki J and Sofo JO 2014. *J. Phys. Condens. Matter* **26**, 244101.

Dewand S, Yeganeh M and Borguet E 2013. *J. Phys. Chem. Lett.* **4**, 1977.

Digne M, Sautet P, Raybaud P, Toulhoat H and Artacho E 2002. *J. Phys. Chem. B* **106**, 5155.

Eftekhari-Bafrooei A and Borguet E 2010. *J. Am. Chem. Soc.* **132**, 3756.

Eisenthal K 2006. *Chem. Rev.* **106**, 1462.

Eisenthal KB 1996 Liquid interfaces probed by second-harmonic and sum-frequency spectroscopy. *Chem. Rev.* **96**, 1343.

Eng P, Trainor T, Brown GE, Jr., Waychunas G, Newville M, Sutton S and Rivers M 2000. *Science* **288**, 1029.

Fenter P, Cheng L, Park C, Zhang H and Sturchio N 2003. *Geochim. Cosmochim. Acta* **67**, 4267.

Fenter P, Geissbuhler P, DiMasi E, Srajer G, Sorensen L and Sturchio N 2000a. *Geochim. Cosmochim. Acta* **64**, 1221.

Fenter P, Teng H, Geissbuhler P, Hanchar J, Nagy K and Sturchio N 2000b. *Geochim. Cosmochim. Acta* **64**, 3663.

Fenter P, Zhang Z, Park C, Sturchio N, Hu X and Higgins S 2007. *Geochim. Cosmochim. Acta* **71**, 566.

Ferrage E, Sakharov BA, Michot LJ, Delville A, Bauer A, Lanson B, Grangeon S, Frapper G, Jiménez-Ruiz M and Cuello GJ 2011 Hydration properties and interlayer organization of water and ions in synthetic Na-smectite with tetrahedral layer charge. Part 2. Toward a precise coupling between molecular simulations and diffraction data. *J. Phys. Chem. C* **115**(5), 1867–1881.

Feynman RP 1939. *Phys. Rev.* **56**, 340.

Gaigeot MP 2010 Theoretical spectroscopy of floppy peptides at room temperature. A DFTMD perspective: gas and aqueous phase. *Phys. Chem. Chem. Phys.* **12**, 3336.

Gaigeot MP and Sprik M 2003 Ab initio molecular dynamics computation of the infrared spectrum of aqueous uracil. *J. Phys. Chem. B* **107**, 10344.

Gaigeot MP, Martinez M and Vuilleumier R 2007. *Mol. Phys.* **105**, 2857.

Gaigeot MP, Sprik M and Sulpizi M 2012. *J. Phys. Condens. Matter* **24**, 124106.

Geiger F 2009. *Annu. Rev. Phys. Chem.* **60**, 61.

Geissbuhler P, Fenter P, DiMasi E, Srajer G, Sorensen L and Sturchio N 2004. *Surf. Sci.* **573**, 191.

Goedecker S, Teter M and Hutter J 1996 Separable dual-space gaussian pseudopotentials. *Phys. Rev. B*, **54**, 1703.

Gopalakrishnan S, Liu D, Allen HC, Kuo M and Shultz MJ 2006. *Chem. Rev.* **106**, 1155.

Greathouse JA and Cygan RT 2005 Molecular dynamics simulation of uranyl(VI) adsorption equilibria onto an external montmorillonite surface. *Phys. Chem. Chem. Phys.* **7**, 3580–3586.

Grimme S, Antony J, Ehrlich S and Krieg H 2010. *J. Chem. Phys.* **132**, 154104.

Hartwigsen C, Goedecker S and Hutter J 1998 Relativistic separable dual-space gaussian pseudopotentials from H to Rn. *Phys. Rev. B* **58**, 3641.

Hayes P, Malin J, Jordan D and Geiger F 2010. *Chem. Phys. Lett.* **499**, 183.

Heaton R and Madden P 2008 Fluctuating ionic polarizabilities in the condensed phase: first-principles calculations of the Raman spectra of ionic melts. *Mol. Phys.* **106**, 1703.

Heaton R, Madden P, Clark S and Jahn S 2006 Condensed phase ionic polarizabilities from plane wave density functional theory calculations. *J. Chem. Phys.* **125**, 144104.

Hellmann H 1937. *Einfuhrung in die Quantenchemie*. Deuticke, Leipzig.

Hiemstra T, Vanema P and Riemsdijk WV 1996. *Colloid Interface Sci.* **184**, 680.

Hohenberg P and Kohn W 1964 Inhomogeneous electron gas. *Phys. Rev.* **136**(3B), 864–871.

Hopkins A, McFearin C and Richmond G 2005. *Curr. Opin. Solid State Mater. Sci.* **9**, 19.

Huang P, Pham T, Gallu G and Schwegler E 2014 Alumina(0001)/water interface: structural properties and infrared spectra from first-principles molecular dynamics simulations. *J. Phys. Chem. C* **118**, 8944.

Hummer G and Szabo A 1996 Calculation of free energy differences from computer simulations of initial and final states. *J. Chem. Phys.* **105**, 2004.

Iftimie R and Tuckerman M 2005. *J. Chem. Phys.* **122**, 214508.

Ishiyama T, Hideaki T and Morita A 2012 Vibrational spectrum at a water surface: a hybrid quantum mechanics/molecular mechanics molecular dynamics approach. *J. Phys. Condens. Matter* **24**, 124107.

Jubb A, Hua W and Allen H 2012. *Annu. Rev. Phys. Chem.* **63**, 107.

Jun YS, Ghose S, Trainor T, Eng P and Martin S 2007. *Environ. Sci. Technol.* **41**, 3918.

Khaliullin R, VandeVondele J and Hutter J 2013. *J. Chem. Theory Comput.* **9**, 4421.

Kim H and Rossky PJ 2006. *J. Chem. Phys.* **125**, 074107.

Kohn W and Sham L 1965 Self-consistent equations including exchange and correlation effects. *Phys. Rev.* **140** (4A), 1133–1138.

Konigsberger E, Konigsberger L and Ilievski D 2011. *Hydrometallurgy* **110**, 33.

Kosmulski M 2002. *J. Colloid Interface Sci.* **253**, 77.

Kremleva A, Krüger S and Rösch N 2011 Uranyl adsorption at (010) edge surfaces of kaolinite: a density functional study. *Geochim. Cosmochim. Acta* **75**, 706–718.

Kubicki J, Xiao Y and Lasaga A 1993. *Geochim. Cosmochim. Acta* **57**, 3847.

Kubo R, Toda M and Hashitsume N 1985 *Statistical physics II-nonequilibrium statistical mechanics* 2 edn. Springer, Heidelberg.

Larentzos JP, Greathouse JA and Cygan RT 2007 An ab initio and classical molecular dynamics investigation of the structural and vibrational properties of talc and pyrophyllite. *J. Phys. Chem. C* **111**(34), 12752–12759.

Lasaga A and Gibbs GV 1990. *Am. J. Sci.* **290**, 263–295.

Lawrence CP and Skinner JL 2005. *Proc. Nat. Acad. Sci.* **102**, 6720.

Lee C, Yang W and Parr R 1988 Development of the Colle-Salvetti correlation-energy formula into a functional of the electron density. *Phys. Rev. B* **37**(2), 785–789.

Leung K and Criscenti L 2012 Predicting the acidity constant of a goethite hydroxyl group from first principles. *J. Phys. Condens. Matter* **24**, 124105.

Leung K, Nielsen IMB and Criscenti LJ 2009. *J. Am. Chem. Soc.* **131**, 18358.

Liptak MD and Shields GC 2001. *J. Am. Chem. Soc.* **123**, 7314.

Lis D, Backus E, Hunger J, Parekh S and Bonn M 2014 Liquid flow along a solid surface reversibly alters interfacial chemistry. *Science* **344**, 1138.

Liu X, Cheng J, Sprik M, Lu X and Wang R 2014. *Geochim. Cosmochim. Acta* **140**, 410.

Liu X, Lu X, Meijer EJ, Wang R and Zhou H 2012 Atomic-scale structures of interfaces between phyllosilicate edges and water. *Geochim. Cosmochim. Acta* **81**, 56–68.

Liu X, Lu X, Sprik M, Cheng J, Meijer E and Wang R 2013 Acidity of edge surface sites of montmorillonite and kaolinite. *Geochim. Cosmochim. Acta* **117**, 118.

Liu X, Lu X, Wang R, Meijer EJ and Zhou H 2011 Acidities of confined water in interlayer space of clay minerals. *Geochim. Cosmochim. Acta* **75**, 4978–4986.

Liu X, Lu X, Wang R, Zhou H and Xu S 2008 Surface complexes of acetate on edge surfaces of 2:1 type phyllosilicate: insights from density functional theory calculation. *Geochim. Cosmochim. Acta* **72**, 5896–5907.

Lockwood G and Garofalini S 2014 Proton dynamics at the water-silica interface via dissociative molecular dynamics. *J. Phys. Chem. C* **118**, 29750–29759.

Machesky ML, Predota M, Wesolowski DJ, Vlcek, L, Cummings PT, Rosenqvist J, Ridley MK, Kubicki JD, Bandura AV, Kumar N and Sofo JO 2008 Surface protonation at the rutile (110) interface: explicit incorporation of solvation structure within the refined music model framework. *Langmuir* **24**(21), 12331–12339.

Mamontov E, Vlcek L, Wesolowski D, Cummings P, Wang W, Anovitz L, Rosenqvist J, Brown C and Sakai VG 2007 Dynamics and structure of hydration water on rutile and cassiterite nanopowders studied by quasielastic neutron scattering and molecular dynamics simulations. *J. Phys. Chem. C* **111**, 4328–4341.

Mangold M, Rolland L, Costanzo F, Sprik M, Sulpizi M and Blumberger J 2011 Absolute pK_a values and solvation structure of amino acids from density functional based molecular dynamics simulation. *J. Chem. Theory Comput.* **7**, 1951.

Marinica C, Grgoire G, Desfranois C, Schermann JP, Borgis D and Gaigeot MP 2006. *J. Phys. Chem. A* **110**, 8802.

Marry V and Turq P 2003 Microscopic simulations of interlayer structure and dynamics in bihydrated heteroionic montmorillonites. *J. Phys. Chem. B* **107**, 1832–1839.

Marry V, Rotenberg B and Turq P 2008 Structure and dynamics of water at a clay surface from molecular dynamics simulation. *Phys. Chem. Chem. Phys.* **10**, 4802–4813.

Marry V, Turq P, Cartailler T and Levesque D 2002 Microscopic simulation for structure and dynamics of water and counterions in a monohydrated montmorillonite. *J. Chem. Phys.* **117**, 3454–3463.

Martinez M, Gaigeot MP, Borgis D and Vuilleumier R 2006. *J. Chem. Phys.* **125**, 144106.

Marzari N and Vanderbilt D 1997. *Phys. Rev. B* **56**, 12847–12865.

Mathias G, Ivanov SD, Witt A, Baer MD and Marx D 2012. *J. Chem. Theory Comput.* **8**, 224.

McQuarrie D 1976 *Statistical mechanics*. Harper-Collins Publishers, New York.

Mercuri F, Costa D and Marcus P 2009. *J. Phys. Chem. C* **113**, 5228.

Morita A 2004 Toward computation of bulk quadrupolar signals in vibrational sum frequency generation spectroscopy. *Chem. Phys. Lett.* **398**, 361.

Morita A and Hynes JT 2002 A theoretical analysis of the SFG spectrum of the water surface. II. Time dependent approach. *J. Phys. Chem. B.* **106**, 673.

Morita A and Ishiyama T 2008 Recent progress in theoretical analysis of vibrational sum frequency generation spectroscopy. *Phys. Chem. Chem. Phys.* **10**, 5801.

Motta A, Gaigeot M and Costa D 2012. *J. Phys. Chem. C* **116**, 12514.

Nihonyanagi S, Ishiyama T, Lee T, Yamaguchi S, Bonn M, Morita A and Tahara T 2011 Unified molecular view of the air/water interface based on experimental and theoretical spectra of an isotopically diluted water surface. *J. Am. Chem. Soc.* **133**, 16875.

Nonella M, Mathias G and Tavan P 2003. *J. Phys. Chem. A* **107**, 8638.

Ong S, Zhao X and Eisenthal KB 1992. *Chem. Phys. Lett.* **191**, 327.

Ostroverkhov V, Waychunas G and Shen Y 2005. *Phys. Rev. Lett.* **94**, 046102.

Pelmenschikov A, Strandh H, Pettersson L and Leszczynski J 2000. *J. Phys. Chem. B* **104**, 5779.

Perdew J, Burke K and M. Ernzerhof 1936. *Phys. Rev. Lett.* **1997**, 78.

Peryea F and Kittrick J 1988. *Clays Clay Miner.* **36**, 391.

Pulay P 1969. *Mol. Phys.* **17**, 197.

Ramirez R, Lopez-Ciudad T, Kumar P and Marx D 2004. *J. Chem. Phys.* **121**, 3973.

Raybaud P, Digne M, Iftimie R, Wellens W, Euzen P and Toulhoat H 2001. *J. Catal.* **201**, 236.

Resta R 1998. *Phys. Rev. Lett.* **80**, 1800.

Rimola A, Costa D, Sodupe M, Lambert J and Ugliengo P 2013. *Chem. Rev.* **113**, 4216.

Rinnert E, Carteret C, Humbert B, Fragneto-Cusani G, Ramsay JDF, Delville A, Robert JL, Bihannic I, Pelletier M and Michot LJ 2005 Hydration of a synthetic clay with tetrahedral charges: a multidisciplinary experimental and numerical study. *J. Phys. Chem. B* **109**(49), 23745–23759.

Rotenberg B, Marry V, Malikova N and Turq P 2010 Molecular simulation of aqueous solutions at clay surfaces. *J. Phys. Condens. Matter* **22**, 284114.

Rotenberg B, Marry V, Vuilleumier R, Malikova N, Simon C and Turq P 2007 Water and ions in clays: unraveling the interlayer/micropore exchange using molecular dynamics. *Geochim. Cosmochim. Acta* **71**, 5089–5101.

Rotenberg B, Morel J, Marry V, Turq P and Morel-Desrosiers N 2009 On the driving force of cation exchange in clays: insights from combined microcalorimetry experiments and molecular simulation. *Geochim. Cosmochim. Acta* **73**, 4034–4044.

Rotenberg B, Patel AJ and Chandler D 2011 Molecular explanation for why talc surfaces can be both hydrophilic and hydrophobic. *J. Am. Chem. Soc.* **133**, 20521–20527.

Sahai N 2000. *Geochim. Cosmochim. Acta* **64**, 3629.

Sahai N 2002. *Environ. Sci. Technol.* **36**, 445.

Salanne M, Vuilleumier R, Madden P, Simon C, Turq P and Guillot B 2008 Polarizabilities of individual molecules and ions in liquids from first principles. *J. Phys. Condens. Matter* **20**, 494207.

Santra B, Klimes J, Tkatchenko A, Alfe D, Slater B, Michaelides A, Car R and Scheffler M 2013 On the accuracy of van der Waals inclusive density-functional theory exchange-correlation functionals for ice at ambient and high pressures. *J. Chem. Phys.* **139**, 154702.

Saracino GAA, Improta R and Barone V 2003. *Chem. Phys. Lett.* **373**, 411.

Schlegel M, Nagy K, Fenter P and Sturchio N 2002. *Geochim. Cosmochim. Acta* **66**, 3037.

Sediki A, Snoek LC and Gaigeot MP 2011. *Int. J. Mass Spectrom.* **308**, 281.

Shen Y 2012 Basic theory of surface sum-frequency generation. *J. Phys. Chem. C* **116**, 15505.

Shen Y 2013. *Annu. Rev. Phys. Chem.* **64**, 129.

Shen YR and Ostroverkhov V 2006 Sum-frequency vibrational spectroscopy on water interfaces: polar orientation of water molecules at interfaces. *Chem. Rev.* **106**, 1140.

Silvestrelli PL 1999 Maximally localized Wannier functions for simulations with supercells of general symmetry. *Phys. Rev. B* **59**(15), 9703–9706.

Silvestrelli PL and Parrinello M 1999a Structural, electronic, and bonding properties of liquid water from first principles. *J. Chem. Phys.* **111**(8), 3572–3580.

Silvestrelli PL and Parrinello M 1999b Water molecule dipole in the gas and in the liquid phase. *Phys. Rev. Lett.* **82**(16), 3308–3311.

Silvestrelli PL, Marzari N, Vanderbilt D and Parrinello M 1998 Maximally-localized Wannier functions for disordered systems: application to amorphous silicon. *Solid State Commun.* **107**(1), 7–11.

Skelton A, Fenter P, Kubicki J, Wesolowski D and Cummings P 2011. *J. Phys. Chem. C* **115**, 2076.

Stebbins J 1987 Identification of multiple structural species in silicate glasses by 29Si NMR. *Nature* **330**, 465.

Sulpizi M and Sprik M 2008 Acidity constants from vertical energy gaps: density functional theory based molecular dynamics implementation. *Phys. Chem. Chem. Phys.* **10**, 5238–5249.

Sulpizi M and Sprik M 2010. *J. Phys. Condens. Matter* **22**, 284116.

Sulpizi M, Gaigeot M and Sprik M 2012 The silica-water interface: how the silanols determine the surface acidity and modulate the water properties. *J. Chem. Theory Comput.* **8**, 1037–1047.

Sulpizi M, Salanne M, Sprik M and Gaigeot M 2013. *J. Phys. Chem. Lett.* **4**, 83.

Sung J, Zhang L, Tian C, Waychunas G and Shen Y 2011. *J. Am. Chem. Soc.* **133**, 3846.

Suter JL, Boek ES and Sprik M 2008 Adsorption of a sodium ion on a smectite clay from constrained ab initio molecular dynamics simulations. *J. Phys. Chem. C* **112**, 18832–18839.

Sverjensky DA 1994. *Geochim. Cosmochim. Acta* **58**, 3123.

Sverjensky DA 1996. *Geochim. Cosmochim. Acta* **60**, 3773.

Tambach TJ, Hensen EJM and Smit B 2004 Molecular simulations of swelling clay minerals. *J. Phys. Chem. B* **108**(23), 7586–7596.

Tangney P 2006. *J. Chem. Phys.* **124**, 044111.

Tanwar K, Catalano J, Petitto S, Ghose S, Eng P and Trainor T 2007. *Surf. Sci.* **601**, L59–L64.

Tazi S, Rotenberg B, Salanne M, Sprik M and Sulpizi M 2012. *Geochim. Cosmochim. Acta* **94**, 1.

Teppen BJ and Miller DM 2006 Hydration energy determines isovalent cation exchange selectivity by clay minerals. *Soil Sci. Soc. Am. J.* **70**(1), 31–40.

Thomas M, Brehm M, Fligg R, Vorhinger P and Kirchner B 2013. *Phys. Chem. Chem. Phys.* **15**, 6608.

Tian C and Shen YR 2014. *Surf. Sci. Rep.* **69**, 105.

Tkatchenko A, DiStasio R, Car R and Scheffler M 2012 Accurate and efficient method for many-body van der Waals interactions. *Phys. Rev. Lett.* **108**, 236402.

Tombácz E and Szekeres M 2004 Colloidal behavior of aqueous montmorillonite suspensions: the specific role of pH in the presence of indifferent electrolytes. *Appl. Clay Sci.* **27**, 75–94.

Tournassat C, Ferrage E, Poinsignon C and Charlet L 2004 The titration of clay minerals. II. Structure-based model and implications for clay reactivity. *J. Colloid Interface Sci.* **273**, 234–246.

Trainor T, Chaka A, Eng P, Newville M, Waychunas G, Catalano J and Brown GE, Jr. 2004. *Surf. Sci.* **573**, 204.

Tunega D, Gerzabek MH and Lischka H 2004 Ab initio molecular dynamics study of a monomolecular water layer on octahedral and tetrahedral kaolinite surfaces. *J. Phys. Chem. B* **108**, 5930–5936.

Tunega D, Haberhauer G, Gerzabek MH and Lischka H 2002 Theoretical study of adsorption sites on the (001) surfaces of 1:1 clay minerals. *Langmuir* **18**, 139–147.

van Duin ACT, Strachan A, Stewman S, Zhang Q, Xu X and Goddard WA 2003 Reaxffsio reactive force field for silicon and silicon oxide systems. *J. Phys. Chem. A* **107**, 3803–3811.

VandeVondele J and Hutter J 2012. *J. Chem. Theory Comput.* **8**, 3565.

VandeVondele J, Krack M, Mohamed F, Parrinello M, Chassaing T and Hutter J 2005. *Comput. Phys. Commun.* **167**, 103–128.

von Lilienfeld O, Tavernelli I, Rothlisberger U and Sebastiani D 2005. *Phys. Rev. B* **71**, 195119.

Wang J, Kalinichev AG and Kirkpatrick R 2006 Effects of substrate structure and composition on the structure, dynamics, and energetics of water at mineral surfaces: a molecular dynamics modeling study. *Geochim. Cosmochim. Acta* **70**, 562–582.

Wang J, Kalinichev AG and Kirkpatrick RJ 2009 Asymmetric hydrogen bonding and orientational ordering of water at hydrophobic and hydrophilic surfaces: a comparison of water/vapor, water/talc, and water/mica interfaces. *J. Phys. Chem. C* **113**, 11077–11085.

White GN and Zelazny LW 1988 Analysis and implications of the edge structure of dioctahedral phyllosilicates. *Clays Clay Miner.* **36**, 14–146.

Xiao Y and Lasaga A 1994. *Geochim. Cosmochim. Acta* **58**, 5379.

Xiao Y and Lasaga A 1996. *Geochim. Cosmochim. Acta* **60**, 2283.

Zhang L, Tian C, Waychunas G and Shen Y 2008. *J. Am. Chem. Soc.* **130**, 7686.

Zhang Q, Cağin T, van Duin A, Goddard WA, Qi Y and Hector LG 2004 Adhesion and nonwetting-wetting transition in the Al/α-Al$_2$O$_3$ interface. *Phys. Rev. B* **69**, 045423.

Zhang Z, Fenter P, Sturchio N, Bedzyk M, Machesky M and Wesolowski D 2007. *Surf. Sci.* **601**, 1129.

Zhang Z, Piatkowski L, Bakker H and Bonn M 2011. *Nat. Chem.* **3**, 888.

Zhao H, Bhattacharjee S, Chow R, Wallace D, Masliyah JH and Xu Z 2008 Probing surface charge potentials of clay basal planes and edges by direct force measurements. *Langmuir* **24**, 12899–12910.

Zhu X, Suhr H and Shen Y 1987. *Phys. Rev. B* **35**, 3047.

9

Biogeochemistry

Weilong Zhao,[1] Zhijun Xu,[1,2] and Nita Sahai[1,3,4]

[1] *Department of Polymer Science, University of Akron, Akron, OH, USA*
[2] *Department of Chemical Engineering, Nanjing University, Nanjing, China*
[3] *Department of Geology, University of Akron, Akron, OH, USA*
[4] *Integrated Bioscience Program, University of Akron, Akron, OH, USA*

9.1 Introduction

Physical–chemical interactions at the mineral–water–organic interface are ubiquitous in a wide range of biogeochemical processes. For example, biomineralization is directed by highly specific organic–mineral interactions (Addadi and Weiner, 1997; De Yoreo and Dove, 2004; George and Veis, 2008; Lovley, 1998; Mann, 2001). From a medical mineralogy perspective, normal and pathological biomineralization in humans are involved in bone and kidney stone formation. Geochemical reactions at mineral surfaces also control element release and uptake and, thus, affect water quality, contaminant transport in groundwater and surface water, soil formation, etc., which are essential processes to the regulation of our environment (Hochella and White, 1990). Many of these environmental processes involve bacterially mediated redox reactions, such as in acid mine drainage. Furthermore, minerals may have played a vital role in prebiotic chemistry and the origins of life on early Earth by concentrating prebiotic organic molecules by adsorption and/or by providing catalytic surfaces for nonenzymatic syntheses of biomolecules (Albery and Knowles, 1976; Bernal, 1951; Ellington and Benner, 1987; Haldane, 1929; James Cleaves et al., 2012), and carbonaceous chondritic meteorites may have served as sources for organic molecule delivery to early Earth. The increase in concentration of molecules at the mineral–water interface can provide a greater thermodynamic driving force compared to bulk solution. However, the role of minerals in interfacial processes is not restricted only to providing surfaces to concentrate molecules by adsorption. Interfacial reaction kinetics may also depend on the conformation of the adsorbed species. A great deal of research effort has, therefore, been invested in unveiling the mechanisms of these physical–chemical interactions.

Molecular Modeling of Geochemical Reactions: An Introduction, First Edition. Edited by James D. Kubicki.
© 2016 John Wiley & Sons, Ltd. Published 2016 by John Wiley & Sons, Ltd.

The extent of adsorption and molecular confirmation of the adsorbate rely on electrostatic interactions, covalent and hydrogen bonds, van der Waals forces, and hydrophobic interactions. The adsorption of amino acids on mineral surface is usually controlled by the ionic and hydrogen bonds between carboxyl or amine groups of the organic molecules and the ions on the surface. Solvation and desolvation free energy barriers of the amino acids and of the surface could also be pivotal, especially in the case where the mineral surface has a small surface charge. The strength of different driving forces can vary significantly, depending on charge density, solubility, topology, and defects, as well as solution pH, ionic strength, water activity, etc. Given these complexities, a complete understanding of the interactions at the mineral–water–organic interface cannot be achieved unless detailed atomic level structural and dynamic information of the adsorbent, adsorbate, and interfacial water is obtained.

Tremendous theoretical and computational advances have been made in recent years to predict the structure, energetics and dynamics of minerals (Cygan et al., 2004; Gibbs, 1982; Stack et al., 2012, 2013; Tossell and Vaughan, 1992), organic molecules (Jorgensen, 2004; McCammon et al., 1977; Warshel and Levitt, 1976), and the complex mineral–water–organic interface (see Chapter 8) (Harding et al., 2008; Latour, 2008; Raiteri et al., 2012; Yang et al., 2011). The rapid development of computational hardware resources and high-efficiency parallel algorithms has enabled the simulation of complex systems to be extended to large spatial ($\sim 10^9$ atoms) and time scales (~ 1 ms) (Das and Baker, 2008; Shaw et al., 2010). Therefore, molecular modeling techniques are being heavily exploited to understand the physical and chemical interactions in biogeochemical processes, such as biomineralization, and are beginning to be applied in the origins of life field (Šponer et al., 2013). *Ab initio* electronic structure methods have also been applied to tackle electron transfer reactions involving proteins and redox-active minerals in bacterially mediated redox reactions. In this chapter, we will focus on the insights provided and challenges posed by the application of atomistic molecular modeling techniques to understanding biomineralization mechanisms.

The creation of structurally delicate and functionally versatile biominerals in invertebrates and vertebrates requires carefully orchestrated interactions between organic molecules, inorganic dissolved ions, and inorganic minerals (Addadi et al., 2003; De Yoreo and Dove, 2004; Mann, 2001). The formation of the highly orientated and uniquely shaped biomineral crystals is controlled by the structure and chemistry of specialized organic matrices, which introduce huge complexities in understanding biomineralization mechanisms. Hence, researchers in the field have adopted a reductionist approach. That is, the complex problem is broken down into experimental and theoretical studies using simplified models of the major components, organic molecules, inorganic ions, and minerals. Despite much progress in this area, two formidable challenges remain in accurately modeling biomineralization. One is the development of force fields (FFs) that can accurately describe the complex free energy landscape of the mineral–water–organic interactions (see Chapter 2), and the other is extensive sampling of conformations for large-scale systems, especially those involving strong electrostatic bonds.

The content of this chapter is organized in the following fashion. We begin with a brief introduction to the basic interactions at the mineral–water–organic interface. We then focus on biomineralization and illustrate how computational modeling can aid the understanding of the pertinent interactions. The challenges to modeling nucleation and growth of biominerals are discussed, and the approaches to these challenges are explored in depth. Finally, to highlight the aforementioned points, we review several modeling studies of biogenic apatite and calcium carbonate ($CaCO_3$) phases. Since this is a book about molecular modeling on geochemical reactions, we shall assume a basic familiarity with techniques such as density functional theory (DFT) calculation (see Chapter 1) and molecular dynamics (MD) and Monte Carlo (MC) simulations (see Chapter 2). Those who intend to acquire detailed aspects about these techniques should consult texts and previous reviews (Frenkel and Smit, 2001; Yang et al., 2014).

9.1.1 Mineral–Water Interactions

The mineral surface has high complexity. For example, when cleaved from a bulk crystal, the atoms on an ideal mineral surface maybe expected to have identical positions to the crystal. In reality, the surface atoms undergo relaxation because of boundary effects, and, thus, their positions are shifted relative to the bulk (Hochella, 1990; Stipp and Hochella, 1991). Kinks, steps, vacancies, and other topological features are also inevitably found on crystal faces, especially high-index surfaces. Furthermore, surfaces exposed to air or aqueous phase are usually subject to chemical reactions such as oxidation and hydroxylation (Stipp, 2002). These deviations from perfectly flat mineral surfaces and changes in surface chemistry complicate the adsorption of molecules (McFadden et al., 1996; Sholl, 1998).

A charged mineral surface in aqueous solution develops an electrical double layer (EDL) that can be described by classical theory (Chapman, 1913; Stern, 1924) and its modern variants (Brown and Parks, 2001; Sahai and Rosso, 2006; Sahai and Sverjensky, 1997; Schindler, 1990; Stumm et al., 1970). Potential-determining ions such as protons, hydroxide ions, and surface-bound ions form a compact layer of charge at the mineral–water interface. These ions are held to the surface (only two-dimensional (2-D) diffusivity along the surface plane) by covalent or H bonding, short-range electrostatic forces, or some combination thereof. Counterbalancing this "compact" charged layer is a diffuse layer of ions of opposite charge and usually composed of background electrolyte ions. These ions are attracted to the surface by long-range electrostatic forces and have some limited diffusion in three dimensions. The concentration of counter ions decreases exponentially away from the surface until it reaches bulk solution values. The compact and the diffuse layers of charge comprise the EDL. The Gouy–Chapman equation describes the change in concentration and potential with distance from the surface (Chapman, 1913). Modifications to this classical EDL include the Stern–Grahame model to describe the potential function for a part of the compact layer (Stern, 1924). From this description of the EDL, it is clear that the surface charge and concentration of adsorbed ions depend on pH, ionic strength, and ion concentrations of the bulk solution. For instance, in the absence of other potential-determining ions, low pH results in surface protonation and, thus, the surface is positively charged, whereas at high pH the surface undergoes hydroxylation and becomes negatively charged. The pH at which surface charge is neutral is called the point of zero charge (PZC).

Due to the effects of electrostatic and H-bonding interactions with the surface, the structure and dynamics of interfacial water differ from bulk water in many ways (Brown and Calas, 2012; Zhao et al., 2014). One example is the different order of magnitude of dielectric constant (ε) between interfacial water (~ 6) and bulk water (~ 78) (Bockris and Jeng, 1990; Bockris and Reddy, 1973). This phenomenon is mainly the result of changes in the normal H-bonding network of water at mineral surface. Direct evidence of such structural changes has been found at the surfaces of corundum (α-Al_2O_3) (Catalano, 2010), calcite (Fenter et al., 2000), mica (Park and Sposito, 2002), and quartz (Du et al., 1994) by using X-ray reflectivity and spectroscopic methods. Results from MD simulation studies have also corroborated the breaking of bulk water H-bonding network for interfacial water on silicon and apatite surfaces at the atomic level (Phan et al., 2012; Zhao et al., 2014). These differences in the structures and dynamics of interfacial versus bulk water affect the interactions of mineral surfaces with adsorbed species. It is, therefore, indispensable to consider these phenomena when studying mineral–water–organic interfaces.

9.1.2 Mineral–Organic Interactions

A large variety of organic molecules are involved in biogeochemical interactions, including amino acids, peptides, soluble and insoluble proteins, polysaccharides, and polyamines. The interactions of hydrophilic organic species with the mineral surface can be classified as physisorption and chemisorption. The former is mediated by interfacial water molecules, protons, or hydroxyls through weak intermolecular forces, and the latter implies the formation of one or more chemical bonds. Generally speaking, depending on whether or not the organic species are able to penetrate the hydration layer on mineral surface, they form inner-sphere or outer-sphere species. In some cases, it is difficult to distinguish between H-bonded and outer-sphere species (Greiner et al., 2014; Jonsson et al., 2009).

In addition to the principles for local physical and chemical bond formation, changes in conformation between bulk solution and the adsorbed state must be considered for organic species with large molecular weight (100s to 10 000s Da), such as proteins. For example, nuclear magnetic resonance (NMR) and circular dichroism (CD) spectroscopies have revealed that the backbone conformation of hydrophilic polypeptides determines their interactions with the aragonite and calcite mineral surfaces (Collino and Evans, 2007; Kim et al., 2004). It has been suggested that polypeptides enhance their binding affinity by adopting flexible unfolded conformations on the surface, thus maximizing the number of interaction sites (Hunter et al., 2010). It is also likely that globular proteins encounter conformational transitions during adsorption. As a consequence of the order–disorder transition of the secondary structure, significant entropic contribution to free energy of adsorption arises in several ways: (i) the release of bound ions and water molecules from protein increases the entropy; (ii) the ordering of protein three-dimensional (3-D) structure decreases the entropy; and (iii) the flexibility of protein binding conformation on the surface increases the entropy. With the current experimental techniques, however, it remains extremely difficult to determine these thermodynamic driving forces.

In the rest of this chapter, we describe how molecular modeling can contribute to resolving interfacial physical–chemical interactions in biomineralization. The challenges and potential solutions in applying molecular modeling techniques will also be discussed and illustrated by specific case studies. Some of the solutions may not be ideal, but they serve as examples to direct future research in the field of biomineralization simulation.

9.2 Challenges and Approaches to Computational Modeling of Biomineralization

9.2.1 Biominerals: Structure, Nucleation, and Growth

Biominerals are composite materials containing organic matrices and minerals, with properties distinct from either phase alone, which maximize the advantages of organic matrices and inorganic minerals while minimizing their weaknesses. Examples include apatite in bone and dentin, aragonite and calcite in mollusk shells and eggshells, silica in sponge and diatom exoskeletons, and magnetite produced by bacteria. Living organisms manufacture biominerals for a variety of functions, such as mechanical support to organs, protection of soft body parts, movement, camouflage, and magnetic field sensing. Bone combines the fracture resistance and light weight of organic molecules with the hardness of inorganic crystals (Fratzl, 2008; Fratzl et al., 2004). The organic part of bone structure is composed of Type I collagen fibrils, which present themselves as extracellular matrix within which hydroxyapatite (HAP, ideal stoichiometry $Ca_5(PO_4)_3OH$) nanocrystals are deposited (Glimcher, 1959, 1989, 2006). The collagen molecules self-assemble to form fibrils and fibers from nano- to micrometer scale, and HAP nanocrystals form a strict structural registry with the collagen molecules (Figure 9.1). The multilevel hierarchical structure of the collagen–HAP composite is responsible for bone's superior mechanical properties (Fratzl et al., 1996; Landis, 1995; Qin et al., 2012). Understanding bone mineralization can assist in both the design of composite biomaterials for orthopedic applications and the determination of biological pathways for treating bone diseases. Thus,

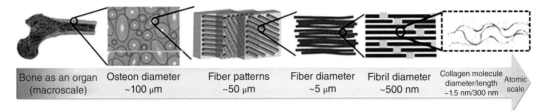

Bone as an organ (macroscale)	Osteon diameter ~100 μm	Fiber patterns ~50 μm	Fiber diameter ~5 μm	Fibril diameter ~500 nm	Collagen molecule diameter/length ~1.5 nm/300 nm	Atomic scale

Figure 9.1 *Hierarchical structure of bone form macro to atomic scale. The yellow rectangles represent HAP nanocrystals between and within the fibrils. See Section 9.3 for further details. Nair et al. (2013). Reproduced with permission of Nature Publishing.*

biomineralization serves as an excellent illustration of the general significance of studying biomineral formation at the molecular level.

The notion that the organic matrix acts as a structural template to promote epitaxial nucleation of biominerals by lattice matching between atomic positions on the organic and the mineral was a commonly held principle for decades (Addadi et al., 1989). This lattice match, for example, was proposed to facilitate the direct nucleation and growth of aragonite in the nacre of the mollusk shell (Weiner et al., 1983). However, based on *in vitro* studies, a different model had been proposed previously for bone biomineralization, in which amorphous calcium phosphate (ACP) clusters were believed to form as precursors to apatite (Termine and Posner, 1966a, 1966b). The ACP solid particles precipitated initially were 30–100 nm in diameter and consisted of randomly packed $Ca^{2+}-PO_4^{3-}/HPO_4^{2-}$ (Pi) clusters, which were themselves approximately 9.5 Å in diameter. It was estimated that these smaller, so-called Posner's clusters contain about nine Ca^{2+} and six Pi ions. The 30–100 nm solid particles lacked long-range order, although they did have some short-range order (Posner and Betts, 1975; Termine and Posner, 1967).

The model of epitaxial nucleation and growth is further being challenged by the application of advanced spectroscopic and microscopy approaches to the early stages of mineralization of tissues *in vivo* and *ex vivo* (Mahamid et al., 2008). By applying a combination of X-ray adsorption near-edge structure (XANES) spectromicroscopy, transmission electron microscopy (TEM), and Fourier transform infrared (FTIR) spectroscopy techniques, Beniash et al. (Beniash et al., 2009) characterized the early stages of enamel mineralization and concluded that the early formed solid phase is ACP.

The nucleation of amorphous precursors before the crystalline phase has been observed for calcareous invertebrate tissues (Gotliv et al., 2003; Nassif et al., 2005; Politi et al., 2004, 2006, 2008). Based on the experimental results, the biomineralization community has questioned the role of organic matrix in acting as a structural template for epitaxial nucleation of biominerals. The pendulum appears now to have swung to the other extreme where the formation of "stable prenucleation clusters (PNCs)" (discussed later) and amorphous phases as precursors to the final crystalline phases is believed to be a universal phenomenon for all mineralized tissues (Navrotsky, 2004).

Recently, the so-called PNCs of calcium carbonate $(Ca-CO_3)$ or calcium phosphate $(Ca-PO_4)$ have been recognized *in vitro* by cryo-TEM analysis of solutions supersaturated with respect to calcite (Gebauer et al., 2008; Pouget et al., 2009). PNCs are described as "stable clusters that are present in solution already before nucleation" (Dey et al., 2010). Gebauer et al. (2008) have recognized that the intermediate step from PNCs to solid ACC was not identified. Subsequently, a considerable amount of effort was invested in determining this intermediate step using cryo-TEM for the $Ca-CO_3$ system in the presence of a stearic acid monolayer as a substrate (Pouget et al., 2009) and for the $Ca-PO_4$ system in the presence of arachidic acid monolayer. PNCs of 1 nm size were identified to form *in solution* in both systems. In the $Ca-CO_3$ system, the PNCs were proposed to assemble *in solution* into solid ACC nuclei (i.e., homogeneous nucleation pathway) of approximately 30–70 nm in size, and the ACC nuclei were then proposed to attach to the monolayer and, finally, transform into crystalline $CaCO_3$ (Pouget et al., 2009). In contrast, a heterogeneous nucleation mechanism for the $Ca-PO_4$ system was proposed, in which the PNCs were proposed to attach directly to the arachidic acid monolayer and form aggregates (Dey et al., 2010). These aggregates were then supposed to densify by partial dehydration and transform into solid ACP nuclei. In another study, cryo-TEM was also employed to investigate *in vitro* $Ca-PO_4$ mineralization of collagen in the presence of polyaspartate (Nudelman et al., 2010). Without collagen, polyaspartate was found to limit the size of the ACP particles. In the presence of collagen and polyaspartate, ACP particles of approximately 10 nm size were observed at 24 h incubation, and some of these particles were located near the early nucleation sites within collagen fibrils.

More recently, various techniques were employed to study a $Ca-PO_4$ solution within a pH range of 7.2–7.4 and without any other additives present (Habraken et al., 2013). It was determined that $Ca(HPO_4)_3^{4-}$ and $Ca(H_2PO_4)(HPO_4)_2^{3-}$ are the dominant (>86%) PNC species and were renamed as "ion-association complexes," which eventually aggregated and formed amorphous ACP nuclei. Thus, the formation of 1-nm-sized clusters or ion-association complexes in solution and solid, amorphous ACC or ACP particles was observed in both the $Ca-CO_3$ and $Ca-PO_4$ systems.

While these are interesting studies, the experimental systems do not reflect the *in vivo* system, especially the fatty acid monolayer substrates. Most importantly, complicated by the presence of PNCs, the pathway of amorphous precursors transforming into crystalline phases remains difficult to establish.

MD simulations have been brought to bear in an attempt to identify the mechanisms for $Ca-CO_3$ and $Ca-PO_4$ nucleation (Demichelis et al., 2011; Wallace et al., 2013; Xu et al., 2015). Demichelis et al. (2011) applied classical MD simulations to study cluster formation in a $Ca-CO_3$ system at high concentrations of Ca^{2+}, HCO_3^-, and CO_3^{2-} ions at different pH. The cluster structures were examined as a function of the radius of gyration, and the Helmholtz free energy for clusters formation was obtained. The solution-phase cluster was named dynamically ordered liquid–liquid oxyanion polymer (DOLLOP). Structure analysis results were interpreted as suggesting the presence of two distinct energy minima, for DOLLOP and for ACC. It was proposed that the minima are separated by a large activation barrier, which prevents spontaneous transitions between these two structures. The free energy calculations were not able to address the thermodynamics associated with the dehydration of DOLLOP; nonetheless, the inferences made from this study were consistent with the results of the original study of Gebauer et al. (2008). Notably, the authors themselves recognized that the term "PNC" is an inappropriate term for the solution-phase cluster (DOLLOP), because they are stable and exist even after nucleation. In a different study, the transition of ions in solution to the dense liquid clusters was examined in reference to the classical liquid–liquid phase separation (Wallace et al., 2013). No energy barrier was obtained up to clusters of 40 atoms, but a barrier was identified for transition from clusters to ACC. The authors described these clusters as "nanoscale droplets," which were proposed to coalesce and density to form ACC, though the simulation did not address this step of the process.

The identification and nomenclature of these various species and phases, such as PNCs, ion-association complexes, DOLLOP, and nanoscale droplets, pose some experimental and conceptual questions. Firstly, based on decades of geochemical research, it is already well known that *aqueous* ion pairs and/or soluble multinuclear complexes may be present in solution before a solid nucleus forms, so the novelty of their identification in the studies summarized earlier is not clear. Gebauer et al. (2008) did, in fact, recognize that these sorts of oligomeric species are already known in the literature for various metals and metalloids. Examples include the dissolved ion complex of calcium carbonate ($CaCO_3^0$) and multinuclear mixed hydroxo–carbonato complexes of uranyl ($(UO_2)_2CO_3(OH)_3^-$), silica, and iron sulfides of various stoichiometry. "Clusters" are also known to be involved in the dissolution or precipitation of other kinds of solids such as polyoxometalates. Well-known examples include the Keggin anion in aluminum oxyhydroxide and hafnium sulfate ionic complexes, which are an intermediate phase during crystallization of Hf_{18} polyoxometalate (Bertsch et al., 1994; Ruther et al., 2014). Based on these considerations, it would appear that the various species of 1 nm size are the visualized *aqueous* multinuclear complexes.

A similar phenomenon has been identified, by the application of synchrotron X-ray absorption spectroscopy, in the adsorption of transition metal ions at the surfaces of metal oxides (Sahai et al., 2000). With increasing concentration of the transition metals, the adsorbed species changed from mononuclear metal ions, to ion pairs, to multinuclear hydroxo and carbonato complexes, and, ultimately, to a complete hydroxide or carbonate surface precipitate. The difference between adsorbed multinuclear complexes and surface precipitate nucleation is difficult to determine experimentally.

Another conceptual problem is that whether clusters of such small sizes can be differentiated between crystalline and amorphous. If the length scale of the visualized object is less than or approximately equal to one unit cell of the crystal, then by the very definition of crystallinity (long range order), such clusters cannot be anything but amorphous. The unit cell parameters of HAP are $a = b = 9.4$ Å and $c = 6.88$ Å. The unit cell parameters of calcite are $a = 4.99$ Å and $c = 17.06$ Å, and aragonite is $a = 4.95$ Å, $b = 7.96$ Å, and $c = 5.74$ Å. Thus, it is obvious that the smallest 1 nm size clusters have to be amorphous, by default.

Finally, the *in vivo* characterization of the stoichiometry and structure of the first nucleated phase remains technically challenging, because of (i) the difficulty in timing the exact "moment" when the first nuclei are formed and when any potential phase transition may occur and (ii) artifacts introduced when preparing the tissue samples for spectroscopic or microscopic analyses, including cryo-TEM (Glimcher, 2006).

In summary, the high spatial and temporal resolution of modern microscopic and spectroscopic methods enables the observation of dissolved ion complexes and clusters. Yet the intermediate phase, in the transition to the solid nuclei, we argue, has to be metastable, by definition.

9.2.2 Conformational Sampling in Modeling Biomineralization

9.2.2.1 Challenges

Using computational methods to study nucleation in biomineralization offers a major advantage, as in theory, the sub-nanometer to 10s of nanometers structural and energetic information of crystallization pathways can be obtained directly from potential-energy-based calculations. The induction time for nucleation in experiments may be longer than can be handled by classical MD simulations, and advanced sampling methods can be applied to overcome this limitation in large part (see Section 9.2.2.2; also see Chapter 11). Phase transitions between dissolved ion complexes and solid-phase clusters or between different polymorphs of a particular stoichiometry can be identified in computer simulation, by properly defining order parameters, such as bond orientation parameter and angular order parameter (Lechner and Dellago, 2008; Leyssale et al., 2005; Steinhardt et al., 1983). The bond orientation parameter is a measure of the relative arrangement of atoms or ions, and the evolution of its probability distribution in a cluster can be used to calculate the free energy surface of nucleation. By performing statistical studies of bond orientation parameter on the results of conventional, longtime MD or rare events methods (e.g., umbrella sampling and metadynamics), previous workers have successfully investigated the structural evolution and the energy landscape of liquid to solid phase transition of water (Matsumoto et al., 2002; Quigley and Rodger, 2008; Radhakrishnan and Trout, 2003). Despite such successes, however, molecular simulation of biomineral nucleation is especially complicated, because of the presence of the organic matrix in addition to the ionic nature of the system.

Complexity of Nucleation Pathways. There are a series of events starting from the assembly of multinuclear complexes to the formation of a solid nucleus, along with the diffusion of individual ions or ion-association complexes to the central nucleus. These dynamic processes are extremely difficult to simulate computationally with a faithful replication of realistic experimental conditions. One promising strategy is to divide the nucleation process into different stages, including initial sequestration of ions by proteins, diffusion of ion-association complexes and multinuclear complexes, dehydration of dissolved clusters, etc. With the aid of high-performance computing, the application of structural analysis and free energy calculation to each stage can provide clues to the complex pathways of nucleation in biomineralization (Raiteri and Gale, 2010; Wallace et al., 2013; Yang et al., 2010).

Organic Matrix. Apart from simulating nucleation pathways, another major hurdle in modeling biomineralization is the general lack of high-resolution (Å-scale), 3-D structures of the proteins comprising the organic matrix (Yang et al., 2014). Taking the case of the bone, several acidic, non-collagenous proteins (ANCPs) play important roles in controlling the HAP formation in addition to insoluble collagen fibrils, by acting as nucleation promoter, inhibitor, or structural template for crystal growth (George and Veis, 2008). These proteins, including bone sialoprotein (BSP), osteopontin (OPN), and matrix extracellular phosphoglycoprotein (MEPE), are highly negatively charged, because of the presence of domains that are rich in side chain residues of glutamic acid (Glu), aspartic acid (Asp), and phosphorylated serine (SerP). Furthermore, many of the ANCPs take flexible, random coil structures in solution (Dunker et al., 2002; Fisher et al., 2001; Tye et al., 2003). This intrinsically disordered nature of ANCPs is believed to be related to their biological functions, in that the flexible, easily accessible local domains can provide multiple binding sites for inorganic ions, mineral, collagen, and cell surface receptors at the same time (Bellahcene et al., 2008; Ganss et al., 1999; Hunter and Goldberg, 1993, 1994; Yang et al., 2011). Without a global energy-minimized protein crystal structure to enable the initial simulation setup, it is inevitable that some arbitrary starting conformations are used (Azzopardi et al., 2010). Sampling the folding of protein from an arbitrary starting point is a problem already complicated in itself because of multiple intramolecular interactions, and the computational cost grows exponentially as the size of the protein increases for each additional amino acid residue (Bowman et al., 2011; Freddolino et al., 2008, 2010). This problem adds further complexities to the simulation of the functions of ANCPs in controlling crystal nucleation and growth. If the conformation at local energy minimum is taken as the interaction motif of the protein with mineral surfaces or dissolved ions, there remains a huge possibility that the free energy barrier of protein folding cannot be overcome to find the lowest energy state. Thus, the results of such local energy minimum ensembles may not be realistic.

The situation is more tractable when the 3-D structures of proteins pertinent to biomineralization are available (Orgel et al., 2006; Reyes-Grajeda et al., 2007). However, even in these cases, a full exploration of the possible conformation changes is hard to achieve. For example, the crystal structures of osteocalcin (OCN) and statherin are available, and their interactions with HAP are proposed to influence the nucleation and growth of the mineral (Flade et al., 2001; Hoang et al., 2003; Long et al., 2001). Most experimental studies of the HAP–protein interactions, however, reveal only partial structural information at the interface. For instance, CD spectra can provide the backbone conformation of the protein, and the side chain residue positions are not revealed, so spatial correlations with HAP surfaces cannot be deduced. Solid-state NMR (ss-NMR) can detect spatial information to local interactions (<1 nm), but this is less than the length scale of side chain, backbone, and mineral interaction. Thus, a single type of experiment or spectroscopic analysis cannot provide the structural relationship between the proteins and the mineral. Therefore, a great deal of effort has been invested in applying MD simulations to understand whether and how the HAP surface affects the 3-D structure of adsorbed proteins (Capriotti et al., 2007; Gray, 2004; Makrodimitris et al., 2007; Masica and Gray, 2009; Yang et al., 2011).

Sampling Electrostatic Interactions. Identifying systematic patterns of conformational changes of proteins from the aqueous state to the surface-bound state is the central objective in these studies. However, because most biominerals are ionic salts, their electrostatic interactions with the partnering proteins produce large energy barriers. The energy barriers easily trap the proteins at local minima, so classical MD or MC methods often cannot sample adequately the conformational changes within limited simulation time scale. This problem is common to both flexible, disordered proteins and those for which 3-D crystal structures are known. Additionally, ionic minerals are hydrophilic and their surfaces are strongly solvated. It is challenging in the simulation for weakly charged proteins to overcome the desolvation energy barrier of the mineral surface and ultimately reach the most favorable binding conformation.

In general, when the correct native conformation of the protein cannot be sampled by MD simulation, any interpretation of its interactions with mineral surfaces, clusters, or ion-association complexes (or multinuclear complexes) would be misleading. Hence, how to establish the most stable 3-D structures of proteins adsorbed at the mineral surface is an essential issue in computational studies of biomineralization (Figure 9.2).

For example, highlighting the issues for determining both protein conformations and nucleation is offered by the formation of $Ca^{2+} - Pi$ in bone. Type I collagen molecules are the dominant organic component of the bone. The molecules self-assemble into a fibril such that the molecules are arranged in a pseudohexagonal array (Figure 9.1). The position of the collagen molecules is staggered with respect to the others, resulting in the formation of "hole" zones within which the mineral phase is initially deposited (Figure 9.1). The detailed nucleation and localization mechanisms of the $Ca^{2+} - Pi$ phase in the intrafibrillar holes remained unknown for the past six decades, because the 3-D crystal structure of collagen fibrils was not determined. The crystal structure has only been obtained recently by synchrotron X-ray diffraction (XRD) and this is a low-resolution structure (5 Å axial and 11 Å equatorial) with only backbone information (Orgel et al., 2006). The fact that the crystal structures of the collagen side chains have not been reported precluded studies about their interactions with water, inorganic ions, and organic molecules. It was only recently that the aforementioned challenges were start to be addressed by the application of advanced sampling methods, including replica exchange MD (REMD) and umbrella sampling (Demichelis et al., 2011; Wallace et al., 2013; Xu et al., 2014, 2015).

9.2.2.2 Advanced Simulation Approaches

In classical MD simulation, the physical properties of the system of interest are reflected by a time-resolved trajectory that records the positions of atoms in a 3-D configuration space. The trajectories are obtained based on Newton's laws of motion by calculating the interatomic potentials and the derived interacting forces at each time step. The structural and dynamic information of the system can be directly obtained by statistical analysis of the trajectory. Thus, classical MD would appear to be a suitable method to study nucleation events in biomineralization. However, this method has a prominent limitation, which is the temporal and spatial scale of the system of interest. Nucleation is usually a rare event, sometimes taking up to days under laboratory conditions and *in vivo*. Confined

(a)

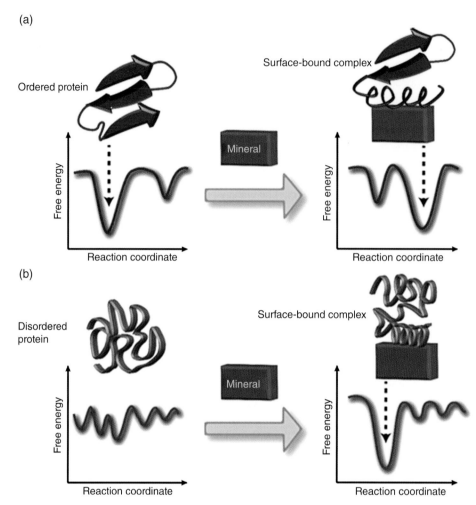

(b)

Figure 9.2 *Schematic illustration of the complex energy landscapes of protein–mineral interactions. (a) Ordered protein shifts its free energy minimum configuration upon binding to the mineral surface. (b) Disordered protein displays a multimodal distribution of free energy minima configurations; it forms a sharp free energy minimum upon binding to the mineral surface. Note that the bound protein may be partially structured as shown or may remain disordered with a random different configuration compared to the solution state.*

by the capacities of computational algorithms and hardware, classical MD simulation is usually terminated at 100 ns, which is much below the time scale of a nucleation event. One of the few exceptions is the study of ice formation. By extending the simulation time to 500 ns, Matsumoto et al. (2002) have observed the phase transition from liquid water to hexagonal ice by classical MD. Nonetheless, performing microsecond level of simulation in an acceptable real-world timeframe is not easily achievable for biological large systems (10^5 to 10^9 atoms) in general. For this reason, research efforts have been dedicated to accelerate the calculation of classical MD simulation. Through the engineering of a specialized high-performance computation facility equipped with a high-efficiency MD calculation algorithm, Shaw and coworkers have reported the folding and unfolding dynamics of protein at millisecond time scale (Piana et al., 2012; Shaw et al., 2010). The calculation speed of

Table 9.1 *A list of molecular simulation methods that can be used for modeling different processes at the mineral–water–organic interface.*

Method	Example of system studied	Advantages	Major limitations
Ab initio MD	Interaction of amino acids with calcium oxalate monohydrate surface	Models chemical bond breaking/making	Extremely expensive for system larger than 1000 atoms
Classical MD	Nucleation of calcium phosphate in the presence of NCP peptide	Directly extracts structural and dynamic information	Inadequate conformational sampling
REMD	Nucleation of calcium phosphate inside collagen matrix	Provides sufficient conformational sampling	Expensive for explicit solvent
RosettaSurface	Adsorbed structure of statherin on HAP surfaces	Provides efficient conformational sampling	No dynamic information; solvent is implicit
Steered MD	Binding free energy of amelogenin peptide with HAP surfaces	Provides reaction pathway and corresponding free energy surface	Inadequate conformational sampling
Umbrella sampling	Binding free energy of amino acids with HAP surfaces	Provides accurate free energy surface	Relevant reaction coordinates limited to one or two
Metadynamics	Nucleation of calcium carbonate in solution	Provides free energy surface of complex pathway	Difficult to achieve convergence

this special MD simulation machine exceeds that of the most advanced supercomputer cluster by a factor of more than 1000, which, in theory, is able to simulate rare events such as crystal nucleation within an affordable real timeframe. In addition, with the rapid development of modern graphics processing units (GPU), a tremendous acceleration of numerically intensive scientific computing has been realized (Stone et al., 2010; Sweet et al., 2013). Implementing GPU and GPU-based algorithms into classical MD method provides a promising route to expand the current spatial and temporal scales of biological simulation. However, limited by the common use of general-purpose MD and CPU-based algorithms, expanding the time scale of the simulation is more approachable by implementing advanced sampling methods (Table 9.1).

Replica Exchange Molecular Dynamics (REMD). REMD is a special technique that allows the system to cross energy barriers efficiently and, hence, can accelerate conformational sampling. The most widely used REMD scheme is temperature REMD (TREMD) (Hansmann, 1997; Sugita and Okamoto, 1999). In TREMD, N replicas of the system are simulated in parallel at different temperatures (Figure 9.3). The first replica, P_0, is coupled to the temperature of interest (usually room temperature), T_0. The last replica, P_n, is coupled to the final temperature, T_n, which is high enough to eliminate any bottlenecks in free energy space. At regular intervals, the conformations of adjacent replicas, P_i and P_j, are exchanged by the simulation software, based on the following Metropolis-fashion acceptance probability:

$$p\left(P_i \leftrightarrow P_j\right) = \min\left(1, \exp\left(\frac{1}{kT_i} - \frac{1}{kT_j}\right)\left(U_i - U_j\right)\right) \tag{9.1}$$

where k is the Boltzmann constant, T is temperature, and U is the potential energy of the system. By using this scheme, the replica that resides at the temperature of interest can travel to higher temperatures and, thus, is able to cross the energy barriers. The canonical distribution at low temperature can be fully recovered to sample energy minima configurations. An optimal TREMD simulation

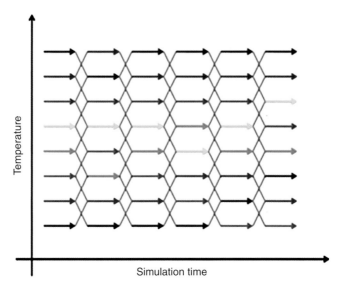

Figure 9.3 *Basic algorithm of TREMD method. N independent replicas are spaced on a temperature "ladder" and are simulated in parallel (in HREMD, potential energies are scaled on the ladder instead of temperature). At fixed time interval, Metropolis test is applied to decide if the two adjacent replicas can be interchanged according to their potential energies (Eq. 9.1). Orozco (2014). Reproduced with permission of Royal Society of Chemistry.*

should be performed in such a way that the temperature interval can result in a uniform exchange probability between adjacent replicas (Patriksson and van der Spoel, 2008; Rathore et al., 2005; Trebst et al., 2006). TREMD enables the efficient sampling of the conformational changes of biological molecules, provided that the replica at the temperature of interest explores the high temperature states extensively (van der Spoel and Seibert, 2006). This approach has led to the successful exploration of free energy landscapes for adsorption of peptides to self-assembled monolayer and metal oxide surfaces (Deighan and Pfaendtner, 2013; Schneider and Colombi Ciacchi, 2011) and should be suitable in biomineralization studies.

Another variant of the replica-exchange scheme is Hamiltonian REMD (HREMD) (Fukunishi et al., 2002). HREMD is far more flexible and usually computationally less expensive (Jang et al., 2003; Liu et al., 2005). In HREMD, each replica is characterized by a different potential energy rather than by a temperature. In its simplest implementation, the potential energies of the replicas differ by a scaling factor, $c^{(m)}$, with $c^{(m)} = 1$ for the original system. The potential energy of the mth replica (U_m) can be decomposed to different energy contributions (u_i), such as van der Waals interaction, electrostatic interaction, bonded interaction, solvent–solvent interaction, solvent–solute interaction, etc., which renders the following expression:

$$U_m = \sum_{i=1}^{k} c_i^{(m)} u_i \tag{9.2}$$

where $c_i^{(m)}$ is the scaling factor for the mth replica. At specified time intervals, replicas are exchanged in a similar fashion as in TREMD:

$$p(c_m \leftrightarrow c_n) = \min\left(1, \exp\left(\frac{1}{kT}(c_m - c_n)(u_m - u_n)\right)\right) \tag{9.3}$$

The ability of HREMD to scale the target component of potential energy (u_i) offers a solution in which specific interactions of the system can be localized and sampled separately, with the remaining parts behaving normally. The large energy barriers produced by strong electrostatics can, therefore, be crossed efficiently without spending huge computational resources on sampling other interactions. HREMD method can be applied to enhance the conformational sampling of protein–ion and ion–ion interactions in biomineralization. By adopting HREMD to account for the interactions between ionic species, our group has recently reported intrafibrillar $Ca^{2+} - Pi$ multinuclear complexes formation within a high-resolution 3-D collagen fibril structure (Xu et al., 2015). This example will be discussed in Section 9.3.1 to demonstrate the advantages of HREMD for modeling nucleation in the presence of an organic matrix.

Bioinformatics. Resolving the adsorption conformation of biomolecules on mineral surface is one of the major goals in biomineralization research, as the information contributes to the understanding of growth modification and morphology control of biominerals (Harding et al., 2008; Hu et al., 2010; Yang et al., 2014). REMD methods lend possible solutions, but the realization still requires a significant amount of computational resources, which may not be available under many circumstances. A recently developed, bioinformatics-based simulation protocol, *RosettaSurface*, offers an alternative approach and has been used to predict the most stable structures of adsorbed proteins at calcite and HAP surface (Makrodimitris et al., 2007; Masica and Gray, 2009; Masica et al., 2011; Pacella et al., 2013). The algorithm of *RosettaSurface* is derived from the applications of protein folding and protein–ligand docking (Bradley et al., 2005; Rohl et al., 2004; Schueler-Furman et al., 2005). One of its significant advantages is its ability to generate rapidly a large ensemble of conformations. In *RosettaSurface*, the solid surface is analogous to the "binding pocket" in traditional protein–ligand docking, and the protein of interest is transferred to the solid surface by optimizing the interaction energies between the protein and the surface. The protocol first runs a fast, backbone-only calculation to generate protein folds and, subsequently, performs all-atom optimization steps to improve predicted structures. The refinement of protein adsorption conformations on the solid surface is based on energy minimization using both physical potentials and database-derived, empirical functions. On a modern workstation laptop, approximately 10 000 adsorption structures can be generated within 1 day; therefore the protein's conformational space can be sampled extensively using only limited computational resources. *RosettaSurface* can be coupled with available structural data of protein–surface interface, such as solid-state NMR, to give the most experimentally relevant, low energy conformations of proteins on the mineral surface (Masica et al., 2010).

However, in order to reduce computational cost, *RosettaSurface* treats the solvent implicitly in the energy calculation. This design feature may become a significant drawback in solid–aqueous interface simulation, as the validity of applying implicit solvent model is not fully justified (Ma and Nussinov, 1999; Verdaguer et al., 2006). Furthermore, the detailed information involving water molecules at the protein–solid interface, such as specific hydrogen bonding, is inevitably lost. An improvement to this approach is to combine *RosettaSurface* with classical MD simulation (Yang et al., 2013). The low energy conformations of adsorbed peptides or proteins on the solid surface can be used as the initial structures with explicit solvation MD simulation. By taking advantage of the extensive conformational sampling of *RosettaSurface*, much greater confidence can be placed in the results from subsequent MD, and global energy minima of peptide/protein structures at the surface can thus be obtained. An example of such a bioinformatics-assisted MD approach is presented in Section 9.3.1.

Metadynamics. Going beyond classical MD, free energy calculation methods, including umbrella sampling (Torrie and Valleau, 1977) and metadynamics (MetaD) (Laio and Gervasio, 2008; Laio and Parrinello, 2002), are now being employed in attempts to model biomineralization mechanisms. These methods lend themselves to an accurate calculation of the free energy surface of a complex process by estimating the potential of mean force (PMF) along specific reaction coordinates (order parameters of the system) of the process. The technical aspects of umbrella sampling and its applications in biomineralization simulation have already been introduced elsewhere by Sahai and coworkers (Yang et al., 2014) and will not be discussed here. Compared to umbrella sampling, MetaD approach is more easily adapted to describing the free energy landscape spanned by complex reaction coordinates and, thus, may be more suitable to model biomineralization mechanisms. In MetaD, a history-dependent, external potential is iteratively and adaptively added to the system to compensate for the underlying free energy surface. The external potential is a sum of Gaussian functions, which are defined on the basis of the system's reaction coordinates:

$$V\left(\vec{s},t\right) = \int_0^t dt' \omega \exp\left(-\sum_{i=1}^d \left(\frac{s_i\left(\vec{R}\right) - s_i\left(\vec{R}\left(t'\right)\right)}{2\sigma_i^2}\right)^2\right) \qquad (9.4)$$

where $V\left(\vec{s},t\right)$ is the biasing potential, t is time, ω is the energy scale of the Gaussian function at each timestep (Gaussian height), $s_i\left(\vec{R}\right)$ is the reaction coordinate of the system in the ith dimension, and σ_i is the Gaussian width corresponding to the ith reaction coordinate. The biasing potential forces the reaction coordinates to evolve toward unfavorable regions in phase space, thus driving the system to escape from local energy minima. Based on different interactions under consideration, the reaction coordinate can be distance, angle, coordination number, potential energy, or root-mean-square deviation (RMSD) of these values, and the combination of several reaction coordinates in one simulation is also feasible (Piana and Laio, 2007; Raiteri et al., 2005). The free energy landscape can finally be reconstructed through the integration of Gaussian functions (Bussi et al., 2006a; Ensing et al., 2006). The capacity of MetaD to extend the sampling to multiple dimensions of reaction coordinates makes it a promising method to investigate nucleation in which multiple-order parameters, such as interatomic distances and bond orientations, are involved (Quigley and Rodger, 2008a, 2008b).

For MetaD simulation, full convergence is a critical premise to reproduce accurately the free energy surface, and it relies upon carefully tuning the phase space explored (Barducci et al., 2008). Implementing REMD method into the MetaD scheme (Bussi et al., 2006b) and considering hidden degrees of freedom explicitly in simulation (Mori et al., 2013) are practical approaches to achieve good convergence in MetaD simulation.

9.2.3 Force Field Benchmarking

The lack of reliable empirical all-atom FFs arises as a significant issue in MD simulation of biomineralization, as stressed in previous publications (see Chapter 2) (Harding et al., 2008; Latour, 2008; Raiteri et al., 2013; Yang et al., 2014). Unlike biological molecules, such as proteins, linear peptides, DNA, and RNA, the accurate parameterization of FFs for minerals in bulk and at the mineral–water interface is absent for routinely used molecular mechanics (MM) FFs, including AMBER, CHARMM, OPLS, and GROMOS. The establishment of FFs to describe biominerals is difficult partly because the relevant experimental information is not readily available, especially in the case of mineral–water and mineral–organic interfaces. Face-specific information is scarce for HAP because of the difficulty in growing single crystals with large surface area (>1 mm^2). The situation is less severe for which large and fresh surfaces can be created along cleavage planes. The calibration between experiments and simulations, however, is limited in both cases.

Another major obstacle is that mineral surfaces are not flat and presents defects, including steps, edges, kinks, vacancies, etc. These local structural features are crucial to crystal growth modulation by biomolecules yet often difficult to measure quantitatively. Chemical modifications to the surface, such as hydroxylation, protonation, and isomorphic substitutions, are not well characterized either. Taking the example of HAP, most studies are limited to average properties measured from bulk crystal powder, thus providing little information on the surface density of modified sites on specific crystal faces, which are important for HAP–protein interactions. These structural variations create tremendous complexities in benchmarking FFs of the mineral–water–organic interface.

Several FFs have been developed to treat HAP (Bhowmik et al., 2007; Hauptmann et al., 2003; Xu et al., 2014), calcite (Freeman et al., 2007; Raiteri et al., 2010), and some general FFs such as CLAYFF (Cygan et al., 2004) and INTERFACE FF (Heinz et al., 2013). Precaution is required when applying some of these FFs, since their parameters are originally fitted for optimizing the bulk structural and mechanical properties of the solid, whereas interfacial terms were not accounted for accurately. The differences between the effective electrostatics and other nonbonded interactions of ions in the bulk crystal and at the mineral–water interface will lead to inaccuracy of the calculated thermodynamic behavior. Moreover, the common approach to treat the van der Waals interactions between the mineral surface and the organic molecule is to employ Lorentz–Berthelot mixing rule; however, only in a few cases has this method lead to success with respect to reproducing the interfacial properties (Xu et al., 2014). Without systematic benchmarks using available experimental information and results from high-level quantum mechanics (QM) calculations, the simulation could be seriously biased toward unphysical results, as seen in some previous studies (Kawska et al., 2008; Pan et al., 2007).

In order to reproduce the thermodynamic properties of the mineral–water–organic interface with high accuracy, several criteria can be used to address whether the FFs are reliable (Stack et al., 2013): (i) Do the structure and dynamic behavior of water molecules resemble those measured in experiments? (ii) Do the FFs predict the solvation and pairing free energies correctly for ions that constitute the minerals? (iii) Are the FFs able to predict the most stable polymorph in a phase-transition simulation? (iv) Can the dissolution constant/solubility product of the mineral (K_{sp}) be reproduced by the FFs? (v) Can the binding free energies of organic molecules be accurately calculated using the FFs? In spite of some recent progress (Fenter et al., 2013; Raiteri et al., 2013; Xu et al., 2014), the answers to these questions remain elusive so far for many applications. In summary, the importance of carefully validating the FFs applied in simulation of mineral–water–organic interface cannot be overstated.

9.2.4 *Ab Initio* MD and Hybrid QM/MM Approaches

QM methods provide an alternative to overcome some of the shortfalls in the development of FFs in classical MD. The electronic structure of the system is calculated by solving Schrödinger's equation, thus providing high-accuracy interaction energies. With the appropriate choice of basis sets, a satisfying balance between accuracy and efficiency can be achieved (see Chapter 1). In the case of peptide or protein adsorption to the mineral surface, the reaction sites on these biomolecules can be reduced to only a few amino acids (Chen et al., 2006; Freeman et al., 2011). Thus, isolating a system to a size of 10s to 100s atoms and applying QM scheme will provide insight about protein–mineral interactions. *Ab initio* MD and hybrid QM/MM are the two QM approaches that have potential for addressing some of the hurdles in modeling biomineralization.

Ab initio MD, such as Car–Parrinello MD (Car and Parrinello, 1985), extends the range of molecular modeling methods. The basic idea underlying *ab initio* MD is to compute the interatomic forces by performing high-level electronic structure calculation, meanwhile including time-evolved dynamics of the system. The combination delicately resolves one limitation of classical MD, which is the simulation of dynamic breaking and making of chemical bonds. Given a suitable approximation of the many-electron orbitals, chemical reactions of the system can be handled by *ab initio* MD and have been successfully applied to tackle problems in a variety of material science and chemistry subjects (Andreoni and Curioni, 2000; Hutter et al., 1996). *Ab initio* MD is a powerful approach to deal with mineral–water–organic interfacial systems on the order of 100–200 atoms. By employing this method, Lischka and coworkers have investigated the adsorption and chemical reactions of water and small organic molecules on the surface of kaolinite clay (see Chapter 6) (Tunega et al., 2001, 2004). The interactions of model amino acids with calcium oxalate monohydrate have also been investigated by *ab initio* MD (Hug et al., 2010).

Despite the success of this approach, the computational cost of *ab initio* MD is considerably expensive for systems larger than approximately 200 atoms, and the simulation time scale is limited to picosecond range, due to the embedding of high-precision electronic structure calculations. Both the spatial and temporal scales are inadequate for describing problems the conformational relaxation of peptides at the mineral–organic interface. Interfacial water dynamics around organic molecules and surface ions surrounding the interaction sites are also difficult to be considered. In this regard, building the modeling system using the hybrid QM/MM framework becomes a better choice.

In the hybrid QM/MM approach, the central atoms of interest are treated in a QM subsystem in which the chemical reactivity is preserved and the remaining atoms are simulated in the MM subsystem with a less sophisticated energy calculation method (Field et al., 1990; Singh and Kollman, 1986; Warshel and Levitt, 1976). Such formalism takes advantage of the accuracy of the *ab initio* method and the efficiency of MM FFs at the same time and is well established in the molecular modeling of biological systems (Riccardi et al., 2006). Adopting this approach to mineral–water–organic interface studies is reasonable in principle, since only a subset of atoms are involved in interfacial chemical reactions and the rest of the atoms merely serve as structural frame. However, because the organic molecules are finite and the inorganic minerals have infinite periodic surface, the theoretical approaches to treat the boundaries between QM and MM regions are different for these two components; therefore, how to properly describe the interactions between QM and MM subsystems in a mineral–organic interface is often not straightforward. Due to the differences in the nature of charge mobility and polarization effects between ionic minerals and covalently bonded organics, there is an inherent incompatibility in treating these effects at the QM/MM boundary (Danyliv et al., 2007; Nasluzov et al., 2001). Rigorous validation is required to produce consistent results of the interaction potentials between QM and MM regions for both mineral and organic systems. This prerequisite should be met before applying the hybrid QM/MM approach to mineral–water–organic interface simulation.

9.3   Case Studies

In this section, we review specific molecular modeling studies of apatite and calcite biominerals, in which the aforementioned advanced approaches were applied and provided realistic insight to the physical–chemical interactions at the mineral–water–organic interface.

9.3.1 Apatite

9.3.1.1 The Potential Roles of ANCPs in Ca^{2+}–Pi Nucleation

The potential roles of ANCPs in Ca^{2+} – Pi nucleation *ab initio* cluster calculations have been applied to explore the nucleation of Ca^{2+} – Pi mineral at inorganic and organic interfaces, although bulk solvation and dynamics could not be explicitly considered because of limited computational resource and computational algorithms at that time (Sahai, 2005; Sahai and Anseau, 2005). Alternatively, classical MD can be employed to study the early stages of nucleation in biomineralization by carefully designing the *in silico* experiment. Sahai and coworkers have previously used this method to explore the potential role of BSP in promoting HAP nucleation, by looking at whether and how one of the acidic domains of BSP sequesters Ca^{2+} and Pi in solution (Yang et al., 2010). Belonging to the family of ANCPs, BSP's acidic Glu-, Asp-, and SerP-rich domain has been proposed to nucleate HAP (Baht et al., 2008; Hunter and Goldberg, 1993). BSP is highly flexible and takes random coil conformation in solution (Fisher et al., 2001), so no crystallographic structure data exists; hence, simulation studies on BSP's interaction with Ca^{2+} and Pi ions or HAP crystal are rare. To ensure that the simulation of nucleation events is both scientifically sound and technically possible, several important conditions were set up in the study. A model peptide composed of two SerP residues and eight Glu was used, starting with different backbone conformations (α-helix and random coil) and three initial configurations for each backbone conformation. The electrostatic bonding interactions of Ca^{2+} to Glu and SerP side chains and subsequent formation of Ca^{2+} – Pi clusters near the model peptide were found to be independent of starting conformations. Of all the conformations, *only in one case* of the α-helix backbone was a stable Ca^{2+} equilateral triangle formed, which resembled the Ca^{2+} ion positions on the HAP (001) face. (In the context of our study, "stable" means that the cluster persisted up to 17 ns in a 25 ns simulation.) In that configuration, Ca^{2+} ions in solution were by coordinated Glu and SerP side chains. However, such a potential epitaxial template did not persist beyond 5 ns of the simulation for other initial configurations of the α-helix backbone and did not form at all for any of the random coil conformations up to 35 ns. It was concluded that the acidic domain of BSP *does not* provide an epitaxial matching template for HAP nucleation. Further, the Ca^{2+} – Pi cluster lacked structure, so it was also conjectured that if BSP plays any role in nucleation, it is more likely to aid ACP rather than direct HAP nucleation (Figure 9.4).

9.3.1.2 The Potential Roles of Collagen in Ca^{2+}–Pi Nucleation

Ca^{2+} – Pi nuclei in bone are formed within an extracellular structural scaffold provided by Type 1 collagen. Individual collagen molecules are approximately 300 nm long and are composed of three peptides in a triple helix structure (Figure 9.1). The molecules are arranged linearly is such a way that the end of one molecule is separated from another resulting in a "hole" zone. The molecules self-assemble into a pseudohexagonal array, which comprises a collagen fibril. The earliest nuclei form within collagen fibrils at these specific hole and neighboring "overlap" zones. Charged amino acid side chain residues of the collagen molecules have long been proposed to surround the hole zone and recruit Ca^{2+} and Pi, ultimately leading to intrafibrillar Ca^{2+}–Pi nucleation (Chapman et al., 1990; Silver and Landis, 2011). However, the hypothesis was never proven because, as described in Section 9.2.1, a major hurdle in modeling intrafibrillar Ca^{2+} – Pi nucleation was the lack of a 3-D structure of collagen.

Previous attempts to model collagen structure and Ca^{2+} – Pi nucleation were limited to a single triple-helical molecule without the whole fibril structure and to simulation approaches that could not

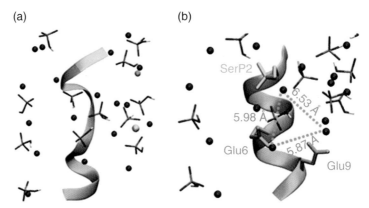

Figure 9.4 *Snapshots that illustrate the interactions between Ca^{2+}, Pi, and the peptide $(SerP)_2$-$(Glu)_8$ in MD simulations. The peptide backbone is rendered in ribbon. (a) The peptide with a random coil structure. No apparent Ca^{2+} or Pi network was observed. (b) The peptide with an α-helix conformation. An equilateral triangle configuration was found only with this helix backbone conformation. The side chain non-hydrogen atoms of SerP2, Glu6, and Glu9, which coordinate to three Ca^{2+} ions in an equilateral triangle (indicated with green dotted lines, distances in Å), are highlighted in licorice representation. Color designation: blue, calcium; cyan, Glu residue; gold, SerP residue; green, sodium; red, oxygen; tan, phosphate; white, hydrogen; yellow, chloride. Yang et al. (2010). Reproduced with permission of American Chemical Society.*

provide adequate sampling. Zahn and coworkers applied an MC-based docking scheme to examine the interactions of Ca^{2+}, Pi, and OH^- or F^- with a single triple-helical collagen molecule to determine whether nuclei of HAP or fluorapatite could form (Kawska et al., 2008; Zahn et al., 2007). It was reported that inorganic ions formed electrostatic bonds with the atoms of collagen, but the collagen triple-helical structure was disrupted by ions and the entire fibril was not modeled. In another study, the potential nucleation of $Ca^{2+}-Pi$ associated with a single triple helical molecule was investigated (Almora-Barrios and De Leeuw, 2011). Again, the study was of limited value, because ion concentrations were 1000 times higher than the normal value in extracellular fluids, the simulations were limited to only 5 ns, and the application of classical MD was far from sufficient to sample adequate configuration relaxation of the ions.

A 3-D crystal structure of an entire fibril consisting of collagen molecules self-assembled in the pseudohexagonal array was determined recently, but it was at low resolution (~ 5 Å axial and ~ 11 Å equatorial) and provided only the protein backbone atom positions (Orgel et al., 2006). Nonetheless, this was a major breakthrough in the field, because it provided the starting point for building a complete, high-resolution structure (1 Å) model and examining intrafibrillar $Ca^{2+}-Pi$ nucleation (Xu et al., 2015). An initial entire fibril structure was constructed based on the low-resolution XRD data of backbone positions of the molecules combined with side chain positions of short collagen-mimic peptides and the structure was optimized to the lowest energy configuration for all atoms (Figure 9.5). Analysis of the simulation results showed that the charged side chains were localized primarily around the hole and overlap zones of the fibril and, crucially, to be oriented toward the hole. When Ca^{2+} and Pi were added to the system, the charged side chains recruited the ions to the hole zone leading to $Ca^{2+}-Pi$ nucleation, thus proving the hypothesis for the first time. HREMD was applied to the nucleation simulation to overcome energy barriers arising from strong electrostatic interactions between ions and between ions and charged side chains.

Longitudinal view   Equatorial view

(a) Low resolution XRD structure provides fibril pseudohexagonal structure, in which only Cα atom positions were available.

(b) Hence, amino acid residues and Cα positions were created using collagen-mimic peptide structures (2 Å resolution) and the well-known triple-helix structure in biology.

(c) Backbone atom positions in (b) were replaced by atom positions from (a) using targeted MD simulation. This creates the initial structure for final structure optimization using classical MD.

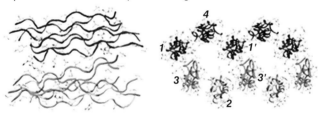

(d) Final optimized self-assembled pseudohexagonal fibril structure from MD.

Figure 9.5 Construction of high-resolution collagen molecules self-assembled in the pseudohexagonal structure of a fibril. Each molecule is composed of three peptides in a triple helix (colored ribbons). The colored lines represent chemical bonds between backbone and side chain atoms. The numbers in the right panel of (d) indicate the designated molecular segments in the original 2-D collagen packing model, inferred from TEM results (Petruska and Hodge, 1964). Xu et al. (2015). Reproduced with permission of Elsevier.

9.3.1.3 The Potential Roles of Small Molecules and Proteins in Mediating HAP Crystal Growth

Compared to the limited number of studies on nucleation, numerous studies have been focused on how the adsorption of amino acids to a growing HAP crystal can potentially affect growth. The amino acids serve as important prototypes to investigate the interactions between proteins and HAP surfaces. In a series of *ab initio* studies using periodic B3LYP calculations, Ugliengo and coworkers investigated the surface reactivity of HAP with up to 10 water molecules and a single molecule of glycine (Gly). The surface coverage was approximately 4.5 water molecules per nm^2. Water was found to dissociative on HAP (100) and (001) faces, resulting in hydroxylation of Ca^{2+}

and protonation of Pi (Corno et al., 2009). The zwitterion form of glycine (Gly) adsorbed to HAP surfaces in the gas phase by interaction of its NH_3^+ and COO^- groups with Pi and Ca^{2+} of HAP, respectively. In the presence of water, such direct binding interaction has to be achieved by replacing one adsorbed water molecule on the HAP surface (Rimola et al., 2008, 2009). These authors also examined an unusual charged state of Gly in which the amine group was uncharged and the carboxyl group was deprotonated (negatively charged). The molecule adsorbed by H bonding between the COO^- group and HPO_4^{2-} on the surface, and these results were consistent with IR spectra (Rimola et al., 2011). It is likely, however, that the Gly molecule was zwitterionic in the experiment so the agreement with computational results for the uncharged Gly molecules is puzzling. Such *ab initio* periodic DFT calculations were subsequently applied to the zwitterionic forms of Glu and lysine (Lys), and the adsorption energy was shown to be higher on (100) than on (001) surface of HAP (Rimola et al., 2012a). These *ab initio* studies provided mechanistic information for HAP–water–organic interactions, but due to the small size of the systems, their fidelity to the real, solvated HAP–protein interface is limited.

A study of SerP and Glu interacting with HAP (100) and (001) surfaces in fully solvated environment with explicit water molecules was conducted by Sahai and coworkers recently, using classical MD combined with umbrella sampling (Xu et al., 2014). The adsorption free energies obtained were in good agreement with separate DFT calculations, confirming the reliability of the FFs used and the convergence of conformational sampling. For both SerP and Glu, the adsorption free energies were higher on the (100) surface compared to (001). This observation was in line with previous *ab initio* calculations for Glu (Rimola et al., 2012a). The preferential binding of amino acids to a certain face of HAP has been proposed to be responsible for the modulation of HAP crystal growth by proteins and small biomolecules (Hunter and Goldberg, 1994; Xie and Nancollas, 2010). Based on the results of the computational studies, it is conceivable, then, that the growth of an HAP crystal in the (Tye et al., 2003) direction will be inhibited relative to the [001] direction resulting in the plate-like HAP crystals formed in the bone.

The aforementioned studies illustrate that the interactions between amino acids and HAP surface are dominated by electrostatic forces between COO^- and Ca^{2+} and/or between NH_3^+ and Pi, and water-mediated H bonds are also observed. Such interactions may also direct protein or peptide adsorption as suggested in previous classical MD studies (Dong et al., 2007; Friddle et al., 2011). According to the results of these studies, the electrostatic bonds between the charged amino acid residues and the surface ions were so strong that a substantial amount of free of energy $(20–100 \text{ kcal.mol}^{-1})$ was required to desorb the protein. Whether or not the limited time scale (<2 ns) of these simulations was capable of overcoming energy barriers and obtaining sufficient conformation sampling is in question. As discussed in Section 9.2.2.2, a bioinformatics-based approach may be able to bypass the difficulty of conformational sampling.

The adsorption of statherin, a protein presented in saliva and responsible for controlling enamel formation, has been explored by Gray and coworkers, using the *RosettaSurface* protocol (Makrodimitris et al., 2007; Masica and Gray, 2009; Masica et al., 2011). Based on the 100 000 adsorbed structures generated by the protocol, different energy contributions to protein adsorption were identified. Again, electrostatic interactions dominated and similar patterns were indicated for adsorption of phosphorylated or non-phosphorylated leucine-rich amelogenin protein (LRAP), where the side chains of SerP, Lys, and arginine (Arg) bonded to the HAP.

Interestingly, Makrodimitris et al. (2007) reported that a molecular motif located at the N-terminus of statherin bound to HAP (001) through four positively charged residues of Lys (Figure 9.6) and a "lattice matching" between these side chains and mineral surface was obtained. The interatomic distances of side chains to surface calculated from this conformation were also in

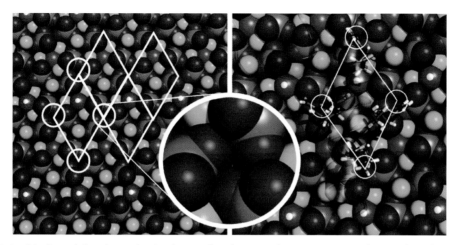

Figure 9.6   *Binding of the charged side chains of statherin to the HAP (001) surface. Left: Schematic drawing of the proposed "lattice matching" pattern from N-terminus of statherin across the surface (white parallelograms). Inset: close-up view. Right: four basic amino acids comprise the recognition motif to the pattern on the surface. Amino acid side chains are represented by licorice model and HAP surface is represented by ball model. Color designation: blue, nitrogen; gray, carbon; green, calcium; orange, phosphate; red, oxygen; white, hydrogen. Makrodimitris et al. (2007). Reproduced with permission of American Chemical Society.*

agreement with distances obtained by ss-NMR spectroscopy (Long et al., 2001). No lattice matching was found between the protein and the (100) face. The idea of lattice matching between proteins and the mineral surfaces has long been raised as a possible mechanism of protein-directed inorganic crystal growth in biomineralization (Hauschka et al., 1989; Hoang et al., 2003; Weiner et al., 1983). We note that the lattice matching observed in the case of statherin–HAP (001) may not be a universal principle for other proteins and biominerals. Indeed, much computational effort has been invested in trying to justify this concept, but no substantial evidence has been found in general. What has been confirmed in common from these studies, instead, is that the major driving force of adsorption is the long-range electrostatic attraction between single or multiple charged side chains to surface ions of random positions (Rimola et al., 2012b; Yang et al., 2011). As discussed in Section 9.2.2.2, a combination of bioinformatics with explicit solvent MD simulation or utilizing advanced MD sampling techniques such as REMD and MetaD will be able to deliver more concrete results (Liao and Zhou, 2014; Yang et al., 2013).

9.3.2   Calcite

$CaCO_3$ minerals are the most common in skeletal tissues of invertebrates. The three crystalline polymorphs in decreasing order of stability, calcite, aragonite, and vaterite, exist in different species. In many species, these polymorphs undergo phase transition from metastable to stable form during various stages of biomineralization (Xu et al., 2006). Due to the availability of large calcite crystals with well-defined facets and cleavage plains, a vast amount of experimental information exists about its nucleation and growth inorganically and in the presence of modifying inorganic ions and organic molecules. Computational modeling of calcite along with its two polymorphs is, therefore, much more extensive compared with that of apatite, as reviewed by Harding et al. (2008).

As the very earliest stages of inorganic calcite nucleation, the phase transitions from dissolved ions to calcium carbonate multinuclear complexes in solution and to a solid phase have drawn significant attention (Gebauer et al., 2014). Tribello et al. (2009) have performed combined MD and umbrella sampling simulations targeting the free energy profiles for association of $Ca^{2+} - CO_3^{2-}$, $Ca^{2+} - CaCO_3^0$, and $CO_3^{2-} - CaCO_3^0$ pairs, as well as adsorption of Ca^{2+} to the calcite (104) surface. The calculation indicated that the association of ion pairs or individual ions to an existing aqueous multinuclear complex was essentially barrierless, while the adsorption of Ca^{2+} or CO_3^{2-} to the calcite surface required crossing a free energy barrier of approximately 5 kcal.mol^{-1}. It was concluded that amorphous clusters were the first to nucleate. Furthermore, the local structural features of the amorphous clusters resembled both the aragonite and vaterite phases rather than the most stable calcite.

By applying a recently developed FF (Raiteri et al., 2010), the formation of liquid-like, CaCO$_3$ "ionic polymers" from solutions (0.06, 0.28, and 0.5 M) was reported (Demichelis et al., 2011). Linear, branched, and ring-shaped CaCO$_3$ oligomers up to 10 mers were observed. The ionic polymers were also immersed into a large water box ($\sim$200 000 molecules) to examine the stability of these structures at low concentration (10 mM) to represent biological conditions more closely. The results of PMF calculation indicated that chain-like topology of CaCO$_3$ was more stable than globular amorphous clusters, which supports the proposed "ionic liquid-state precursor" theory for CaCO$_3$ nucleation (Gebauer et al., 2008). In a more detailed study, Wallace et al. employed REMD to enhance the sampling of CaCO$_3$ cluster formation in supersaturated solution (0.2 M) (Figure 9.7)

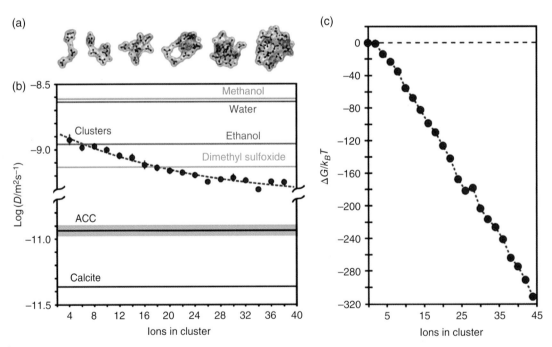

Figure 9.7 *Structural, dynamical, and energetic properties of CaCO$_3$ clusters. (a) Snapshots taken from REMD showing the stoichiometric evolution of polymeric cluster configurations. (b) Plot showing the diffusion coefficient (D) of Ca^{2+} ions within the clusters compared with calcite and ACC (dashed curve, from simulation) and the self- diffusion coefficients (solid lines, experimental) of several common solvents. (c) The free energy of the solvated ions as a function of cluster size determined at $[Ca^{2+}] = [CO_3^{2-}] = 0.015\,mol.l^{-1}$. Wallace et al. (2013). Reproduced with permission of AAAS.*

(Wallace et al., 2013). The diffusion coefficient was calculated for Ca^{2+} ions in the cluster, and the value fell into the regime between water- and solid-like ACC but was much closer to that of water. This was taken to indicate the liquid-like nature of the cluster. Notably, the free energy of the $CaCO_3$ cluster decreased monotonically as the size of the cluster grew, as determined by the method of Lin et al. (Lin et al., 2010). Similar results were obtained by Raiteri and Gale, showing that the initial formation of ACC phase is preferred over direct nucleation of calcite or aragonite (Raiteri and Gale, 2010). In these studies, the authors attempted to resolve the issue of sampling nucleation events at ion concentration that is relevant to *in vitro* or *in vivo* biomineralization, and indeed they have provided inspiring examples for the field to follow.

The evolution of anhydrous ACC to $CaCO_3$ crystalline polymorphs has been previously studied (Quigley and Rodger, 2008b; Quigley et al., 2011). Due the complex structural transformation in this process, biasing particular order parameters that can describe the packing structure of atoms is a possible approach. By using MetaD to sample the conformational space of bond orientation parameters, the free energy landscape of ACC to calcite phase transition for a 2-nm-sized cluster has reported. Subsequently, the approach was extended to study larger sized $CaCO_3$ clusters (3–4 nm), and free energy surfaces were acquired for the phase transition between the three $CaCO_3$ polymorphs. The method of employing MetaD to bias relevant order parameters allows the rapid generation of trajectories for amorphous–crystalline transition, by reducing the computational cost of overcoming associated free energy barriers. Therefore, in principle, this method can be applied to simulate heterogeneous nucleation as well, such as $CaCO_3$ phase transition induced by proteins or self-assembled monolayers (Freeman et al., 2010; Quigley et al., 2009). It is imperative, however, to ensure that the exploration of conformational space by multidimensional MetaD is converged at the end of the simulation, which is a prerequisite of any valid free energy calculations.

The simulations of calcite–organic interactions have largely relied on molecular docking energy minimization (Elhadj et al., 2006; Orme et al., 2001; Wolf et al., 2007). In these studies, the adsorbed structure was obtained merely by minimizing the interaction energies between the organic molecules and the surfaces. The goal of these studies was to provide structural insight into the step growth mechanisms of calcite by combining the results of simulations with experiments; however, adsorption energies and kinetics cannot be obtained by molecular docking, and progress has been made since then by using classical MD and DFT approaches. The first widely used method to obtain the energetic information of calcite–organic interactions was to calculate the "adsorption energy," which was taken as the potential energy difference before and after the adsorption of the organic molecule and oligopeptides to the calcite surface (Harding and Duffy, 2006). The adsorption energy values reported were on the order of $-100\,kJ.mol^{-1}$ (Filgueiras et al., 2006). The simulation of large system (>1000 atoms) by this approach is limited, because attaining the convergence of potential energy (which usually means fluctuation of energy is less than 10^{-4} of the energy value itself) is very difficult, especially for the electrostatics contributions. Besides, only the potential energy is calculated and the entropic contribution to free energy cannot be addressed.

Since free energy can be expressed as a function of probability distribution, researchers have attempted to calculate the free energy of adsorption by counting the probability density of adsorbed species along the direction perpendicular to the surface (Raut et al., 2005). By employing this method, free energy values of ~ -5 and $\sim -2.5\,kJ.mol^{-1}$ have been obtained for ethanol and water adsorption to the (104) surface, respectively (Cooke et al., 2010; Kerisit and Parker, 2004). This direct free energy calculation approach, however, has a limited efficiency of sampling enough conformations along the adsorption or desorption reaction coordinate.

With the development of computational power, more sophisticated calculations have been performed for the calcite–organic interface. Adsorption free energies of -15.5 and $-9\,kJ.mol^{-1}$, respectively, have been reported for ethanol interactions at the terrace and obtuse step on the (104) face, by using DFT calculation (Andersson and Stipp, 2012). Shen et al. (2013) used umbrella sampling approach to study the interactions between styrene sulfonate oligomer and the (001) or (104) surface and the corresponding free energy profiles. It was shown that the oligomer adsorbed to the (001) surface with a free energy of $-60\,kJ.mol^{-1}$; in contrast, the value was $-5\,kJ.mol^{-1}$ for the (104) surface. This face-selective adsorption behavior is consistent with observations in calcite growth experiments and also resembles the preferential binding of single amino acids with HAP surfaces (Wang et al., 2005; Xu et al., 2014). Through reconstruction of the free energy landscape, kinetic information about the process can subsequently be obtained by transition path sampling, which may allow direct comparison to organic-mediated crystal growth experiment (Best and Hummer, 2005; Bolhuis et al., 2000; Stack et al., 2012).

9.4 Concluding Remarks and Future Perspectives

In this chapter, we have attempted to describe the challenges and approaches to modeling physical–chemical interactions at the mineral–water–organic interface that are pertinent to the topic of biomineralization. Similar considerations apply to modeling processes potentially involved in the prebiotic chemistry and the origins of life, such as the nonenzymatic, mineral surface-mediated polymerization of simple organic molecules into functional biomolecules. Two major issues arise, namely, sufficient conformational sampling of rare events such as crystal nucleation and growth and the validity of FFs that can faithfully reproduce the inorganic–organic interfacial properties. Advanced computational methodologies and hardware that have been developed for biological and material simulations can be adapted to biomineralization modeling toward addressing the first issue. Tackling the second question requires proper benchmarking between different levels of approaches (QM, MD, and coarse grained) and close comparisons with experimental data. An individual simulation method alone often cannot provide accurately the interactions between proteins or peptides with mineral surfaces and approaches with different levels of sampling efficiency and accuracy should be combined. Furthermore, for the purpose of modeling structure–property relationships of proteins or peptides in biomineralization and prebiotic chemistry, computational methodologies that bridge multiple length scales from angstrom to micron are yet to be developed.

Even if the premises of sufficient computational power and carefully validated FFs are achieved, solving the mysteries of biomineralization mechanisms through molecular modeling faces significant hurdles. Linking the vast amount of data from simulation with plausible physical principles calls for the establishment of theories and such efforts are just beginning to be made. Taking apatite as an example, while most computational studies focus on stoichiometric and pristine HAP surface, the biological counterparts in the bone and dentin are ion substituted and nonstoichiometric. Such apatite phases are virtually impossible to be modeled with current atomistic simulations. In this case, a coarse-grained model, which preserves the fundamental physical properties of mineral surface, may be able to capture the essence of its interactions with organic species. Furthermore, the underlying physics of the adsorption of flexible peptides and ANCPs to biomineral surfaces is analogous to polyelectrolytes adsorption to charged solid surfaces in salt solution, the principles of which have already been developed (Dobrynin, 2001; Sens and Joanny, 2000). Application of such theories to the field of biomineralization relies on the extent of details we can learn from molecular modeling and experiments.

It is also critical to note that many differences exists in conditions between *in vivo* biomineralization and *in vitro* experiments and computational models or differences in biomineralization mechanisms between different species or between different tissues within the same species. Thus, a universal principle or model for biomineralization may not exist, further highlighting the importance of revealing detailed mechanisms for different systems. Furthermore, expanding the current focus of mineral–water–protein interactions to extracellular and cellular processes, such as collagen–NCP interaction, ligand–receptor binding, inter- and intracellular transportation of mineral, is an ultimate challenge yet a great opportunity for biomineralization modeling research.

As a final remark, running classical molecular modeling software is relatively easy, but knowing how to get reliable and realistic results is a highly demanding task, which requires state-of-the-art approaches and critical discrimination in recognizing the quality of the results.

Acknowledgments

The authors would like to acknowledge funding from NSF CAREER Award (EAR 0346689) and NSF DMR ARRA (DMR 0906817) and start-up funds from University of Akron to N. S., Frank Kelley Fellowship to W. Z., and Robert E. Helm Jr. Post-Doctoral Fellowship to Z. X. The authors thank Prof. William J. Landis, Prof. Qiang Cui, Prof. Yang Yang, Prof. Jeffrey J. Gray, Prof. Mesfin Tsige, Dr. Ling Chen, Mr. Michael Pacella, and Ms. Ziqiu Wang for fruitful discussions. The computational resources generously provided by Ohio Supercomputer Center Grant PBS0286 and University of Wisconsin-Madison, High Throughput Computing are greatly appreciated.

References

L. Addadi, S. Weiner, *Nature* 1997, **389**, 912.
L. Addadi, A. Berman, J. M. Oldak, S. Weiner, *Connect. Tissue Res.* 1989, **21**, 127.
L. Addadi, S. Raz, S. Weiner, *Adv. Mater.* 2003, **15**, 959.
W. J. Albery, J. R. Knowles, *Biochemistry* 1976, **15**, 5631.
N. Almora-Barrios, N. H. De Leeuw, *Cryst. Growth Des.* 2011, **12**, 756.
M. P. Andersson, S. L. S. Stipp, *J. Phys. Chem. C* 2012, **116**, 18779.
W. Andreoni, A. Curioni, *Parallel Comput.* 2000, **26**, 819.
P. V. Azzopardi, J. O'Young, G. Lajoie, M. Karttunen, H. A. Goldberg, G. K. Hunter, *PLoS One* 2010, **5**, e9330.
G. S. Baht, G. K. Hunter, H. A. Goldberg, *Matrix Biol.* 2008, **27**, 600.
A. Barducci, G. Bussi, M. Parrinello, *Phys. Rev. Lett.* 2008, **100**, 020603.
A. Bellahcene, V. Castronovo, K. U. Ogbureke, L. W. Fisher, N. S. Fedarko, *Nat. Rev. Cancer* 2008, **8**, 212.
E. Beniash, R. A. Metzler, R. S. Lam, P. U. Gilbert, *J. Struct. Biol.* 2009, **166**, 133.
J. D. Bernal, *The Physical Basis of Life*, Routledge and Paul, London 1951.
P. M. Bertsch, D. B. Hunter, S. R. Sutton, S. Bajt, M. L. Rivers, *Environ. Sci. Technol.* 1994, **28**, 980.
R. B. Best, G. Hummer, *Proc. Natl. Acad. Sci. U. S. A.* 2005, **102**, 6732.
R. Bhowmik, K. S. Katti, D. Katti, *Polymer* 2007, **48**, 664.
J. O. M. Bockris, K. T. Jeng, *Adv. Colloid Interface Sci.* 1990, **33**, 1.
J. O. M. Bockris, A. K. N. Reddy, *Modern Electrochemistry*, Vol. **2**, Springer, New York 1973.
P. G. Bolhuis, C. Dellago, D. Chandler, *Proc. Natl. Acad. Sci. U. S. A.* 2000, **97**, 5877.
G. R. Bowman, V. A. Voelz, V. S. Pande, *Curr. Opin. Struct. Biol.* 2011, **21**, 4.
P. Bradley, K. M. Misura, D. Baker, *Science* 2005, **309**, 1868.
G. E. Brown, G. Calas, *Geochem. Perspect.* 2012, **1**, 483.
G. E. Brown, G. A. Parks, *Int. Geol. Rev.* 2001, **43**, 963.
G. Bussi, A. Laio, M. Parrinello, *Phys. Rev. Lett.* 2006a, **96**, 090601.
G. Bussi, F. L. Gervasio, A. Laio, M. Parrinello, *J. Am. Chem. Soc.* 2006b, **128**, 13435.
L. A. Capriotti, T. P. Beebe, Jr., J. P. Schneider, *J. Am. Chem. Soc.* 2007, **129**, 5281.
R. Car, M. Parrinello, *Phys. Rev. Lett.* 1985, **55**, 2471.
J. G. Catalano, *J. Phys. Chem. C* 2010, **114**, 6624.
D. L. Chapman, *Philos. Mag.* 1913, **25**, 475.
J. A. Chapman, M. Tzaphlidou, K. M. Meek, K. E. Kadler, *Electron Microsc. Rev.* 1990, **3**, 143.

X. Chen, Q. Wang, J. Shen, H. Pan, T. Wu, *J. Phys. Chem. C* 2006, **111**, 1284.

S. Collino, J. S. Evans, *Biomacromolecules* 2007, **8**, 1686.

D. J. Cooke, R. J. Gray, K. K. Sand, S. L. S. Stipp, J. A. Elliott, *Langmuir* 2010, **26**, 14520.

M. Corno, C. Busco, V. Bolis, S. Tosoni, P. Ugliengo, *Langmuir* 2009, **25**, 2188.

R. T. Cygan, J.-J. Liang, A. G. Kalinichev, *J. Phys. Chem. B* 2004, **108**, 1255.

O. Danyliv, L. Kantorovich, F. Corá, *Phys. Rev. B* 2007, **76**, 045107.

R. Das, D. Baker, *Annu. Rev. Biochem.* 2008, **77**, 363.

J. J. De Yoreo, P. M. Dove, *Science* 2004, **306**, 1301.

M. Deighan, J. Pfaendtner, *Langmuir* 2013, **29**, 7999.

R. Demichelis, P. Raiteri, J. D. Gale, D. Quigley, D. Gebauer, *Nat. Commun.* 2011, **2**, 590.

A. Dey, P. H. Bomans, F. A. Muller, J. Will, P. M. Frederik, G. de With, N. A. Sommerdijk, *Nat. Mater.* 2010, **9**, 1010.

A. V. Dobrynin, *Phys. Rev. E Stat. Nonlin. Soft Matter Phys.* 2001, **63**, 051802.

X. Dong, Q. Wang, T. Wu, H. Pan, *Biophys. J.* 2007, **93**, 750.

Q. Du, E. Freysz, Y. R. Shen, *Phys. Rev. Lett.* 1994, **72**, 238.

A. K. Dunker, C. J. Brown, J. D. Lawson, L. M. Iakoucheva, Z. Obradovic, *Biochemistry* 2002, **41**, 6573.

S. Elhadj, J. J. De Yoreo, J. R. Hoyer, P. M. Dove, *Proc. Natl. Acad. Sci. U. S. A.* 2006, **103**, 19237.

A. D. Ellington, S. A. Benner, *J. Theor. Biol.* 1987, **127**, 491.

B. Ensing, M. De Vivo, Z. Liu, P. Moore, M. L. Klein, *Acc. Chem. Res.* 2006, **39**, 73.

P. Fenter, P. Geissbühler, E. DiMasi, G. Srajer, L. B. Sorensen, N. C. Sturchio, *Geochim. Cosmochim. Acta* 2000, **64**, 1221.

P. Fenter, S. Kerisit, P. Raiteri, J. D. Gale, *J. Phys. Chem. C* 2013, **117**, 5028.

M. J. Field, P. A. Bash, M. Karplus, *J. Comput. Chem.* 1990, **11**, 700.

M. R. T. Filgueiras, D. Mkhonto, N. H. de Leeuw, *J. Cryst. Growth* 2006, **294**, 60.

L. W. Fisher, D. A. Torchia, B. Fohr, M. F. Young, N. S. Fedarko, *Biochem. Biophys. Res. Commun.* 2001, **280**, 460.

K. Flade, C. Lau, M. Mertig, W. Pompe, *Chem. Mater.* 2001, **13**, 3596.

P. Fratzl, *Nat. Mater.* 2008, **7**, 610.

P. Fratzl, O. Paris, K. Klaushofer, W. J. Landis, *J. Clin. Invest.* 1996, **97**, 396.

P. Fratzl, H. S. Gupta, E. P. Paschalis, P. Roschger, *J. Mater. Chem.* 2004, **14**, 2115.

P. L. Freddolino, F. Liu, M. Gruebele, K. Schulten, *Biophys. J.* 2008, **94**, L75.

P. L. Freddolino, C. B. Harrison, Y. Liu, K. Schulten, *Nat. Phys.* 2010, **6**, 751.

C. L. Freeman, J. H. Harding, D. J. Cooke, J. A. Elliott, J. S. Lardge, D. M. Duffy, *J. Phys. Chem. C* 2007, **111**, 11943.

C. L. Freeman, J. H. Harding, D. Quigley, P. M. Rodger, *Angew. Chem. Int. Ed. Engl.* 2010, **49**, 5135.

C. L. Freeman, J. H. Harding, D. Quigley, P. M. Rodger, *J. Phys. Chem. C* 2011, **115**, 8175.

D. Frenkel, B. Smit, *Understanding Molecular Simulation: From Algorithms to Applications*, Vol. **1**, Academic Press, San Diego 2001.

R. W. Friddle, K. Battle, V. Trubetskoy, J. Tao, E. A. Salter, J. Moradian-Oldak, J. J. De Yoreo, A. Wierzbicki, *Angew. Chem. Int. Ed.* 2011, **50**, 7541.

H. Fukunishi, O. Watanabe, S. Takada, *J. Chem. Phys.* 2002, **116**, 9058.

B. Ganss, R. H. Kim, J. Sodek, *Crit. Rev. Oral Biol. Med.* 1999, **10**, 79.

D. Gebauer, A. Völkel, H. Cölfen, *Science* 2008, **322**, 1819.

D. Gebauer, M. Kellermeier, J. D. Gale, L. Bergstrom, H. Colfen, *Chem. Soc. Rev.* 2014, **43**, 2348.

A. George, A. Veis, *Chem. Rev.* 2008, **108**, 4670.

G. V. Gibbs, *Am. Mineral.* 1982, **67**, 421.

M. J. Glimcher, *Rev. Mod. Phys.* 1959, **31**, 359.

M. J. Glimcher, *Anat. Rec.* 1989, **224**, 139.

M. J. Glimcher, in *Medical Mineralogy and Geochemistry*, Vol. **64** (Eds: N. Sahai, M. A. A. Schoonen), Mineralogical Society of America, Chantilly, VA 2006.

B. A. Gotliv, L. Addadi, S. Weiner, *ChemBioChem* 2003, **4**, 522.

J. J. Gray, *Curr. Opin. Struct. Biol.* 2004, **14**, 110.

E. Greiner, K. Kumar, M. Sumit, A. Giuffre, W. Zhao, J. Pedersen, N. Sahai, *Geochim. Cosmochim. Acta* 2014, **133**, 142.

W. J. Habraken, J. Tao, L. J. Brylka, H. Friedrich, L. Bertinetti, A. S. Schenk, A. Verch, V. Dmitrovic, P. H. Bomans, P. M. Frederik, J. Laven, P. van der Schoot, B. Aichmayer, G. de With, J. J. DeYoreo, N. A. Sommerdijk, *Nat. Commun.* 2013, **4**, 1507.

J. B. S. Haldane, *Rationalist Annu.* 1929, **3**, 3.

U. H. E. Hansmann, *Chem. Phys. Lett.* 1997, **281**, 140.

J. H. Harding, D. M. Duffy, *J. Mater. Chem.* 2006, **16**, 1105.

J. H. Harding, D. M. Duffy, M. L. Sushko, P. M. Rodger, D. Quigley, J. A. Elliott, *Chem. Rev.* 2008, **108**, 4823.

S. Hauptmann, H. Dufner, J. Brickmann, S. M. Kast, R. S. Berry, *Phys. Chem. Chem. Phys.* 2003, **5**, 635.

P. V. Hauschka, J. B. Lian, D. E. Cole, C. M. Gundberg, *Physiol. Rev.* 1989, **69**, 990.

H. Heinz, T. J. Lin, R. K. Mishra, F. S. Emami, *Langmuir* 2013, **29**, 1754.

Q. Q. Hoang, F. Sicheri, A. J. Howard, D. S. Yang, *Nature* 2003, **425**, 977.

M. F. Hochella Jr., in *Mineral-Water Interface Geochemistry*, Vol. **23** (Eds: M. F. Hochella Jr., A. F. White), Mineralogical Society of America, Chantilly, VA 1990, 87.

M. F. Hochella Jr., A. F. White (Eds.) *Mineral-Water Interface Geochemistry*, Vol. **23**, Mineralogical Society of America, Chantilly, VA 1990.

Y. Y. Hu, A. Rawal, K. Schmidt-Rohr, *Proc. Natl. Acad. Sci. U. S. A.* 2010, **107**, 22425.

S. Hug, G. K. Hunter, H. Goldberg, M. Karttunen, *Phys. Procedia* 2010, **4**, 51.

G. K. Hunter, H. A. Goldberg, *Proc. Natl. Acad. Sci. U. S. A.* 1993, **90**, 8562.

G. K. Hunter, H. A. Goldberg, *Biochem. J.* 1994, **302**, 175.

G. K. Hunter, J. O'Young, B. Grohe, M. Karttunen, H. A. Goldberg, *Langmuir* 2010, **26**, 18639.

J. Hutter, P. Carloni, M. Parrinello, *J. Am. Chem. Soc.* 1996, **118**, 8710.

H. James Cleaves II, A. Michalkova Scott, F. C. Hill, J. Leszczynski, N. Sahai, R. Hazen, *Chem. Soc. Rev.* 2012, **41**, 5502.

S. Jang, S. Shin, Y. Pak, *Phys. Rev. Lett.* 2003, **91**, 058305.

C. M. Jonsson, C. L. Jonsson, D. A. Sverjensky, H. J. Cleaves, R. M. Hazen, *Langmuir* 2009, **25**, 12127.

W. L. Jorgensen, *Science* 2004, **303**, 1813.

A. Kawska, O. Hochrein, J. Brickmann, R. Kniep, D. Zahn, *Angew. Chem. Int. Ed.* 2008, **47**, 4982.

S. Kerisit, S. C. Parker, *J. Am. Chem. Soc.* 2004, **126**, 10152.

I. W. Kim, D. E. Morse, J. S. Evans, *Langmuir* 2004, **20**, 11664.

A. Laio, F. L. Gervasio, *Rep. Prog. Phys.* 2008, **71**, 126601.

A. Laio, M. Parrinello, *Proc. Natl. Acad. Sci. U. S. A.* 2002, **99**, 12562.

W. J. Landis, *Bone* 1995, **16**, 533.

R. A. Latour, *Biointerphases* 2008, **3**, FC2.

W. Lechner, C. Dellago, *J. Chem. Phys.* 2008, **129**, 114707.

J. M. Leyssale, J. Delhommelle, C. Millot, *J. Chem. Phys.* 2005, **122**, 104510.

C. Liao, J. Zhou, *J. Phys. Chem. B* 2014, **118**, 5843.

S.-T. Lin, P. K. Maiti, W. A. Goddard, *J. Phys. Chem. B* 2010, **114**, 8191.

P. Liu, B. Kim, R. A. Friesner, B. J. Berne, *Proc. Natl. Acad. Sci. U. S. A.* 2005, **102**, 13749.

J. R. Long, W. J. Shaw, P. S. Stayton, G. P. Drobny, *Biochemistry* 2001, **40**, 15451.

D. R. Lovley, *Science* 1998, **280**, 54.

B. Ma, R. Nussinov, *Proteins: Struct., Funct., Bioinf.* 1999, **37**, 73.

J. Mahamid, A. Sharir, L. Addadi, S. Weiner, *Proc. Natl. Acad. Sci. U. S. A.* 2008, **105**, 12748.

K. Makrodimitris, D. L. Masica, E. T. Kim, J. J. Gray, *J. Am. Chem. Soc.* 2007, **129**, 13713.

S. Mann, *Biomineralization Principles and Concepts in Bioinorganic Materials Chemistry*, Oxford University Press, New York 2001.

D. L. Masica, J. J. Gray, *Biophys. J.* 2009, **96**, 3082.

D. L. Masica, J. T. Ash, M. Ndao, G. P. Drobny, J. J. Gray, *Structure* 2010, **18**, 1678.

D. L. Masica, J. J. Gray, W. J. Shaw, *J. Phys. Chem. C* 2011, **115**, 13775.

M. Matsumoto, S. Saito, I. Ohmine, *Nature* 2002, **416**, 409.

J. A. McCammon, B. R. Gelin, M. Karplus, *Nature* 1977, **267**, 585.

C. F. McFadden, P. S. Cremer, A. J. Gellman, *Langmuir* 1996, **12**, 2483.

T. Mori, R. J. Hamers, J. A. Pedersen, Q. Cui, *J. Chem. Theory Comput.* 2013, **9**, 5059.

A. K. Nair, A. Gautieri, S. W. Chang, M. J. Buehler, *Nat. Commun.* 2013, **4**, 1724.

V. A. Nasluzov, V. V. Rivanenkov, A. B. Gordienko, K. M. Neyman, U. Birkenheuer, N. Rösch, *J. Chem. Phys.* 2001, **115**, 8157.

N. Nassif, N. Pinna, N. Gehrke, M. Antonietti, C. Jager, H. Colfen, *Proc. Natl. Acad. Sci. U. S. A.* 2005, **102**, 12653.

A. Navrotsky, *Proc. Natl. Acad. Sci. U. S. A.* 2004, **101**, 12096.

F. Nudelman, K. Pieterse, A. George, P. H. Bomans, H. Friedrich, L. J. Brylka, P. A. Hilbers, G. de With, N. A. Sommerdijk, *Nat. Mater.* 2010, **9**, 1004.

J. P. R. O. Orgel, T. C. Irving, A. Miller, T. J. Wess, *Proc. Natl. Acad. Sci. U. S. A.* 2006, **103**, 9001.

C. A. Orme, A. Noy, A. Wierzbicki, M. T. McBride, M. Grantham, H. H. Teng, P. M. Dove, J. J. DeYoreo, *Nature* 2001, **411**, 775.

M. Orozco, *Chem. Soc. Rev.* 2014, **43**, 5051.

M. S. Pacella, C. E. Koo da, R. A. Thottungal, J. J. Gray, *Methods Enzymol.* 2013, **532**, 343.

H. Pan, J. Tao, X. Xu, R. Tang, *Langmuir* 2007, **23**, 8972.

S.-H. Park, G. Sposito, *Phys. Rev. Lett.* 2002, **89**, 085501.

A. Patriksson, D. van der Spoel, *Phys. Chem. Chem. Phys.* 2008, **10**, 2073.

J. A. Petruska, A. J. Hodge, *Proc. Natl. Acad. Sci. U. S. A.* 1964, **51**, 871.

A. Phan, T. A. Ho, D. R. Cole, A. Striolo, *J. Phys. Chem. C* 2012, **116**, 15962.

S. Piana, A. Laio, *J. Phys. Chem. B* 2007, **111**, 4553.

S. Piana, K. Lindorff-Larsen, D. E. Shaw, *Proc. Natl. Acad. Sci. U. S. A.* 2012, **109**, 17845.

Y. Politi, T. Arad, E. Klein, S. Weiner, L. Addadi, *Science* 2004, **306**, 1161.

Y. Politi, Y. Levi-Kalisman, S. Raz, F. Wilt, L. Addadi, S. Weiner, I. Sagi, *Adv. Funct. Mater.* 2006, **16**, 1289.

Y. Politi, R. A. Metzler, M. Abrecht, B. Gilbert, F. H. Wilt, I. Sagi, L. Addadi, S. Weiner, P. U. Gilbert, *Proc. Natl. Acad. Sci. U. S. A.* 2008, **105**, 17362.

A. S. Posner, F. Betts, *Acc. Chem. Res.* 1975, **8**, 273.

E. M. Pouget, P. H. H. Bomans, J. A. C. M. Goos, P. M. Frederik, G. de With, N. A. Sommerdijk, *Science* 2009, **323**, 1455.

Z. Qin, A. Gautieri, A. K. Nair, H. Inbar, M. J. Buehler, *Langmuir* 2012, **28**, 1982.

D. Quigley, P. M. Rodger, *J. Chem. Phys.* 2008a, **128**, 154518.

D. Quigley, P. M. Rodger, *J. Chem. Phys.* 2008b, **128**, 221101.

D. Quigley, P. M. Rodger, C. Freeman, J. Harding, D. Duffy, *J. Chem. Phys.* 2009, **131**, 094703.

D. Quigley, C. L. Freeman, J. H. Harding, P. M. Rodger, *J. Chem. Phys.* 2011, **134**, 044703.

R. Radhakrishnan, B. L. Trout, *J. Am. Chem. Soc.* 2003, **125**, 7743.

P. Raiteri, J. D. Gale, *J. Am. Chem. Soc.* 2010, **132**, 17623.

P. Raiteri, A. Laio, F. L. Gervasio, C. Micheletti, M. Parrinello, *J. Phys. Chem. B* 2005, **110**, 3533.

P. Raiteri, J. D. Gale, D. Quigley, P. M. Rodger, *J. Phys. Chem. C* 2010, **114**, 5997.

P. Raiteri, R. Demichelis, J. D. Gale, M. Kellermeier, D. Gebauer, D. Quigley, L. B. Wright, T. R. Walsh, *Faraday Discuss.* 2012, **159**, 61.

P. Raiteri, R. Demichelis, J. D. Gale, *Methods Enzymol.* 2013, **532**, 3.

N. Rathore, M. Chopra, J. J. de Pablo, *J. Chem. Phys.* 2005, **122**, 024111.

V. P. Raut, M. A. Agashe, S. J. Stuart, R. A. Latour, *Langmuir* 2005, **21**, 1629.

J. P. Reyes-Grajeda, L. Marin-Garcia, V. Stojanoff, A. Moreno, *Acta Crystallogr. Sect. F Struct. Biol. Cryst. Commun.* 2007, **63**, 987.

D. Riccardi, P. Schaefer, Y. Yang, H. Yu, N. Ghosh, X. Prat-Resina, P. Konig, G. Li, D. Xu, H. Guo, M. Elstner, Q. Cui, *J. Phys. Chem. B* 2006, **110**, 6458.

A. Rimola, M. Corno, C. M. Zicovich-Wilson, P. Ugliengo, *J. Am. Chem. Soc.* 2008, **130**, 16181.

A. Rimola, M. Corno, C. M. Zicovich-Wilson, P. Ugliengo, *Phys. Chem. Chem. Phys.* 2009, **11**, 11662.

A. Rimola, Y. Sakhno, L. Bertinetti, M. Lelli, G. Martra, P. Ugliengo, *J. Phys. Chem. Lett.* 2011, **2**, 1390.

A. Rimola, M. Corno, J. Garza, P. Ugliengo, *Philos. Transact. A Math. Phys. Eng. Sci.* 2012a, **370**, 1478.

A. Rimola, M. Aschi, R. Orlando, P. Ugliengo, *J. Am. Chem. Soc.* 2012b, **134**, 10899.

C. A. Rohl, C. E. Strauss, K. M. Misura, D. Baker, *Methods Enzymol.* 2004, **383**, 66.

R. E. Ruther, B. M. Baker, J. H. Son, W. H. Casey, M. Nyman, *Inorg. Chem.* 2014, **53**, 4234.

N. Sahai, *Am. J. Sci.* 2005, **305**, 661.

N. Sahai, M. Anseau, *Biomaterials* 2005, **26**, 5763.

N. Sahai, K. Rosso, in *Surface Complexation Modeling*, Vol. **11** (Ed: J. Lutzenkirchen), Elsevier, Amsterdam 2006.

N. Sahai, D. A. Sverjensky, *Geochim. Cosmochim. Acta* 1997, **61**, 2801.

N. Sahai, S. A. Carroll, S. Roberts, P. A. O'Day, *J. Colloid Interface Sci.* 2000, **222**, 198.

P. W. Schindler, in *Mineral-Water Interface Geochemistry*, Vol. **23** (Eds: M. F. Hochella Jr., A. F. White), Mineralogical Society of America, Chantilly, VA 1990, 281.

J. Schneider, L. Colombi Ciacchi, *J. Am. Chem. Soc.* 2011, **134**, 2407.

O. Schueler-Furman, C. Wang, P. Bradley, K. Misura, D. Baker, *Science* 2005, **310**, 638.

P. Sens, J. Joanny, *Phys. Rev. Lett.* 2000, **84**, 4862.

D. E. Shaw, P. Maragakis, K. Lindorff-Larsen, S. Piana, R. O. Dror, M. P. Eastwood, J. A. Bank, J. M. Jumper, J. K. Salmon, Y. Shan, W. Wriggers, *Science* 2010, **330**, 341.

J.-W. Shen, C. Li, N. F. A. van der Vegt, C. Peter, *J. Phys. Chem. C* 2013, **117**, 6904.

D. S. Sholl, *Langmuir* 1998, **14**, 862.

F. H. Silver, W. J. Landis, *Connect. Tissue Res.* 2011, **52**, 242.

U. C. Singh, P. A. Kollman, *J. Comput. Chem.* 1986, **7**, 718.

J. Šponer, J. E. Šponer, A. Mládek, P. Banáš, P. Jurečka, M. Otyepka, *Methods* 2013, **64**, 3.

A. G. Stack, P. Raiteri, J. D. Gale, *J. Am. Chem. Soc.* 2012, **134**, 11.

A. G. Stack, J. D. Gale, P. Raiteri, *Elements* 2013, **9**, 211.

P. J. Steinhardt, D. R. Nelson, M. Ronchetti, *Phys. Rev. B* 1983, **28**, 784.

O. Stern, *Zeitschrift fur Electrochemie* 1924, **30**, 508.

S. L. S. Stipp, *Mol. Simul.* 2002, **28**, 497.

S. L. Stipp, M. F. Hochella Jr., *Geochim. Cosmochim. Acta* 1991, **55**, 1723.

J. E. Stone, D. J. Hardy, I. S. Ufimtsev, K. Schulten, *J. Mol. Graph. Model.* 2010, **29**, 116.

W. Stumm, C. P. Huang, S. R. Jenkins, *Croat. Chem. Acta* 1970, **42**, 223.

Y. Sugita, Y. Okamoto, *Chem. Phys. Lett.* 1999, **314**, 141.

J. C. Sweet, R. J. Nowling, T. Cickovski, C. R. Sweet, V. S. Pande, J. A. Izaguirre, *J. Chem. Theory Comput.* 2013, **9**, 3267.

J. D. Termine, A. S. Posner, *Science* 1966a, **153**, 1523.

J. D. Termine, A. S. Posner, *Nature* 1966b, **211**, 268.

J. D. Termine, A. S. Posner, *Calcif. Tissue Res.* 1967, **1**, 8.

G. M. Torrie, J. P. Valleau, *J. Comput. Phys.* 1977, **23**, 187.

J. A. Tossell, D. J. Vaughan, *Theoretical Geochemistry: Application of Quantum Mechanics in the Earth and Mineral Science*, Oxford University Press, New York, Oxford 1992.

S. Trebst, M. Troyer, U. H. Hansmann, *J. Chem. Phys.* 2006, **124**, 174903.

G. A. Tribello, F. Bruneval, C. Liew, M. Parrinello, *J. Phys. Chem. B* 2009, **113**, 11680.

D. Tunega, G. Haberhauer, M. H. Gerzabek, H. Lischka, *Langmuir* 2001, **18**, 139.

D. Tunega, M. H. Gerzabek, H. Lischka, *J. Phys. Chem. B* 2004, **108**, 5930.

C. E. Tye, K. R. Rattray, K. J. Warner, J. A. Gordon, J. Sodek, G. K. Hunter, H. A. Goldberg, *J. Biol. Chem.* 2003, **278**, 7949.

D. van der Spoel, M. M. Seibert, *Phys. Rev. Lett.* 2006, **96**, 238102.

A. Verdaguer, G. M. Sacha, H. Bluhm, M. Salmeron, *Chem. Rev.* 2006, **106**, 1478.

A. F. Wallace, L. O. Hedges, A. Fernandez-Martinez, P. Raiteri, J. D. Gale, G. A. Waychunas, S. Whitelam, J. F. Banfield, J. J. De Yoreo, *Science* 2013, **341**, 885.

T. Wang, H. Colfen, M. Antonietti, *J. Am. Chem. Soc.* 2005, **127**, 3246.

A. Warshel, M. Levitt, *J. Mol. Biol.* 1976, **103**, 227.

S. Weiner, Y. Talmon, W. Traub, *Int. J. Biol. Macromol.* 1983, **5**, 325.

S. E. Wolf, N. Loges, B. Mathiasch, M. Panthöfer, I. Mey, A. Janshoff, W. Tremel, *Angew. Chem. Int. Ed.* 2007, **46**, 5618.

B. Xie, G. H. Nancollas, *Proc. Natl. Acad. Sci. U. S. A.* 2010, **107**, 22369.

X. Xu, J. T. Han, H. Kim do, K. Cho, *J. Phys. Chem. B* 2006, **110**, 2764.

Z. Xu, Y. Yang, Z. Wang, D. Mkhonto, C. Shang, Z.-P. Liu, Q. Cui, N. Sahai, *J. Comput. Chem.* 2014, **35**, 70.

Z. Xu, Y. Yang, W. Zhao, Z. Wang, W. J. Landis, Q. Cui, N. Sahai, *Biomaterials* 2015, **39**, 59.

Y. Yang, Q. Cui, N. Sahai, *Langmuir* 2010, **26**, 9848.

Y. Yang, D. Mkhonto, Q. Cui, N. Sahai, *Cells Tissues Organs* 2011, **194**, 182.

Y. Yang, Z. Xu, Q. Cui, N. Sahai, *Biophys. J.* 2013, **104**, 230a.

Y. Yang, Z. Xu, Q. Cui, N. Sahai, in *Biomineralization Sourcebook* (Eds: E. DiMasi, L. B. Gower), CRC Press, Taylor & Francis Group, Boca Raton, FL 2014, 265.

D. Zahn, O. Hochrein, A. Kawska, J. Brickmann, R. Kniep, *J. Mater. Sci.* 2007, **42**, 8966.

W. Zhao, Z. Xu, Y. Yang, N. Sahai, *Langmuir* 2014, **30**, 13283.

10

Vibrational Spectroscopy of Minerals Through *Ab Initio* Methods

Marco De La Pierre,[1] Raffaella Demichelis,[1] and Roberto Dovesi[2]

[1]*Nanochemistry Research Institute, Curtin Institute for Computation, and Department of Chemistry, Curtin University, Perth, Western Australia, Australia*
[2]*Dipartimento di Chimica, Università degli Studi di Torino and NIS Centre of Excellence "Nanostructured Interfaces and Surfaces", Torino, Italy*

10.1 Introduction

In the past years, spectroscopic investigations, that is, studies of how a radiation of some type interacts with matter, have found increasingly broad application in geochemistry. In fact, the variety and the accuracy of the several techniques that are nowadays available allow for the study of almost any characteristics of rocks and minerals: average and local structure, chemical composition, bonding, oxidation states, elasticity, electronic/magnetic structure and optical properties. The importance of these techniques in geochemistry is such that the Mineralogical Society of America has recently published one volume of the *Reviews in Mineralogy and Geochemistry* (number 78, 2014) entirely devoted to the topic. Notably, in Chapter 17 of that volume, Jahn and Kowalski (2014) summarise the role of theoretical methods in modelling vibrational, electronic and magnetic spectroscopies.

In this chapter, we will focus on the details of vibrational spectroscopies, namely, IR (infrared), Raman and related techniques, as modelled through first principles methods. We address the readers to the aforementioned publication for a more general overview on modelling spectroscopic properties through a variety of computational methods.

Vibrational spectroscopies have a unique capability of getting insights on the features of the electronic potential, thus providing effective fingerprints for mineralogical materials. They can be used on both crystalline and amorphous samples and are suitable for laboratory, *in situ* and remote measurements. Therefore, it is not surprising that they have become widespread means for characterisation and analytical purposes, with applications including geochemical analysis (Dubessy et al. 2012),

Molecular Modeling of Geochemical Reactions: An Introduction, First Edition. Edited by James D. Kubicki.
© 2016 John Wiley & Sons, Ltd. Published 2016 by John Wiley & Sons, Ltd.

geological mapping (Yamaguchi et al. 1998), space exploration (Carlson et al. 1992) and cultural heritage (Dubessy et al. 2012).

Due to the high sensitivity of vibrational techniques to small variations in chemical composition and structure, the main difficulty in their use is the correct assignment of peaks to the proper compounds. Large databases of experimental data have been constructed over the years to help researchers overcome the issue (e.g. Farmer (1974), RRUFF database,[1] Handbook of Minerals Raman Spectra[2]) by cumulating knowledge on the natural variations in spectral features. Computer simulation brings here one of its major advantages, by providing powerful tools not only to readily assign peaks but also to understand their nature and relationship with the microscopic structure. A second advantage, common to any use of simulation, is the possibility to explore conditions that are hardly accessible by experiments, such as high pressures and temperatures, and specific chemical and isotopic compositions.

Methods for computing vibrational properties for solids started to be developed in the 1980s (Baroni et al. 1987; Yin and Cohen 1982; Zein 1984) and then spread in the following decade, when improved levels of theory were explored (for a review, see Baroni et al. (2001)). Hybrid functionals, which attempt to overcome the lack of pure density functional theory (DFT) in describing systems where electrons are localised, such as minerals, were firstly used to compute accurate vibrational data only in recent times (Pascale et al. 2004b; Zicovich-Wilson et al. 2004), with the CRYSTAL code providing the first public implementation for solids in 2006. Besides, a substantial performance improvement in computer hardware has occurred in the past decades, which has proven decisive in reducing the required computational costs down to affordable time scales. As a result of these evolutions, present state-of-the-art *ab initio* simulations of solids are finally a mature instrument for computational spectroscopy.

The aim of this chapter is to illustrate the key points that make nowadays *ab initio* simulation an ideal complimentary aid for vibrational spectroscopies applied to minerals. While the methods highlighted in the following sections could in principle be applied to investigate structural features and properties of any kind of material, including disordered systems such as glasses, solid solutions and species adsorbed on surfaces (De La Pierre et al. 2013b; Kubicki et al. 2012; Rimola et al. 2010; Skyes and Kubicki 1996), we will here focus on the case of pure crystalline materials. Simple examples will be used to demonstrate the levels of accuracy in the calculation of vibrational properties, to introduce tools for the peak interpretation and to discuss current shortcomings related to the presence of hydrogen. Also, the use of vibrational calculations in the task of structure determination will be outlined. Prior to introducing these topics, the following section provides details on the adopted methods.

10.2   Theoretical Background and Methods

In the last few years, first principles calculation of vibrational spectra of solid state systems has become a well-known standard procedure and has been implemented in many computer codes, based on a variety of quantum mechanical and classical approaches (e.g. Dovesi et al. (2014), Gale and Rohl (2003), Giannozzi et al. (2009), Hafner (2008), Refson et al. (2006), Soler et al. (2002)).

At present, and to the best of our knowledge, all these implementations rely on the assumption that the potential energy surface (PES) of the system is harmonic close to its minimum. This

---

[1] Online at http://rruff.info/
[2] Online at http://www.ens-lyon.fr/LST/Raman/

approximation can be confidently applied to estimate the vibrational properties of minerals, as numerous studies show that the anharmonic contribution is as small as $10\,cm^{-1}$ or less in most packed solids.[3]

The only major limit of this so-called harmonic approximation is represented by a nearly systematic inaccuracy in predicting the vibrational properties of modes involving hydrogen atoms. Because of the light mass of hydrogen, these modes usually have a non-negligible anharmonic character, which can bring both to a vibrational frequency shifted by as much as $200\,cm^{-1}$ compared to the harmonic value and to vibrational mode displacements that need to be described as linear combinations of the harmonic normal coordinates.

For molecules, various schemes for computing anharmonic frequencies based on the vibrational self-consistent field (VSCF) approach have been implemented in electronic structure codes and successfully applied to a range of problems (e.g. Barone (2005)). However, the complexity inherent in the treatment of solids is such that a first principles investigation of anharmonicity effects in vibrational frequencies for the condensed phase has first appeared only in 2013 (Monserrat et al. 2013) for simple systems with small unit cells, such as diamond and lithium hydride. In the context of this chapter, a simple *a posteriori* method will be outlined, yielding good estimations for the anharmonic constants associated with X–H stretching modes, with X being a generic atom (most frequently O in minerals). The harmonic approximation will still be retained for other modes related to H (bendings, librations).

In this section, we will summarise the harmonic approximation and the basic principles and equations that are used to model the vibrational spectra of minerals, including the estimation of IR and Raman intensities. We will outline the *frozen phonon* (or *finite difference*) method for computing frequencies, which relies on computing second derivatives of the crystal energy by applying finite displacements to each atom in the unit cell. This approach is implemented in many codes and is the simplest way to compute frequencies at the Γ point, that is, the ones of interest for spectroscopic purposes. Other methods exist, such as density functional perturbation theory, which become more effective when studying the entire dispersion spectrum of a solid (Baroni et al. 2001).

A full overview of quantum mechanical methods, including Hartree–Fock (HF) and DFT, is available in Chapters 1 and 3. Here we will focus our efforts in reviewing the performance of some different choices of basis set and of some of the most popular DFT schemes in predicting vibrational properties.

Most of our examples in this and in the next sections will be based on calculations performed through the CRYSTAL code (Dovesi et al. 2014), introduced in Chapter 3, to which we address the readers for a detailed description of the algorithms for energy calculation and minimisation. This code allows for a very accurate analysis of mineral vibrational properties, through the use of automatic and well-tested analysis tools that will be described here and in the next sections (e.g. symmetry, isotopic substitution, anharmonicity, fragment analysis, scanning along imaginary modes, IR and Raman intensities). Actually, in the last 10 years, the proposed method has been successfully applied in tens of studies on IR and Raman spectra of minerals belonging to numerous groups, including silicates (from orthosilicates to tectosilicates, e.g. Dovesi et al. (2011), Pascale et al. (2002), Prencipe et al. (2012), Smirnov et al. (2010), Tosoni et al. (2006)), carbonates (De La Pierre et al. 2014a; Valenzano et al. 2007), phosphates (Ulian et al. 2013b), oxides (Canepa et al. 2012; Montanari et al. 2006), hydroxides (Demichelis et al. 2011a), titanates (Evarestov and Bandura 2012), nitrates (Bourahla et al. 2014), halides (Evarestov and Losev 2009), germanates (Leonidov et al. 2014) and phospho-silica glasses (Corno and Pedone 2009). Excellent performances have also

---

[3] Here, and in the remaining of this chapter, we will refer to wavenumbers $\tilde{\nu}$, in cm^{-1}, as per tradition in vibrational spectroscopy. Their relation to frequencies ν is $\tilde{\nu} = \nu/c$, with c being the speed of light. We will also refer to 'frequencies' expressed in wavenumber unit.

been proved for systems containing Fe^{2+} ions, non-trivial to simulate due to the d^6 electronic configuration; for example, see almandine (Ferrari et al. 2009) and fayalite (Noël et al. 2012).

10.2.1   Calculation of Vibrational Frequencies

In principle, given the translational invariance of a crystalline solid, its full spectrum of vibrations takes the form of a dispersion relation as a function of a reciprocal space vector $\vec{k}$. Under periodic boundary conditions (PBC), a simulation box containing several replicas of the crystal unit cell is required, to sample the dispersion curve by means of appropriate Fourier transforms between direct and reciprocal spaces (Born and Huang 1954). However, the only vibrational frequencies that account for IR and Raman spectra of solids are those lying close to the Γ point in the reciprocal space, that is, $\vec{k} \approx (0,0,0)$. Restricting our description to the Γ point makes the model much easier: the simulation box can be restricted to the unit cell, the reciprocal space does not need to be introduced, and no Fourier transforms are required. From now on, our discussion will consider the single crystal unit cell as the reference. The extension to the other points of the reciprocal lattice is discussed in Chapter 3.

As mentioned earlier, a common basic assumption when computing vibrational frequencies in solids is the harmonic shape of the PES, $V(\vec{x})$, as a function of the atomic displacements, $\vec{x}$, when close to minimum energy (i.e. around the equilibrium position), where higher-order terms can be neglected:

$$V(\vec{x}) = V_0 + \frac{1}{2}\vec{x}^T \mathbf{V}\vec{x} \tag{10.1}$$

Here V_0 is the potential energy at the equilibrium position $(\vec{x} = \vec{0})$, $\vec{x}$ has $3N$ components, where N is the number of atoms in the unit cell, and $\mathbf{V}$ is the *Hessian matrix*, whose elements are the second derivatives of the PES with respect to the atomic displacements around the equilibrium:

$$\mathbf{V}_{ij} = \left[\frac{\partial^2 V(\vec{x})}{\partial x_i \partial x_j}\right]_0 \tag{10.2}$$

We can now define the weighted displacement coordinates $\vec{q} = \mathbf{M}^{\frac{1}{2}}\vec{x}$ with $\mathbf{M}$ being a diagonal matrix whose iith element is the mass of the atom associated with the ith atomic displacement. Correspondingly, we can define the *dynamical matrix*, or mass-weighted Hessian matrix, at the Γ point as $\mathbf{W} = \mathbf{M}^{-\frac{1}{2}}\mathbf{V}\mathbf{M}^{-\frac{1}{2}}$, whose elements have the dimensions of inverse squared time.[4] The harmonic PES can then be recast as

$$V(\vec{x}) = V_0 + \frac{1}{2}\vec{x}^T \mathbf{M}^{\frac{1}{2}}\mathbf{M}^{-\frac{1}{2}}\mathbf{V}\mathbf{M}^{-\frac{1}{2}}\mathbf{M}^{\frac{1}{2}}\vec{x} = V_0 + \frac{1}{2}\vec{q}^T \mathbf{W}\vec{q} \tag{10.3}$$

By diagonalising the dynamical matrix, $\mathbf{U}^{\dagger}\mathbf{W}\mathbf{U} = \mathbf{\Lambda}$, the PES can be expressed as follows:

$$V(\vec{x}) = V_0 + \frac{1}{2}\vec{q}^T \mathbf{U}\mathbf{U}^{\dagger}\mathbf{W}\mathbf{U}\mathbf{U}^{\dagger}\vec{q}$$
$$= V_0 + \frac{1}{2}\left[\vec{q}^T \mathbf{U}\right]\mathbf{U}^{\dagger}\mathbf{W}\mathbf{U}\left[\mathbf{U}^{\dagger}\vec{q}\right] = V_0 + \frac{1}{2}\vec{Q}^T \mathbf{\Lambda}\vec{Q} \tag{10.4}$$

---

[4] If the reciprocal space outside Γ were considered, the definition of $\mathbf{W}$ would require a Fourier transform.

Here, the new coordinate system (*normal coordinates*) is $\vec{Q} = \mathbf{U}^{\dagger}\vec{q} = \mathbf{U}^{\dagger}\mathbf{M}^{\frac{1}{2}}\vec{x}$, $\mathbf{U}$ is a unitary matrix, whose columns represent the harmonic *normal modes*, and the matrix of eigenvalues $\mathbf{\Lambda}$ contains the *vibrational frequencies* v_n:

$$\mathbf{\Lambda}_{nm} = 4\pi^2 v_n^2 \delta_{nm} \tag{10.5}$$

In this way, we have successfully decomposed our lattice dynamics problem in a set of $3N$ independent collective motions or normal modes.

Once the Hessian matrix $\mathbf{V}$ is known, the effect of the *isotopic substitution* can be readily computed by constructing a matrix $\mathbf{M}$ using the desired masses. $\mathbf{M}$ can be used to build the dynamical matrix through Equation 10.3, which is then diagonalised as in Equation 10.4.

The key quantity to compute the vibrational frequencies is the Hessian matrix $\mathbf{V}$, whose elements are defined in Equation 10.2. In solid-state calculations, given the complexity of the exact formulation for high-order derivatives of the PES, $\mathbf{V}$ is usually built by computing the analytical first derivatives $v_j = \partial V(\vec{x})/\partial x_j$, and then resorting to a numerical evaluation of the second derivatives. The latter can be performed by means of either right finite difference (each atom is displaced once by a positive quantity along each of the three axes):

$$\mathbf{V}_{ij} = \left[\frac{\partial v_j}{\partial x_i}\right]_0 \approx \frac{v_j(0,\ldots,x_i,0,\ldots) - v_j(0,\ldots,0,0,\ldots)}{x_i} \tag{10.6}$$

or a more accurate central finite difference (each atom is displaced also backwards):

$$\mathbf{V}_{ij} = \left[\frac{\partial v_j}{\partial x_i}\right]_0 \approx \frac{v_j(0,\ldots,x_i,0,\ldots) - v_j(0,\ldots,-x_i,0,\ldots)}{2x_i} \tag{10.7}$$

Note that term $v_j(0, \ldots, 0, 0, \ldots)$ in formula (10.6) in principle is equal to zero at the equilibrium (minimum energy) geometry. However, its inclusion permits to account for the finite numerical accuracy of the optimisation algorithm (typical thresholds for gradient components are in the order of 10^{-4} atomic units per unit cell).

Concerning other relevant computational parameters, the displacement x_i (or *step*) must be small enough to allow for an accurate sampling of the PES around its minimum (i.e. infinitesimal displacement). Optimal values are in the order of a few thousandths of angstroms (Pascale et al. 2004b); in the literature displacements 10–20 times larger have often been adopted, leading to relatively large errors (though fortuitous and nearly systematic in many cases) on those modes that have a non-negligible anharmonic component (such as modes involving H atoms). These hopefully tiny displacements applied to each single atom at a time result in small increases in the total energy. In order to retain accuracy in the construction of the Hessian matrix, it is crucial that computed energy values are converged compared to a threshold far below these energy variations. Suggested values for the energy threshold are in the order of 10^{-10} to 10^{-11} atomic units per unit cell.

The central finite difference approach in Equation 10.7 is certainly more accurate but also computationally more expensive as the number of energy and gradient calculations is essentially doubled with respect to Equation 10.6. In fact, $3N + 1$ and $6N + 1$ calculations of energy and gradients are required when using Equations 10.6 and 10.7, respectively, to build the Hessian matrix $\mathbf{V}$ for a solid with N atoms in the unit cell. However, the application of group theory can drastically reduce this number, thus making the calculation of vibrational frequencies affordable also for large unit cell systems (Dovesi et al. 2007; Stanton 1991).

In fact, whenever a set of atoms related by symmetry is identified in the unit cell, energy and gradient need to be computed only for displacements relative to a single, symmetry irreducible atom within this set. As an example, in garnets there are 80 atoms in the unit cell, but only four of them are irreducible; therefore, a set of $3 \cdot 4 + 1 = 13$ (or $6 \cdot 4 + 1 = 25$ using central finite difference) energy and gradient calculations is required rather than $3 \cdot 80 + 1 = 241$ (or $6 \cdot 80 + 1 = 481$). Additional symmetry properties can be exploited, such as translational invariance and line symmetry, that can further reduce these numbers. The latter case occurs when atoms lie on symmetry operators (e.g. Si, the divalent and the trivalent cations in garnets), so that displacements along x, y and z are no more independent from each other. In garnets, only 9 (17) calculations are eventually required rather than 241 (481).

10.2.2   Splitting of the Longitudinal Optical (LO) and Transverse Optical (TO) Modes

Atomic Born charge tensors are the key quantities for the calculation of LO frequencies, infrared intensities and vibrational contributions to the dielectric function. They represent the linear response of atomic motions to an applied electric field $\vec{F}$ (Born and Huang 1954):

$$ \mathbf{Z}^*_{i,\alpha j} = \frac{\partial}{\partial x_{\alpha j}} \left(\frac{\partial V(\vec{x})}{\partial F_i} \right) = \frac{\partial}{\partial x_{\alpha j}} \mu_i \tag{10.8} $$

where the indices i and j stand for the Cartesian components of the electric field and of the atomic displacement, respectively; α indicates an atom in the unit cell; and $\vec{\mu}$ is the cell dipole moment. Note that $\mathbf{Z}^*$ is a $3 \times 3N$ matrix.

According to the modern theory of polarisation (Resta 2000), $\vec{\mu}$ is not a bulk property, and it cannot be determined as its value depends on the (arbitrary) choice of the unit cell. There exists however a corresponding quantity that can be computed, the polarisation per unit cell, which is the dipole moment difference between close geometries of the same system. Partial derivatives in formula (10.8) can be estimated numerically from the polarisation generated by the same small atomic displacements from equilibrium that are exploited in the calculation of the Hessian matrix elements (see Section 10.2.1). Various strategies have been designed for the calculation of polarisation: numerical approaches exploit either localised Wannier functions or the Berry phase (Dall'Olio et al. 1997). Recently, a coupled perturbed HF/Kohn–Sham (CPHF/KS) fully analytical approach has been devised (Maschio et al. 2012).

It is often useful to represent Born charges in the basis of normal modes, defining the mode effective Born charge vectors $\bar{\mathbf{Z}}_n$ as the rows of the following $3N \times 3$ matrix:

$$ \bar{\mathbf{Z}} = \mathbf{U}^\dagger \mathbf{M}^{-\frac{1}{2}} (\mathbf{Z}*)^T \tag{10.9} $$

In polar compounds, TO and LO modes in the long-wavelength limit (i.e. close to Γ) have different frequencies, as only the latter couple with the macroscopic electric field arising from the long-range character of the restoring Coulomb forces (Born and Huang 1954). The electronic potential related to this macroscopic field is not lattice periodic and is thus neglected in standard calculations due to the adoption of PBC. Therefore, in order to describe LO modes properly, the dynamical matrix close to Γ for this type of compounds must be rewritten as

$$ \mathbf{W}(\vec{k} \to \vec{0}) = \mathbf{W}(\vec{k} = \vec{0}) + \mathbf{W}^{\mathrm{NA}}(\vec{k} \to \vec{0}) \tag{10.10} $$

The former term, that is, the *analytical contribution*, is the one discussed in Section 10.2.1. The latter term is the so-called non-analytical contribution, which arises from momentum ($\vec{k}$) exchange with the radiation field (Born and Huang 1954):

$$\mathbf{W}^{NA}\left(\vec{k}\rightarrow\vec{0}\right) = \frac{4\pi}{\Omega}\frac{\left[\mathbf{M}^{-\frac{1}{2}}(\mathbf{Z}^*)^T\vec{k}\right]\otimes\left[\mathbf{M}^{-\frac{1}{2}}(\mathbf{Z}^*)^T\vec{k}\right]}{\vec{k}^T\,\boldsymbol{\epsilon}^\infty\,\vec{k}} \qquad (10.11)$$

where Ω is the unit cell volume and $\boldsymbol{\epsilon}^\infty$ is the high-frequency electronic (clamped nuclei) dielectric tensor, which can be evaluated by means of a CPHF/KS treatment (Ferrero et al. 2008). When rotating the dynamical matrix to the normal coordinate system, formula (10.11) turns to

$$\boldsymbol{\Lambda}^{NA}\left(\vec{k}\rightarrow\vec{0}\right) = \frac{4\pi}{\Omega}\frac{\left[\bar{\mathbf{Z}}\vec{k}\right]\otimes\left[\bar{\mathbf{Z}}\vec{k}\right]}{\vec{k}^T\,\boldsymbol{\epsilon}^\infty\,\vec{k}} \qquad (10.12)$$

where $\bar{\mathbf{Z}}$ is the matrix of mode effective Born charge vectors, each row of which gives the response of a normal mode to an applied electric field (see definition (10.9)). When the exchanged momentum $\vec{k}$ is orthogonal to the Born charge vector of the nth mode, $\bar{\mathbf{Z}}_n\cdot\vec{k} = 0$, the corresponding non-analytical term vanishes and we have a TO mode frequency. On the opposite, when $\vec{k}$ is parallel to $\bar{\mathbf{Z}}_n$, the corresponding non-analytical term has a finite value and diagonalisation of the full dynamical matrix gives a frequency for an LO mode.

10.2.3   Calculation of Infrared (IR) and Raman Peak Intensities and of the IR Dielectric Function

Vibrational transitions are active in IR and Raman spectroscopies when the corresponding atomic motions generate a variation in the system dipole moment and polarisability, respectively. Therefore, the derivatives of these two quantities with respect to the normal mode coordinates are the keys to estimate the intensities of IR- and Raman-active modes.

The isotropic IR intensity of the nth mode, A_n, is proportional to the absolute square of the cell dipole moment first derivative with respect to the normal mode coordinate, which can be turned in terms of the corresponding mode effective Born charge vector (Eq. 10.9):

$$A_n \propto \left|\frac{\partial\vec{\mu}}{\partial Q_n}\right|^2 \propto \sum_i \bar{\mathbf{Z}}_{n,i}^2 \qquad (10.13)$$

This relation implies that IR-active modes are identified by non-zero Born charge vectors.[5] Single crystal IR intensities, or *oscillator strengths*, are tensor quantities defined as

$$\mathbf{f}_{n,ij} = \frac{1}{4\pi\epsilon_0}\frac{4\pi}{\Omega}\frac{\bar{\mathbf{Z}}_{n,i}\bar{\mathbf{Z}}_{n,j}}{v_n^2} \qquad (10.14)$$

---

[5] Degenerate modes are pairs ($g = 2$) or triplets ($g = 3$) of modes having the same v and A values by symmetry. When analysing isotropic IR spectra, they are usually reported as a single mode with frequency v and intensity $g \cdot A$.

where ϵ_0 is the vacuum dielectric permittivity ($1/4\pi\epsilon_0 = 1$ atomic unit). An expression for the IR dielectric function can be written exploiting the oscillator strengths, according to the classical Drude–Lorentz model:

$$\epsilon(v)_{ij} = \epsilon_{ij}^\infty + \sum_n \frac{\mathbf{f}_{n,ij} v_n^2}{v_n^2 - v^2 - iv\gamma_n} \tag{10.15}$$

Here, γ_n is the damping factor of the nth mode, which is related to the corresponding phonon lifetime. As the model adopted in our calculations is unable to compute the damping factors, examples reported in this chapter will make use of the experimental values. In case no experimental data is available for γ, a unique empirical value for all modes can be adopted with very good results (e.g. see the case of garnets in Dovesi et al. (2011)). Note that in a reflectance spectrum, this quantity does not affect the peak position, width or shape but only the peak height. From definition (10.15), the static ($v = 0$) dielectric tensor comes out as

$$\epsilon_{ij}^0 = \epsilon_{ij}^\infty + \sum_n \mathbf{f}_{n,ij} \tag{10.16}$$

A simple relation holds between reflectance $R(v)$ and the dielectric function (here θ is the angle between the incident light and the normal to the surface; see Born and Huang (1954)):

$$R(v) = \left| \frac{\sqrt{\epsilon(v) - \sin^2\theta} - \cos\theta}{\sqrt{\epsilon(v) - \sin^2\theta} + \cos\theta} \right|^2 \tag{10.17}$$

The calculation of non-resonant Raman intensities is based on Placzek's approximation (Placzek 1934), where the key quantities are the partial derivatives of the polarisability tensor α with respect to the atomic displacements, forming a $3 \times 3 \times 3N$ matrix:

$$\rho_{ij,\alpha k} = \frac{\partial}{\partial x_{\alpha k}} \left(\frac{\partial^2 V(\vec{x})}{\partial F_i \partial F_j} \right) = \frac{\partial}{\partial x_{\alpha k}} \alpha_{ij} \tag{10.18}$$

This matrix can be computed by means of a fully analytical CPHF/KS approach (Maschio et al. 2013), similar to the one mentioned earlier for IR intensities (Maschio et al. 2012). Like for Born charges, we can redefine these quantities in the normal coordinate basis rather than in the Cartesian atomic basis, obtaining the Raman susceptibility tensors:

$$\chi_{n,ij} = \frac{\partial}{\partial Q_n} \left(\frac{\partial^2 V(\vec{x})}{\partial F_i \partial F_j} \right) = \frac{\partial}{\partial Q_n} \alpha_{ij} \tag{10.19}$$

Single crystal Raman intensities are then calculated according to Prosandeev et al. (2005):

$$II_{n,ij} \propto C \cdot \Omega \cdot \left(\chi_{n,ij} \right)^2 \tag{10.20}$$

where the prefactor C gives the dependence on temperature T and frequency of laser v_L:

$$C \propto (v_L - v_n)^4 \frac{1 + n_B(v_n)}{30 v_n} \tag{10.21}$$

with the Bose occupancy factor being $1 + n_B(v_n) = \left[1 - \exp(-hv_n/k_B T) \right]^{-1}$

For a powder poly-crystalline sample, the isotropic Raman intensity of the nth mode can be obtained using the following formula (Prosandeev et al. 2005):

$$II_n^{\text{tot}} \propto C \cdot \Omega \cdot \left[10 G_n^{(0)} + 5 G_n^{(1)} + 7 G_n^{(2)} \right] \tag{10.22}$$

where the terms $G_n^{(k)}$ are the rotational invariants defined through

$$
\begin{aligned}
G_n^{(0)} &= \frac{1}{3} \left(\chi_{n,xx} + \chi_{n,yy} + \chi_{n,zz} \right)^2 \\
G_n^{(1)} &= \frac{1}{2} \left[\left(\chi_{n,xy} - \chi_{n,yx} \right)^2 + \left(\chi_{n,xz} - \chi_{n,zx} \right)^2 + \left(\chi_{n,yz} - \chi_{n,zy} \right)^2 \right] \\
G_n^{(2)} &= \frac{1}{2} \left[\left(\chi_{n,xy} + \chi_{n,yx} \right)^2 + \left(\chi_{n,xz} + \chi_{n,zx} \right)^2 + \left(\chi_{n,yz} + \chi_{n,zy} \right)^2 \right] \\
&\quad + \frac{1}{3} \left[\left(\chi_{n,xx} - \chi_{n,yy} \right)^2 \right] + \left(\chi_{n,xx} - \chi_{n,zz} \right)^2 + \left(\chi_{n,yy} - \chi_{n,zz} \right)^2 \right]
\end{aligned}
\tag{10.23}
$$

10.2.4   Estimation of the Anharmonic Constant for X–H Stretching Modes

As mentioned at the beginning of Section 10.2, vibrational frequencies calculated within the harmonic approximation are generally only a few cm^{-1} far from the experimental values. However, modes involving H atoms show a large and non-negligible anharmonic character.

For X–H stretching modes (X being a generic atom, e.g. O, N, C, S), a simple *a posteriori* approach has been devised to estimate the corresponding anharmonic constant $\omega_e \chi_e$ (Pascale et al. 2004a; Ugliengo et al. 1990; Ugliengo P, 1989, ANHARM: a program to solve monodimensional nuclear Schrödinger equation, unpublished). The fundamental assumption is that the X–H distance can be considered as a normal coordinate fully decoupled from all others. Then, the following steps are taken:

1. The PES is explored as a function of the X–H distance, by displacing H along this coordinate around its equilibrium position.
2. A polynomial curve of sixth degree is used to best fit the energy points, while keeping the root-mean-square error below 10^{-6} atomic units.
3. The one-dimensional nuclear Schrödinger equation is solved numerically, following the method proposed by Lindberg (1988), based on a matrix formulation of Numerov's method. The lowest three eigenvalues of the equation are computed, E_0, E_1 and E_2.

Knowledge of the eigenvalues permits to obtain the frequency of the first (fundamental, v_{01}) and second (overtone, v_{02}) vibrational transitions and then the anharmonic constant $\omega_e \chi_e$:

$$v_{01} = E_1 - E_0; \quad v_{02} = E_2 - E_0; \quad \omega_e \chi_e = \frac{(2 v_{01} - v_{02})}{2} \tag{10.24}$$

The corresponding anharmonic vibrational frequencies can be obtained from the harmonic ones upon subtraction of $2\omega_e \chi_e$. This estimation is reasonable provided that the modes do not couple with any other, that is, H atoms are not involved in strong hydrogen bonds or other interactions. At present, no simple and general scheme is available to estimate the anharmonicity of other kinds of modes.[6]

---

[6] If not differently specified, all computed frequencies will refer to the harmonic model in the following discussion, tables and figures.

10.2.5   Accuracy of Basis Set and Hamiltonian

The method proposed in this chapter is based on the use of localised Gaussian-type basis set functions to define the wave function (more details in Chapters 1 and 3). The main factors affecting the accuracy of localised basis sets are the number of functions associated with valence electrons and the presence of higher angular momentum 'polarisation' functions.[7]

To give an idea of the role of the basis set in determining the accuracy of vibrational calculations, we will here take calcite, the most stable $CaCO_3$ polymorph, as a reference system, and consider four different basis sets. The first three, A, B and C, are derived from Pople basis sets and correspond to BSA, BSC and BSD in Valenzano et al. (2006). BS A uses (8s)-(6511sp)-(3d) for Ca, (6s)-(21sp)-(1d) for C and (8s)-(411sp)-(1d) for O. BS B uses a modified (6s)-(311sp)-(1d) set for C, with an additional third valence shell consisting of a single Gaussian function, and a modified set for Ca with the d polarisation being split into two shells with respect to BS A. BS C adds a second d polarisation shell to both C and O. The fourth basis set, BS D, has been recently derived from the Alrichs def2-TZVP molecular basis set (Peintinger et al. 2013). It has three valence and one polarisation shells for each element and uses different exponents for s and p valence Gaussian functions.

Statistics on the comparison of calculated and experimental vibrational properties of calcite using different basis sets are reported in Table 10.1, top part. Structural data are not reported, as all the adopted basis sets give deviations from the experimental volume at 298 K (Antao et al. 2009) that are consistent with the systematic volume overestimation affecting the adopted functional.

Overall, data in Table 10.1 (top part) demonstrate how the accuracy of the calculations can improve when using richer basis sets. For all the considered properties, the mean absolute deviation $|\bar{\Delta}|$ decreases in going from BS A to BS D. For frequencies, the largest improvement for $|\bar{\Delta}|$ is found from BS B to BS C, that is, when double polarisation functions are added to C and O. For IR oscillator strengths, the largest improvement, by about 10%, is achieved when passing from BS C to BS D, that is, when switching to Gaussian functions with distinct optimised exponents for s and p

Table 10.1 *Calculated versus experimental vibrational properties of calcite using various basis sets (top part, B3LYP functional) and functionals (bottom, BS D).*

	v			f(IR)		II (Raman)	
	$|\bar{\Delta}|$	$\bar{\Delta}$	$|\Delta|_{max}$	$|\bar{\Delta}|$	$|\Delta|_{max}$	$|\bar{\Delta}|$	$|\Delta|_{max}$
BS A	9.7	+5.6	29.7	0.457	2.116	95	422
BS B	8.7	−1.7	26.9	0.432	2.036	76	253
BS C	7.0	+0.6	27.6	0.450	2.073	80	285
BS D	5.9	+0.3	19.2	0.391	1.915	28	58
SVWN	26.2	+5.0	55.0	1.123	4.838	43	148
PBE	21.8	−20.6	48.7	0.220	1.325	33	59
PBEsol	17.4	−9.7	47.8	0.273	1.383	38	90
PBE0	12.7	+12.5	49.4	0.278	1.385	26	54
B3LYP	5.9	+0.3	19.2	0.391	1.915	28	58

Experimental values from Long et al. (1993) (IR) and De La Pierre et al. (2014a) (Raman). Typical values of IR and Raman properties of calcite are reported in Tables 10.3 and 10.4, respectively, in Section 10.3.1.
v, vibrational frequencies (cm^{-1}); f, oscillator strengths (dimensionless); II, single crystal Raman intensities (arbitrary units, A.U.);
$\bar{\Delta}$, $|\bar{\Delta}|$, $|\Delta|_{max}$, mean, mean absolute, and maximum absolute difference, respectively.

---

[7] Other kinds of basis set are available in solid-state codes, whose accuracy may be based on other parameters. For instance, when using a pseudopotential/plane-wave code, the main issues to consider are the energy cut-off for the plane-wave functions and the choice of the effective core potential. See Chapter 1 for further discussion.

valence shells. However, the most remarkable improvement over the entire data set is found for Raman intensities, when considering BS D. Here $|\bar{\Delta}|$ reduces by a factor around 3.

The maximum absolute deviation, $|\Delta|_{max}$, has a similar trend as $|\bar{\Delta}|$. Again, the largest improvement is found when considering BS D for the calculation of Raman intensities: here $|\Delta|_{max}$ decreases from 285 A.U. (BS C) to 58 A.U., by a factor larger than 4.

It is not surprising that Raman intensities are the most sensitive to the choice of the basis set: as discussed in Section 10.2.3, they are related to derivatives of the crystal polarisability, which is in turn the second derivative of the potential energy with respect to an applied field.

We now briefly discuss the effect of the adopted Hamiltonian (or functional) on the accuracy of vibrational calculations. We will consider five functionals as representative of three wider classes: SVWN (Slater 1951; Vosko et al. 1980), the most commonly used local density approximation (LDA) functional; PBE (Perdew et al. 1996) and PBEsol (Perdew et al. 2008), two flavours of the generalised gradient approximation (GGA), the latter corresponding to a re-parametrisation of PBE for solids; and PBE0 (Adamo and Barone 1999) and B3LYP (Becke 1993), two popular forms of the global hybrid HF/DFT approach.[8]

The corresponding data for calcite are shown in Table 10.1, bottom part. The unit cell volume (not reported in the table) has a well-known trend, with SVWN largely underestimating it, PBE and B3LYP overestimating it and PBE0 and PBEsol providing deviations that are in between these two extremes.

The main improvement for v values is observed when using hybrid functionals instead of pure DFT, with B3LYP giving the best agreement. The mean deviation $\bar{\Delta}$ indicates that PBE and PBEsol tend to underestimate frequencies, whereas PBE0 and SVWN tend to overestimate them. B3LYP does not show any pronounced trend here. For IR oscillator strengths and Raman intensities, it is clear from Table 10.1 that SVWN gives by far the worst results. GGA and hybrid functionals have comparable performance in reproducing IR spectra, whereas hybrid functionals appear to be slightly better for Raman intensities.

While the analysis performed on calcite can be considered representative for hydrogen-free minerals, with GGA being a less accurate but still acceptable choice with respect to hybrid functionals, the inclusion of some exact exchange is crucial for calculating the vibrational properties of minerals containing hydrogen in their structure. Structural and vibrational data related to hydrogen in $Mg(OH)_2$ brucite are shown in Table 10.2. Here, a relevant structural parameter is the O—H bond length: SVWN gives the largest deviation from the experimental value, whereas hybrid functionals provide much closer estimations. The calculation of O—H stretching frequencies (corrected by anharmonicity) follows the same trend, with SVWN, PBEsol and PBE giving deviations from the experiment in the order of hundreds of cm^{-1}.

Overall, when studying vibrational properties of solids, LDA functionals such as SVWN usually give very poor results compared to other functionals. Even if they provide a good description for metals, they should be avoided when investigating solids with insulating electronic structure, like minerals.

Between GGA and global hybrid functionals, the latter give improved performance in computing frequencies, especially when hydrogen atoms are involved. In particular, as already proved in a number of publications (e.g. De La Pierre et al. (2011), Demichelis et al. (2010b)), which also provide a general overview of the DFT performance in predicting vibrational properties of aluminosilicates,

---

[8] Another increasingly popular class is that of dispersion-corrected functional (DFT-D) derived from the ideas of Grimme (2004). They are crucial for a proper description of structure and energetics of systems where dispersion forces are relevant. However, assessment of their performance in computing vibrational properties, relative to the original DFT methods, is still a matter of study; the main complication is the dependence of the results on the parameterisation of the dispersion term. Typical examples of applications involve layered minerals (Ugliengo et al. 2009; Ulian et al. 2013a).

Table 10.2 *Calculated versus experimental structures and O—H stretching frequencies of brucite with various functionals.*

	$\Delta V\%$	Δd_{OH}	$\Delta v(A_{1g})$	$\Delta v(A_{2u})$
SVWN	−10.6	+0.024	−298	−256
PBE	+2.1	+0.014	−151	−149
PBEsol	−5.0	+0.021	−243	−217
PBE0	−0.8	+0.003	+36	+33
B3LYP	+3.6	+0.004	+14	−5

Basis set from Pascale et al. (2004a) and computed data from Demichelis et al. (2010a).
$\Delta V\%$, relative difference with respect to the experimental volume (Catti et al. 1995); Δd_{OH}, difference with respect to the experimental O—H bond distance (Å) (Catti et al. 1995); Δv, difference between the calculated anharmonic and experimental O—H stretching frequencies (cm^{-1}) (Lutz et al. 1994).

B3LYP gives the best performance for frequencies, at the price of slightly worse IR oscillator strengths. More generally, all hybrid functionals are expected to provide very comparable performance. For H-free minerals, also GGA functionals represent a reasonable choice for predicting vibrational properties. Often, the choice within these ranges of functionals can be made also on the basis of other factors, such as uniformity with other studies, or improved description of other kinds of properties.

10.3 Examples and Applications

The aim of this section is to show how the tools and methods described earlier can be successfully applied to investigate the vibrational properties of mineral systems. Minerals belonging to diverse groups (carbonates, pyroxenes, oxides and hydroxides) will be considered, so as to include a variety of compositions, structural motifs and physical properties.

Carbonates are commonly found in the Earth's upper crust and represent important tracers of chemical, physical and biological activity throughout geological times. In the last decade, carbonates have been at the centre of intensive research, due to their nucleation and crystal growth occurring via pathways that differ from those envisaged within the classical nucleation theory (Cartwright et al. 2012; Demichelis et al. 2011b; Gebauer et al. 2008; Wallace et al. 2013). Calcium and magnesium carbonate phases are arguably the most studied, due to their crucial role in biomineralisation (see Chapter 9) and to their undesired precipitation in industrial and domestic devices in the form of scale.

Pyroxenes are a group of inosilicates that, together with olivines and amphiboles, represent the major constituents of the Earth's mantle and of many extraterrestrial planetary bodies. Therefore, they are involved in the majority of geochemical and geophysical phenomena occurring on Earth and in space (Müntener 2010). Pyroxenes, especially ortho-enstatite ($MgSiO_3$), are also the most frequent silicate compounds found in interstellar and circumstellar dusts (Spoon et al. 2006; Wooden et al. 2004).

The oxide and hydroxide groups include a variety of minerals that are more or less widely present in the Earth's crust and mantle, as well as in extraterrestrial environments. Among these, we will focus on brucite, $Mg(OH)_2$, a mineral commonly formed as a product of serpentinisation (Guillot and Hattori 2013), and on diaspore, α-AlOOH, an aluminium oxyhydroxide that is present as a main component in bauxite ores, together with boehmite (its layered polymorph) and gibbsite (Hart 1990; Wefers and Misra 1987). Both these minerals play important technological and geochemical roles and are often used as a reference to model the properties of oxyhydroxides (diaspore) and layered hydroxides (brucite).

In all these contexts, first principles methods are able to provide accurate datasets that can serve as a main reference for understanding structural features and composition of real samples, contributing then to validate hypotheses in regard to physical processes and chemical reaction mechanisms.

10.3.1 Vibrational Properties of Calcium and Magnesium Carbonates

In this section, we will focus on the simple case of calcite ($CaCO_3$, $R\bar{3}c$ space group), magnesite ($MgCO_3$, $R\bar{3}c$) and dolomite ($MgCa(CO_3)_2$, $R\bar{3}$). Vibrational properties of calcite as modelled with the B3LYP functional and the BS D basis sets (see Section 10.2.5) will be fully analysed and then compared with those of magnesite and dolomite. For the sake of brevity, we will be unable to analyse the remaining $CaCO_3$ polymorphs, aragonite and vaterite, for which further details can be found in Carteret et al. (2013) and De La Pierre et al. (2014a, b).

The 27 fundamental vibrational modes of calcite and magnesite at the Γ point can be classified according to the irreducible representations of the $\bar{3}m$ point group as follows: $\Gamma_{total} = 1A_{1g} \oplus 2A_{1u} \oplus 3A_{2g} \oplus 3A_{2u} \oplus 4E_g \oplus 5E_u$. A_{2u} and E_u modes are IR active, A_{1g} and E_g modes are Raman active, and the remaining ones are silent.

10.3.1.1 Calcite: Frequencies and Intensities

IR data can be extracted from experimental single crystal reflectance spectra through best fit, by exploiting Equations 10.15 and 10.17: TO frequencies and oscillator strengths are explicit in these formulas, whereas LO frequencies are identified as the positions of the minima in the imaginary part of $1/\epsilon(v)$. Table 10.3 presents results obtained in this way by Hellwege et al. (1970), together with values from our calculations. The agreement on frequencies is excellent, with a mean absolute deviation of $8\,cm^{-1}$ for both TO and LO modes. Only 2 TO and 3 LO modes have absolute differences larger than $10\,cm^{-1}$. In 4 out of 5 such cases, the affected modes lie below $140\,cm^{-1}$, in a spectral region involving cation soft motions. This region has shown large discrepancies between simulation and experiment also for other minerals, such as aragonite (Carteret et al. 2013) and pyrope (Dovesi

Table 10.3 *The IR spectrum of calcite.*

	v_{TO}		f		v_{LO}	
	Exp.	Calc. (Δ)	Exp.	Calc. (Δ)	Exp.	Calc. (Δ)
A_{2u}	92	+22	4.962	−2.236	136	+13
	303	+5	1.394	−0.118	387	+12
	872	−9	0.088	−0.006	890	−8
E_u	102	+16	2.980	−1.018	123	+12
	223	+2	1.026	−0.332	239	−3
	297	−3	1.655	−0.037	381	−3
	712	−6	0.010	−0.000	715	−7
	1407	−0	0.550	−0.035	1549	+2
$|\bar{\Delta}|$		8		0.473		8
$\bar{\Delta}$		+3		−0.473		+2
$|\Delta|_{max}$		22		2.236		13

Experimental values from Hellwege et al. (1970). Reproduced with permission of Springer.
v_{TO}, v_{LO}, TO and LO frequencies (cm^{-1}); Δ, difference between calculated and experimental data. The rest as in Table 10.1.

Table 10.4 *The Raman spectrum of calcite.*

	ν		X^2			Y^2		
	Exp.	Calc. (Δ)	Exp.	Calc. (Δ)	X	Exp.	Calc. (Δ)	Y
A_{1g}	1086.5	−4.8	1000	0	a	176	−38	b
E_g	155.4	−2.5	25	−7	c	147	+36	d
E_g	282.0	−5.2	71	−58	c	428	−24	d
E_g	712.5	−7.5	56	+49	c	20	−13	d
E_g	1436.5	+2.0	11	+21	c	6	−6	d
$\overline{|\Delta|}$		4.4					28	
$\overline{\Delta}$		−3.6					−4	
$|\Delta|_{max}$		7.5					58	

Experimental data from De La Pierre et al. (2014a). Reproduced with permission of AIP.
a^2–d^2, single crystal Raman intensities (X^2 and Y^2 used as placeholders, for the sake of compactness), re-normalised so that the a^2 element of mode at 1086.5 cm^{-1} is equal to 1000. The rest as in Tables 10.1 and 10.3.

et al. 2011). Oscillator strengths are very well reproduced, the only two large discrepancies being again found in the aforementioned region.

Raman frequencies and intensities, as obtained from both calculation and experiment (De La Pierre et al. 2014a), are given in Table 10.4. Frequencies are in striking agreement, the largest absolute deviation being 7.5 cm^{-1}. We here follow the convention of normalising intensities, so that the largest value is set to 1000. As the corresponding mean absolute deviation is 28, principal Raman peaks with intensity larger than 100 are in general well reproduced, whereas relatively large discrepancies occur for minor features with intensity in the order of few tens (see also aragonite in De La Pierre et al. (2014a)).

Overall, these data show that computer simulation is capable of properly identifying and characterising all the fundamental modes in both IR and Raman spectra of a mineral.

10.3.1.2 Calcite: IR and Raman Spectra

We have constructed the two polarised IR reflectance spectra of single crystal calcite using the computed frequencies and intensities and compared them with the experimental counterparts in Figure 10.1. As discussed in Section 10.2.3, we have adopted damping factors γ (Eq. 10.15) taken from the experiment.

The correspondence between simulated and measured curves is remarkable. Certain reflectance bands show an asymmetric profile, such as the one at 303 cm^{-1} in the A_{2u} spectrum. In this case, a better description could be achieved, by using a four-parameter model for the dielectric function (Berreman and Unterwald 1968), instead of the Drude–Lorentz expression (10.15), as it has been done for forsterite, grossular and ortho-enstatite in De La Pierre et al. (2013a).

Even in those cases where all the fundamental peaks have been characterised, sometimes there are other minor interesting features that can be investigated through computer simulation. In the present case, the peak around 848 cm^{-1} in the A_{2u} spectrum is usually interpreted as a combination mode (Donoghue et al. 1971). Periodic codes implementing crystal symmetry, such as that adopted here, readily permit to characterise combination modes and overtones, by analysing all possible combinations of fundamentals and assigning them the appropriate irreducible representation. Here, it turns out that only two combinations provide values close to the observed frequency, namely, $102 \; (E_u) + 712.5 (E_g) = 814.5 \, \text{cm}^{-1}$ and $155.4 \; (E_g) + 712 (E_u) = 867.4 \, \text{cm}^{-1}$. The latter is 20 cm^{-1}

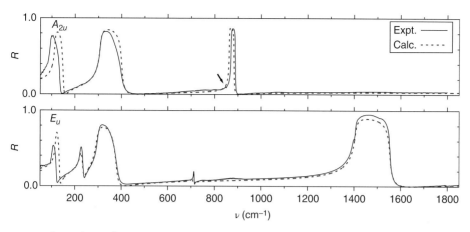

Figure 10.1   *Polarised IR reflectance spectra of single crystal calcite. The two datasets correspond to the electric field of the incident radiation being parallel and perpendicular to the c axis, respectively. Arrow points to the experimental extra peak at 848 cm^{-1} (see text).*

far from the observed one and is the most likely, the difference being probably accounted for by a Fermi resonance with the mode at 872 cm^{-1} (Donoghue et al. 1971).

Computed and experimental polarised Raman spectra are shown in Figure 10.2. Here simulation helps in characterising the peak at 1436.5 cm^{-1}, found in both $a^2 + c^2$ and d^2 spectra, as a fundamental mode. Two additional minor features appear at 1066.8 and 1749.2 cm^{-1}. They are only present in the $a^2 + c^2$ and b^2 spectra; therefore they must have A_{1g} symmetry. The former can be identified as the ^{18}O satellite peak of the intense peak at 1086.5 cm^{-1}. In fact, by simulating the isotopic substitution of one ^{16}O atom in one CO_3^{2-} unit with a ^{18}O species, a mode appears that is 20.2 cm^{-1} lower than the main A_{1g} mode (19.7 cm^{-1} in the experiment). Analysis of the eigenvectors confirms that the lowest frequency mode corresponds to ^{18}O—C stretching and the highest frequency mode to ^{16}O—C stretching.

As proposed by Gillet et al. (1996), the second additional peak at 1749.2 cm^{-1} (well beyond the highest fundamental frequency) is an overtone mode. Symmetry analysis confirms that the most likely combination is $2 \times 872(A_{2u}) = 1744$ cm^{-1}.

A more detailed computational-aided analysis of these extra features in the Raman spectra of calcite is presented in De La Pierre et al. (2014a).

10.3.1.3   Calcite: Mode Analysis and Isotopic Substitution

Computer simulation provides additional tools for characterising and interpreting the vibrational properties of minerals. The eigenvectors obtained diagonalising the dynamical matrix contain information on the contribution of the various atoms to each mode. These contributions, which may be hard to understand just by looking at the eigenvectors (a large table of numbers), can be post-processed to obtain graphical animations. The latter permit to easily perform a qualitative and intuitive analysis of the nature of vibrational modes.[9]

The simulation of isotopic mass substitution is another way of exploring the nature of the modes, allowing for a quantitative description of the contribution of each chemical species to the normal

---

[9] Interactive animation for calcite available at http://www.crystal.unito.it/vibs/calcite/

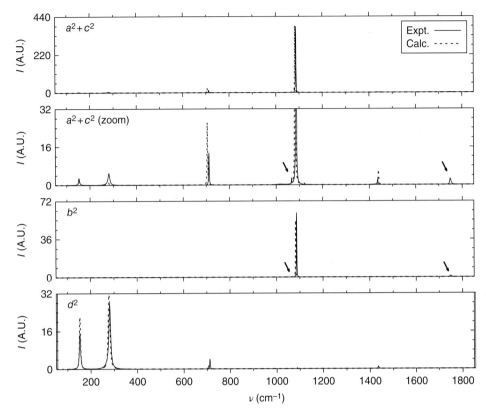

Figure 10.2   *Polarised Raman spectra of single crystal calcite. The three datasets correspond to XX, ZZ and XZ polarisations, respectively. Damping factors γ here correspond to the peak full width at half maximum (FWHM). Arrows point to the experimental extra peaks discussed in the text. Experimental data from De La Pierre et al. (2014a). Reproduced with permission of AIP.*

modes. It can be computed at zero cost once the Hessian matrix is built (see Eq. 10.3) and can be used to a much larger extent than in the experiments. Indeed, the latter are affected by many limitations, including the existence of a limited range of isotopes and the inability to substitute isotopes to specific sites. From this point of view, computer simulation offers the possibilities of selecting any mass and of substituting isotopes with a specific distribution in the lattice.

This versatility can be useful to investigate several kinds of features. One example, the ^{18}O contribution to the Raman spectrum of calcite, has been described in the previous paragraph. Other examples could involve the need of analysing isolated fragments of a structure (e.g. O—H bonds on a mineral surface or subunits of a bulk structure), of obtaining equilibrium isotopic fractionation factors or of understanding what are the chemical species that contribute to a particular spectral peak (Daramola et al. 2010; Demichelis et al. 2011a; Dovesi et al. 2011; Imrie et al. 2013; Orlando et al. 2006; Schauble et al. 2006). A further advantage is the possibility of choosing any desired mass (e.g. freezing all atoms but a fragment by attributing them infinite masses and investigating the dynamical properties of that specific fragment, as in Orlando et al. (2006)).

Here we show how isotopic substitution can be exploited to systematically analyse the contribution of each chemical species to each normal mode. A percent variation to the element masses is applied, rather than substituting with naturally occurring isotopes. It is then assumed that, when varying the mass of a given species, the resulting isotopic shift of a mode frequency is proportional to the contribution of that species to that mode. In this case, the masses of Ca, C and O species in calcite have been in turn increased by 10%.

High-frequency modes (700–1500 cm^{-1}) are related to the internal motions of the CO_3^{2-} units (bending and stretching), as confirmed by graphical animation. Isotopic shifts reported in Table 10.5 substantiate this point, showing that changing the mass of Ca atoms has no effect on these frequencies. The shifts also indicate that in-plane bendings and symmetric stretchings mostly involve oxygen motions, whereas out-of-plane bendings and asymmetric stretchings mostly involve carbon.

Low-frequency modes correspond to external motions of the CO_3^{2-} anions (librations and translations) and to translations of the Ca^{2+} ions. Their description is complex due to their mutual coupling. A few possibilities are ruled out by symmetry, so that all gerade 'g' modes have no contribution from Ca atoms, and the only A_{1u} mode is a pure Ca translation. In the other cases, the contribution of each species to a given mode can be quantitatively (though arbitrarily) estimated by referring to the corresponding percent shift in frequency. Notably, the shifts provide a way to distinguish CO_3^{2-} translations and librations, as the former involve carbon motions to a much larger extent. More details on isotopic analysis of calcite can be found in Valenzano et al. (2006).

10.3.1.4 Calcite, Magnesite and Dolomite: IR Spectra

Compared to calcite and magnesite (see beginning of Section 10.3.1), dolomite has lower space group symmetry and therefore also has a different classification of vibrational modes according to irreducible representations: $\Gamma_{\text{total}} = 4A_g \oplus 5A_u \oplus 4E_g \oplus 5E_u$, where u and g modes are IR and Raman

Table 10.5 *Simulated isotopic analysis of vibrational modes in calcite.*

		$\Delta v\%$			
	v	Ca	C	O	Description
A_{2u}	113.8	−0.9	−0.1	−3.7	l_{in} + Ca t_{out}
E_u	117.7	−1.3	−0.1	−3.5	l_{out} + Ca t_{in}
E_g	152.9	0.0	−0.8	−4.0	l_{out} + t_{in}
A_{2g}	180.3	0.0	−0.3	−4.4	l_{in} + t_{out}
E_u	225.2	−4.2	0.0	−0.4	Ca t_{in}
E_g	276.8	0.0	−0.2	−4.5	l_{out}
A_{1u}	288.9	−4.6	0.0	0.0	Ca t_{out}
E_u	293.7	−2.0	−0.3	−2.4	l_{out} + Ca t_{in}
A_{2g}	306.3	0.0	−0.8	−3.9	l_{in} + t_{out}
A_{2u}	308.4	−1.9	−0.3	−2.4	l_{in} + Ca t_{out}
E_g	705.0	0.0	−0.4	−4.3	b_{in}
E_u	706.3	0.0	−0.4	−4.3	b_{in}
A_{2u}	862.9	0.0	−3.7	−0.9	b_{out}
A_{2g}	869.2	0.0	−3.5	−1.1	b_{out}
A_{1g}	1081.7	0.0	0.0	−4.7	s_{sym}
A_{1u}	1082.8	0.0	0.0	−4.6	s_{sym}
E_u	1406.9	0.0	−3.4	−1.2	s_{asym}
E_g	1438.5	0.0	−3.4	−1.2	s_{asym}

$\Delta\%$, percent isotopic shifts computed for each species by increasing its mass by 10%; v, frequencies (cm^{-1}); s, C—O stretching; b, $\widehat{OCO}$ bending; l, CO_3 libration; t, CO_3 translation; Ca t, Ca translation; sym and asym, symmetric and asymmetric; in and out, in-plane and out of plane.

active, respectively. At variance with the two end member carbonates, there are 2 (3) additional IR (Raman)-active modes, corresponding to the A_{1u} (A_{2g}) silent modes in the end members. Valenzano et al. (2007) carried out a detailed analysis of the vibrational spectra of these carbonates, comparing calculated and experimental data. Here, we focus on the computed IR spectra to provide an example of how simulation can help interpreting specificities and trends along a mineral series.

Frequencies of the IR-active modes for the three compounds are reported in Table 10.6; we make use again of simulated isotopic substitution and graphical animation to characterise the corresponding normal modes.[10] Apart from the occurrence of Mg rather than Ca ions, the nature of vibrational modes in magnesite is nearly the same as in calcite. Isotopic shifts for dolomite reported in the table show how Ca and Mg ions contribute to the six low-frequency modes: three of them are dominated by the former and two of them by the latter, whereas mode at 260.0 cm^{-1} has mixed contributions from both species.

Table 10.6 *Calculated IR frequencies (v, cm^{-1}) of three carbonates.*

		Calcite	Dolomite						Magnesite
					$\Delta v\%$				
		v	v	Ca	Mg	C	O	Ions	v
1	$A_{(2)u}$	113.8	153.7	−1.6	−0.1	−0.1	−3.0	Ca	244.4
2	E_u	117.7	166.4	−1.8	−0.1	−0.1	−2.9	Ca	241.8
3	E_u	225.2	260.0	−1.8	−1.0	0.0	−1.9	Ca + Mg	304.0
*	$A_{(1)u}$	(288.9)	354.2	−0.1	−3.8	−0.2	−0.5	Mg	(360.8)
4	E_u	293.7	342.9	−0.1	−3.0	−0.1	−1.3	Mg	346.2
5	$A_{(2)u}$	308.4	301.1	−2.0	−0.2	−0.1	−2.4	Ca	344.8
6	E_u	706.3	721.0	0.0	0.0	−0.4	−4.3	—	737.8
7	$A_{(2)u}$	862.9	863.9	0.0	0.0	−3.6	−1.0	—	860.7
#	$A_{(1)u}$	(1082.8)	1093.6	0.0	0.0	0.0	−4.6	—	(1092.8)
8	E_u	1406.9	1419.2	0.0	0.0	−3.4	−1.2	—	1419.8

Silent (i.e. IR-inactive) frequencies for calcite and magnesite in brackets. '*' and '#' indicate modes that are silent in calcite and magnesite but IR active in dolomite. Percent isotopic shifts ($\Delta\%$) and contributing ions reported for dolomite. Isotopic shifts computed for each species by increasing the corresponding mass by 10%.

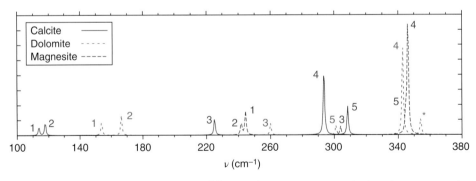

Figure 10.3 *Calculated adsorption IR spectra of three carbonates: zoom on the low-frequency region. Peaks labelled according to Table 10.6.*

---

[10] Interactive animation available at http://www.crystal.unito.it/vibs/carbonates/

The adsorption IR spectra in the 100–$400\,\mathrm{cm}^{-1}$ range are shown in Figure 10.3. A remarkable alteration in the sequence is found for modes 4 and 5 of dolomite, both corresponding to combined ion translations and carbonate librations. However, the former is dominated by Mg and the latter by Ca motions, so that their frequencies fall closer to the matching magnesite (higher) and calcite (lower) values. Dolomite also has an additional IR peak (A_u symmetry, labelled as '*') at $354.2\,\mathrm{cm}^{-1}$. It is a nearly pure out-of-plane translation of Mg ions and corresponds to IR-inactive A_{1u} modes in calcite and magnesite, which in turn are pure out-of-plane translations of Ca and Mg ions, respectively.

Analysing frequencies along the series, values in the low-frequency range increase by 130 (mode 1) to 40 (mode 5) cm^{-1} when going from calcite to magnesite. When considering dolomite as the intermediate compound, these steps are not homogeneous, so that, for example, for mode 1 frequency increases by $40\,\mathrm{cm}^{-1}$ from calcite to dolomite and by a further $90\,\mathrm{cm}^{-1}$ to get to magnesite. Mode analysis helps in recognising that the reason for this behaviour is the contributing cation for each dolomite mode, whose frequency is closer to that of the end member with the matching species: the smallest differences are with calcite when Ca dominates and with magnesite when Mg dominates, whereas differences are nearly equal for the only mode with mixed contribution from both species.

The high-frequency region is much simpler, as cations do not contribute to the corresponding modes. Mode analysis is the same for the three compounds. Note that a relatively small increase in frequency along the series is found for 3 out of 4 modes, which can be mainly related to the shortening of cell parameters.

10.3.2   A Complex Mineral: The IR Spectra of Ortho-enstatite

In order to convey the idea that the computational analysis illustrated so far can be applied to minerals that are much more complex than calcite, we briefly discuss the case of $MgSiO_3$ ortho-enstatite (*Pbca* space group, 80 atoms in the unit cell). Three distinct IR reflectance spectra can be collected along the three inequivalent crystallographic directions of its orthorhombic lattice, each spectrum featuring 29 IR-active fundamental modes.

Demichelis et al. (2012) performed a combined experimental and theoretical study of the IR reflectance spectra of single crystal ortho-enstatite. Accurate experimental curves were collected at the synchrotron radiation facility SPring-8 in Japan. Simulations made use of the PBE0 hybrid functional[11] and the following basis sets: (8s)-(511sp)-(1d) for Mg, (8s)-(6311sp)-(1d) for Si and (8s)-(411sp)-(1d) for O. Analysis of the three measured spectra resulted in the identification of 65 peaks, out of the 87 expected by symmetry analysis and obtained from the simulation. Modes identified in the simulation but not in the experiment were recognised either to have low calculated intensity or to lie close to intense peaks. Comparison between calculated and experimental data are in excellent agreement, so that, for example, the mean absolute deviation for frequencies is 7–$8\,\mathrm{cm}^{-1}$. A visible proof of such consistency is given in Figure 10.4, showing the reflectance spectrum along the c axis.

Building on the increasingly demonstrated accuracy of calculations in the domain of vibrational properties, a later study by De La Pierre et al. (2013a) thoroughly investigated the feasibility of an approach making use of simulation as the principal tool for characterising experimental IR spectra of minerals. Computed frequencies and intensities were there used as starting values to identify peaks by best fit against a model dielectric function, such as the one in Equation 10.15. Ortho-enstatite was chosen as one of the test systems for the richness of its spectra; compared to the purely empirical characterisation adopted by Demichelis et al. (2012), this computational-aided analysis identified 14 additional fundamental modes over the three experimental curves, as well as three combination modes.

---

[11] Similar results can be obtained with B3LYP or other hybrid functionals.

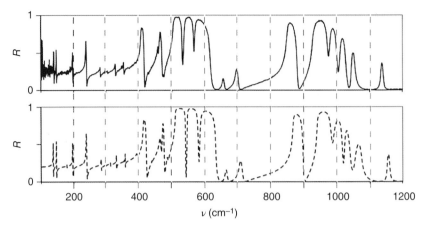

Figure 10.4 *Polarised IR reflectance spectrum along the c axis of single crystal ortho-enstatite. Experimental (full line) and calculated (dashed line) curves both from figure 3 Demichelis et al. (2012). Reproduced with permission of John Wiley & Sons.*

10.3.3 Treatment of the O—H Stretching Modes: The Vibrational Spectra of Brucite and Diaspore

As discussed in the previous sections, the presence of hydrogen atoms in minerals adds a further level of complexity in computing their vibrational properties. While the harmonic approximation is sufficiently accurate to describe the fundamental vibrational transitions in H-free systems, such as those presented in Sections 10.3.1 and 10.3.2, the anharmonic contribution to modes involving the X—H group (X = oxygen in this case) is not negligible.

On top of this, while pure GGA functionals are able to provide a realistic prediction of the vibrational features of H-free systems, the use of hybrid functionals is mandatory in the presence of hydrogen atoms, as shown in Table 10.2 and discussed in Section 10.2.5. In this section, structural optimisation and vibrational calculation have been run using the B3LYP functional, together with the following basis sets: (8s)-(511sp)-(1d) for Mg, (8s)-(621sp)-(1d) for Al, (8s)-(411sp)-(1d) for O and (211s)-(1p) for H (Demichelis et al. 2011a; Pascale et al. 2004a).

The structures of brucite and diaspore are represented in Figure 10.5 and summarised in Table 10.7. The former is a layered system, with layers interacting with each other through van der Waals forces. The fact that DFT and hybrid functionals lack the description of such forces results in an overestimation of the c lattice parameter, which is partly compensated by basis set superposition error (BSSE) due to the use of a finite localised basis set. It is possible to evaluate these two effects separately by correcting our calculation for the BSSE, through the counterpoise method by Boys and Bernardi (1970). BSSE contributes to $\Delta\%$ by -1%, so that the overestimation of c in the limit of the infinite basis set would amount to $+3.3\%$ (rather than $+2.3\%$ in Table 10.7; see Pascale et al. (2004a) for the full discussion).

Diaspore is a much denser and strongly bonded structure, made of double chains of AlO_6 octahedra that form small cavities containing hydrogen atoms. Strong hydrogen bonds (HBs) are present in diaspore, with an $O \cdots H$ distance of 1.68 Å and an $\widehat{OHO}$ angle of $161°$. Brucite and diaspore represent the two extremes in the range of applicability of the harmonic approximation and of the anharmonic correction as proposed in Section 10.2.4.

The IR and Raman spectra of brucite are extremely simple, due to its high symmetry and small unit cell (containing five atoms). There are in total $2E_u + 2A_{2u}$ IR-active and $2E_g + 2A_{1g}$ Raman-active modes, whose frequencies are reported in Table 10.8 and compared with the experimental

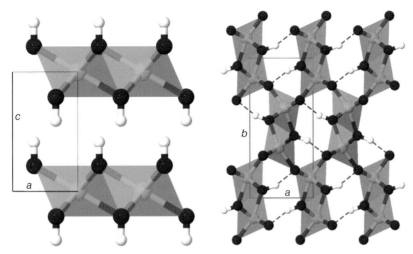

Figure 10.5 *Polyhedra, ball and stick representation of Mg(OH)$_2$ brucite (left) and α-AlOOH diaspore (right). Mg atoms are coloured in green, Al in grey, O in red and H in white. Green dot lines represent hydrogen bonds in diaspore.*

Table 10.7 *The structure of brucite (P$\bar{3}$m1, left) and diaspore (Pbnm, right).*

	Catti et al. (1995)	Calc. (Δ%)	Hill (1979)	Calc. (Δ%)
a	3.150	+0.5	4.401	+0.7
b	=a	=a	9.425	+0.6
c	4.770	+2.3	2.845	+0.9
O—H	0.958	+0.4	0.989	+0.7
X···H	1.999	+1.4	1.676	+0.7
$\widehat{OHX}$	114.5	+1.0	160.7	+0.3

Experimental values and relative differences between calculation and experiment (Δ%) for lattice parameters a, b, c (Å), $X \cdots H$ and O—H bond distances (Å), $\widehat{OHX}$ angle (°). X is hydrogen second nearest atom (H in brucite; O in diaspore, with the formation of a strong hydrogen bond).

Table 10.8 *Calculated versus experimental (Lutz et al. 1994) vibrational frequencies (cm^{-1}) of brucite.*

Sym	Calc.	Exp.	Δ	Sym	Calc.	Exp.	Δ
E_u	360.9	365	−4	E_g	277.8	280	−2
E_u	439.4	415	+24	A_{1g}	451.2	444	+7
A_{2u}	496.6	455	+42	E_g	785.8	725	+61
A_{2u}	3852.1	3698	+154	A_{1g}	3824.6	3654	+171
	(3672.1)		(−26)		(3644.6)		(−9)

Anharmonic values for the high-frequency modes shown in brackets ($\omega_e \chi_e = 90$ cm^{-1}).

data. The anharmonic constant for the O—H stretching modes as estimated through the approach proposed in Section 10.2.4 is $\omega_e \chi_e = 90$ cm^{-1}, resulting in a correction of -180 cm^{-1} to the harmonic frequencies. Due to the poor interactions of hydrogen with its neighbourhood, this estimation is very accurate, as shown through the satisfactory agreement between calculated (anharmonic) and experimental frequencies for the O—H stretching modes.

Graphical visualisation of the eigenvectors permits to characterise the atomic motion associated with the low-frequency modes, as discussed in Pascale et al. (2004a). The computed harmonic values for $\widehat{MgOH}$ bending modes differ from the experiment by +24 and +61 cm^{-1}. These large deviations indicate that the anharmonic contribution is not negligible, though smaller than for the O–H stretching modes. Unfortunately, at the moment there is no tool available to estimate the anharmonicity of bending modes in solids. The remaining four modes involve translations of the Mg and OH units, either within the xy plane (360.9 and 277.8 cm^{-1}) or along the z axis (496.6 and 451.2 cm^{-1}); apart from the third one (where the atomic motion brings the layers closer to each other and anharmonicity would be required to account for mode coupling), they are in excellent agreement with experimental values.

Diaspore exhibits 17 IR-active $\left(3B_{1u} + 7B_{2u} + 7B_{3u}\right)$ and 24 Raman-active $\left(8B_{1g} + 4B_{2g} + 4B_{3g} + 8A_g\right)$ modes. The analysis of normal modes, through animations[12] and isotopic substitution, allows identifying three distinct regions: the region of O–H stretching modes (2900–3000 cm^{-1}), the region of $\widehat{AlOH}$ bending modes (1050–1300 cm^{-1}) and the region of AlO$_6$ octahedral stretching, bending and lattice modes (<800 cm^{-1}). As an example of how isotopic substitution can be used to identify these regions, Table 10.9 shows the different effects of deuterium on the O–H stretching and the $\widehat{AlOH}$ bending regions ($\Delta v < 30$ cm^{-1} in the AlO$_6$ modes region) and also the null contribution of aluminium to these modes.

A full analysis of diaspore normal modes can be found in Demichelis et al. (2007). Here we rather focus on the region of $\widehat{AlOH}$ bending and O–H stretching modes shown in Table 10.9. This region is particularly complex for diaspore, due to the presence of broad bands in the experimental spectrum (Delattre et al. 2012; Stegmann et al. 1973) that make peak assignment a non-trivial task. For example, there are modes that have been recorded by a number of authors (Delattre et al. 2012; Farmer 1974, pp. 146–151; Frost et al. 1999; Ruan et al. 2002; Stegmann et al. 1973) and that do not have a

Table 10.9 IR- and Raman-active $\widehat{AlOH}$ bending (top part) and O–H stretching (bottom) modes of diaspore.

	v Calc.	v Exp.	Δ	Δv ^{29}Al	Δv D		v Calc.	v Exp.	Δ	Δv ^{29}Al	Δv D
A_u	1047.9	Silent	—	0.0	284.8	B_{2g}	1124.5	1046	+78	0.0	313.1
B_{1u}	1049.4	1078	−29	0.0	288.7	B_{3g}	1126.4	1045	+81	0.0	316.2
B_{3u}	1125.7	1153	−27	0.1	288.5	A_g	1277.0	1188	+89	0.7	325.0
B_{2u}	1235.2	n.d.	—	1.9	282.2	B_{1g}	1282.7	1192	+91	0.6	315.2
B_{3u}	3159.0 (2727.8) 2959.0	2920	+239 (−192) +39	0.1	853.6	A_g	3137.2 (2706.2) 2937.2	2920 b	+217 (−214) +17	0.1	833.5
B_{2u}	3165.5 (2734.3) 2965.5	2990	+175 (−256) −24	0.1	851.0	B_{1g}	3155.3 (2724.1) 2955.3	2920 b	+235 (−196) +35	0.1	840.1

A line separates bending from stretching modes. Calculated and the most recent experimental (Delattre et al. 2012) frequencies v and their difference Δ (cm^{-1}); D and ^{29}Al isotopic shifts Δv (cm^{-1}). n.d., not detected; b, broad band. Anharmonic stretching modes in brackets $\left(\omega_e\chi_e = 215.6 \text{ cm}^{-1}\right)$; values corrected using $\omega_e\chi_e = 100$ cm^{-1} are in italic (see text for details).

---

[12] Interactive animation available at http://www.crystal.unito.it/vibs/diaspore/

calculated counterpart. Despite suggested in some of the aforementioned studies, it is only through modelling the IR spectrum that Demichelis et al. (2007) were able to confirm that these modes correspond to either overtones (around 960 and 2000 cm^{-1}) or to water molecules present in the samples ($>3200\ cm^{-1}$).

The anharmonic constant for the O—H stretching modes here is $\omega_e\chi_e = 215.6\ cm^{-1}$, resulting in a correction to the harmonic frequencies as large as $-431.2\ cm^{-1}$. This is a huge overestimation of $\omega_e\chi_e$ for the O—H stretching mode, whose value has been estimated to range around $100\ cm^{-1}$ for numerous minerals (see table 9 in Demichelis et al. (2011a) and references therein), and it is due to two main reasons. The former is that H in diaspore is involved in a very strong HB, so that our assumption of an isolated and fully decoupled O—H oscillator is no longer valid. The latter is that the $\widehat{OHO}$ angle is $161°$ rather than $180°$ (with H$\cdots$O as small as 1.68 Å), and then displacing the H atom along the linear O—H direction (rather than along a curved trajectory) adds a further level of inaccuracy to our model. Table 10.9 shows that a reasonable estimation of the stretching frequencies can indeed be obtained by assuming $\omega_e\chi_e = 100\ cm^{-1}$.

$\widehat{AlOH}$ bending modes suffer from similar problems related to shortcomings in the harmonic approximation, so that a proper description would involve anharmonic modes described as couplings (i.e. linear combinations) of the harmonic eigenvectors. Strong HBs and the proximity of several OH groups pointing towards the same direction and getting closer to each other while vibrating are the main reasons for the discrepancies shown in Table 10.9.

As mentioned earlier, the method proposed here has been successfully applied in tens of studies on minerals, many of which contained hydrogen atoms in their structure (e.g. Orlando et al. (2006), Prencipe et al. (2009), Tosoni et al. (2006)), with HB strength and geometry not as extreme as in diaspore. This latter case still represents one of the most challenging systems, given the present lack of general tools for modelling of anharmonicity in solids.

10.4 Simulation of Vibrational Properties for Crystal Structure Determination

In the past few years, quantum mechanical *ab initio* schemes have been successfully applied to make structural predictions and to unambiguously determine the atomic structure of minerals and solid-state systems (e.g. Brázdová et al. (2004), Demichelis et al. (2009, 2013), Stixrude and Peacori (2002), Tielens et al. (2008), Ugliengo et al. (2008), Zhou et al. (2012)). In this context, computing the vibrational spectrum provides invaluable information about the topography of the PES around the equilibrium position, as normal modes are obtained directly from second derivatives.

Stationary points of the PES (i.e. points where all the gradient components are null) have a physical meaning, as they correspond either to equilibrium structures (minima) or to transition states (saddle points, maxima). Whereas for equilibrium structures all the eigenvalues of the Hessian matrix (see Section 10.2.1 for the definition) are positive, transition states are characterised by the presence of one or more negative eigenvalues, corresponding to 'imaginary' normal modes (see Chapter 11 and Schlegel (1998)).

Though computing second derivatives is crucial in understanding whether the optimised structure represents a minimum energy structure or a transition state, it is often assumed that geometry optimisation algorithms are able to minimise the gradient components following the directions along which the energy decreases (i.e. they lead the structure to the nearest minimum), and no frequency calculation is performed. However, if the PES is particularly flat (i.e. displacing the atoms leads to energy and gradient variations that are lower than the adopted thresholds) or the directions towards a minimum energy configuration are symmetry forbidden (i.e. in case of symmetry constraints), the result of geometry optimisation could be a transition state rather than a minimum energy structure.[13]

Problems of flat PES can be addressed by setting more strict thresholds. Besides, problems related to symmetry constraints could be addressed through symmetry removal. However, keeping symmetry constraints has a series of advantages. From a computational point of view, symmetry allows for a significant reduction of the calculation costs, as described in Section 10.2.1 and in Dovesi et al. (2014) (and references therein). From a scientific point of view, accounting for symmetry provides additional information on the system structure and properties (e.g. classification of normal modes, degeneracy, equivalent sites for substitutions in the presence of solid solutions, equivalent components of tensor properties).

The general approach that will be adopted here for structure determination involves performing geometry optimisation and then frequency calculation by keeping the symmetry constraints. In the presence of negative eigenvalues, for each one of these: (i) the nature of the imaginary mode is examined; (ii) the corresponding symmetry constraint (and only this one) is removed; (iii) a scan of the geometry along the corresponding eigenvector is performed; (iv) the new geometry is re-optimised and vibrational frequencies are re-computed; and (v) the strategy is repeated from point (i) until there are no negative eigenvalues left. Note that the main features of geometry optimisation and vibrational frequency calculation algorithms are discussed in Chapter 3 and in Section 10.2, respectively.

Experimental structure determinations of minerals are usually extremely accurate and provide atomic positions, lattice parameters, and space group that represent an accurate guess for geometry optimisation and properties calculations. However, there are many cases where structures cannot be fully determined due to many reasons, including the presence of hydrogen atoms and water molecules, disorder, impurities, rarity of the material, polytypism, and small size of single crystals. In many of these cases, it is likely that experimental techniques provide average positions rather than the true structure.

Two of these examples are vaterite (a $CaCO_3$ metastable polymorph often obtained as a result of biomineralisation) and boehmite (diaspore's polymorph). After a long debate on the nature of their structures, lasted for decades and involving investigations through several experimental techniques, an accurate combination of geometry optimisation and vibrational properties calculation has provided new perspectives and more realistic structural models for these minerals (Demichelis et al. 2013; Noël et al. 2009). While addressing the readers to the original literature for more details about the delicate task of understanding the structure of vaterite (Demichelis et al. 2013; Kabalah-Amitai et al. 2013; Wehrmeister et al. 2010 and references therein), we present here the simpler example of boehmite.

10.4.1   Proton Disorder in γ-AlOOH Boehmite

The structure of boehmite, or γ-AlOOH (Figure 10.6), consists of AlO_6 irregular octahedra organised into layers through sharing edges; layers are stacked along the y Cartesian direction and strongly interact with each other through hydrogen bonds (HBs). Whereas the lattice parameters and the position of aluminium and oxygen atoms in the structure have been determined through

---

[13] Note that, on top of this, there is the fact that the nearest minimum may not be the absolute minimum on the PES. We will not analyse this case here, as the search for the global minimum requires the use of advanced optimisation tools that are beyond the aim of this chapter. In general, at least average atomic positions are available for the bulk structure of most minerals, which allows making a reasonable starting guess for the structure prior to the geometry optimisation; as a consequence, in the majority of the cases, the optimised structure is reasonably expected to correspond to either a global minimum or a transition state close to it.

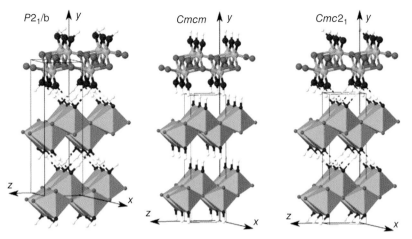

Figure 10.6 *Ball and stick and polyhedral representation of boehmite structures. The Cmcm arrangement is shown, together with the arrangements exhibiting parallel (Cmc2$_1$) and antiparallel (P2$_1$/b) HB chains. Al atoms are coloured in grey and H atoms in white. O atoms involved in HBs are shown in dark red and the others in light red. From figure 1 Noël et al. (2009). Reproduced with permission of Springer.*

Table 10.10   *The structure of boehmite.*

	P2$_1$/b	Cmcm	Cmc2$_1$	Corbato et al. (1985)
a	+0.8	+1.4	+0.8	5.7362[a]
b	+0.1	+4.2	+0.1	12.2336
c	+0.8	+1.1	+0.8	3.6923
V	+1.8	+6.9	+1.7	259.10[a]
O—H	+1.9	−0.7	+2.0	0.97
O···H	−1.0	+31.7	−1.1	1.738
$\widehat{OHO}$	−1.1	−30.0	−1.3	179

Percent relative differences between calculated and experimental lattice parameters *a, b, c* (Å), O—H bonds and HB O···H distances (Å), unit cell volume *V*(Å^3) and $\widehat{OHO}$ angle (°).
[a] To facilitate the comparison, these parameters have been doubled in *Cmcm*, in *Cmc2$_1$* and in the experimental structure.

X-ray and neutron diffraction experiments (Bokhimi et al. 2002; Christensen et al. 1982; Corbato et al. 1985; Hill 1981; Liu et al. 1998), the position of hydrogen atoms and the interlayer HB pattern, necessary to fully define the structure and the space group, have remained ambiguous for many years. A certain number of space groups have been proposed, based either on the aforementioned experiments or on measurements with various spectroscopic techniques, including IR, Raman and NMR (Farmer 1980; Kiss et al. 1980; Russell and Farmer 1978; Slade and Halstead 1980; Stegmann et al. 1973). One of the commonly accepted hypotheses is that the structure belongs to the *Cmcm* (or *Amam*) space group (Corbato et al. 1985). Partial occupancy of the hydrogen sites is often assumed, and the generic term 'proton disorder' is frequently adopted.

In 2009, Noël et al. have examined some of the most popular space groups for boehmite by applying the scheme proposed in this section (with the B3LYP functional and the same basis set used in Section 10.3.3 for diaspore), providing both a model for the structure of boehmite and a mechanism for proton disorder.

Table 10.10 clearly shows that the *Cmcm* model is not suitable for this structure. Geometry optimisation of the experimental structures by Corbato et al. (1985) leads to a calculated unit cell volume that is 7% higher than the experimental value, due to overestimation of the b lattice parameter (i.e. the direction along which layers are stacked). Also, all the parameters related to hydrogen atom positions are in strong disagreement with the experiment. Vibrational frequency calculation of this structure reveals one imaginary mode with frequency $-608\,\text{cm}^{-1}$, indicating that the *Cmcm* model is a transition state.

Following the procedure presented earlier, it turns out that the imaginary mode corresponds to a forbidden bending of the $\widehat{\text{AlOH}}$ angle.[14] Scanning along this mode leads to a minimum energy structure belonging to the *Cmc*2_1 space group, which is $18\,\text{kJ}\,\text{mol}^{-1}$ per formula unit more stable than *Cmcm* ($16.2\,\text{kJ}\,\text{mol}^{-1}$ if we include the zero point energy correction). This represents the simplest modification possible to *Cmcm*, with the inversion operator removed and all H atoms allowed to bend towards the same direction (parallel HB chain, Figure 10.6).

A second possible modification is allowing for the OH groups in the cell to have independent orientations. Of all the space groups compatible with this additional degree of freedom, groups $P2_1/b$ and $P2_1ab$ are possible candidates. They both correspond to a minimum on the PES, are almost isoenergetic with respect to *Cmc*2_1 and have HB chains with antiparallel orientation (Figure 10.6; $P2_1ab$ is not shown as it is very similar to $P2_1/b$). As a result of allowing for two independent orientations of the HB chain, the size of the unit cell of these models doubles to 32 atoms.

Table 10.10 shows that the structures optimised within these model constraints are in agreement with the experimental findings. The presence of at least two almost isoenergetic minimum energy structures and an activation barrier for the proton to flip between two limit configurations (left to right and *vice versa*) of $16.2\,\text{kJ}\,\text{mol}^{-1}$ provide two strong points in favour of a mechanism for proton disorder that involves structure interconversion through $\widehat{\text{AlOH}}$ bending, also called hydrogen flipping in Noël et al. (2009). The mechanism for this process can be represented as in Figure 10.7.

From the full set of vibrational frequencies, the vibrational partition function for each structure, Z_i, can be calculated:

$$Z_i = \prod_{l,\vec{k}} \frac{\exp\left(-\dfrac{hv_{l,\vec{k}}}{2k_{\mathrm{B}}T}\right)}{1-\exp\left(-\dfrac{hv_{l,\vec{k}}}{k_{\mathrm{B}}T}\right)} \qquad (10.25)$$

where h is Planck's constant, $v_{l,\vec{k}}$ is the lth vibrational frequency in point $\vec{k}$ of the reciprocal space lattice, k_{B} is the Boltzmann constant and T is the absolute temperature (the imaginary mode for

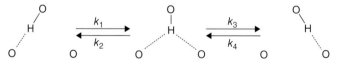

Figure 10.7 *Schematic representation of hydrogen flipping between two positions through the Cmcm configuration. Kinetic constants for each reaction are shown. Only atoms involved in the process are represented.*

---

[14] Interactive animation available at http://www.crystal.unito.it/vibs/boehmite/

Cmcm is not included in the sum). Through Eyring's equation, we can then estimate the kinetic constant K of the process represented in Figure 10.7:

$$K = \frac{k_1 k_3}{k_2 k_4} = \frac{k_B T Z_j}{h Z_i} e^{-\frac{\Delta E_{ij}}{k_B T}} \tag{10.26}$$

where j and i refer to structures *Cmcm* and *Cmc2$_1$*, respectively. At room temperature, $K = 4.9 \cdot 10^{10}$ s^{-1}, meaning that H atoms flip between the two different configurations several times per ns (further details on the calculation of kinetic constants in Chapter 11).[15]

This constant and the energy barrier associated with this transformation relate to a mechanism where all protons in the structure flip from the *Cmc2$_1$* configuration to another *Cmc2$_1$* configuration (all protons at the left side flip towards the right side, or *vice versa*), but they do not tell how likely structures with parallel HB chains interconvert into structures with antiparallel HB chains, and *vice versa*, or how likely and how fast a single proton or a subgroup of protons can move independently from the others. However, since the different HB chains do not interact significantly between each other, we can assume that the proposed values for the activation barrier and the kinetic constant provide realistic estimations also for a more general mechanism.

It is clear, at this point, that one could in principle increase the size of the unit cell to allow for more and more combinations of HB chains and that all these structures together would probably be able to fully describe the proton disorder of boehmite. However, here we have limited our analysis to the two limit configurations *Cmc2$_1$* and *P2$_1$/b*.

In the past, other mechanisms for proton disorder have been proposed: proton tunnelling, proton flipping between two layers (through stretching along the HB direction) and protons totally free to move (Fripiat et al. 1967; Kiss et al. 1980). The last hypothesis can be disregarded based on the vibrational frequencies of the O—H stretching modes, which fall in the region around 3000–3300 cm^{-1} (Kiss et al. 1980; Ruan et al. 2001; Stegmann et al. 1973), typical of many other minerals (e.g. α-AlOOH diaspore). The likelihood of the second hypothesis will not be explored here, though this mechanism is probably secondary to the one highlighted in the discussion earlier, as the latter has low barrier that allows hydrogen flipping at a rate of $K = 4.9 \cdot 10^{10}$ s^{-1}. Finally, from the computed vibrational spectrum of the transition state and the value of the activation barrier, we can estimate the temperature above which proton tunnelling becomes negligible, the 'tunnelling crossover temperature' T_x, as follows (Fermann and Auerbach 2000; Sierka and Sauer 2001):

$$T_x = \frac{\hbar |\omega_{ij}| \Delta E_{ij}}{k_B \left(2\pi \Delta E_{ij} - \hbar |\omega_{ij}| \ln 2 \right)} \tag{10.27}$$

In this equation, $\omega_{ij} = 2\pi \nu_{ij}$ is the transition frequency between state i and state j (in our case, it is 2π times the imaginary mode frequency, with $i = $ *Cmc2$_1$* and $j = $ *Cmcm*), which gives the curvature of the barrier, and ΔE_{ij} is the height of the barrier to cross between state i and state j. For boehmite, with $\nu_{ij} = -608$ cm^{-1} and $\Delta E_{ij} = 16.2$ kJ mol^{-1}, the tunnel crossover temperature is $T_x = 147$ K. This implies that at room temperature tunnelling is unlikely and that proton disorder is mostly due to hydrogen flipping around the Al—O bond.

---

[15] Value for K has been recalculated in this chapter and is slightly different from the one reported in Noël et al. (2009), $9.4 \cdot 10^9$ s^{-1}, though of the same order of magnitude.

10.5 Future Challenges

Throughout this chapter, we have provided examples that demonstrate the usefulness of *ab initio* simulation in analysing the lattice dynamics and vibrational spectroscopic response of minerals. Quantitative comparison with experiments has substantiated the high level of accuracy that can be achieved through calculations. Besides, tools have been presented that can be used to interpret the spectra in terms of relations with the microscopic structure. Together with the availability of powerful computer hardware at relatively low cost, these capabilities define computational methods as the ideal complement for researchers involved in the investigation of minerals using vibrational spectroscopy.

One of the biggest current open issues in the *ab initio* simulation of vibrational properties of minerals is the treatment of anharmonicity, which presumes getting over the harmonic approximation for the electronic potential. This is the requirement to accurately describe phenomena such as anharmonic correction to modes involving hydrogen and to soft modes (e.g. low-frequency cation modes with large vibration amplitudes), quantitative description of combination modes, overtones and Fermi resonances (in fact, purely harmonic normal modes do not couple with each other), accurate prediction of thermal expansion, thermodynamics and phase transitions. As mentioned in Section 10.2, no definitive solution has been designed for solids yet. This is due to the higher complexity related to the infinite periodic model, which implies both algorithmic intricacies and computational costs. However, a variety of successful implementations exist for molecular studies, which are currently representing the starting point for solid state approaches: VSCF (Bowman 1978; Ratner and Gerber 1986), vibrational perturbation theory (Norris et al. 1996), vibrational configuration interaction (Bowman et al. 1979) and vibrational coupled cluster (Christiansen 2004). Recent attempts of devising working strategies for anharmonicity in solids include the works by Monserrat et al. (2013), who first published *ab initio* results on anharmonic frequencies of simple solids, and by Hirata et al. (2010), who presented generalisations of molecular methods to satisfy the criterion of size extensiveness (a crucial question when dealing with extended systems such as solids). A completely alternative way out is offered by *ab initio* molecular dynamics techniques (Car and Parrinello 1985) that natively take into account anharmonic effects by solving the classic equations of motions of the nuclei under the action of the quantum electronic potential. Examples of application include systems such as ice (Putrino and Parrinello 2002) and naphtalene crystal (Pagliai et al. 2008).

Acknowledgements

The authors would like to thank Cédric Carteret at Université de Lorraine (France) for providing us with digitised copies of the experimental spectra for calcite and Piero Ugliengo at Università degli Studi di Torino (Italy) for careful reading of this chapter. R. Demichelis would like to acknowledge Curtin University for funding parts of this work through a Curtin Research Fellowship. The Pawsey Centre and the Australian National Computational Infrastructure are acknowledged for the provision of computer time.

References

Adamo C and Barone V 1999 Toward reliable density functional methods without adjustable parameters: the PBE0 model. *J. Chem. Phys.* **110**, 6158–6170.

Antao SM, Hassan I, Mulder WH, Lee PL and Toby BH 2009 *In situ* study of the $R\bar{3}c \rightarrow R\bar{3}m$ orientational disorder on calcite. *Phys. Chem. Miner.* **36**, 159–169.

Barone V 2005 Anharmonic vibrational properties by a fully automated second order perturbative approach. *J. Chem. Phys.* **122**, 014108.

Baroni S, de Gironcoli S, Dal Corso A and Giannozzi P 2001 Phonons and related crystal properties from density-functional perturbation theory. *Rev. Mod. Phys.* **73**, 515–562.

Baroni S, Giannozzi P and Testa A 1987 Green's-function approach to linear response in solids. *Phys. Rev. Lett.* **58**, 1861–1864.

Becke AD 1993 Density functional thermochemistry. III. The role of exact exchange. *J. Chem. Phys.* **98**, 5648–5652.

Berreman DW and Unterwald FC 1968 Adjusting poles and zeros of dielectric dispersion to fit reststrahlen of $PrCl_3$ and $LaCl_3$. *Phys. Rev.* **174**, 791–799.

Bokhimi X, Sanchez-Valente J and Pedraza F 2002 Crystallization of sol-gel boehmite via hydrothermal annealing. *J. Solid State Chem.* **166**, 182–190.

Born M and Huang K 1954 *Dynamical Theory of Crystal Lattices*. Oxford University Press, Oxford.

Bourahla S, Benamara AA and Moustefai SK 2014 Infrared spectra of inorganic aerosols: *ab initio* study of $(NH_4)_2SO_4$, NH_4NO_3, and $NaNO_3$. *Can. J. Phys.* **92**, 216–221.

Bowman JM 1978 Self-consistent field energies and wavefunctions for coupled oscillators. *J. Chem. Phys.* **68**, 608–610.

Bowman JM, Christoffel K and Tobin F 1979 Application of SCF-SI theory to vibrational motion in polyatomic molecules. *J. Phys. Chem.* **83**, 905–912.

Boys SF and Bernardi F 1970 The calculation of small molecular interactions by the differences of separate total energies. Some procedures with reduced errors. *Mol. Phys.* **19**, 553–566.

Brázdová V, Ganduglia-Pirovano MV and Sauer J 2004 Periodic density functional study on structural and vibrational properties of vanadium oxide aggregates. *Phys. Rev. B* **69**, 165420.

Canepa P, Ugliengo P and Alfredsson M 2012 Elastic and vibrational properties of alpha- and beta-PbO. *J. Phys. Chem. C* **116**, 21514–21522.

Car R and Parrinello M 1985 Unified approach for molecular dynamics and density-functional theory. *Phys. Rev. Lett.* **55**, 2471–2474.

Carlson RW, Weissman PR, Smythe WD, Mahoney JC, Aptaker I, Bailey G, Baines K, Burns R, Carpenter E, Curry K, Danielson G, Encrenaz T, Enmark H, Fanale F, Gram M, Hernandez M, Hickok R, Jenkins G, Johnson T, Jones S, Kieffer H, Labaw C, Lockhart R, Macenka S, Marino J, Masursky H, Matson D, Mccord T, Mehaffey K, Ocampo A, Root G, Salazar R, Sevilla D, Sleigh W, Smythe W, Soderblom L, Steimle L, Steinkraus R, Taylor F and Wilson D 1992 Near-infrared mapping spectrometer experiment on Galileo. *Space Sci. Rev.* **60**, 457–502.

Carteret C, De La Pierre M, Dossot M, Pascale F, Erba A and Dovesi R 2013 The vibrational spectrum of $CaCO_3$ aragonite: a combined experimental and quantum-mechanical investigation. *J. Chem. Phys.* **138**, 014201.

Cartwright JHE, Checa AG, Gale JD, Gebauer D and Sainz-Díaz CI 2012 Calcium carbonate polyamorphism and its role in biomineralization: how many amorphous calcium carbonates are there? *Angew. Chem. Int. Ed.* **51**, 11960–11970.

Catti M, Ferraris G, Hull S and Pavese A 1995 Static compression and H disorder in brucite, $Mg(OH)_2$, to 11 GPa: a powder neutron diffraction study. *Phys. Chem. Miner.* **22**, 200–206.

Christensen AN, Lehmann MS and Convert P 1982 Deuteration of crystalline hydroxides. Hydrogen bonds of γ-AlOO(H,D) and γ-FeOO(H,D). *Acta Chem. Scand.* **36A**, 303–308.

Christiansen O 2004 Vibrational coupled cluster theory. *J. Chem. Phys.* **120**, 2149–2159.

Corbato CE, Tettenhorst RT and Christoph GG 1985 Structure refinement of deuterated boehmite. *Clays Clay Miner.* **33**, 71–75.

Corno M and Pedone A 2009 Vibrational features of phospho-silicate glasses: periodic B3LYP simulations. *Chem. Phys. Lett.* **476**, 218–222.

Dall'Olio S, Dovesi R and Resta R 1997 Spontaneous polarization as a berry phase of the Hartree–Fock wave function: the case of $KNbO_3$. *Phys. Rev. B* **56**, 10105–10114.

Daramola DA, Muthuvel M and Bott GG 2010 Density functional theory analysis of Raman frequency modes of monoclinic zirconium oxide using Gaussian basis sets and isotopic substitution. *J. Phys. Chem. B* **114**, 9323–9329.

De La Pierre M, Carteret C, Maschio L, André E, Orlando R and Dovesi R 2014a The Raman spectrum of $CaCO_3$ polymorphs calcite and aragonite. A combined experimental and computational study. *J. Chem. Phys.* **140**, 164509.

De La Pierre M, Carteret C, Orlando R and Dovesi R 2013a Use of *ab initio* methods for the interpretation of the experimental IR reflectance spectra of crystalline compounds. *J. Comput. Chem.* **34**, 1476–1485.

De La Pierre M, Demichelis R, Wehrmeister U, Jacob DE, Raiteri P, Gale JD and Orlando R 2014b Probing the multiple structures of vaterite through combined computational and experimental Raman spectroscopy. *J. Phys. Chem. C* **118**, 27493–27501.

De La Pierre M, Noël Y, Mustapha S, Meyer A, D'Arco P and Dovesi R 2013b The infrared vibrational spectrum of andradite-grossular solid solutions: a quantum mechanical simulation. *Am. Mineral.* **98**, 966–976.

De La Pierre M, Orlando R, Maschio L, Doll K, Ugliengo P and Dovesi R 2011 Performance of six functionals (LDA, PBE, PBESOL, B3LYP, PBE0, and WC1LYP) in the simulation of vibrational and dielectric properties of crystalline compounds. The case of forsterite Mg_2SiO_4. *J. Comput. Chem.* **32**, 1775–1784.

Delattre S, Balan E, Lazzeri M, Blanchard M, Guillaumet M, Beyssac O, Haussühl E, Winkler B, Salje EKS and Calas G 2012 Experimental and theoretical study of the vibrational properties of diaspore (α-AlOOH). *Phys. Chem. Miner.* **39**, 93–102.

Demichelis R, Catti M and Dovesi R 2009 Structure and stability of the $Al(OH)_3$ polymorphs doyleite and nordstrandite: a quantum mechanical *ab initio* study with the CRYSTAL06 code. *J. Phys. Chem. C* **113**, 6785–6791.

Demichelis R, Civalleri B, D'Arco P and Dovesi R 2010a Performance of 12 DFT functionals in the study of crystal systems. Al_2SiO_5 orthosilicates and Al hydroxides as a case study. *Int. J. Quantum Chem.* **110**, 2260–2273.

Demichelis R, Civalleri B, Ferrabone M and Dovesi R 2010b On the performance of eleven DFT functionals in the description of the vibrational properties of aluminosilicates. *Int. J. Quantum Chem.* **110**, 406–415.

Demichelis R, Noël Y, Civalleri B, Roetti C, Ferrero M and Dovesi R 2007 The vibrational spectrum of α-AlOOH diaspore: an *ab initio* study with the CRYSTAL code. *J. Phys. Chem. B* **111**, 9337–9346.

Demichelis R, Noël Y, Ugliengo P, Zicovich-Wilson CM and Dovesi R 2011a Physico-chemical features of aluminum hydroxides as modeled with the hybrid B3LYP functional and localized basis functions. *J. Phys. Chem. C* **115**, 13107–13134.

Demichelis R, Raiteri P, Gale JD and Dovesi R 2013 The multiples structure of vaterite. *Cryst. Growth Des.* **13**, 2247–2251.

Demichelis R, Raiteri P, Gale JD, Quigley D and Gebauer D 2011b Stable prenucleation mineral clusters are liquid-like ionic polymers. *Nat. Commun.* **2**, 590.

Demichelis R, Suto H, Noël Y, Sogawa H, Naoi T, Koike C, Chihara H, Shimobayashi N, Ferrabone M and Dovesi R 2012 The infrared spectrum of ortho-enstatite from reflectance experiments and first-principle simulations. *Mon. Not. R. Astron. Soc.* **420**, 147–154.

Donoghue M, Hepburn PH and Ross SD 1971 Factors affecting the infrared spectra of planar anions with D_3h, symmetry V: the origin of the splitting of the out-of-plane bending mode in carbonates and nitrates. *Spectrochim. Acta A* **27**, 1065–1072.

Dovesi R, De La Pierre M, Ferrari AM, Pascale F, Maschio L and Zicovich-Wilson CM 2011 The IR vibrational properties of six members of the garnet family: a quantum mechanical *ab initio* study. *Am. Mineral.* **96**, 1787–1798.

Dovesi R, Orlando R, Erba A, Zicovich-Wilson CM, Civalleri B, Casassa S, Maschio L, Ferrabone M, De La Pierre M, DArco P, Noël Y, Causà M, Rérat M and Kirtman B 2014 CRYSTAL14: a program for the *ab initio* investigation of crystalline solids. *Int. J. Quantum Chem.* **114**, 1287–1317.

Dovesi R, Pascale F and Zicovich-Wilson C 2007 The *ab initio* calculation of the vibrational spectrum of crystalline compounds; the role of symmetry and related computational aspects. In *Beyond Standard Quantum Chemistry: Applications from Gas to Condensed Phases* (ed. Hernández-Lamoneda R) Transworld Research Network, Trivandrum, pp. 117–138.

Dubessy J, Caumon MC and Rull F 2012 *Raman Spectroscopy Applied to Earth Sciences and Cultural Heritage* vol EMU Notes in Mineralogy **12**. Mineralogical Society of Great Britain & Ireland, London.

Evarestov RA and Bandura AV 2012 First-principles calculations on the four phases of $BaTiO_3$. *J. Comput. Chem.* **33**, 1123–1130.

Evarestov RA and Losev MV 2009 All-electron LCAO calculations of the LiF crystal phonon spectrum: influence of the basis set, the exchange-correlation functional, and the supercell size. *J. Comput. Chem.* **30**, 2645–2655.

Farmer VC 1974 *Infrared Spectra of Minerals*. The Mineralogical Society, London.

Farmer VC 1980 Raman and IR spectra of boehmite (γ-AlOOH) are consistent with D_{2h}^{16} or C_{2h}^5 symmetry. *Spectrochim. Acta* **36A**, 585–586.

Fermann JT and Auerbach S 2000 Modeling proton mobility in acidic zeolite clusters: II. Room temperature tunneling effects from semiclassical rate theory. *J. Chem. Phys.* **112**, 6787–6794.

Ferrari AM, Valenzano L, Meyer A, Orlando R and Dovesi R 2009 Quantum-mechanical *ab initio* simulation of the Raman and IR spectra of $Fe_3Al_2Si_3O_{12}$ almandine. *J. Phys. Chem. A* **113**, 11289–11294.

Ferrero M, Rérat M, Orlando R and Dovesi R 2008 The calculation of static polarizabilities of periodic compounds. The implementation in the CRYSTAL code for 1D, 2D and 3D systems. *J. Comput. Chem.* **29**, 1450–1459.

Fripiat JJ, Bosmans H and Rouxhet PG 1967 Hydrogenic vibration modes and proton delocalization in boehmite. *J. Phys. Chem.* **71**, 1097–1111.

Frost RL, Kloprogge JT, Russel SC and Szetu J 1999 Dehydroxylation and the vibrational spectroscopy of aluminum (oxo)hydroxides using infrared emission spectroscopy. Part III: diaspore. *Appl. Spectrosc.* **53**, 829–835.

Gale JD and Rohl AL 2003 The general utility lattice program. *Mol. Simul.* **29**, 291–341.

Gebauer D, Völkel A and Cölfen H 2008 Stable prenucleation calcium carbonate clusters. *Science* **322**, 1819–1822.

Giannozzi P, Baroni S, Bonini N, Calandra M, Car R, Cavazzoni C, Ceresoli D, Chiarotti GL, Cococcioni M, Dabo I, Dal Corso A, de Gironcoli S, Fabris S, Fratesi G, Gebauer R, Gerstmann U, Gougoussis C, Kokalj A, Lazzeri M, Martin-Samos L, Marzari N, Mauri F, Mazzarello R, Paolini S, Pasquarello A, Paulatto L, Sbraccia C, Scandolo S, Sclauzero G, Seitsonen AP, Smogunov A, Umari P and Wentzcovitch RM 2009 QUANTUM ESPRESSO: a modular and open-source software project for quantum simulations of materials. *J. Phys. Condens. Matter* **21**, 395502.

Gillet P, McMillan P, Schott J, Badro J and Grzechnik A 1996 Thermodynamic properties and isotopic fractionation of calcite from vibrational spectroscopy of ^{18}O-substituted calcite. *Geochim. Cosmochim. Acta* **60**, 3471–3485.

Grimme S 2004 Accurate description of van der Waals complexes by density functional theory including empirical corrections. *J. Comput. Chem.* **25**, 1463–1473.

Guillot S and Hattori K 2013 Serpentinites: essential roles in geodynamics, arc volcanism, sustainable development, and the origin of life. *Elements* **9**, 95–98.

Hafner J 2008 *Ab initio* simulations of materials using VASP: density-functional theory and beyond. *J. Comput. Chem.* **29**, 2044–2078.

Hart LD 1990 *Alumina Chemicals: Science and Technology Handbook*. The American Ceramic Society, Westerville.

Hellwege K, Lesch W, Plihal M and Schaack G 1970 Zwei-phononen-absorptionsspektren und dispersion der schwingungszweige in kristallen der kalkspatstruktur. *Z. Physik.* **232**, 61–86.

Hill RJ 1979 Crystal structure refinement and electron density distribution in diaspore. *Phys. Chem. Miner.* **5**, 179–200.

Hill RJ 1981 Hydrogen atoms in boehmite: a single crystal X-ray diffraction and molecular orbital study. *Clays Clay Miner.* **29**, 435–445.

Hirata S, Keçeli M and Yagi K 2010 First-principles theories for anharmonic lattice vibrations. *J. Chem. Phys.* **133**, 034109.

Imrie FE, Corno M, Ugliengo P and Gibson IR 2013 *Computational Studies of Magnesium and Strontium Substitution in Hydroxyapatite* vol. **529–530** Key Engineering Materials. Trans Tech Publications Inc., Zürich-Dürnten, pp. 123–128.

Jahn S and Kowalski PM 2014 Theoretical approaches to structure and spectroscopy of earth materials. In *Spectroscopic Methods in Mineralogy and Material Sciences* (ed. Henderson GS, Neuville DR and Downs RT) vol. **78** Reviews of Mineralogy and Geochemistry. Mineralogical Society of America, Chantilly, pp. 691–743.

Kabalah-Amitai L, Mayzel B, Kauffmann Y, Fitch AN, Bloch L, Gilbert PU and Pokroy B 2013 Vaterite crystals contain two interspersed crystal structures. *Science* **340**, 454–456.

Kiss AB, Keresztury G and Farkas L 1980 Raman and IR spectra of boehmite (γ-AlOOH). Evidence for the recently discarded D_{2h}^{17} space group. *Spectrochim. Acta* **36A**, 653–658.

Kubicki JD, Paul KW, Kabalan L, Zhu Q, Mrozik MK, Aryanpour M, Pierre-Louis AM and Strongin DR 2012 ATR-FTIR and density functional theory study of the structures, energetics, and vibrational spectra of phosphate adsorbed onto goethite. *Langmuir* **28**, 14573–14587.

Leonidov II, Petrov VP, Chernyshev VA, Nikiforov AE, Vovkotrub EG, Tyutyunnik AP and Zubkov VG 2014 Structural and vibrational properties of the ordered $Y_2CaGe_4O_{12}$ germanate: a periodic *ab initio* study. *J. Phys. Chem. C* **118**, 8090–8101.

Lindberg B 1988 A new efficient method for calculation of energy eigenvalues and eigenstates of the one-dimensional Schrödinger equation. *J. Chem. Phys.* **88**, 3805–3810.

Liu P, Kendelewicz T, Brown GE, Nelson EJ and Chambers SA 1998 Reaction of water vapor with α-Al_2O_3(0001) and α-Fe_2O_3(0001) surfaces: synchrotron X-ray photoemission studies and thermodynamic calculations. *Surf. Sci.* **417**, 53–65.

Long L, Querry M, Bell R and Alexander R 1993 Optical properties of calcite and gypsum in crystalline and powdered form in the infrared and far-infrared. *Infrared Phys.* **34**, 191–201.

Lutz HD, Moller H and Schmidt M 1994 Lattice vibration spectra. Part LXXXII. Brucite-type hydroxides $M(OH)_2$ (M = Ca, Mn, Co, Fe, Cd) IR and Raman spectra, neutron diffraction of $Fe(OH)_2$. *J. Mol. Struct.* **328**, 121–132.

Maschio L, Kirtman B, Orlando R and Rérat M 2012 *Ab initio* analytical infrared intensities for periodic systems through a coupled perturbed Hartree–Fock/Kohn–Sham method. *J. Chem. Phys.* **137**, 204113.

Maschio L, Kirtman B, Rérat M, Orlando R and Dovesi R 2013 *Ab initio* analytical Raman intensities for periodic systems through a coupled perturbed Hartree–Fock/Kohn–Sham method in an atomic orbital basis I. Theory. *J. Chem. Phys.* **139**, 164102.

Monserrat B, Drummond N and Needs R 2013 Anharmonic vibrational properties in periodic systems: energy, electron-phonon coupling, and stress. *Phys. Rev.* **B 87**, 144302.

Montanari B, Civalleri B, Zicovich-Wilson CM and Dovesi R 2006 Influence of the exchange-correlation functional in all-electron calculations of the vibrational frequencies of corundum (α-Al_2O_3). *Int. J. Quantum Chem.* **106**, 1703–1714.

Müntener O 2010 Serpentine and serpentinization: a link between planet formation and life. *Geology* **38**, 959–960.

Noël Y, De La Pierre M, Maschio L, Rérat M, Zicovich-Wilson CM and Dovesi R 2012 The Electronic structure, dielectric properties and infrared vibrational spectrum of fayalite: an *ab initio* simulation with an all-electron gaussian basis set and the B3LYP functional. *Int. J. Quantum Chem.* **112**, 2098–2108.

Noël Y, Demichelis R, Pascale F, Ugliengo P, Orlando R and Dovesi R 2009 *Ab initio* quantum mechanical study of γ-AlOOH boehmite: structure and vibrational spectrum. *Phys. Chem. Miner.* **36**, 47–59.

Norris LS, Ratner MA, Roitberg AE and Gerber RB 1996 Möller–Plesset perturbation theory applied to vibrational problems. *J. Chem. Phys.* **105**, 11261–11267.

Orlando R, Torres FJ, Pascale F, Ugliengo P, Zicovich-Wilson C and Dovesi R 2006 Vibrational spectrum of katoite $Ca_3Al_2[(OH)_4]_3$: a periodic *ab initio* study. *J. Phys. Chem.* **B 110**, 692–701.

Pagliai M, Cavazzoni C, Cardini G, Erbacci G, Parrinello M and Schettino V 2008 Anharmonic infrared and Raman spectra in Car–Parrinello molecular dynamics simulations. *J. Chem. Phys.* **128**(22), 224514.

Pascale F, Tosoni S, Zicovich-Wilson C, Ugliengo P, Orlando R and Dovesi R 2004a Vibrational spectrum of brucite, $Mg(OH)_2$: a periodic *ab initio* quantum mechanical calculation including OH anharmonicity. *Chem. Phys. Lett.* **396**, 308–315.

Pascale F, Ugliengo P, Civalleri B, Orlando R, D'Arco P and Dovesi R 2002 Hydrogarnet defect in chabazite and sodalite zeolites: a periodic Hartree–Fock and B3-LYP study. *J. Chem. Phys.* **117**, 5337–5346.

Pascale F, Zicovich-Wilson CM, Gejo FL, Civalleri B, Orlando R and Dovesi R 2004b The calculation of the vibrational frequencies of the crystalline compounds and its implementation in the CRYSTAL code. *J. Comput. Chem.* **25**, 888–897.

Peintinger MF, Oliveira DV and Bredow T 2013 Consistent Gaussian basis sets of triple-zeta valence with polarization quality for solid-state calculations. *J. Comput. Chem.* **34**, 451–459.

Perdew J, Ruzsinsky A, Csonka GI, Vydrov OA, Scuseria GE, Constantin LA, Zhou X and Burke K 2008 Restoring the density-gradient expansion for exchange in solids and surfaces. *Phys. Rev. Lett.* **100**, 136406.

Perdew JP, Burke K and Ernzerhof M 1996 Generalized gradient approximation made simple. *Phys. Rev. Lett.* **77**, 3865–3868.

Placzek G 1934 *Handbuch der Radiologie* vol. **6**. Akademische Verlagsgesellschft, Leipzig, p. 208.

Prencipe M, Mantovani L, Tribaudino M, Bersani D and Lottici PP 2012 The Raman spectrum of diopside: a comparison between *ab initio* calculated and experimentally measured frequencies. *Eur. J. Mineral.* **24**, 457–464.

Prencipe M, Noël Y, Bruno M and Dovesi R 2009 The vibrational spectrum of lizardite-1T [$Mg_3Si_2O_5(OH)_4$] at the Γ point: a contribution from an *ab initio* periodic B3LYP calculation. *Am. Mineral.* **94**, 986–994.

Prosandeev S, Waghmare U, Levin I and Maslar J 2005 First-order Raman spectra of $AB'_{1/2}B''_{1/2}O_3$ double perovskites. *Phys. Rev.* **B 71**, 214307.

Putrino A and Parrinello M 2002 Anharmonic Raman spectra in high-pressure ice from *ab initio* simulations. *Phys. Rev. Lett.* **88**(17), 176401.

Ratner MA and Gerber RB 1986 Excited vibrational states of polyatomic molecules: the semiclassical self-consistent field approach. *J. Phys. Chem.* **90**, 20–30.

Refson K, Tulip PR and Clark SJ 2006 Variational density-functional perturbation theory for dielectrics and lattice dynamics. *Phys. Rev. B* **73**, 155114.

Resta R 2000 Manifestations of Berry's phase in molecules and condensed matter. *J. Phys. Condens. Matter* **12**, R107.

Rimola A, Civalleri B and Ugliengo P 2010 Physisorption of aromatic organic contaminants at the surface of hydrophobic/hydrophilic silica geosorbents: a B3LYP-D modeling study. *Phys. Chem. Chem. Phys.* **12**, 6357–6366.

Ruan HD, Frost RL and Kloprogge JT 2001 Comparison of Raman spectra in characterizing gibbsite, bayerite, diaspore and boehmite. *J. Raman Spectrosc.* **32**, 745–750.

Ruan HD, Frost RL, Kloprogge JT and Duong L 2002 Far-infrared spectroscopy of alumina phases. *Spectrochim. Acta* **58A**, 265–272.

Russell JD and Farmer VC 1978 Lattice vibrations of boehmite (γ-AlOOH): evidence for C_{2v}^{12} rather than D_{2h}^{17} space group. *Spectrochim. Acta* **34A**, 1151–1153.

Schauble EA, Ghosh P and Eiler JM 2006 Preferential formation of $^{13}C^{18}O$ bonds in carbonate minerals, estimated using first-principles lattice dynamics. *Geochim. Cosmochim. Acta* **70**, 2510–2529.

Schlegel HB 1998 Geometry optimization. In *Encyclopedia of Computational Chemistry* (ed. von Rague Schleyer P) vol. **2**. John Wiley & Sons, Ltd, Chichester, pp. 1136–1142.

Sierka M and Sauer J 2001 Proton mobility in chabazite, faujasite and ZSM-5 zeolite catalysts comparison based on *ab initio* calculations. *J. Phys. Chem. B* **105**, 1603–1613.

Skyes D and Kubicki JD 1996 Four-membered rings in silica and aluminosilicates glasses. *Am. Mineral.* **81**, 265–272.

Slade RC and Halstead TK 1980 Evidence for proton pairs in γ-AlOOH (boehmite) from NMR absorption spectra. *J. Solid State Chem.* **32**, 119–122.

Slater JC 1951 A simplification of the Hartree–Fock method. *Phys. Rev.* **81**, 385–390.

Smirnov MB, Sukhomlinov SV and Smirnov K 2010 Vibrational spectrum of reidite $ZrSiO_4$ from first principles. *Phys. Rev. B* **82**, 094307.

Soler JM, Artacho E, Gale JD, García A, Junquera J, Ordejón P and Sánchez-Portal D 2002 The Siesta method for *ab initio* order-N materials simulation. *J. Phys. Condens. Matter* **14**, 2745–2779.

Spoon HWW, Tielens AGGM, Armus L, Sloan GC, Sargent B, Cami J, Charmandaris V, Houck JR and Soifer BT 2006 The detection of crystalline silicates in ultraluminous infrared galaxies. *Astrophys. J.* **638**, 759–765.

Stanton JF 1991 Point group symmetry and Cartesian force constant redundancy. *Int. J. Quantum Chem.* **39**, 19–29.

Stegmann MC, Vivien D and Mazieres C 1973 Étude des modes de vibration infrarouge dans les oxyhydroxides d'aluminium boehmite et diaspore. *Spectrochim. Acta* **29A**, 1653–1663.

Stixrude L and Peacori R 2002 First-principles study of illite–smectite and implications for clay mineral systems. *Nature* **420**, 165–168.

Tielens F, Gervais C, Lambert JF, Mauri F and Costa D 2008 *Ab initio* study of the hydroxylated surface of amorphous silica: a representative model. *Chem. Mater.* **20**, 3336–3344.

Tosoni S, Doll K and Ugliengo P 2006 Hydrogen bond in layered materials: structural and vibrational properties of kaolinite by periodic B3LYP approach. *Chem. Mater.* **18**, 2135–2143.

Ugliengo P, Saunders V and Garrone E 1990 Silanol as a model for the free hydroxyl of amorphous silica: *ab initio* calculations of the interaction with water. *J. Phys. Chem.* **94**, 2260–2267.

Ugliengo P, Sodupe M, Musso F, Bush IJ, Orlando R and Dovesi R 2008 Realistic models of hydroxylated amorphous silica surfaces and MCM-41 mesoporous material simulated by large-scale periodic B3LYP calculations. *Adv. Mater.* **20**, 4579–4583.

Ugliengo P, Zicovich-Wilson CM, Tosoni S and Civalleri B 2009 Role of dispersive interactions in layered materials: a periodic B3LYP and B3LYP-D* study of $Mg(OH)_2$, $Ca(OH)_2$ and kaolinite *J. Mater. Chem.* **19**, 2564–2572.

Ulian G, Tosoni S and Valdrè G 2013a Comparison between Gaussian-type orbitals and plane wave *ab initio* density functional theory modeling of layer silicates: talc [$Mg_3Si_4O_{10}(OH)_2$] as model system. *J. Chem. Phys.* **139**, 204101.

Ulian G, Valdre G, Corno M and Ugliengo P 2013b The vibrational features of hydroxylapatite and type A carbonated apatite: a first principle contribution. *Am. Mineral.* **98**, 752–759.

Valenzano L, Noël Y, Orlando R, Zicovich-Wilson CM, Ferrero M and Dovesi R 2007 *Ab initio* vibrational spectra and dielectric properties of carbonates: magnesite, calcite and dolomite. *Theor. Chem. Accounts* **117**, 991–1000.

Valenzano L, Torres FJ, Doll K, Pascale F, Zicovich-Wilson C and Dovesi R 2006 *Ab initio* study of the vibrational spectrum and related properties of crystalline compounds: the case of CaCO$_3$ calcite. *Z. Phys. Chem.* **220**, 893–912.

Vosko SH, Wilk L and Nusair M 1980 Accurate spin-dependent electron liquid correlation energies for local spin density calculations: a critical analysis. *Can. J. Phys.* **58**, 1200–1211.

Wallace AF, Hedges LO, Fernandez-Martinez A, Raiteri P, Gale JD, Waychunas GA, Whitelam S, Banfield JF and De Yoreo J 2013 Microscopic evidence for liquid-liquid separation in supersaturated CaCO$_3$ solutions. *Science* **341**, 885–889.

Wefers K and Misra C 1987 *Oxides and Hydroxides of Aluminium*. ALCOA Technical Report vol. **19**. ALCOA Laboratories, Pittsburgh.

Wehrmeister U, Soldati L, Jacob DE, Häger T and Hofmeister W 2010 Raman spectroscopy of synthetic, geological and biological vaterite: a Raman spectroscopic study. *J. Raman Spectrosc.* **41**, 193–201.

Wooden DH, Woodward CE and Harker DE 2004 Discovery of crystalline silicates in comet C/2001 Q4 (NEAT). *Astrophys. J. Lett.* **612**, 77.

Yamaguchi Y, Kahle AB, Tsu H, Kawakami T and Pniel M 1998 Overview of advanced spaceborne thermal emission and reflection radiometer (ASTER). *IEEE Trans. Geosci. Remote Sens.* **36**, 1062–1071.

Yin MT and Cohen ML 1982 Theory of lattice-dynamical properties of solids: application to Si and Ge. *Phys. Rev. B* **26**, 3259–3272.

Zein NE 1984 Density functional calculations of elastic moduli and phonon spectra of crystals. *Sov. Phys. Solid State* **26**, 1825–1828. [Zein NE 1984. *Fiz. Tverd. Tela (Leningrad)* **26**, 3028–3034.]

Zhou XF, Oganov A, Qian GR and Zhu Q 2012 First-principles determination of the structure of magnesium borohydride. *Phys. Rev. Lett.* **109**, 245503.

Zicovich-Wilson CM, Pascale F, Roetti C, Saunders VR, Orlando R and Dovesi R 2004 Calculation of the vibrational frequencies of α-quartz: the effect of Hamiltonian and basis set. *J. Comput. Chem.* **25**, 1873–1881.

11

Geochemical Kinetics via Computational Chemistry

James D. Kubicki[1] and Kevin M. Rosso[2]

[1]*Department of Geological Sciences, University of Texas at El Paso, El Paso, TX, USA*
[2]*Physical Sciences Division, Pacific Northwest National Laboratory, Richland, WA, USA*

11.1 Introduction

Organic chemists have the ability to synthesize a complex array of compounds guided by fundamental principles that control reactions. Upon examination of an organic chemistry textbook, one can find reaction sequences composed of numerous steps that result in specific, detailed structures. In contrast, for many years geochemistry was not concerned to any great extent with reaction mechanisms. The high temperatures and long time scales associated with the geochemical processes of interest diminished the idea that kinetics was a critical topic in geochemistry. With an increased interest in lower-temperature biogeochemical processes, geochemical kinetics (rates and mechanisms) became an area of greater concern. With lower temperatures and shorter time scales, equilibrium was often not achieved. This recognition also caused researchers to reexamine assumptions of equilibrium in higher-temperature systems, and disequilibrium could be found even in igneous and metamorphic rocks that had been reacted at high temperature.

Today geochemical kinetics is a common focus of published studies. The need to predict time scales of geologic processes such as aqueous-phase exchange reactions (Balogh et al., 2007; Evans et al., 2008; Grant and Jordan, 1981; Harley et al., 2011; Jin et al., 2011a; Johnson et al., 2011; Panasci et al., 2012) is a significant driver of this interest. The Ostwald step rule (i.e., precipitation of metastable phases prior to formation of more stable solid phases) is observed in action across a wide variety of environments, so understanding nucleation, crystal growth, and nanoparticle chemistry is imperative (Anh et al., 2014; Hummer et al., 2009; Jensen et al., 2014; Kadota et al., 2014; Lee and Kubicki, 1993). These metastable phases generally convert to more crystalline, thermodynamically favored phases over time (e.g., Torn et al. 1997) which makes the process of solid phase

---

Molecular Modeling of Geochemical Reactions: An Introduction, First Edition. Edited by James D. Kubicki.
© 2016 John Wiley & Sons, Ltd. Published 2016 by John Wiley & Sons, Ltd.

transformation a relevant study (Bazilevskaya et al., 2011; Kwon et al., 2013). For example, concerns about anthropogenic climate change have piqued interest in isotope signatures of paleoclimates (McInerney and Wing, 2011) which means that we need to be able to account for kinetic isotope effects (KIE) regarding isotopic fractionations (Cohen and Gaetani, 2010; DePaolo, 2011; Dreybrodt and Scholz, 2011; Immenhauser et al., 2010). Three related areas are weathering of minerals affecting the C cycle (Berner, 2009), mineral aerosol chemistry influencing the Earth's radiation budget and precipitation (Alexander et al., 2013), and CO_2 behavior in deep saline aquifers as part of a climate change mitigation strategy (DePaolo and Cole, 2013).

The chemical weathering of minerals (or dissolution) and adsorption/desorption on minerals have been the most extensively studied processes using computational chemistry techniques. Whether in the deep subsurface, in soils and sediments at the Earth's near surface, or among mineral particles in the atmosphere, the most relevant interface is that between the mineral solid and liquid or vapor H_2O. Starting with Casey et al. (1990), these studies have grown in sophistication over the past 25 years to include better representations of real mineral–water interfaces (Adeagbo et al., 2008; Bandura et al., 2011; Bhandari et al., 2010; Bickmore et al., 2008; Kubicki et al., 1993, 1996, 2012a; Lockwood and Garofalini, 2009, 2010, 2012; Morrow et al., 2010; Pelmenschikov et al., 2000, 2001; Walsh et al., 2000; Xiao and Lasaga, 1994, 1996). Mineral dusts have multiple effects on atmospheric chemistry and climate (Choobari et al., 2014), and these minerals undergo complex chemical reactions within soils before being suspended (Park and Kim, 2014). Hence, knowledge of their interaction with water and adsorption mechanisms and the effects on optical and surficial properties is important (Baltrusaitis et al., 2012; Baptista et al., 2008). The kinetics of these adsorption/desorption reactions is poorly understood. Considering the competition among myriad species for surface sites, rates and mechanisms must be quantified to better account for this atmospheric component.

In addition to the ability to predict how climate will change, minimizing the buildup of atmospheric CO_2 and its impacts has also been an area of active study (Boot-Handford et al., 2014). One option is to capture man-made CO_2 emissions, pressurize and pump them as a supercritical fluid into deep saline aquifers. Long-term storage of C can occur via mineral sequestration wherein Ca^{2+} and Mg^{2+} minerals dissolve and react with aqueous HCO_3^- to precipitate carbonate minerals, sequestering the CO_2 in the form of stable solid precipitates (Lal, 2008). The dissolution, nucleation, and crystal growth kinetics are all critical in this process, particularly because some of these transformations are very slow, so computational chemistry can provide a strong knowledge base complementing laboratory- and field-scale studies. The details computational chemistry can provide about reaction mechanisms beyond what is easily measureable could be helpful in designing safer and more efficient C sequestration projects (Loring et al., 2011; Tenney and Cygan, 2014). The engineering of geochemical systems also comes out of the desire to construct remediation projects for acid mine drainage (Gammons et al., 2010; Heidel and Tichomirowa, 2010, 2011; Heidel et al., 2011) and other types of terrestrial contamination (Johnson et al., 2013; Kim et al., 2009; Kubicki et al., 2009; Lucena et al., 2014).

Understanding geochemical kinetics is required for higher-temperature processes as well. The T and P conditions of many geochemical processes make them difficult to study experimentally (Karki et al., 2001). Alternatively, some geochemical reactions are slow on human time scales, so experiments are run at elevated temperatures in order to drive the reaction forward in a reasonable time frame (Wieclaw et al., 2010; Yakob et al., 2012). In either case, knowledge of the activation energy (ΔE_a) and mechanism will allow for extrapolation to T and P conditions outside the experimentally accessible range. For example, Doltsinis et al. (2007) studied reactions in supercritical solutions of $SiO_2 + H_2O$ where in situ observation of species is problematic. Another example is the application of laboratory observed dissolution rates to field weathering rates

(Fischer et al., 2012). Mechanistic information can help resolve the discrepancy between these two rates.

An exemplary problem in computational geochemistry is ionic diffusion in minerals. Diffusion through mineral lattices can be exceeding slow (on the order of 10^{-19} cm^2/s), and the environments in which ionic diffusion in most minerals becomes significant are at high temperature, such as in the Earth's mantle, along with concomitant high pressure. Consequently, a relatively large number of simulation studies have been performed to estimate ion diffusion rates (see Ammann et al. (2010) and references therein for a thorough review). These simulations may be performed with either classical or quantum mechanical methods. A recent series of papers that explain issues with simulating solid-state diffusion and interesting geochemical implications can be found in Ammann et al. (2009, 2011, 2012). Ammann et al. (2010) discuss the choices of method for diffusion simulations (i.e., classical vs. quantum, LDA vs. GGA DFT methods—see Chapter 1) and how to calculate the attempt frequencies, defect probabilities, and diffusive hop enthalpy barriers. In one example (Ammann et al., 2010), DFT methods are used to calculate Si^{4+} and Mg^{2+} diffusion rates in postperovskite $MgSiO_3$. The diffusion rate is related to the mineral viscosity via the Nabarro–Herring expression for diffusion creep; hence, the simulated diffusion constants provide an estimate for the D'' layer viscosity. The large diffusion anisotropy found in the model calculations serves to explain lateral viscosity variations near the core–mantle boundary.

As mentioned earlier, silica hydrolysis and silicate dissolution have garnered significant attention. The main hypotheses regarding silicate dissolution are that H^+-transfer to a bridging O atom (O_{br}) in an Si—O—Si linkage weakens the Si—O bonds under acidic conditions (Casey et al., 1990) and that under basic conditions, nucleophilic attack (see Section 11.2.6) by OH^- leads to a 5-coordinate Si with weaker Si—O bonds that break more readily than the normal Si—O bonds in tetrahedral coordination (Kubicki et al., 1993; Xiao and Lasaga, 1996).

This seemingly simple problem has been difficult to solve for a number of reasons. First, H^+-transfer and OH^- nucleophilic attack are strongly influenced by the solvation environment. H^+-transfer usually occurs via a concerted mechanism involving a number of H_2O molecule intermediates (Kumar et al., 2011). Most of the early studies involved a limited number of H_2O molecules, so solvation effects were not completely accounted for. Second, the mineral surface must be reproduced in detail. Although some recent studies (Adeagbo et al., 2008; Kubicki et al., 2012a) have used periodic silicate surfaces in their calculations, mineral dissolution may commonly occur at steps on the surface (Dove et al., 2005; Kurganskaya et al., 2012) and, to our knowledge, no molecular modeling study of silicate dissolution has been conducted with this type of defect present on the silica surface. Most quantum calculations have overestimated the ΔE_a of quartz dissolution which is approximately 72 kJ/mol (Dove, 1999), and this could be due to inadequate model construction and methodological errors. Early quantum studies by Xiao and Lasaga (1994, 1996) using the MP2 method produced reasonably accurate activation energies with only simple molecules such as $H_3SiOSiH_3$ and $(OH)_3SiOSi(OH)_3$ representing the quartz surface. Recently, Kagan et al. (2014) have used a reactive force field classical mechanical approach and obtained excellent agreement with observed ΔE_a values for amorphous SiO_2 dissolution. By using a reactive classical force field, the authors were able to employ a potential mean force (PMF) approach and examine a variety of surfaces and reactions. The relatively large model size (5988 atoms) and long-duration simulations (100s of picoseconds) were pivotal in realistically modeling Si—O—Si hydrolysis and dissolution. These authors found that variations in the amorphous SiO_2 surface structure (i.e., Si—O—Si constraints, strained Si—O—Si angles, and surface concavity) all influenced the simulated ΔE_a.

KIE are common in geochemical reactions but difficult to study experimentally. Computational geochemistry techniques complement natural sample analyses and laboratory experiments to obtain estimates of KIE and isotopic fractionation mechanisms. Eiler et al. (2014) have published a detailed review of recent isotope geochemistry research with numerous examples of computational geochemistry providing key information. Ionic diffusion in aqueous solution can be a source of isotopic fractionation (Nielsen and DePaolo, 2013; Saenger et al., 2013). Bourg and coworkers (Bourg and Sposito, 2007; Bourg et al., 2010) have performed MD simulations demonstrating the fractionation process and explaining the variation based on the long-term coupling of solvent–solute motions. These aqueous diffusion isotope effects can influence the observed isotopic fractionation in crystals grown in these solutions (Bourg et al., 2010). One significant factor controlling isotopic fractionation is the desolvation rate (i.e., the breaking of the cation–H_2O bond) which depends in part on the isotopic mass (Hofmann et al., 2012).

An intriguing study is that of Lacks et al. (2012) who have performed classical MD simulations to explain the observed isotopic fractionation along a temperature gradient in silicate melts (Richter et al., 2009). The anomalously more rapid diffusion of the heavier isotopes in these melts is explained in terms of the atomic momentum during diffusive collisions. The heavier isotopes continue moving in the same direction after a collision more than lighter isotopes resulting in a net increased diffusion coefficient. Further work on this topic would be worthwhile. Although Lacks et al. (2012) performed long (68 ns or 68×10^6 time steps) on a reasonably large system (2160 atoms), the heavy isotopes of Si, Mg, and O in these simulations were four times larger than the light isotopes rather than the natural 7, 8, and 13% mass differences between light and heavy isotopes.

KIE have been investigated with quantum as well as classical mechanical methods. In a series of papers, Felipe and coworkers (2003, 2004, 2005) calculated reaction pathways and KIE for H and O isotope exchange on H_4SiO_4 and ^{13}C exchange between HCO_3^- and CH_3COO^-. The former two studies are applicable to aqueous silica (e.g., Zotov and Keppler, 2002) and to silicate–water interface isotopic exchange (e.g., Clayton et al., 1988; Webb and Longstaffe, 2006). The third study was designed to explain the observation of ^{13}C exchange between organic and inorganic components in oil field brine studies (Dias et al., 2002). Felipe et al. (2003) explain the theory behind calculating equilibrium fractionation factors (α) and KIE. The reactions were modeled using a combination of a small number of explicit H_2O molecules of solvation combined with a polarized continuum model of solvation (integral equation formalism polarized continuum model = IEFPCM (Cancès et al., 1997; Mennucci et al., 1997)). H/D and $^{18}O/^{16}O$ exchange on aqueous H_4SiO_4 were determined to be sufficiently rapid that equilibrium would likely be achieved in most natural environments. However, the $^{13}C/^{12}C$ exchange between HCO_3^- and CH_3COO^- had a large enough barrier that at low temperatures exchange would be slow. At higher temperatures associated with petroleum generation (i.e., 200°C), the exchange reaction would be fast enough to be observed in experiments (i.e., 72 h (Dias et al., 2002)). These calculations predicted a reaction mechanism involving an enol intermediate species as opposed to the initially hypothesized S_N2 mechanism.

The KIE for H/D fractionation in organic molecules has also been modeled and can be used to estimate thermal maturities of natural gas (Ni et al., 2011). Tang and coworkers (Ni et al., 2011; Tang et al., 2005; Tan et al., 2010) have published detailed studies of molecular-specific isotopic fractionation of natural hydrocarbon molecules. One complicating factor in organic H/D fractionation is the distribution of exchange site α values that occur within a single molecule (Wang et al., 2009a, 2009b, 2013). The rates of exchange at these sites can also vary significantly (Kubicki et al., 2008), so interpreting observed H isotopes in organic mixtures becomes an extremely complex problem.

Another noteworthy case where kinetics can be important is in redox processes. For example, the cycling of redox-active metals such as Fe and Mn in surface waters depends in part on their rates of oxidation by dissolved O_2. Such reactions occur as these reduced aqueous metals encounter a positive gradient in $O_2(aq)$, such as when sediment layers on the bottom of lakes are exposed to more oxygenated waters from the air–water interface (Stumm and Morgan, 1996). Steep redox gradients can also occur in acid mine drainage systems (Benner et al., 2000) and at the boundaries of polluted groundwater plumes (McGuire et al., 2000) and injected CO_2 plumes (Lammers et al., 2015). Where not catalyzed by biologic activity or mineral surfaces (e.g., Sung and Morgan (1981) and Emerson et al. (1982)), the rates of oxidation in homogeneous aqueous solution can be orders of magnitude slower for Mn(II) than Fe(II) (Diem and Stumm, 1984). Slow kinetics of electron transfer arise because these reactions are predicated on both the close encounter of donor and acceptor species (just a few angstroms) and then the random vibrational motions of nuclei to attain the transition state nuclear configuration. Because homogeneous electron transfer reactions with O_2 can be slow, redox equilibrium may never be reached in some settings (Keating and Bahr, 1998). There are many other examples where the behavior of geochemical and biogeochemical redox reactions is strongly influenced by electron transfer kinetics, such as pyrite oxidation (Lowson, 1982), the transport of redox-active radionuclides and toxic metals (Cui and Eriksen, 1996; Peterson et al., 1997; Scott and Morgan, 1996), the reductive/oxidative degradation of organic solvents and pesticides (Amonette et al., 2000; Elsner et al., 2004; Pecher et al., 2002; Strathmann and Stone, 2003), and the reductive dissolution of Fe(III)-oxide and oxyhydroxide minerals (Maurice et al., 1995; Stone and Morgan, 1987), including enzymatically by dissimilatory metal-reducing bacteria (Breuer et al., 2014; Kerisit et al., 2007).

11.2 Methods

11.2.1 Potential Energy Surfaces

Now that some applications of computational geochemical kinetics have been discussed to motivate the reader, this section will reiterate information that has been in previous chapters. But here emphasis will be placed on issues that are of particular importance when modeling reaction pathways and kinetics as compared to equilibrium or metastable states. The reader is referred to previous chapters for more detail.

Accurately calculating a potential energy surface (PES; Figure 11.1) for even simple systems is not a trivial task (Truhlar, 2007). Achieving this goal for large, complex geochemical systems is a daunting challenge. Although Schrodinger's equation guides us, the approximations made in calculating the wave function or electron probability density have a great effect on the level of accuracy. In molecular orbital methods, one can increase the basis set (see Chapter 1) up to the complete basis set limit, but this is not always possible in practice. Furthermore, the choice of basis set is connected, in DFT, to the choice of exchange–correlation functionals and method of dispersion correction (see Chapter 1). For example, Mardirossian and Head-Gordon (2013) have pointed out that the inhomogeneity correction factor (ICF) in the M06-L exchange functional affects the basis set superposition error (BSSE).

Systematic benchmarking studies are guides to preferable choices. De Jong et al. (2005) tested 24 DFT functionals against higher-level methods. Typically, experimental and/or high-level ab initio calculations are used as the standard of comparison. These authors used coupled-cluster singles, doubles, and triples (CCSD(T)) calculations as the benchmark. Excellent agreement for the energy differences between reactant and transition state among DFT methods and the CCSD(T) calculations were achieved except for the commonly used B3LYP (Becke, 1993; Perdew et al., 1988) hybrid functional. BSSE was approximately zero when the doubly polarized triple-ζ (TZ2P) and quadruply polarized quadruple-ζ (TZ2P) basis sets were used. The general strategy is to select a test reaction that can be modeled at a high level of theory that is similar to the reaction of interest and then to test increasingly accurate basis set/exchange–correlation functionals to search for convergence on the

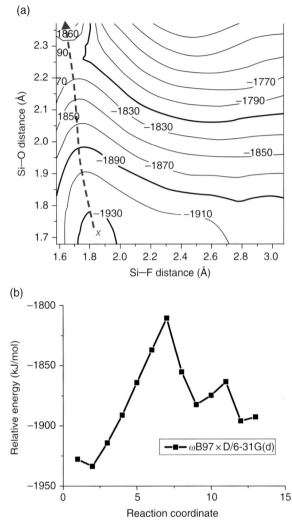

Figure 11.1   *(a) Potential energy surface (PES—contours in kilojoule per mole) and (b) cross section for*
Si(OH)$_4$ + F$^-$ → Si(OH)$_3$F + OH$^-$ illustrate how reactions may proceed via the minimum-energy pathway
between two states. In this case, the Si—F distance (i.e., the reaction coordinate) is decreased by constraining
the Si and F atomic coordinates during the energy minimization and the Si—O distance is the dependent
parameter that relaxes as the Si—F bond is formed. The difference between the stable configurations (i.e.,
reactants and products) and the highest energy point along the reaction coordinate (i.e., the transition state) is
an estimate of the activation energy (ΔE$_a$) of the reaction.

high-level result. Depending on the size of the system of interest, one cannot always use the most accurate method and is forced to accept a larger uncertainty. These tests are valuable in providing an estimate for the uncertainty to the user which should be reported in any publication. In addition, if one is
comparing various reaction mechanisms, in some cases an error of ±10 kJ/mol may be acceptable, for
instance, if the difference in calculated ΔE_a among possible mechanisms is much larger than this uncertainty. Other examples of note are Estreicher et al. (2011) who focused on diffusion and Ribeiro et al.
(2010) who studied the biologically relevant hydrolysis of phosphodiester bonds.

Gomes et al. (2012) performed a similar method test focusing on the ethane methylation reaction in the zeolite MFI. This study required a larger system size than the Pd–CH_4 model system of de Jong et al. (2005), so hybrid quantum mechanical/molecular mechanical (QM/MM; Chapter 1) methods were chosen. The benchmarks used were the experimental activation enthalpy barrier and PBC–DFT:MP2 calculations (a combination of periodic DFT and MP2 (Svelle et al., 2009)). Good agreement was obtained with the ωB97x-D/6-311++G(3df,3pd)//ωB97x-D/6-31G (d,p) method (note that "//" indicates a single-point calculation to obtain more accurate energies with the method on the left on a structure energy minimized with the method on the right). Hence, QM/MM can be an efficient method for simulating complex system reactions because it allows the researcher to incorporate the reaction environment while modeling the TS accurately. The user must be cautious with QM/MM. Care in selecting the QM region and the method of handling the boundary region as well as basis set/exchange–correlation functional choices is imperative.

In conjunction with choosing an appropriate basis set/exchange–correlation method, simultaneous consideration of solvation models should be performed. There are two basic approaches: explicit and implicit solvation (see Chapter 4). Explicit solvation means including H_2O molecules in the model system; implicit solvation is generally using a polarizable continuum to surround the solute (e.g., IEFPCM (Cances et al., 1997); COSMO (Klamt et al., 1998; Reinisch and Klamt, 2014; Saukkoriipi and Laasonen, 2008)). When strong H-bonding is likely (e.g., between oxyanions such as HCO_3^- and water; Figure 11.2), explicit solvation is recommended, but adding more atoms to the QM calculation can increase energy minimization times dramatically because each energy calculation is slower and more steps are necessary to optimize the H_2O positions. If H-bonding is weaker to the solute, implicit solvation can be an efficient method to approximate the effects of water on solute structure and energy. The methods are not mutually exclusive and one can include explicit solvation and surround the solute plus H_2O molecules in the polarizable continuum.

Aguilar and Rocha (2011) investigated solvation effects for aqueous Ru ligand exchange reactions. Their approach was to use a QM/MM method where H_2O molecules are included as effective fragment potentials (DFT/EFP (Chen and Gordon, 1996; Day et al., 1996)). These EFPs mimic the solvent effect without needing to solve for the electron density of the H_2O molecules themselves. In this case, the B3LYP/cc-pVDZ/EFP method gave good agreement with experiment for the H_2O–Cl exchange reaction (99 kJ/mol calculated vs. 97 kJ/mol experimental (Broomhead et al., 1964)), and switching from B3LYP to MP2 did not significantly affect the model energy barriers for the other reactions that were studied.

A further complication that arises when modeling aqueous reactions is that the reaction pathway may not be well represented by a single configuration from reactants to TS to products. Gallet et al. (2012) applied DFT–MD with path metadynamics (MTD) to model CO_2 hydration (i.e., $CO_2 + nH_2O \rightarrow H_2CO_3$ where $n = 1, 2,$ or 3). This method with appropriate choices for path-collective variables allows for a more flexible determination of reaction path than methods such as nudged elastic band (NEB; see section III). Multiple possible reaction pathways were modeled, and quasidegenerate configurations for the TS were found.

For electron transfer reactions the relevant PESs are those based on the collective nuclear coordinates (q) of the donor (D) and acceptor (A) species in an encounter complex, both before and after electron transfer (Figure 11.3). The ET treatment discussed here is based on Marcus' two PES model (Marcus, 1956). Before electron transfer, the reactants could be, for example, a pair of solution species in an encounter complex, or neighboring metal sites in a solid phase, or a surface complex at a solid–liquid interface. The donor and acceptor need not be bridged by bonds, but their close proximity is implied. The initial electronic state (i.e., the "reactant state") can be referred to as $\psi_A = D \cdots A$, and the final electronic state (i.e., the "product state") as $\psi_B = D^+ \cdots A^-$. At thermal

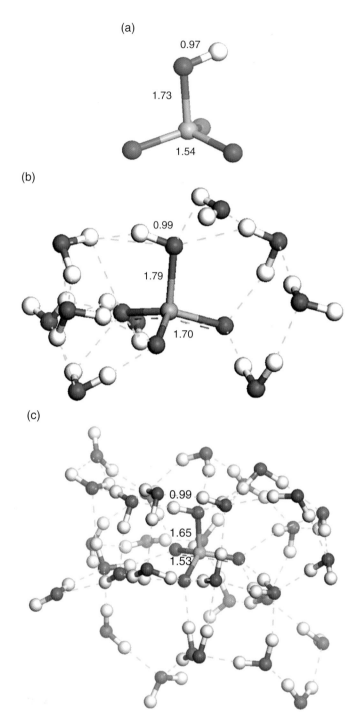

Figure 11.2 (a) Implicit (i.e., polarized continuum) and (b) explicit (i.e., addition of H_2O or other solvent molecules) solvation methods result in significantly different bond distances for HPO_4^{2-}. The continuum solvation method is a better approximation for low dielectric constant organic solvents, but when strong H-bonds form, the continuum approximation does not account as well for these specific interactions. These changes affect calculated structures, energies, and spectroscopic properties. (c) Addition of H_2O molecules to include second or third solvation shells can make the calculations significantly more time-consuming, but comparison of (b) and (c) shows that the P—O bond distances change with addition of the extra H_2O molecules. Molecules drawn with Materials Studio 7 (Accelrys Inc., San Diego, CA). (H, white; H-bonds, dashed blue lines; O, green; P, pink).

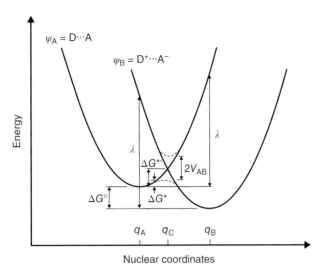

Figure 11.3 *Potential energy surfaces for an electron transfer process from donor to acceptor (D···A), where the initial (ψ_A) and final (ψ_B) electronic states are parabolic with respect to collective nuclear coordinates (q). $\Delta G°$ is the energy difference between the initial state in its equilibrium nuclear configuration (q_A) and that (q_B) of the final state. λ is the reorganization energy required to distort the nuclear coordinates from q_A to q_B while keeping the electron on the donor. $\Delta G^{*'}$ prime is the diabatic activation energy to distort nuclear coordinates from q_A to that of the crossing point configuration q_C, whereas ΔG^* is that energy reduced by the magnitude of the electronic coupling (V_{AB}) between ψ_A and ψ_B at q_C.*

equilibrium in either state, the free energy difference between them is $\Delta G°$, the thermodynamic driving force for the electron transfer reaction.

One important aspect of electron transfer reactions is that they obey the Franck–Condon principle, that is, transfer occurs at approximately fixed nuclear positions, which can only be satisfied if the electron transfer takes place on a time scale that is short relative to that of the nuclear motion (Nitzan, 2006). No bonds are formed or broken during an electron transfer reaction; through random fluctuations the nuclear coordinates of the reactant state must achieve a transition state configuration in which the energy is equivalent to that of the product state. The distortion of the reactant configuration (donor and acceptor at equilibrium before electron transfer) to the configuration at which electron transfer can take place (i.e., the transition state) and the relaxation to the product configuration (donor and acceptor at equilibrium after electron transfer) are represented by two intersecting parabolas, which are free energy surfaces of the reactant and product states. The reorganization energy (λ) is the energy to distort the configuration of the reactants into that of the products, or vice versa, without changing the electronic density distribution. Electron transfer requires degeneracy of donor and acceptor orbitals, a condition that depends on random fluctuations in nuclear coordinates to achieve the appropriate configuration. Hence, changes in the nuclear coordinates of the donor and acceptor species and the accompanying rearrangement of the solvent molecules have to occur prior to the actual electron transfer.

Marcus' use of a continuum dielectric model (linear response theory) initially yielded the parabolic shape of the free energy surfaces (Marcus and Sutin, 1985). In other words, the changes in the total collective energy along the reaction coordinate satisfy the harmonic approximation. In solution, Marcus theory separates the relevant nuclear coordinates into those involving changes in the bond lengths and angles of the species within the first solvation sphere (internal reorganization energy, λ_I) and the reorientation of the solvent molecules beyond the first solvation sphere (external reorganization energy, λ_E) (Newton and Sutin, 1984). In addition, both factors are taken as being additive ($\lambda = \lambda_I + \lambda_E$). Therefore, in the Marcus model, typically only the solvent molecules in the first shell are treated explicitly and those beyond the first shell are often described by a dielectric continuum (Newton, 2007). In the continuum treatment, the solvent reorganization is characterized by two time scales: fast (electronic response) and slow (nuclear and simultaneous electronic response), whereby the former is characterized by the optical dielectric constant (ε_{opt}), while the latter is described by the static dielectric constant (ε_s).

11.2.2 Choice of Solvation Methods

Solvation methods have been discussed in detail in Chapter 4 of this volume, so this chapter will not repeat that information. The points made in Chapter 4 are echoed here, and they are amplified when applied to solvating the TS complex. The structures and energies of molecules are altered by solvation effects in water, so these effects need to be accounted for in modeling reaction mechanisms. Figure 11.2 shows how bonding changes around the HPO_4^{2-} ion under various solvation schemes. For reactions involving charged species in aqueous solution, it is hard to imagine when an implicit solvation scheme by itself will be sufficient. H^+-transfers are often cooperative among H_2O molecules in the vicinity (e.g., Stirling and Papai, 2010) so explicit treatment of at least the first solvation shell is imperative. A second or third shell is desirable if practical. If not, then one solvation shell and a polarized continuum such as IEFPCM (Cances et al., 1997; Mennucci et al., 1997) or COSMO-RS (Klamt et al., 1998) will suffice.

With respect to electron transfer reactions between two hydrated aqueous ions with a small overlap of electronic orbitals, Marcus employed a strategy similar to the Born solvation model (i.e., spherical cavity with radius r) to calculate the external reorganization energy. Using two arbitrarily defined spheres with radius r_D and r_A for the donor and acceptor sites in the electron transfer complex, separated at a distance r_{AD}, immersed in continuum dielectric, the expression is

$$\lambda_E = \gamma_p e^2 \left(\frac{1}{2r_A} + \frac{1}{2r_D} - \frac{1}{r_{AD}} \right), \quad \text{where } \gamma_p = \frac{1}{\varepsilon_{opt}} - \frac{1}{\varepsilon_s} \tag{11.1}$$

where ε_{opt} and ε_s are optical and static dielectric constants of the solvent, respectively. This treatment involves calculating the change in the solvation of an elementary charge during reorganization from the reactant state at equilibrium to the charge distribution in the transition state. Because $\varepsilon_s \gg \varepsilon_{opt}$ for this example system, the magnitude of the reorganization energy is governed by ε_{opt}, and it is rather insensitive to the low-frequency solvent response. This model generally overestimates the solvent contribution and loses applicability upon large overlap of the spheres. In this regard, an improvement over the Marcus model can be achieved by employing more elaborate cavity shapes, multipolar expansions, or by replacing the classical model with a quantum self-consistent reaction field model, but these approaches also suffer from arbitrary definitions of the solute cavities.

Such models also rely on the assumption that a polar solvent such as water exhibits a linear response with respect to electrostatic forces. That is to say that the solvent polarization response to changes in the electrostatic field exerted by the solute is linear and characterized by a single dielectric constant. Marcus' parabolic free energy surfaces for electron transfer derive from the linear response assumption; Gaussian statistics of energy gap fluctuations (i.e., the free energy difference between reactant and product electron transfer states) is one of the consequences of the linearity in solvent response (Ayala and Sprik, 2008; Chandler, 1998). Researchers have shown this assumption to be a good approximation for ionic solutes (Aqvist and Hansson, 1996).

However, cases for which the assumption of linear response was inappropriate have also been illustrated, such as when there is strong electrostatic coupling of the solute and solvent molecules (Ghorai and Matyushov, 2006), when local binding/unbinding events take place (Martin and Matyushov, 2008), or when significant variations in ion coordination occur with a change in ion charge (solvent density fluctuations) (Ayala and Sprik, 2008). Despite such exceptions showing that the solvent response is not strictly linear, the Marcus approximation can still be used to estimate the reorganization energy (Ayala and Sprik, 2008). However, assessing the extent to which linear response theory is satisfied, such as by analyzing the statistics of the energy fluctuations along the reaction coordinate, is an important accompanying step in many cases of modeling electron transfer reactions. In addition, analysis of dipole moment and density fluctuations allows for the estimation of their contribution to the overall entropy, which is expected to be positive (Ghorai and Matyushov, 2006).

Taken together this provides additional insight into the mechanism and driving force for electron transfer reactions. Some contributions of aqueous solvation to the reorganization energy such as H-bonding and dipole moment fluctuations cannot be fully addressed with continuum solvent models. Furthermore, because the radius of solute cavities may change as a consequence of changing the charge (electrostriction), explicit treatment of solvent can be a necessary approach. Explicit solvation based on classical molecular dynamics or Monte Carlo (MC) simulations can account for these effects. For example, to determine the PESs for an electron transfer reaction in aqueous solution, including an explicit evaluation of the solvent contributions to the free energy, one may use classical molecular dynamics simulations. To calculate the free energy curves of the reactant and product states, we define the reaction coordinate ΔE (also referred to as the polarization coordinate or energy gap), as $\Delta E = E_B - E_A$, where E_B and E_A are the potential energies of the product and reactant states, respectively. The system evolves on the reactant free energy curve when it is in the normal charge state and on the product curve when it is in the reversed charge state. The free energy function of the reactant state can be obtained from the following relationship:

$$\Delta G_R(X) = -RT \ln\left[\frac{P(X)}{P(\langle \Delta E \rangle)}\right], \tag{11.2}$$

where $\langle \Delta E \rangle$ is the ensemble average of the energy gap when the system evolves on the reactant free energy curve and $P(X)$ is the probability of finding the system in a configuration for which $\Delta E = X$. The same relationship is true for the reversed charge state.

The diabatic activation energy $\Delta G^{*\prime}$ is related to the probability of being at the transition state, that is, when the two states are degenerate and thus $\Delta E = 0$:

$$\Delta G^{*\prime} = -RT \ln\left[\frac{P(0)}{P(\langle \Delta E \rangle)}\right]. \tag{11.3}$$

Due to the large free energy barriers typically involved, the probability of being at the transition state in a simulation is exceedingly small. Thus, extremely long simulation times would be required to obtain reliable statistics in the regions of phase space where ΔE is significantly different from $\langle \Delta E \rangle$. Therefore, an alternative sampling method is necessary (Hwang and Warshel, 1987; Kerisit and Rosso, 2005; King and Warshel, 1990; Kuharski et al., 1988). For example, using umbrella sampling (see Chapter 4), the potential of the system is gradually changed from that of the reactants to that of the products to favorably bias the formation of configurations which would normally have a very low probability, enabling sampling of the reaction coordinate in the vicinity of $\Delta E = 0$. This is achieved by varying θ from 0 to 1 in the following equation to obtain a new potential from a linear combination of the reactant and product state potentials:

$$V_\theta = (1-\theta)V_A + \theta V_B. \tag{11.4}$$

The effect of the biasing potential is accounted for, after the simulation has been run, in the expression of free energy:

$$\Delta G(X) = -RT \ln\left[\frac{P(X)_\theta}{P(\langle \Delta E \rangle)}\right] - \theta X + C, \tag{11.5}$$

where C is a normalization constant and $P(X)_\theta$ is the probability of finding the system at $\Delta E = X$ when its potential is V_θ.

The approach described earlier has been used successfully to calculate the PESs for ferrous–ferric aqueous ion electron transfer (Rustad et al., 2004a) and for small polaron hopping in iron oxides (Kerisit and Rosso, 2005, 2006; Kerisit et al., 2015; Zarzycki et al., 2015). However, fast relaxation associated with solvent electronic degrees of freedom (the optical part) is available only in polarizable solvent models, which can become computationally expensive for a large system. Furthermore, careful attention must be given to the choice of force fields for the donor and acceptor species and how they change in different charge states (Kerisit and Rosso, 2005). An approach that avoids these kinds of parameterization problems is to use umbrella sampling within an AIMD scheme (Ayala and Sprik, 2008; Blumberger and Sprik, 2005; Vladimirov et al., 2008), which is the most comprehensive treatment but because of computational expense is practical only for very small systems. This method also suffers technical challenges for simulating electron transfer reactions, such as the fact that proper charge distributions required to obtain the reactant and product electron transfer states are often difficult to impose or require arbitrary parameterization. If these aspects can be managed, the AIMD approach is clearly the most optimal prospect for accuracy and completeness.

11.2.3 Activation Energies and Volumes

Experimentally observed activation energy barriers (ΔE_a) and activation volumes (ΔV_a) can provide useful clues about potential mechanisms of geochemical reactions. The clues can be vague however, so computation can be used to fill in important details. If the simulated ΔE_a and ΔV_a are similar to observed values, then one may have attained a reasonable model of the reaction mechanism *consistent with experiment*. One must not be tempted to state that he or she has determined *the* reaction mechanism because the previously stated agreement is not proof, merely one successful test of the hypothesized mechanism.

11.2.3.1 Reaction Pathways and Constrained Energy Minimizations and MD

According to the Arrhenius equation,

$$k = A\exp\left(\frac{-\Delta E_a}{RT}\right), \tag{11.6}$$

the rate constant, k, depends exponentially on the activation energy, ΔE_a. Hence, small errors in ΔE_a translate into order of magnitude errors in k. For example, assuming $A = 100\ \text{s}^{-1}$ and $\Delta E_a = 100\ \text{kJ/mol}$,

$$k = 100 \ \text{s}^{-1} \exp\left[\frac{-100 \ \text{kJ/mol}}{(8.314 \times 10^{-3} \ \text{kJ/mol/K})(298 \ \text{K})}\right] = 3 \times 10^{-18} \ \text{s}^{-1}$$

but

$$k = 100 \ \text{s}^{-1} \exp\left[\frac{-110 \ \text{kJ/mol}}{(8.314 \times 10^{-3} \ \text{kJ/mol/K})(298 \ \text{K})}\right] = 5 \times 10^{-20} \ \text{s}^{-1}.$$

Many DFT methods will have discrepancies with experiment on the order of ± 10 kJ/mol for complex aqueous or oxide systems even for the stable reactant and product states. Accurate determination of the TS structure and energy is more difficult. Given this situation, it may seem futile to attempt to simulate reaction paths and ΔE_a, but there can still be useful information gained in reaction mechanism modeling.

First, in most reactions there is likely to be a small number of probable reaction mechanisms. If one can determine which of these has the lowest ΔE_a that is in best agreement with experiment, then a reasonable approximation for the reaction mechanism can be posited. Second, in many cases, an estimate of k with two orders of magnitude may be sufficient to answer a given question. In the aforementioned example, it may not matter if k is 10^{-18} or 10^{-20} s^{-1}. The 10^{-18} s^{-1} value translates into a time frame of 10^{10} years, so for a given time frame of a geologic process, whether the reaction takes 10^8 or 10^{12} years does not matter because the reaction can be considered infinitely slow for many purposes. Third, T and P conditions may be simulated that are beyond the reach of experiment to predict trends in rate constants. For instance, Mookherjee and Stixrude (2006) computed the ΔE_a for H$^+$-transfer between sites in brucite (Mg(OH)$_2$) as a function of pressure. They predicted that ΔE_a increases with pressure causing a transition from H$^+$ dynamic to static disorder at high pressure.

Other examples of the utility of reaction path and ΔE_a calculations in low T and P geochemistry are found in Kwon et al. (2013) and Shi et al. (2013). Kwon et al. (2013) calculated the energy barriers of various transition metal migrations between TCS (interlayer triple-corner-sharing inner-sphere surface complexes) and INC (vacancy site) and found that the ΔE_a decreased by 70 kJ/mol when the number of H$^+$ bonded in a vacancy decreased to a fully deprotonated condition. This is consistent with the observation that the INC/(TS + INC) ratio for Ni^{2+} increases from pH 4 to 7 (Manceau et al., 2007; Peacock and Sherman, 2007; Pena et al., 2010). Shi et al. (2013) calculated the activation Gibbs free energy, ΔG_a, to estimate the logarithm of the water-exchange rate constant, $\log(k_{ex})$. By obtaining $\log(k_{ex})$ values for various Al–salicylate complexes, they were able to determine that the water exchange at the Al^{3+} ion was more consistent with the model value for the 1:1 bidentate complex ($\log(k_{ex}) = 4.0$ vs. experiment = 3.7 (Sullivan et al., 1999)) than for the 1:1 monodentate complex ($\log(k_{ex}) = 1.3$–1.9).

The activation volume, ΔV_a, can also be calculated and can be used to distinguish among various water-exchange reaction mechanisms such as associative, dissociative, or interchange (Rotzinger (2005) and references therein). Calculation of the reactant, transition state, and product volumes is relatively straightforward. In quantum calculations, the volume can be defined as some electron density contour (i.e., where the number of electrons per unit volume falls below a specified value such as 0.01 e$^-$/bohr^3). Another method for calculating ΔV_a from MD simulations is to run simulations as a function of pressure (at constant T) and use the equation

$$\ln(k) = (\Delta V_a)\left(-\frac{P}{RT}\right) + (\ln A_2) \tag{11.7}$$

then calculate the slope of a plot of $\ln(k)$ versus P. Kubicki and Lasaga (1988) used this method to obtain an activation volume for Si diffusion (D_{Si}) in SiO_2 melts that reproduced the anomalously negative ΔV_a observed for D_{Si} in silicate melts (Kushiro, 1983).

Of course, this level of uncertainty will not always be acceptable, so research efforts have and will continue to focus on improving TS structure and energy calculations. This entails the aforementioned efforts to more accurately calculate molecular structures and energies as discussed previously and in Chapter 1 in this volume with special emphasis on diffuse electron density, improved methods for transition state searches, and more realistic modeling of the environment in which the reaction occurs.

For electron transfer reactions the activation energy ΔG^* is based on Marcus' single excess-electron description of an electron transfer process (May and Kuhn, 2003), occurring at a transition state nuclear configuration where the reactant and product localized electronic states are degenerate (Figure 11.3). At sufficiently close distances, due to donor–acceptor orbital interaction, the wave functions of the two states are coupled, and their coupling reduces the activation energy for electron transfer. The coupling is quantitatively described by the electronic coupling matrix element V_{AB}, which is the energy difference between the hypothetical zero-interaction (diabatic) crossing point of the Marcus parabolas and the adiabatic energy surface. The electronic coupling matrix element may be obtained using the general formula (Brunschwig et al., 1980; Farazdel et al., 1990; Newton, 1980):

$$V_{AB} = \frac{\left| H_{AB} - \frac{1}{2}S_{AB}(H_{AA} + H_{BB}) \right|}{1 - S_{AB}^2}, \quad \text{where } H_{ij} = \langle \psi_i | H | \psi_j \rangle, S_{ij} = \langle \psi_i | \psi_j \rangle \quad (11.8)$$

where H and S are the interaction energy and overlap, respectively, between the reactants and products states. This approach has been implemented in, for example, the free computational chemistry software package NWChem.

The electron transfer rate constant has different expressions depending on whether the charge transfer is adiabatic ($V_{AB} \geq kT$), in which case the system is treated as evolving on a single double-well PES and the rate is Arrhenius-like, or diabatic ($V_{AB} \ll kT$), where weak coupling dominates the rate. For adiabatic electron transfer the rate expression is similar to Equation 11.6:

$$k_{et} = n v_n \exp\left(\frac{-\Delta G^*}{k_B T} \right), \quad (11.9)$$

where v_n is a typical frequency for nuclear motion along the reaction coordinate q, n is the multiplicity of pathways to equivalent electron transfer products, and the activation energy ΔG^* is given by:

$$\Delta G^* = \frac{(\lambda + \Delta G^\circ)^2}{4\lambda} - V_{AB} \quad (11.10)$$

For weak coupling the rate is given by

$$k_{et} = \frac{2\pi}{\hbar} |V_{AB}|^2 \frac{1}{\sqrt{4\pi\lambda k_B T}} \exp\left[-\frac{(\Delta G^\circ + \lambda)^2}{4\lambda k_B T} \right]. \quad (11.11)$$

11.2.3.2   *Vibrational Frequencies and ZPE*

Another equation for obtaining the rate constant is based on the Gibbs free energy change, ΔG_a, of reaction

$$k = \left(\frac{k_B T}{\kappa h}\right) \exp\left(\frac{-\Delta G_a}{RT}\right) \qquad (11.12)$$

where k_B is Boltzmann's constant, κ is a transmission coefficient, and h is Planck's constant. A critical factor in estimating ΔG_a or ΔE_a is calculating the vibrational frequencies and zero-point vibrational energy (ZPE; (Felipe et al., 2001)). For example, Ma et al. (2008) found that the ZPE is the dominant controlling factor on the temperature dependence of H and C isotope fractionation in methane (see also Chapter 7). Because ZPE is the primary cause for normal mass-dependent isotope effects, predicting equilibrium and KIE effects depends strongly on accurate determination of frequencies (Domagal-Goldman and Kubicki, 2008; Eiler et al., 2014; Felipe et al., 2001, 2003, 2004, 2005; Schauble et al., 2001; Zeebe, 2010).

Accurate computational methods for vibrational frequencies are discussed in detail in Chapter 10 of this volume, so a detailed discussion is not presented here. The methodology is the same for stationary and transitions states, but like most properties, the transition state vibrational frequencies are particularly sensitive to errors in the basis set, exchange–correlation functionals, etc. This is exacerbated by the fact that vibrational spectra of the transition state complex are not generally available for comparison.

Determining the rate of adiabatic electron transfer requires knowing the frequency of motion along the reaction path. Different approaches have been used to determine the nuclear frequency ν_n. For instance, for Fe^{2+} electron transfer to Fe^{3+}, either as hydrated ions in aqueous solution or as cations in an iron oxide crystal, a large component of the nuclear reaction coordinate is contraction of the first coordination shell Fe—O distances in the donor Fe^{2+} site and expansion of the same in the acceptor Fe^{3+}. The most relevant frequencies are those associated with these site volume oscillation modes. By volume oscillation, we mean the compression of the volume of the donor species and the simultaneous decompression of that of the acceptor (e.g., a breathing mode vibration). Thus, in the study of polaronic hopping processes in iron oxides, ν_n could be approximated by the highest infrared active longitudinal optic mode phonon (Farmer, 1974). For the same system in solution, the average frequency of the Fe—O stretching mode could be chosen (Kerisit and Rosso, 2005). Unfortunately, neither the experimental phonon spectrum nor the isolated vibrational modes strictly correspond to the vibrational motion along the reaction path. In particular, none of the iron dimer harmonic modes correspond exactly to the asymmetric stretching of the Fe—O bonds in the two monomers. More completely, ν_n should be taken as the sum of the weighted contributions of all normal modes that contribute to taking the system from the reactants to the transition state geometry (Marcus and Sutin, 1985).

From molecular dynamics simulations of the electron transfer PESs, such as using umbrella sampling as described previously, one can "work backward" to compute ν_n for dominant vibrational motions. The parabolic shape of the PESs is a consequence of the underlying harmonic constraints for the free energy, so a harmonic oscillator model is justified. Although the volume oscillation example should be considered as a three-dimensional oscillator, we can reduce it to a one-dimensional problem by focusing on one chosen geometry descriptor (e.g., the Fe—O distance). From computed PESs one already has the potential curve associated with this vibrational motion, so the force constant and frequency can be determined from Hooke's law ($k = d^2 V(r_{Fe-O})/dr^2_{Fe-O|eq}$). This harmonic analysis corresponds to a temperature of 0 K, but assuming a lack of correlation with the other intramolecular motions, it can be scaled to room temperature.

This approach also allows us to look into the effects of isotopic mass on the ET rate, which is important for understanding isotopic fractionations arising from geochemical redox reactions. For example, in the Fe^{2+}/Fe^{3+} donor–acceptor example, the effects of changing the Fe isotopes on the enthalpy, entropy, and frequency of motion along the reaction coordinate are relatively large.

If both Fe atoms are substituted by heavier isotopes, a shift in ν_n on the order of a wavenumber is found (Zarzycki et al., 2011). The associated isotopic effect can be assessed using the reduced partition function ratio (β) (Rustad and Zarzycki, 2008), which will generally show that the Fe encounter complexes comprised of heavier isotopes have a lower propensity for electron transfer, or in other words the heavier Fe isotopes will remain in their oxidation states longer than lighter ones. Computational molecular modeling of redox-based isotopic signatures is an important and growing area under development.

11.2.4 Transition States and Imaginary Frequencies

In classical transition state theory, the transition state corresponds to a saddle point on the PES (Figure 11.1 (Felipe et al., 2001)). By definition, the saddle point has only one negative second derivative, and the second derivatives of the PES are related to vibrational frequencies (Chapter 10). Hence, the TS should have one and only one negative vibrational frequency associated with it. The vibrational mode giving rise to this negative or imaginary (because the square root of a negative number is imaginary) frequency should correspond to atomic motions that lead toward reactants and products (Figure 11.4). This is a necessary condition for a TS, but it is not sufficient. A model configuration at the top of the PES between reactants and products with one and only one imaginary frequency can be claimed to be an approximation to the actual TS, but these qualities do not prove

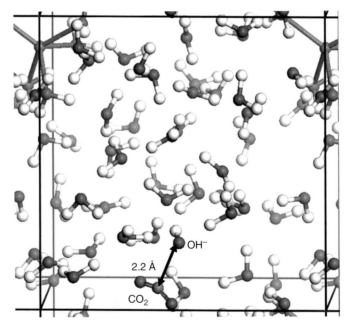

Figure 11.4 *The imaginary mode (i.e., the vibrational mode associated with a negative frequency) of the reaction $CO_2 + OH^- \rightarrow HCO_3^-$ is illustrated. A configuration from the highest energy point along the reaction pathway defined by the $C \cdots O$ distance was subjected to a frequency analysis. The presence of one and only one imaginary mode at the highest point along the reaction coordinate helps to identify the configuration as a transition state. In addition, the atomic motions defined by this mode should lead back and forth between the reactants and products. In this example, as the $C \cdots OH^-$ distance extends and contracts, the products, HCO_3^-, and reactants ($CO_2 + OH^-$) are formed, respectively. Molecules drawn with Materials Studio 7 (Accelrys Inc., San Diego, CA). (C, gray; H, white; Na, purple; O, green).*

one has determined the actual TS. By comparing calculated versus observed ΔE_a's and/or KIE, the accuracy of the model TS can be judged more stringently (Felipe et al., 2005).

As mentioned in Section 11.2.1, Gallet et al. (2012) have performed a rigorous analysis of the gas-phase $nH_2O + CO_2 \rightarrow H_2CO_3 \cdot (n - 1)H_2O$ reactions. A number of important advances were included in this study. The use of MTD allowed for a statistically more robust determination of the TS. The use of flexible collective variables removes the constraints imposed by methods such as constrained energy minimizations or MD simulations with atomic distances defining the reaction coordinate frozen. Furthermore, the collective variables were validated using committor probabilities to determine whether configurations along the reaction coordinate belong to reactant or product states. Lastly, the activation entropy, ΔS_a, was determined using

$$F_\Omega = -k_B T \ln \frac{\int_\Omega \exp(-\beta F(s,z))dsdz}{\int \exp(-\beta F(s,z))dsdz} \tag{11.13}$$

$$H_\Omega = -k_B T \ln \frac{\int_\Omega E_{DFT}(s,z)\exp(-\beta F(s,z))dsdz}{\int \exp(-\beta F(s,z))dsdz} \tag{11.14}$$

and

$$T\Delta S_\Omega = \Delta H_\Omega - \Delta F_\Omega \tag{11.15}$$

where F is the Helmholtz free energy, $\beta = 1/k_B T$, and E_{DFT} is the total energy. This method removes the assumption that the vibrational, rotational, and translational contributions to the barrier entropy can be calculated separately and the harmonic approximation applied to the vibrational entropy.

11.2.5   Rate Constants

11.2.5.1   Calculating k Using TST and Variational TST

From Equation 11.6, one can see that after calculating ΔE_a, determining the preexponential factor A is necessary to calculate a rate constant. Felipe et al. (2001) provide a commonly used expression:

$$k = \left(\frac{k_B T}{h}\right) \left(\frac{Q_{TS}}{Q_{Reactants} Q_{Products}}\right) \exp\left(\frac{-\Delta E_a}{k_B T}\right) \tag{11.16}$$

where ΔE_a is the zero-point corrected energy difference between reactants and TS. Q represents the partition functions for vibrations, rotations, and translations. When the rigid-rotor/harmonic oscillator approximation is assumed, these Q can be separated into q_{vib}, q_{rot}, and q_{trans}. (Note: The nuclear partition function can usually be ignored and the electronic partition function is folded into ΔE_a.) Equations for partition functions can be found in statistical mechanics textbooks and in Chapter 3. Hence,

$$A = \left(\frac{k_B T}{h}\right) \left(\frac{Q_{TS}}{Q_{Reactants} Q_{Products}}\right) \tag{11.17}$$

Note that $A = (k_B T/h)\exp(\Delta S_a/RT)$ as well. Campbell et al. (2013) have cautioned that the TST approach for calculating A based on DFT calculations may overestimate the preexponential factor by 3 orders of magnitude for molecular adsorption. Via empirical derivation of ΔS_a values from desorption rates of adsorbates with residence times of less than 1000 s, Campbell et al. (2013) found that the A was much smaller and that the DFT-derived ΔS_a were too small (177 vs. 130 J/mol/K).

11.2.5.2   Tunneling

An issue of concern in many geochemical reactions is quantum tunneling. This effect is more important for lighter nuclei such as H and Li. Reactions involving H^+-transfer or H–D exchange are of special concern. A common method for estimating quantum tunneling effects is the Wigner treatment (Truhlar et al., 1982):

$$\kappa = 1 + \left(\frac{1}{24}\right)\left|\frac{h\nu^*}{2\pi k_B T}\right|^2 \tag{11.18}$$

where κ is the correction factor and ν^* is the imaginary frequency associated with the TS, but this approximation is valid only when the T is high relative to ΔE_a. Another method is to calculate the rate constant with a Marcus-like model (see Roston et al. (2013) and references therein):

$$k = \left(\frac{|V|^2}{\hbar}\right)\left(\frac{\pi}{\lambda k_B T}\right)\exp\left(\frac{-(\Delta G^\circ + \lambda)^2}{4 k_B T \lambda}\right)\int_0^\infty F(m, \text{DAD})\exp\left(\frac{-E(\text{DAD})}{k_B T}\right)d\text{DAD} \tag{11.19}$$

where V is the reactant/product electron coupling, λ the reorganization energy, and ΔG° the Gibbs free energy change of the reaction. DAD is the donor–acceptor distance. The $\exp(-E(\text{DAD})/k_B T)$ term is the Boltzmann distribution probability of a given DAD existing. The main assumption in this approach is that the heavy and light atom motion can be treated separately. Roston et al. (2013) suggest that observed KIE that cannot be explained via normal fractionation can be caused by tunneling effects.

11.2.5.3   Relating Rate Constants to Rates of Reactions

The novice computational geochemist should be warned not to conflate the rate constant, k, with the rate of reaction, R. This is worth mentioning here because geochemical kinetics is not heavily emphasized in geology curricula and because the latter has been substituted for the former in some early geochemical kinetics papers (see Casey and Sposito (1992) for a discussion of this problem). Generally, the rate of reaction is expressed as

$$R = k[A]^\alpha [B]^\beta .... [Z]^\gamma \tag{11.20}$$

where the "[...]" symbols indicate a reactant, product, or reactive intermediate concentration and the α, β, ... γ symbols are the exponential rate dependence (i.e., rate order) of the reaction with respect to the given species (see Brantley and Conrad (2008) for a thorough discussion of rate laws in geochemistry). Most papers on geochemical kinetics do not derive a complete rate law; instead a conditional rate constant is determined as a function of one or more components. The rate constants determined in this manner are dependent upon the experimental conditions, so comparing the experimentally reported k to computationally derived k for an elementary reaction is not appropriate. The recommended approach is to search for reactions that have complete rate law determinations or to perform such experiments in conjunction with simulation of the reaction path.

The reader is reminded that k is calculated based on the model ΔE_a or ΔG_* (Eqs. 11.6 and 11.9), both of which have significant errors. Due to the exponential dependence of k on these parameters, it is difficult to calculate k extremely accurately as mentioned previously. Often studies of rates in low-temperature geochemistry do not determine ΔE_a values or they are determined inaccurately due to temperature constraints. This situation makes it problematic to verify model reaction mechanisms and rate constants. One can distinguish among probable pathways, however, as long as the modeler keeps the level of uncertainty in the calculations in mind.

Another problem is that many geochemical reactions occur in numerous elementary steps. The computational geochemist must either model all possible steps or intuit which step or steps may be the rate-controlling step in a series of elementary reactions. Thus, even when experimental ΔE_a values are available, unless there is one clear rate-controlling elementary step in a reaction, the observed ΔE_a could be the result of numerous activation energy barriers during the overall reaction (Fischer et al., 2012).

For an electron transfer reaction in the simplest case involving a single electron transfer between a donor and acceptor in solution, macroscopically observable rates are based on the following basic steps:

1. Diffusion of the reactants together to form the precursor complex
2. Electron transfer within the precursor complex to form the successor complex
3. Dissociation of the successor complex

Using the steady-state approximation and assuming the reaction (Equation 11.20) is not diffusion limited and that dissociation is fast, the net rate that is usually observed in experiment is due only to the equilibrium constant for the formation of the precursor complex (K_{pre}) and the rate of electron transfer (k_{et}):

$$k_{obs} = K_{pre}k_{et} \tag{11.21}$$

where K_{pre} can be modeled as

$$K_{pre} = 4\pi N R^2 dR \exp\left(\frac{-w}{RT}\right)$$

where N is Avogadro's number, R is the effective separation of the reactants at their closest approach, dR is the effective reaction zone thickness, and w is the electrostatic work to bring the reactants together. The electrostatic work depends on the charges of the reactants, the separation of the reactants R, and the dielectric properties of the solvent corrected for ionic strength. For low ionic strength, Debye–Hückel theory provides flexibility and sufficient accuracy for modeling the work term as

$$w = \left(\frac{1}{4\pi\epsilon_0}\right)Z_1 Z_2 e^2 N / \epsilon_s R_m \left(1 + BR_{\text{Å}}\mu^{1/2}\right) \tag{11.22}$$

where Z_1 and Z_2 are the charges on the donor and acceptor ions, the factor $1/4\pi\epsilon_0$ equals 8.988×10^9 Nm^2/C^2, R is expressed in meters, B is the Debye–Hückel parameter for R expressed in angstroms, and μ is the ionic strength (Rosso and Rustad, 2000).

11.2.6 Types of Reaction Mechanisms

Much of the previous discussion has focused on the reasons why one would be interested in obtaining models of the reaction mechanism and methods for simulating the reaction pathway and rate. This section discusses some of the basic reaction mechanism types one can expect to obtain.

Two common reaction types are based on nucleophilic substitution reactions. "Nucleophilic substitution" means an electronegative species attraction to a positive center replaces a weaker electronegative species. These reaction mechanisms are designated S_N1 and S_N2 for first-order and second-order nucleophilic substitution. S_N1 reactions depend on the concentration of one reactant and its ability to form and reactive intermediate (i.e., transition state, TS), whereas S_N2 reactions depend on the concentration of two reactants (a bimolecular reaction) to form a TS. The first-order reaction S_N1 means that the rate law only depends on the reactant containing the original nucleophile (i.e., $R = k[A]$). For example, if the reaction – $Si(OH)_4 + F^- \rightarrow Si(OH)_3F + OH^-$ (Figure 11.1) only depended upon the concentration of $Si(OH)_4$, this would be S_N1. This first-order dependence occurs when the Si—(OH) bond dissociates before the F^- replaces it. On the other hand, the reaction $CH_3F + OH^- \rightarrow CH_3OH + F^-$ is S_N2 because the OH^- begins to bond to the C atoms as the C—F bond breaks, producing a $CH_3F(OH)^-$ TS, and making it an S_N2 reaction (Chen et al., 2014). For this example of an S_N2 reaction, the second-order rate law would be $R = k[CH_3F][OH^-]$. Related types of reactions occur in aqueous solution termed associative (A), dissociative (D), and interchange (I). "I" indicates that a reactive intermediate species forms. The I reactions can be further classified as having associative (I_a) or dissociative (I_d) TS (Helm and Merbach, 2005). Stack et al. (2005) and Qian et al. (2010) modeled H_2O exchange on an Al polyoxocation and Mg^{2+}, respectively, and concluded an I_d mechanism was prevalent based on the calculated activation volume, ΔV_a, but this method of calculating aqueous reaction mechanisms is controversial due to the involvement of H_2O molecules with reactant, TS, and product (W.H. Casey, personal communication).

Another critical reaction mechanism is proton-coupled electron transfer (PCET). Many electron transfer reactions involve a concomitant rearrangement of atoms as the system relaxes as a consequence of the new (electron transfer product) charge distribution. In aqueous systems in particular changes in proton distribution often accompany electron transfer (Alexandrov and Rosso, 2015; Mayer, 2004). The normal situation is that while an electron transfer can cause proton transfer, the electron transfer process itself is not dependent upon it. However, cases exist where these two processes are coupled, and the coupling must be accounted for to properly model the kinetics. In the limit of PCET is when a reaction occurs via an H radical transfer (e.g., Kubicki et al. (2009)). This should be considered an extreme case because the H^+ and e^- are 100% coupled and transferred as a unit. Much of geochemistry occurs where aqueous solutions are present, so free radical mechanisms are less important than ion-based mechanisms. However, in situations where water is limited, such as in petroleum chemistry (Chapter 7) and organic matter-rich soils (Chapter 6), free radical mechanisms should not be ignored. Descriptions of addition, elimination, substitution, and rearrangement reactions can be found in organic chemistry textbooks.

11.3 Applications

11.3.1 Diffusion

Diffusion is an important geochemical process even though it operates on limited spatial scales. Concentration gradients frozen into melts and glasses can be used to estimate diffusion coefficient (D) values and constrain time frames of geologic processes (e.g., Cruz-Uribe et al. (2014)). Computational techniques are used to estimate D outside the realm of experimentally accessible P and T conditions and to provide model diffusion mechanisms.

11.3.1.1 Melts and Aqueous Solutions

One of the first applications of molecular modeling in geochemical kinetics focused on diffusion in melts (Angell et al., 1982, 1987; Kubicki and Lasaga, 1988; Rustad et al., 1990; Woodcock et al., 1976). More recently, classical MD simulations have been applied to cation diffusion in aqueous solutions and thermal diffusion in melts (Lacks et al., 2012). With increasingly available computer power and more efficient DFT codes, DFT–MD simulations have become possible.

Karki and coworkers (Karki and Stixrude, 2010; Karki et al., 2013; Lacks et al., 2012; Verma et al., 2012) have applied DFT–MD to model diffusion and viscosity of silicate melts. In Karki and Stixrude (2010) the effect of H_2O on diffusion in SiO_2 melts was examined. This study found that addition of H_2O ($20SiO_2 + 6H_2O$ vs. $24SiO_2$ for the anhydrous model melt) increased diffusion coefficients and decreased viscosity by an order of magnitude. They also predicted anomalous pressure dependence of Si diffusion in silica melts and explained this observation due to the formation of fivefold coordinate Si consistent with the experimental observation of Kushiro (1983) and earlier classical MD simulations (Angell et al., 1982; Kubicki and Lasaga, 1988). In Karki et al. (2013), the viscosity changes of melts along the $MgO–SiO_2$ join were predicted as a function of temperature and pressure. As expected, viscosity decreases significantly with MgO content due to increasing configurational entropy. Limitations of DFT–MD for modeling liquid-state diffusion are the number of atoms that can be treated in the simulation cell and the duration of the simulation. The short-duration (i.e., 1 ns or less) force simulation temperatures to be much higher than typical geophysical conditions, except for perhaps during bolide impacts (Karki et al., 2013). Approaches toward stretching the current limits are discussed in the Section 11.4.

11.3.1.2 Minerals

DFT studies of diffusion in solids are more common (e.g., Ammann et al., 2009, 2010, 2011, 2012; Saadoune and de Leeuw, 2009; Verma and Karki, 2009a, b). Solid-state diffusion in minerals can be difficult to study because it is a slow process except at high temperatures. Smaller simulation cells can be used for diffusion in minerals because one is attempting to calculate the energy barrier and hopping frequency between an occupied and a vacant site. If the mineral structure and vacancy type and concentration are known (e.g., Stashans and Flores, 2013; Umemoto et al., 2011), then the system is better constrained than for the cases of aqueous solutions and silicate melts where the disordered nature and large number of configurations make DFT calculations more time-consuming. For example, Ammann et al. (2011) used DFT to model Fe diffusion across the high- to low-spin transition for ferropericlase in order to estimate the effect of this transition on mantle viscosities. These authors used 64 and 108 atom simulation cells with $2 \times 2 \times 2$ k-point sampling and 1000 eV cutoff energies. For comparison, liquid-state studies such as Karki et al. (2013) use simulation cells of 72–228 atoms with 1 k-point (the Γ point in the Brillouin zone) and 400 eV cutoff energies. However, Ammann et al. (2011) were able to use harmonic transition state theory and the climbing-image nudged elastic band method (ci-NEB (Henkelman et al., 2000)) rather than running MD simulations for a million time steps. Furthermore, direct comparison to diffusion constants as a function of temperature is simpler because the calculations do not need to be run at overheated temperatures. At lower temperatures, DFT–MD can be applied to model diffusion of H_2O through minerals as well. For example, Molina-Montes et al. (2013) used CPMD metadynamics to simulate dehydration of pyrophyllite.

11.3.2 Ligand Exchange Aqueous Complexes

Water or ligand exchange reactions for aqueous solutes have been investigated using computational chemistry methods in a variety of studies (e.g., Jin et al., 2011a, b; Morrow et al., 2010; Panasci et al.,

2012; Shi et al., 2013). A recent study by Amaro-Estrada et al. (2014) examined the aqueous chemistry of the environmentally important element Hg. These authors used MP2 (Moller and Plesset, 1934), the B3PW91 exchange–correlation functional (Becke, 1993; Lee et al., 1988), Stuttgart–Köln relativistic effective core potentials (Bergner et al., 1993; Kuchle et al., 1991) for Hg and Cl, and the 6-311G** basis set (Francl et al., 1982; Frisch et al., 1984; Krishnan et al., 1980; McLean and Chandler, 1980) for H and O in the Gaussian 09 program (Gaussian Inc.) to model HgClOH—$(H_2O)_n$ clusters. Born–Oppenheimer MD (BO-MD) simulations were carried out using the Geraldyn2.1 code (Raynaud et al., 2004). These authors note that the 10 ps BO-MD simulations required 58 CPU days using 32 processors. These simulations predict the thermal stability of the HgClOH complex, systematically examine explicit and implicit solvation effects, and predict the rate of H_2O exchange to be approximately $10^{11}\,s^{-1}$.

11.3.3 Adsorption

Numerous studies have employed DFT to model surface complexes (Kubicki, 2005; Roques et al., 2009; Tan et al., 2010), but surprisingly few have attempted to model adsorption/desorption reaction mechanisms in geochemistry (He et al., 2011; Watts et al., 2014). He et al. (2011) modeled arsenate ($H_2AsO_4^-$) sorption kinetics on anatase (100) using a $[Ti_4O_{18}H_{24}]^{4+}$ cluster. IEFPCM was used to model solvation effects, and constrained energy minimizations were used to determine the reaction pathway. Activation energy barriers from an H-bonded to a monodentate and from monodentate to bidentate complexes were +102 and +55 kJ/mol, respectively. The authors concluded that the bidentate complex was most thermodynamically stable but that it would form via a monodentate intermediate. These are fairly large energy barriers (especially the +102 kJ/mol), so the predicted adsorption rate should be slow, but the authors do not calculate a rate constant or rate based on their model. The high charge (+4) on the cluster will have a strong influence on the calculated adsorption energy because $H_2AsO_4^-$ will be overstabilized compared to a less highly charged cluster (Kwon and Kubicki, 2004). Another concern with the models used in He et al. (2011) is that there were no H_2O molecules of solvation other than two H_2O molecules liberated by exchange of $H_2AsO_4^-$ in a bidentate configuration with 2 Ti—OH_2 groups. The IEFPCM model in this case cannot adequately mimic solvation of the oxyanion, and it also calculates solvation on the bottom of the $[Ti_4O_{18}H_{24}]^{4+}$ cluster which would not occur on a real surface (Bandura et al., 2004).

Watts et al. (2014) modeled adsorption/desorption from goethite (α-FeOOH) using a neutral cluster and H_2O molecules of solvation around the $HAsO_4^{2-}$ ion as well as periodic DFT calculations. Based on energy-minimized model surface complexes for a monodentate configuration, the Fe···As distance was increased incrementally and the other atoms were allowed to relax. In the periodic DFT calculations, short (i.e., 3 ps) MD simulations at 300 K were performed at each Fe···As distance followed by energy minimizations. The cluster and periodic models were in disagreement with respect to their calculated activation energy barriers. Even the smaller periodic DFT activation energy barrier of +70 kJ/mol was considered too large in this case compared to experiment. The discrepancy was hypothesized to be due to arsenate adsorption onto the (210) surface of goethite rather than the (010) surface modeled in this study.

11.3.4 Dissolution

Application of computational chemistry to dissolution of minerals has a relatively long history in this field starting with Casey et al. (1990). As discussed in Section 11.1, numerous papers have been published based on model systems of increasing complexity. Recent studies have begun to examine the role of the silicate–water interface in dissolution kinetics. In one case, Nangia and Garrison (2010a) suggested that on cristobalite surfaces intrasurface H-bonding of Q^2 terrace groups inhibits dissolution and that dissolution at non-H-bonded Q^3 sites is preferred site (Washton et al., 2008). In another study, Kubicki et al. (2012a) hypothesized that the addition of salts to the solution favors intrasurface H-bonding on the (101) surface of α-quartz which promotes H^+-transfer to bridging O atoms and dissolution (Dove and Nix, 1997).

A series of papers by Nangia and Garrison developed rate constants for Si—O—Si hydrolysis as a function of pH predicted by quantum mechanics (Nangia and Garrison, 2009a; Nangia et al., 2007) and then used these rate constants to perform MC simulations of silica dissolution at larger scales (Nangia and Garrison, 2009b, 2010b). This approach using the combined reactive MC and configurational bias MD (RxMC–CBMC) was able to reproduce the experimental trend of silica dissolution rate (Knauss and Wolery, 1988) as a function of pH (Nangia and Garrison, 2010b). Kundin et al. (2010) have used a similar approach to model the effect of oxalic acid on dissolution of Ca—Mg silicate glasses.

Luttge and coworkers (e.g., Kurganskaya and Luttge, 2013a, b and references therein) have combined vertical scanning interferometry (VSI) with KMC simulations of mineral dissolution. VSIM and atomic force microscopy (AFM) are capable of observing growth rates of etch pits that control mineral dissolution in many cases, especially quartz and calcite (Dove and Han, 2007). Kinetic MC simulations can be complex, such as RxMC–CBMC mentioned earlier, but the essence of the method is to predict the probability (P) of an event (i.e., an elementary reaction) occurring using a Boltzmann distribution:

$$P = \exp\left(\frac{-\Delta E_\mathrm{a}}{kT}\right) \text{ or } P_n = \exp\left(\frac{-n\Delta E_\mathrm{a}}{kT}\right) \tag{11.23}$$

where ΔE_a is the activation energy of the reaction and n is the number of bonds to be broken for a given surface atom to dissolve. KMC results based on ΔE_a and k derived from quantum mechanical calculations are able to model dissolution on larger spatial scales and longer time scales. The caveat is that the ΔE_a and k need to be as accurate as possible in order to model real-time rates of mineral dissolution. If relative ΔE_a and k are reasonably accurate, then the KMC can reproduce observed mineral surface dissolution textures and provide insight into dissolution mechanisms (Kurganskaya and Luttge, 2013b). Thus, KMC and DFT can be used in tandem—DFT providing initial ΔE_a and k values for KMC and KMC providing larger-scale, longer-duration perspective upon which more realistic DFT calculations can be based.

Electron transfer reactions can be pivotal in the dissolution of metal oxides such as hematite (α-Fe_2O_3) and birnessite (MnO_2) because the reduced forms of the metal cations are more soluble than the oxidized forms (e.g., Fe^{2+} vs. Fe^{3+} and Mn^{2+} vs. Mn^{4+}). Note that oxidative dissolution is also possible for phases such as metal sulfides (e.g., pyrite FeS_2) and uraninite (UO_{2-x}). Reductive dissolution often occurs mediated by dissimilatory Fe-reducing bacteria. Organic compounds can also adsorb to Fe-hydroxide surfaces and transfer electrons that lead to dissolution (Debnath et al., 2010). In the presence of sunlight, solar radiation can also provide the energy necessary to overcome the ΔE_a for reduction and dissolution (Kwon et al., 2009). In Kwon et al. (2009) the role of Mn vacancies in birnessite on reducing the band gap and increasing the photoactivity of the mineral for producing electrons to reduce Mn^{4+} to Mn^{2+} was explored. This example is rare in geochemistry because it explores photochemistry which is typically ignored because much of geochemistry occurs in the absence of sunlight. Atmospheric and surface water reactions may be strongly influenced by solar radiation, especially UV light (Kubicki, 2005), so the effect of sunlight on surficial geochemical process should be studied more often than it is.

11.3.5 Nucleation

The Ostwald step rule is often observed to apply in geochemical systems such that metastable solids form before the more thermodynamically stable phases (see Torn et al. (1997) for a good example of the implications of this principle). The general principle behind this observation (i.e., the trade-off between bulk and interfacial Gibbs free energies) is well understood (Boerio-Goates et al., 2013; Navrotsky, 2011), but the details of how and why this occurs in any given system are still the subject of investigation (see Gebauer et al. (2014) and references therein for a recent review). Classical nucleation and crystal growth theories assumes that these processes occur unit-by-unit where a "unit" is an atom or molecule (Figure 11.5). However, speciation in aqueous solution is a continuum from individual dissolved ions, ion pairs and dimers (Schroedle et al., 2007; Zhu et al., 2013), molecular clusters (Gebauer et al., 2008; Helz et al., 2014), and nanoparticles (Banfield et al., 2000; Conrad et al., 2007; Hochella et al., 2008; Maurice and Hochella, 2008). Thus, nucleation and growth may occur through aggregation of oligomeric units and nanoparticles of various sizes. Computational work on dimers and molecular clusters (e.g.,; Rustad et al., 2004b; Stack et al., 2005; Wander et al., 2013; Zhu et al., 2013) and nanoparticles (e.g., Gilbert et al., 2003; Hummer et al., 2009, 2013; Kerisit et al., 2005; Kubicki et al., 2012b; Pinney et al., 2009; Zhang and Banfield, 2004) that provide insights into the nucleation process has been increasing, but few studies have tackled nucleation directly (Lee and Kubicki, 1993; Matsumoto et al., 2002).

Gale and coworkers have published a number of interesting studies in this area mainly on amorphous calcium carbonate (ACC; (Demichelis et al., 2011; Jones et al., 2008; Piana and Gale, 2006; Piana et al., 2005; Raiteri et al., 2010, 2012). The papers include analysis of prenucleation clusters that can form in classical MD simulations and take the form of ionic polymers (Demichelis et al., 2011).

Figure 11.5 *A TiO$_2$ nanoparticle with a Ti^{4+}(H$_2$O)$_6$ ion is illustrated to provide an example of the classical crystal growth theory where macroscopic crystals are assumed to grow via addition of monomers from a supersaturated solution. Recently, this theory has been shown to be incorrect in some cases where oligomers or molecular clusters form in the supersaturated solution, and these become the building blocks for crystal growth. The difference in size of the monomer and nanocluster shown is intended to provide a graphical representation of why the monomeric growth model can be problematic. Even this small nanoparticle is made up of over 100 atoms, so it becomes hard to imagine that the system is bimodally distributed between monomers and nanoparticles or critical nuclei. A distribution of molecular sizes is more reasonable in most cases.*

A key aspect of this work is that the force field has been accurately parameterized using the Generalized Utility Lattice Program (GULP; Gale, 1996, 2005; Gale and Rohl, 2003) and benchmarked against observed thermodynamic data (Raiteri and Gale, 2010). Furthermore, a flexible model for water, SPC/Fw (Wu et al., 2006), was chosen to better represent the critical H_2O–Ca^{2+}–CO_3^{2-} interactions (Raiteri and Gale, 2010). These authors concluded that incorporation of H_2O into ACC nanoparticles varies as a function of nanoparticle size which allows the nucleating entity to minimize its interfacial energy in solution (Raiteri and Gale, 2010). One area for further development is in the creation of the ACC nanoparticles themselves. Rather than being built in aqueous solution from smaller units, model calcite structures were heated to 3000 K to disrupt the crystal structure. Simulating nucleation from smaller units is now a significant challenge (see Section 11.4).

11.4 Future Challenges

The previous discussion has focused on the basics of modeling kinetics of geochemical reactions with a selective review of previous studies. In this section, possibilities for future research are explored with an eye toward including factors in the calculations that have not generally been considered in computational geochemistry. Five main issues are discussed:

1. Connection of computational chemistry with femtosecond spectroscopy to better elucidate the transition state
2. Improving H-bonding accuracy within DFT to better account for solvation effects
3. Modeling reaction mechanisms with roaming to loosen the constraints on the reaction path imposed by the modeler
4. Larger size and longer-duration quantum molecular dynamics simulations
5. Development of reactive force fields to expand the temporal and spatial scales of model systems while retaining the ability to model chemical reactions

This list is not intended to cover all potential areas for improvement, but these are some key areas that are ready to be developed and tested.

11.4.1 Femtosecond Spectroscopy

"Femtochemistry" usually combines rapid, time-dependent spectroscopic techniques with high-level quantum mechanical approaches in order to obtain information on states between reactants and products. The impact of this field was stated succinctly by Zewail (2000):

> ...the transition state, the cornerstone of reactivity, could be clocked as a molecular species $TS^{\ddagger}$, providing a real foundation to the hypothesis of Arrhenius, Eyring, and Polanyi for ephemeral species $[TS]^{\ddagger}$, and leading the way to numerous new studies. Extensions will be made to study transition-state dynamics in complex systems, but the previous virtual status of the transition state has now given way to experimental reality;....

Thus, the level of complementarity between experiment and theory previously reserved for reactants and products may now exist for transition states and reaction mechanisms. These techniques are just beginning to be applied to geochemical problems. Cohen et al. (2010) have studied the dynamics of a dye adsorbed to amorphous silica and found that adsorption affects the dynamics of the dye. The authors suggested the possibility of an excited state H^{+}-transfer between the dye and the amorphous silica. Katz et al. (2012) used pump-probe spectroscopy to study the dynamics of electrons introduced into iron(III) (oxyhydr) oxide nanoparticles via ultrafast interfacial electron transfer. Using time-resolved x-ray spectroscopy and *ab initio* calculations, they observed the formation of reduced and structurally distorted metal sites consistent with small polarons. Geochemists have interest in adsorption and H^+-transfer reactions at silicate surfaces so transferring this approach to geochemical applications is worth exploring. Mohammed et al. (2010) used femtochemistry methods to study solvation dynamics. These authors found two time scales (13 and 35–60 ps) of relaxation related to charge-transfer reactions in solution. This type of knowledge could define how long simulations would need to be to model a given reaction. Not only would this approach be applicable to aqueous reactions, but it could also be adapted to mineral–water interfaces. Yang et al. (2013) have used scanning ultrafast electron microscopy (S-UEM) to observe surface dynamics of different solvents on CdSe surfaces. S-UEM will allow for femtochemistry studies at spatial resolutions less than the diffraction limit of visible light (Yang et al., 2013) which would be extremely useful for quantifying chemical reactions at specific, reactive sites on mineral surfaces.

11.4.2 H-Bonding

If one is attempting to model solvation and mineral–water interfaces, then simulating H-bonding and H^+-transfer reactions accurately are necessities. Standard DFT methods result in significant discrepancies with experimental observations, so refining DFT methods will be a critical step forward in computational geochemistry. One way to solve this is through the use of coupled-cluster or MPn calculations, but these methods can be impractical for geochemical systems containing 10s or 100s of atoms (Boese, 2013). This means that high-level calculations on small clusters may not be adequate for bulk phases. Gillan et al. (2013) have determined the causes of these discrepancies. Commonly used functionals, such as BLYP, underestimate nonbonded forces other than H-bonding so the H-bonding interaction is overemphasized. PBE errors on the other hand are the result of problems reproducing 2-body and multibody interactions. Consequently, modeling H_2O in dimers, small clusters, and condensed phases exposes different inadequacies such that comparisons of a DFT method to experiments on dimers may not mean the results are applicable to bulk water. After the application of Gaussian approximation potentials, significant errors remain (Gillan et al., 2013).

Confounding the H-bonding problem is the fact that van der Waals forces are in effect over similar intermolecular distances. This means that deconvoluting discrepancies between DFT and observation into vdW and H-bonding terms is not trivial. Methods for adding dispersion corrections to DFT have been developed (Dahlke et al., 2008; Dion et al., 2004; Hujo and Grimme, 2011; Xu and Goddard, 2004; Zhao and Truhlar, 2008), but if these methods are parameterized in a manner that compensates for H-bonding errors, then it is not clear how the methods will function under all circumstances. These contributions to the energy are particularly critical for transition states (DiLabio et al., 2013). Recent progress in this area using dispersion-correcting potentials (DCPs) is encouraging. DiLabio and Koleini (2014) have shown improvement in nonbonded and covalent bond enthalpy results using LC-ωPBE/6-31+G(2d,2p) calculations to obtain mean absolute errors of 6.7 kJ/mol.

11.4.3 Roaming

Reactions occurring through a "roaming" mechanism (Townsend et al., 2004) have been associated with photochemical processes. These unimolecular dissociation reactions are alternative pathways to those obtained through conventional transition state theory (Andrews et al., 2013). Studies of reactions that occur via roaming mechanisms are commonly found in atmospheric applications (e.g., Grubb et al., 2012; North, 2011). Because high-level quantum calculations such as CASSCF are necessary to model these reactions, systems of larger number of atoms have not been studied. Exploration of roaming as an alternative to TST and VTST is encouraged here for two reasons. First, computational geochemists have not yet worked extensively in the area of photochemistry.

This is because the calculations and theory are complex and because most geochemistry occurs without light being a significant factor. However, the role of mineral dusts as atmospheric aerosols is critical to atmospheric chemistry and climate change (Choobari et al., 2014). Some studies have begun to work in this area (Baltrusaitis and Grassian, 2010; Baltrusaitis et al., 2007a, b; Pierre-Louis et al., 2013; Tribe et al., 2012) and few have included photochemistry (Nanayakkara et al., 2014; Rubasinghege et al., 2010). The second reason is that roaming may be thermally activated as well as photochemically activated especially for alkanes (Harding and Klippenstein, 2010). Hence, high-temperature and organic geochemists should consider this alternate pathway to reaction.

Another approach is transition path sampling (Kerisit and Rosso, 2009; Dellago et al., 1998). TPS avoids the problem of user-selected reaction coordinates by generating an ensemble of potential reaction pathways. The most probable reaction coordinate is then selected from this ensemble. TPS has been employed in biochemistry (e.g., Knott et al., 2014) and materials science (e.g., Tranca et al., 2012). In geochemistry, Kerisit and Rosso (2009) used TPS to examine rates and mechanisms of water exchange for Na^+ and Fe^{2+} aqueous ions, using the calculated pressure dependence of the water exchange rate to determine activation volumes. The technique shows promise for attacking a wide range of geochemical problems that would require rare event sampling with poorly constrained reaction paths.

11.4.4 Large-Scale Quantum Molecular Dynamics

The system size and temporal limitations of quantum molecular dynamics are a significant bottleneck toward progress in computational geochemistry. One possible way around this problem is to parallelize the simulation in time as well as in the number of atoms. Bylaska et al. (2013) have made progress with this approach and tested it on 1000 Si atom and $HCl + 4H_2O$ models. The method is effective for ab initio molecular dynamics simulations and simulations can be run on separate computers simultaneously.

Another strategy is the divide–conquer–recombine (DCR) algorithm. Shimojo et al. (2014) have used DCR to model over 50 million atoms on the IBM Blue Gene/Z computer, but this simulation took advantage of over 780 000 cores. DCR is an integration of quantum and classical MD simulations with finite-element dynamics. The method is adaptive such that the quantum region selected within the model varies depending on the reaction dynamics. An important aspect of this type of simulation is the availability of accurate, reactive classical force fields to bridge the gap between the quantum and finite-element simulations. Combined with KMC, DCR could be a route to the large-scale, long-duration simulations and embedding scaling required for complex geochemical systems.

11.4.5 Reactive Force Fields

Optimization of parameters for use in classical molecular mechanical force fields is usually performed by fitting to experimental data and results of quantum mechanical calculations (see Chapter 2). The ability to reproduce experimental observables (e.g., structures, cohesive energies, vibrational spectra) is a necessary condition for a force field, but it is not sufficient to explore chemical kinetics. Information on states far from the equilibrium positions is necessary to describe bond-making and bond-breaking processes. Quantum mechanics can supply this information, but the accuracy of the quantum results for complex systems is limited as discussed previously. One would like to improve the accuracy of the quantum results and the fitting of force field parameters (Farah et al., 2012). Ideally, reoptimizing these parameters to reproduce benchmark activation energies and rate constants on relatively simple test reactions would then be carried out.

Some of the earliest simulations in geochemistry utilized "reactive force fields" because the equations involved only included Coulombic, short-range repulsive, and van der Waals terms associated with each ion (Angell et al., 1982, 1987; Kubicki and Lasaga, 1988; Rustad et al., 1991a, b, c; Woodcock et al., 1976) such as Angell et al. (1987):

$$V(r) = \frac{q_i q_i}{r} + A \exp\left[\frac{(\sigma_i + \sigma_j - r)}{\rho}\right] \tag{11.24}$$

where $V(r)$ is the potential energy, q is the charge on the ion, r is the distance between two ions, σ is a size parameter for each ion, and A and ρ are fit constants. Recently, Butenuth et al. (2012) have used a similar type of force field to model H_2O adsorption on SiO_2. These authors achieve an excellent reproduction of quantum mechanical results for the H_2O–SiO_2 interaction over a range of differences, but note that the force field parameterization varies significantly depending on the quantum mechanical method employed (e.g., PBE vs. MP2).

A more explicit attempt to derive a reactive force field for geochemical applications was published by Rustad and coworkers (1996a, b). The Stillinger and David (1978) model for water was used as a starting point. Formal charges could be used which allows for a straightforward handling of H^+-transfer reactions; the authors modified the force field by cutting off the charge–dipole interaction to reproduce the gas-phase dipole moment of H_2O (Halley et al., 1993). Rustad and coworkers used this force field to model aqueous Fe^{3+} hydrolysis (1995) and surface charging behavior of goethite (Felmy and Rustad, 1998; Rustad et al., 1996b).

Another approach to the H^+ charge problem was used by Garofalini and coworkers. Lockwood and Garofalini (2013) used a constant charge of $+0.4$ e^- for each H^+ (with -0.8 e^- for O atoms). These charges are closer to those derived from quantum mechanical calculations, but when using these charges for dissociation reaction such as $H_2O \rightarrow H^+$ and OH^-, there is not a $+1$ charge on the H^+ nor -1 charge on the OH^-. Despite this, MD simulations using this force field have reproduced numerous experimentally observable properties. The simplicity of this approach has been effective for studying silicate–water interfaces (Lockwood and Garofalini, 2009, 2010, 2012; Kagan et al., 2014).

A contrasting approach has been developed by van Duin and coworkers (e.g., van Duin et al., 2003). The "ReaxFF" approach has been to attempt to model every element with a set of equations and parameters that mimics all possible chemical states including changes in oxidation state. Thus, ReaxFF for the Si–O–H system not only describes SiO_2–H_2O but also Si, H_2, and O_2 species. ReaxFF is complex because it contains numerous energy terms and higher-order equations for each term. The energy is the combination of terms:

$$E_{sys} = E_{bond} + E_{over} + E_{under} + E_{lp} + E_{val} + E_{pen} + E_{tors} + E_{conj} + E_{vdWalls} + E_{Coul} \tag{11.25}$$

where $E_{bond} + E_{over} + E_{under} + E_{lp} + E_{val} + E_{pen} + E_{tors} + E_{conj} + E_{vdWalls} + E_{Coul}$ are covalent bonding, over and under coordination penalties, lone-pair interactions, valence angle, penalty energy, torsion angle, conjugation, van der Waals, and Coulombic terms. With terms such as E_{val}, there are cubic polynomials that describe the bond stretch and bond angle bend terms. Hence, a typical ReaxFF parameterization has numerous parameters that are fit to quantum mechanical results. ReaxFF has been applied by a vast number of groups to many materials (see Liang et al., 2013 for a review). One geochemical example is the work of Kulkarni et al. (2013) who fit ReaxFF parameters to M06-L and M06-2X (Zhao and Truhlar, 2008) DFT calculations to model the Si–O–H system. Parameterizations of ReaxFF are sometimes less accurate than force fields dedicated to a particular model chemistry (e.g., Skelton et al., 2011), but there is always the possibility of training ReaxFF solely on results from a single oxidation state and making it more accurate for a particular application.

In conclusion, although there are many challenges facing computational geochemical kinetics, the advances in computational power and software development occurring today leave one optimistic that these challenges can be met. Furthermore, collaborations among experimental and computational geochemists are allowing for increased synergy between approaches to solving problems. The student should keep in mind that each geochemical question requires a tool or set of tools that is appropriate for obtaining the information necessary to answer that question and that each tool has its strengths and limitations, so the combination of measurements and calculations that will best provide a comprehensive picture of geochemical reactions must be carefully chosen in each instance. The ability to incorporate a wide variety of results from numerous methods will be the key to progress in geochemical kinetics.

References

Adeagbo, W.A., et al., Transport processes at alpha-quartz-water interfaces: Insights from first-principles molecular dynamics simulations. *Chemphyschem*, 2008. **9**(7): p. 994–1002.

Aguilar, C.M. and W.R. Rocha, Ligand exchange reaction involving Ru(III) compounds in aqueous solution: A hybrid quantum mechanical/effective fragment potential study. *Journal of Physical Chemistry B*, 2011. **115**(9): p. 2030–2037.

Alexander, J.M., et al., A combined laboratory and modeling study of the infrared extinction and visible light scattering properties of mineral dust aerosol. *Journal of Geophysical Research-Atmospheres*, 2013. **118**(2): p. 435–452.

Alexandrov, V. and K.M. Rosso, Ab initio modeling of Fe(II) adsorption and interfacial electron transfer at goethite (alpha-FeOOH) surfaces. *Physical Chemistry Chemical Physics*, 2015. **17**(22): p. 14518–14531.

Amaro-Estrada, J.I., L. Maron, and A. Ramirez-Solis, Aqueous solvation of HgClOH. Stepwise DFT solvation and Born–Oppenheimer molecular dynamics studies of the $HgClOH\text{-}(H_2O)_{(24)}$ complex. *Physical Chemistry Chemical Physics*, 2014. **16**(18): p. 8455–8464.

Ammann, M.W., J.P. Brodholt, and D.P. Dobson, DFT study of migration enthalpies in $MgSiO_3$ perovskite. *Physics and Chemistry of Minerals*, 2009. **36**(3): p. 151–158.

Ammann, M.W., et al., First-principles constraints on diffusion in lower-mantle minerals and a weak D'' layer. *Nature*, 2010. **465**(7297): p. 462–465.

Ammann, M.W., J.P. Brodholt, and D.P. Dobson, Ferrous iron diffusion in ferro-periclase across the spin transition. *Earth and Planetary Science Letters*, 2011. **302**(3–4): p. 393–402.

Ammann, M.W., J.P. Brodholt, and D.P. Dobson, Diffusion of aluminium in MgO from first principles. *Physics and Chemistry of Minerals*, 2012. **39**(6): p. 503–514.

Amonette, J.E., et al., Dechlorination of carbon tetrachloride by Fe(II) associated with goethite. *Environmental Science & Technology*, 2000. **34**(21): p. 4606–4613.

Andrews, D.U., S.H. Kable, and M.J.T. Jordan, A phase space theory for roaming reactions. *Journal of Physical Chemistry A*, 2013. **117**(32): p. 7631–7642.

Angell, C.A., P.A. Cheeseman, and S. Tamaddon, Pressure enhancement of ion mobilities in liquid silicates from computer-simulation studies to 800-kilobars. *Science*, 1982. **218**(4575): p. 885–887.

Angell, C.A., P.A. Cheeseman, and R.R. Kadiyala, Diffusivity and thermodynamic properties of diopside and jadeite melts by computer-simulation studies. *Chemical Geology*, 1987. **62**(1–2): p. 83–92.

Anh, P., D.R. Cole, and A. Striolo, Aqueous methane in slit-shaped silica nanopores: High solubility and traces of hydrates. *Journal of Physical Chemistry C*, 2014. **118**(9): p. 4860–4868.

Aqvist, J. and T. Hansson, On the validity of electrostatic linear response in polar solvents. *Journal of Physical Chemistry*, 1996. **100**(22): p. 9512–9521.

Ayala, R. and M. Sprik, A classical point charge model study of system size dependence of oxidation and reorganization free energies in aqueous solution. *Journal of Physical Chemistry B*, 2008. **112**(2): p. 257–269.

Balogh, E., et al., Rates of ligand exchange between $>Fe^{III}\text{-}OH_2$ functional groups on a nanometer-sized aqueous cluster and bulk solution. *Inorganic Chemistry*, 2007. **46**(17): p. 7087–7092.

Baltrusaitis, J. and V.H. Grassian, Carbonic acid formation from reaction of carbon dioxide and water coordinated to Al(OH)(3): A quantum chemical study. *Journal of Physical Chemistry A*, 2010. **114**(6): p. 2350–2356.

Baltrusaitis, J., et al., FTIR spectroscopy combined with quantum chemical calculations to investigate adsorbed nitrate on aluminium oxide surfaces in the presence and absence of co-adsorbed water. *Physical Chemistry Chemical Physics*, 2007a. **9**(36): p. 4970–4980.

Baltrusaitis, J., et al., Surface reactions of carbon dioxide at the adsorbed water-oxide interface. *Journal of Physical Chemistry C*, 2007b. **111**(40): p. 14870–14880.

Baltrusaitis, J., E.V. Patterson, and C. Hatch, Computational studies of CO_2 activation via photochemical reactions with reduced sulfur compounds. *Journal of Physical Chemistry A*, 2012. **116**(37): p. 9331–9339.

Bandura, A.V., et al., Adsorption of water on the TiO_2 (rutile) (110) surface: A comparison of periodic and embedded cluster calculations. *Journal of Physical Chemistry B*, 2004. **108**(23): p. 7844–7853.

Bandura, A.V., J.D. Kubicki, and J.O. Sofo, Periodic density functional theory study of water adsorption on the alpha-quartz (101) surface. *Journal of Physical Chemistry C*, 2011. **115**(13): p. 5756–5766.

Banfield, J.F., et al., Aggregation-based crystal growth and microstructure development in natural iron oxyhydroxide biomineralization products. *Science*, 2000. **289**(5480): p. 751–754.

Baptista, L., E.C. da Silva, and G. Arbilla, Theoretical investigation of the gas phase oxidation mechanism of dimethyl sulfoxide by OH radical. *Journal of Molecular Structure (THEOCHEM)*, 2008. **851**(1–3): p. 1–14.

Bazilevskaya, E., et al., Aluminum coprecipitates with Fe (hydr)oxides: Does isomorphous substitution of Al^{3+} for Fe^{3+} in goethite occur? *Geochimica et Cosmochimica Acta*, 2011. **75**(16): p. 4667–4683.

Becke, A.D., Density-functional thermochemistry. 3. The role of exact exchange. *Journal of Chemical Physics*, 1993. **98**(7): p. 5648–5652.

Benner, S.G., W.D. Gould, and D.W. Blowes, Microbial populations associated with the generation and treatment of acid mine drainage. *Chemical Geology*, 2000. **169**(3–4): p. 435–448.

Bergner, A., et al., Ab-initio energy-adjusted pseudopotentials for elements of groups 13–17. *Molecular Physics*, 1993. **80**(6): p. 1431–1441.

Berner, R.A., Phanerozoic atmospheric oxygen: New results using the geocarbsulf model. *American Journal of Science*, 2009. **309**(7): p. 603–606.

Bhandari, N., et al., Photodissolution of ferrihydrite in the presence of oxalic acid: An in situ ATR-FTIR/DFT study. *Langmuir*, 2010. **26**(21): p. 16246–16253.

Bickmore, B.R., et al., Reaction pathways for quartz dissolution determined by statistical and graphical analysis of macroscopic experimental data. *Geochimica et Cosmochimica Acta*, 2008. **72**(18): p. 4521–4536.

Blumberger, J. and M. Sprik, Ab initio molecular dynamics simulation of the aqueous Ru^{2+}/Ru^{3+} redox reaction: The Marcus perspective. *Journal of Physical Chemistry B*, 2005. **109**(14): p. 6793–6804.

Boerio-Goates, J., et al., Characterization of surface defect sites on bulk and nanophase anatase and rutile TiO_2 by low-temperature specific heat. *Journal of Physical Chemistry C*, 2013. **117**(9): p. 4544–4550.

Boese, A.D., Assessment of coupled cluster theory and more approximate methods for hydrogen bonded systems. *Journal of Chemical Theory and Computation*, 2013. **9**(10): p. 4403–4413.

Boot-Handford, M.E., et al., Carbon capture and storage update. *Energy & Environmental Science*, 2014. **7**(1): p. 130–189.

Bourg, I.C. and G. Sposito, Molecular dynamics simulations of kinetic isotope fractionation during the diffusion of ionic species in liquid water. *Geochimica et Cosmochimica Acta*, 2007. **71**(23): p. 5583–5589.

Bourg, I.C., et al., Isotopic mass dependence of metal cation diffusion coefficients in liquid water. *Geochimica et Cosmochimica Acta*, 2010. **74**(8): p. 2249–2256.

Brantley, S.L., Conrad, C.F., Analysis of rates of geochemical reactions, in *Kinetics of Water-Rock Interaction*, S.L. Brantley, Kubicki, J.D., and White, A.F., Editor. 2008, New York: Springer. p. 1–37.

Breuer, M., K.M. Rosso, and J. Blumberger, Electron flow in multiheme bacterial cytochromes is a balancing act between heme electronic interaction and redox potentials. *Proceedings of the National Academy of Sciences of the United States of America*, 2014. **111**(2): p. 611–616.

Broomhead, J.A., F. Basolo, and R.G. Pearson, Kinetics of acid + bases hydrolyses of chloropentaammineruthenium(3) ion. *Inorganic Chemistry*, 1964. **3**(6): p. 826–832.

Brunschwig, B.S., et al., A semi-classical treatment of electron-exchange reactions: Application to the hexaaquoiron(II)-hexaaquoiron(III) system. *Journal of the American Chemical Society*, 1980. **102**(18): p. 5798–5809.

Butenuth, A., et al., Ab initio derived force-field parameters for molecular dynamics simulations of deprotonated amorphous-SiO_2/water interfaces. *Physica Status Solidi B-Basic Solid State Physics*, 2012. **249**(2): p. 292–305.

Bylaska, E.J., J.Q. Weare, and J.H. Weare, Extending molecular simulation time scales: Parallel in time integrations for high-level quantum chemistry and complex force representations. *Journal of Chemical Physics*, 2013. **139**(7): p. 074114.

Campbell, C.T., L. Arnadottir, and J.R.V. Sellers, Kinetic prefactors of reactions on solid surfaces. *Zeitschrift Fur Physikalische Chemie-International Journal of Research in Physical Chemistry & Chemical Physics*, 2013. **227**(9–11): p. 1435–1454.

Cances, E., B. Mennucci, and J. Tomasi, A new integral equation formalism for the polarizable continuum model: Theoretical background and applications to isotropic and anisotropic dielectrics. *Journal of Chemical Physics*, 1997. **107**(8): p. 3032–3041.

Casey, W.H. and G. Sposito, On the temperature-dependence of mineral dissolution rates. *Geochimica et Cosmochimica Acta*, 1992. **56**(10): p. 3825–3830.

Casey, W.H., A.C. Lasaga, and G.V. Gibbs, Mechanisms of silica dissolution as inferred from the kinetic isotope effect. *Geochimica et Cosmochimica Acta*, 1990. **54**(12): p. 3369–3378.

Chandler, D., Electron transfer in water and other polar environments, how it happens, in *Computer Simulation of Rare Events and Dynamics of Classical and Quantum Condensed-Phase Systems: Classical and Quantum Dynamics in Condensed Phase Simulations*, B.J. Berne, G. Ciccotti, and D.F. Coker, Editors. 1998, Singapore: World Scientific. p. 25–49.

Chen, W. and M.S. Gordon, The effective fragment model for solvation: Internal rotation in formamide. *Journal of Chemical Physics*, 1996. **105**(24): p. 11081–11090.

Chen, J., Y. Xu, and D. Wang, A multilayered representation, quantum mechanical and molecular mechanics study of the CH3F+OH- reaction in water. *Journal of Computational Chemistry*, 2014. **35**(6): p. 445–450.

Choobari, O.A., P. Zawar-Reza, and A. Sturman, The global distribution of mineral dust and its impacts on the climate system: A review. *Atmospheric Research*, 2014. **138**: p. 152–165.

Clayton, R.N., et al., Oxygen isotope fractionation factors among rock-forming minerals at high-temperatures. *Chemical Geology*, 1988. **70**(1–2): p. 183.

Cohen, A.L. and G.A. Gaetani, Ion partitioning and the geochemistry of coral skeletons: Solving the mystery of the vital effect, in *Ion Partitioning in Ambient-Temperature Aqueous Systems*, M. Prieto and H. Stoll, Editors. 2010, Twickenham: European Mineralogical Union. p. 377–397.

Cohen, B., et al., Femtosecond fluorescence dynamics of a proton-transfer dye interacting with silica-based nanomaterials. *Journal of Physical Chemistry C*, 2010. **114**(14): p. 6281–6289.

Conrad, C.F., et al., Modeling the kinetics of silica nanocolloid formation and precipitation in geologically relevant aqueous solutions. *Geochimica et Cosmochimica Acta*, 2007. **71**(3): p. 531–542.

Cruz-Uribe, A.M., et al., Metamorphic reaction rates at similar to 650–800°C from diffusion of niobium in rutile. *Geochimica et Cosmochimica Acta*, 2014. **130**: p. 63–77.

Cui, D.Q. and T.E. Eriksen, Reduction of pertechnetate in solution by heterogeneous electron transfer from Fe(II)-containing geological material. *Environmental Science & Technology*, 1996. **30**(7): p. 2263–2269.

Dahlke, E.E., et al., Assessment of the accuracy of density functionals for prediction of relative energies and geometries of low-lying isomers of water hexamers. *Journal of Physical Chemistry A*, 2008. **112**(17): p. 3976–3984.

Day, P.N., et al., An effective fragment method for modeling solvent effects in quantum mechanical calculations. *Journal of Chemical Physics*, 1996. **105**(5): p. 1968–1986.

Debnath, S., et al., Reductive dissolution of ferrihydrite by ascorbic acid and the inhibiting effect of phospholipid. *Journal of Colloid and Interface Science*, 2010. **341**(2): p. 215–223.

Dellago, C., et al., Transition path sampling and the calculation of rate constants. *Journal of Chemical Physics*, 1998. **108**(5): p. 1964–1977.

Demichelis, R., et al., Stable prenucleation mineral clusters are liquid-like ionic polymers. *Nature Communications*, 2011. **2**: p. 590.

DePaolo, D.J., Surface kinetic model for isotopic and trace element fractionation during precipitation of calcite from aqueous solutions. *Geochimica et Cosmochimica Acta*, 2011. **75**(4): p. 1039–1056.

DePaolo, D.J. and D.R. Cole, Geochemistry of geologic carbon sequestration: An overview, in *Geochemistry of Geologic CO_2 Sequestration*, D.J. DePaolo, et al., Editors. 2013, Chantilly: Mineralogical Society of America. p. 1–14.

Dias, R.F., et al., Delta C-13 of low-molecular-weight organic acids generated by the hydrous pyrolysis of oil-prone source rocks. *Geochimica et Cosmochimica Acta*, 2002. **66**(15): p. 2755–2769.

Diem, D. and W. Stumm, Is dissolved Mn^{-2+} being oxidized by O_2 in absence of Mn-bacteria or surface catalysts. *Geochimica et Cosmochimica Acta*, 1984. **48**(7): p. 1571–1573.

DiLabio, G.A. and M. Koleini, Dispersion-correcting potentials can significantly improve the bond dissociation enthalpies and noncovalent binding energies predicted by density-functional theory. *Journal of Chemical Physics*, 2014. **140**(18): p. 18A542.

DiLabio, G.A., M. Koleini, and E. Torres, Extension of the B3LYP-dispersion-correcting potential approach to the accurate treatment of both inter- and intra-molecular interactions. *Theoretical Chemistry Accounts*, 2013. **132**(10): p. 1389.

Dion, M., et al., Van der Waals density functional for general geometries. *Physical Review Letters*, 2004. **92**(24): p. 246401.

Doltsinis, N.L., et al., Ab initio molecular dynamics study of dissolved SiO_2 in supercritical water. *Journal of Theoretical & Computational Chemistry*, 2007. **6**(1): p. 49–62.

Domagal-Goldman, S.D. and J.D. Kubicki, Density functional theory predictions of equilibrium isotope fractionation of iron due to redox changes and organic complexation. *Geochimica et Cosmochimica Acta*, 2008. **72**(21): p. 5201–5216.

Dove, P.M., The dissolution kinetics of quartz in aqueous mixed cation solutions. *Geochimica et Cosmochimica Acta*, 1999. **63**(22): p. 3715–3727.

Dove, P.M. and N. Han, Kinetics of mineral dissolution and growth as reciprocal microscopic surface processes across chemical driving force, in *Perspectives on Inorganic, Organic, and Biological Crystal Growth: From Fundamentals to Applications*, M. Skowronski, J.J. DeYoreo, and C.A. Wang, Editors. 2007, Melville: American Institute of Physics. p. 215–234.

Dove, P.M. and C.J. Nix, The influence of the alkaline earth cations, magnesium, calcium, and barium on the dissolution kinetics of quartz. *Geochimica et Cosmochimica Acta*, 1997. **61**(16): p. 3329–3340.

Dove, P.M., N.Z. Han, and J.J. De Yoreo, Mechanisms of classical crystal growth theory explain quartz and silicate dissolution behavior. *Proceedings of the National Academy of Sciences of the United States of America*, 2005. **102**(43): p. 15357–15362.

Dreybrodt, W. and D. Scholz, Climatic dependence of stable carbon and oxygen isotope signals recorded in speleothems: From soil water to speleothem calcite. *Geochimica et Cosmochimica Acta*, 2011. **75**(3): p. 734–752.

van Duin, A.C.T., et al., ReaxFF(SiO) reactive force field for silicon and silicon oxide systems. *Journal of Physical Chemistry A*, 2003. **107**(19): p. 3803–3811.

Eiler, J.M., et al., Frontiers of stable isotope geoscience. *Chemical Geology*, 2014. **372**: p. 119–143.

Elsner, M., et al., Mechanisms and products of surface-mediated reductive dehalogenation of carbon tetrachloride by Fe(II) on goethite. *Environmental Science & Technology*, 2004. **38**(7): p. 2058–2066.

Emerson, S., et al., Environmental oxidation rate of manganese(II): Bacterial catalysis. *Geochimica et Cosmochimica Acta*, 1982. **46**(6): p. 1073–1079.

Estreicher, S.K., et al., Activation energies for diffusion of defects in silicon: The role of the exchange-correlation functional. *Angewandte Chemie-International Edition*, 2011. **50**(43): p. 10221–10225.

Evans, R.J., J.R. Rustad, and W.H. Casey, Calculating geochemical reaction pathways: Exploration of the inner-sphere water exchange mechanism in $Al(H_2O)_6^{3+}$(aq) + nH_2O with ab initio calculations and molecular dynamics. *Journal of Physical Chemistry A*, 2008. **112**(17): p. 4125–4140.

Farah, K., F. Muller-Plathe, and M.C. Bohm, Classical reactive molecular dynamics implementations: State of the art. *Chemphyschem*, 2012. **13**(5): p. 1127–1151.

Farazdel, A., et al., Electric-field induced intramolecular electron-transfer in spiro pi-electron systems and their suitability as molecular electronic devices: A theoretical-study. *Journal of the American Chemical Society*, 1990. **112**(11): p. 4206–4214.

Farmer, V.C., *The Infrared Spectra of Minerals*. 1974, London: Mineralogical Society.

Felipe, M.A., Y.T. Xiao, and J.D. Kubicki, Molecular orbital modeling and transition state theory in geochemistry, in *Molecular Modeling Theory: Applications in the Geosciences*, R.T. Cygan and J.D. Kubicki, Editors. 2001, Washington, D.C.: Mineralogical Society of America. p. 485–531.

Felipe, M.A., J.D. Kubicki, and D.M. Rye, Hydrogen isotope exchange kinetics between H_2O and H_4SiO_4 from ab initio calculations. *Geochimica et Cosmochimica Acta*, 2003. **67**(7): p. 1259–1276.

Felipe, M.A., J.D. Kubicki, and D.M. Rye, Oxygen isotope exchange kinetics between H_2O and H_4SiO_4 from ab initio calculations. *Geochimica et Cosmochimica Acta*, 2004. **68**(5): p. 949–958.

Felipe, M.A., J.D. Kubicki, and K.H. Freeman, A mechanism for carbon isotope exchange between aqueous acetic acid and CO_2/HCO_3^-: An ab initio study. *Organic Geochemistry*, 2005. **36**(6): p. 835–850.

Felmy, A.R. and J.R. Rustad, Molecular statics calculations of proton binding to goethite surfaces: Thermodynamic modeling of the surface charging and protonation of goethite in aqueous solution. *Geochimica et Cosmochimica Acta*, 1998. **62**(1): p. 25–31.

Fischer, C., R.S. Arvidson, and A. Luettge, How predictable are dissolution rates of crystalline material? *Geochimica et Cosmochimica Acta*, 2012. **98**: p. 177–185.

Francl, M.M., et al., Self-consistent molecular-orbital methods. 23. A polarization-type basis set for 2nd-row elements. *Journal of Chemical Physics*, 1982. **77**(7): p. 3654–3665.

Frisch, M.J., J.A. Pople, and J.S. Binkley, Self-consistent molecular-orbital methods. 25. Supplementary functions for Gaussian-basis sets. *Journal of Chemical Physics*, 1984. **80**(7): p. 3265–3269.

Gale, J.D., Empirical potential derivation for ionic materials. *Philosophical Magazine B-Physics of Condensed Matter Statistical Mechanics Electronic Optical and Magnetic Properties*, 1996. **73**(1): p. 3–19.

Gale, J.D., GULP: Capabilities and prospects. *Zeitschrift Fur Kristallographie*, 2005. **220**(5–6): p. 552–554.

Gale, J.D. and A.L. Rohl, The general utility lattice program (GULP). *Molecular Simulation*, 2003. **29**(5): p. 291–341.

Gallet, G.A., F. Pietrucci, and W. Andreoni, Bridging static and dynamical descriptions of chemical reactions: An ab initio study of CO_2 interacting with water molecules. *Journal of Chemical Theory and Computation*, 2012. **8**(11): p. 4029–4039.

Gammons, C.H., et al., Geochemistry and stable isotope investigation of acid mine drainage associated with abandoned coal mines in central Montana, USA. *Chemical Geology*, 2010. **269**(1–2): p. 100–112.

Gebauer, D., A. Voelkel, and H. Coelfen, Stable prenucleation calcium carbonate clusters. *Science*, 2008. **322** (5909): p. 1819–1822.

Gebauer, D., et al., Pre-nucleation clusters as solute precursors in crystallisation. *Chemical Society Reviews*, 2014. **43**(7): p. 2348–2371.

Ghorai, P.K. and D.V. Matyushov, Solvent reorganization entropy of electron transfer in polar solvents. *Journal of Physical Chemistry A*, 2006. **110**(28): p. 8857–8863.

Gilbert, B., et al., Special phase transformation and crystal growth pathways observed in nanoparticles. *Geochemical Transactions*, 2003. **4**: p. 20–27.

Gillan, M.J., et al., First-principles energetics of water clusters and ice: A many-body analysis. *Journal of Chemical Physics*, 2013. **139**(24): p. 244504.

Gomes, J., et al., Accurate prediction of hydrocarbon interactions with zeolites utilizing improved exchange-correlation functionals and QM/MM methods: Benchmark calculations of adsorption enthalpies and application to ethene methylation by methanol. *Journal of Physical Chemistry C*, 2012. **116**(29): p. 15406–15414.

Grant, M. and R.B. Jordan, Kinetics of solvent water exchange on iron(III). *Inorganic Chemistry*, 1981. **20**(1): p. 55–60.

Grubb, M.P., M.L. Warter, and S.W. North, Stereodynamics of multistate roaming. *Physical Chemistry Chemical Physics*, 2012. **14**(19): p. 6733–6740.

Halley, J.W., J.R. Rustad, and A. Rahman, A polarizable, dissociating molecular-dynamics model for liquid water. *Journal of Chemical Physics*, 1993. **98**(5): p. 4110–4119.

Harding, L.B. and S.J. Klippenstein, Roaming radical pathways for the decomposition of alkanes. *Journal of Physical Chemistry Letters*, 2010. **1**(20): p. 3016–3020.

Harley, S.J., C.A. Ohlin, and W.H. Casey, Geochemical kinetics via the Swift-Connick equations and solution NMR. *Geochimica et Cosmochimica Acta*, 2011. **75**(13): p. 3711–3725.

He, G., G. Pan, and M. Zhang, Studies on the reaction pathway of arsenate adsorption at water-TiO_2 interfaces using density functional theory. *Journal of Colloid and Interface Science*, 2011. **364**(2): p. 476–481.

Heidel, C. and M. Tichomirowa, The role of dissolved molecular oxygen in abiotic pyrite oxidation under acid pH conditions: Experiments with O-18-enriched molecular oxygen. *Applied Geochemistry*, 2010. **25**(11): p. 1664–1675.

Heidel, C. and M. Tichomirowa, Galena oxidation investigations on oxygen and sulphur isotopes. *Isotopes in Environmental and Health Studies*, 2011. **47**(2): p. 169–188.

Heidel, C., M. Tichomirowa, and C. Breitkopf, Sphalerite oxidation pathways detected by oxygen and sulfur isotope studies. *Applied Geochemistry*, 2011. **26**(12): p. 2247–2259.

Helm, L. and A.E. Merbach, Inorganic and bioinorganic solvent exchange mechanisms. *Chemical Reviews*, 2005. **105**(6): p. 1923–1959.

Helz, G.R., B.E. Erickson, and T.P. Vorlicek, Stabilities of thiomolybdate complexes of iron; implications for retention of essential trace elements (Fe, Cu, Mo) in sulfidic waters. *Metallomics*, 2014. **6**(6): p. 1131–1140.

Henkelman, G., B.P. Uberuaga, and H. Jonsson, A climbing image nudged elastic band method for finding saddle points and minimum energy paths. *Journal of Chemical Physics*, 2000. **113**(22): p. 9901–9904.

Hochella, M.F., Jr., et al., Nanominerals, mineral nanoparticles, and Earth systems. *Science*, 2008. **319**(5870): p. 1631–1635.

Hofmann, A.E., I.C. Bourg, and D.J. DePaolo, Ion desolvation as a mechanism for kinetic isotope fractionation in aqueous systems. *Proceedings of the National Academy of Sciences of the United States of America*, 2012. **109**(46): p. 18689–18694.

Hujo, W. and S. Grimme, Comparison of the performance of dispersion-corrected density functional theory for weak hydrogen bonds. *Physical Chemistry Chemical Physics*, 2011. **13**(31): p. 13942–13950.

Hummer, D.R., et al., Origin of nanoscale phase stability reversals in titanium oxide polymorphs. *Journal of Physical Chemistry C*, 2009. **113**(11): p. 4240–4245.

Hummer, D.R., et al., Single-site and monolayer surface hydration energy of anatase and rutile nanoparticles using density functional theory. *Journal of Physical Chemistry C*, 2013. **117**(49): p. 26084–26090.

Hwang, J.K. and A. Warshel, Microscopic examination of free-energy relationships for electron-transfer in polar-solvents. *Journal of the American Chemical Society*, 1987. **109**(3): p. 715–720.

Immenhauser, A., et al., Magnesium-isotope fractionation during low-Mg calcite precipitation in a limestone cave: Field study and experiments. *Geochimica et Cosmochimica Acta*, 2010. **74**(15): p. 4346–4364.

Jensen, K.R., et al., Water condensation: A multiscale phenomenon. *Journal of Nanoscience and Nanotechnology*, 2014. **14**(2): p. 1859–1871.

Jin, X., et al., DFT study on the mechanism for the substitution of F- into Al(III) complexes in aqueous solution. *Dalton Transactions*, 2011a. **40**(3): p. 567–572.

Jin, X., et al., Density functional theory study on aqueous aluminum-fluoride complexes: Exploration of the intrinsic relationship between water-exchange rate constants and structural parameters for monomer aluminum complexes. *Environmental Science & Technology*, 2011b. **45**(1): p. 288–293.

Johnson, R.L., et al., Multinuclear NMR study of the pressure dependence for carbonate exchange in the UO2(CO3)3 4-(aq) ion. *Chemphyschem*, 2011. **12**(16): p. 2903–2906.

Johnson, R.L., et al., Dynamics of a nanometer-sized uranyl cluster in solution. *Angewandte Chemie-International Edition*, 2013. **52**(29): p. 7464–7467.

Jones, F., S. Piana, and J.D. Gale, Understanding the kinetics of barium sulfate precipitation from water and water-methanol solutions. *Crystal Growth & Design*, 2008. **8**(3): p. 817–822.

de Jong, G.T., et al., DFT benchmark study for the oxidative addition of CH4 to Pd. Performance of various density functionals. *Chemical Physics*, 2005. **313**(1–3): p. 261–270.

Kadota, K., et al., Effect of surface properties of calcium carbonate on aggregation process investigated by molecular dynamics simulation. *Journal of Materials Science*, 2014. **49**(4): p. 1724–1733.

Kagan, M., G.K. Lockwood, and S.H. Garofalini, Reactive simulations of the activation barrier to dissolution of amorphous silica in water. *Physical Chemistry Chemical Physics*, 2014. **16**(20): p. 9294–9301.

Karki, B.B. and L. Stixrude, First-principles study of enhancement of transport properties of silica melt by water. *Physical Review Letters*, 2010. **104**(21): p. 215901.

Karki, B.B., L. Stixrude, and R.M. Wentzcovitch, High-pressure elastic properties of major materials of Earth's mantle from first principles. *Reviews of Geophysics*, 2001. **39**(4): p. 507–534.

Karki, B.B., J. Zhang, and L. Stixrude, First principles viscosity and derived models for MgO-SiO2 melt system at high temperature. *Geophysical Research Letters*, 2013. **40**(1): p. 94–99.

Katz, J.E., et al., Electron small polarons and their mobility in iron (Oxyhydr)oxide nanoparticles. *Science*, 2012. **337**(6099): p. 1200–1203.

Keating, E.H. and J.M. Bahr, Reactive transport modeling of redox geochemistry: Approaches to chemical disequilibrium and reaction rate estimation at a site in northern Wisconsin. *Water Resources Research*, 1998. **34**(12): p. 3573–3584.

Kerisit, S. and K.M. Rosso, Charge transfer in FeO: A combined molecular-dynamics and ab initio study. *Journal of Chemical Physics*, 2005. **123**(22): p. 224712.

Kerisit, S. and K.M. Rosso, Computer simulation of electron transfer at hematite surfaces. *Geochimica et Cosmochimica Acta*, 2006. **70**(8): p. 1888–1903.

Kerisit, S. and K.M. Rosso, Transition path sampling of water exchange rates and mechanisms around aqueous ions. *Journal of Chemical Physics*, 2009. **131**(11): p. 114512.

Kerisit, S., et al., Molecular dynamics simulations of the interactions between water and inorganic solids. *Journal of Materials Chemistry*, 2005. **15**(14): p. 1454–1462.

Kerisit, S., et al., Molecular computational investigation of electron-transfer kinetics across cytochrome-iron oxide interfaces. *Journal of Physical Chemistry C*, 2007. **111**(30): p. 11363–11375.

Kerisit, S., P. Zarzycki, and K.M. Rosso, Computational molecular simulation of the oxidative adsorption of ferrous iron at the hematite (001)-water interface. *Journal of Physical Chemistry C*, 2015. **119**(17): p. 9242–9252.

Kim, J.-H., et al., Modeling the reductive dechlorination of polychlorinated dibenzo-p-dioxins: Kinetics, pathway, and equivalent toxicity. *Environmental Science & Technology*, 2009. **43**(14): p. 5327–5332.

King, G. and A. Warshel, Investigation of the free-energy functions for electron-transfer reactions. *Journal of Chemical Physics*, 1990. **93**(12): p. 8682–8692.

Klamt, A., et al., Refinement and parametrization of COSMO-RS. *Journal of Physical Chemistry A*, 1998. **102**(26): p. 5074–5085.

Knauss, K.G. and T.J. Wolery, The dissolution kinetics of quartz as a function of pH and time at 70°C. *Geochimica et Cosmochimica Acta*, 1988. **52**(1): p. 43–53.

Knott, B.C., et al., The mechanism of cellulose hydrolysis by a two-step, retaining cellobiohydrolase elucidated by structural and transition path sampling studies. *Journal of the American Chemical Society*, 2014. **136**(1): p. 321–329.

Krishnan, R., et al., Self-consistent molecular-orbital methods. 20. Basis set for correlated wave-functions. *Journal of Chemical Physics*, 1980. **72**(1): p. 650–654.

Kubicki, J.D., Computational chemistry applied to studies of organic contaminants in the environment: Examples based on benzo[a]pyrene. *American Journal of Science*, 2005. **305**(6–8): p. 621–644.

Kubicki, J.D. and A.C. Lasaga, Molecular-dynamics simulations of SiO_2 melt and glass: Ionic and covalent models. *American Mineralogist*, 1988. **73**(9–10): p. 941–955.

Kubicki, J.D., Y. Xiao, and A.C. Lasaga, Theoretical reaction pathways for the formation of $[Si(OH)_5]^{1-}$ and the deprotonation of orthosilicic acid in basic solution. *Geochimica et Cosmochimica Acta*, 1993. **57**(16): p. 3847–3853.

Kubicki, J.D., G.A. Blake, and S.E. Apitz, Ab initio calculations on aluminosilicate Q(3) species: Implications for atomic structures of mineral surfaces and dissolution mechanisms of feldspars. *American Mineralogist*, 1996. **81**(7–8): p. 789–799.

Kubicki, J.D., C.C. Trout, and K.H. Freeman, Quantum mechanical calculation of hydrogen isotope exchange thermodynamics and kinetics on organic compounds. *Geochimica et Cosmochimica Acta*, 2008. **72**(12): p. A500–A500.

Kubicki, J.D., et al., Quantum mechanical calculation of aqueous uranium complexes: Carbonate, phosphate, organic and biomolecular species. *Chemistry Central Journal*, 2009. **3**: p. 10–29.

Kubicki, J.D., et al., A new hypothesis for the dissolution mechanism of silicates. *Journal of Physical Chemistry C*, 2012a. **116**(33): p. 17479–17491.

Kubicki, J.D., et al., Quantum mechanical calculations on Fe–O–H nanoparticles. *Geoderma*, 2012b. **189**: p. 236–242.

Kuchle, W., et al., Ab initio pseudopotentials for HG through RN. 1. Parameter sets and atomic calculations. *Molecular Physics*, 1991. **74**(6): p. 1245–1263.

Kuharski, R.A., et al., Molecular-model for aqueous ferrous ferric electron-transfer. *Journal of Chemical Physics*, 1988. **89**(5): p. 3248–3257.

Kulkarni, A.D., et al., Oxygen interactions with silica surfaces: Coupled cluster and density functional investigation and the development of a new ReaxFF potential. *Journal of Physical Chemistry C*, 2013. **117**(1): p. 258–269.

Kumar, N., et al., Faster proton transfer dynamics of water on SnO_2 compared to TiO_2. *Journal of Chemical Physics*, 2011. **134**(4): p. 044706.

Kundin, J., et al., Simulation of adsorption processes on the glass surface in aqueous solutions containing oxalic acid. *Computational Materials Science*, 2010. **49**(1): p. 88–98.

Kurganskaya, I. and A. Luttge, Kinetic Monte Carlo simulations of silicate dissolution: Model complexity and parametrization. *Journal of Physical Chemistry C*, 2013a. **117**(47): p. 24894–24906.

Kurganskaya, I. and A. Luttge, A comprehensive stochastic model of phyllosilicate dissolution: Structure and kinematics of etch pits formed on muscovite basal face. *Geochimica et Cosmochimica Acta*, 2013b. **120**: p. 545–560.

Kurganskaya, I., et al., Does the stepwave model predict mica dissolution kinetics? *Geochimica et Cosmochimica Acta*, 2012. **97**: p. 120–130.

Kushiro, I., Effect of pressure on the diffusivity of network-forming cations in melts of jadeitic compositions. *Geochimica et Cosmochimica Acta*, 1983. **47**(8): p. 1415–1422.

Kwon, K.D. and J.D. Kubicki, Molecular orbital theory study on surface complex structures of phosphates to iron hydroxides: Calculation of vibrational frequencies and adsorption energies. *Langmuir*, 2004. **20**(21): p. 9249–9254.

Kwon, K.D., K. Refson, and G. Sposito, On the role of Mn(IV) vacancies in the photoreductive dissolution of hexagonal birnessite. *Geochimica et Cosmochimica Acta*, 2009. **73**(14): p. 4142–4150.

Kwon, K.D., K. Refson, and G. Sposito, Understanding the trends in transition metal sorption by vacancy sites in birnessite. *Geochimica et Cosmochimica Acta*, 2013. **101**: p. 222–232.

Lacks, D.J., et al., Isotope fractionation by thermal diffusion in silicate melts. *Physical Review Letters*, 2012. **108**(6): p. 065901.

Lal, R., Sequestration of atmospheric CO_2 in global carbon pools. *Energy & Environmental Science*, 2008. **1**(1): p. 86–100.

Lammers, L.N., et al., Sedimentary reservoir oxidation during geologic CO_2 sequestration. *Geochimica et Cosmochimica Acta*, 2015. **155**: p. 30–46.

Lee, W.J. and J.D. Kubicki, Molecular-dynamics simulations of periclase crystallization. *Geophysical Research Letters*, 1993. **20**(19): p. 2103–2106.

Lee, C.T., W.T. Yang, and R.G. Parr, Development of the Colle-Salvetti correlation-energy formula into a functional of the electron-density. *Physical Review B*, 1988. **37**(2): p. 785–789.

Liang, T. et al. Reactive potential for advanced atomistic simulations. *Annual Review of Materials Research*, 2013. **43**: 109–129.

Lockwood, G.K. and S.H. Garofalini, Bridging oxygen as a site for proton adsorption on the vitreous silica surface. *Journal of Chemical Physics*, 2009. **131**(7): p. 074703.

Lockwood, G.K. and S.H. Garofalini, Effect of moisture on the self-healing of vitreous silica under irradiation. *Journal of Nuclear Materials*, 2010. **400**(1): p. 73–78.

Lockwood, G.K. and S.H. Garofalini, Reactions between water and vitreous silica during irradiation. *Journal of Nuclear Materials*, 2012. **430**(1–3): p. 239–245.

Lockwood, G.K. and S.H. Garofalini, Lifetimes of excess protons in water using a dissociative water potential. *Journal of Physical Chemistry B*, 2013. **117**(15): p. 4089–4097.

Loring, J.S., et al., In situ infrared spectroscopic study of forsterite carbonation in wet supercritical CO_2. *Environmental Science & Technology*, 2011. **45**(14): p. 6204–6210.

Lowson, R.T., Aqueous oxidation of pyrite by molecular-oxygen. *Chemical Reviews*, 1982. **82**(5): p. 461–497.

Lucena, A.F., et al., Oxo-exchange of gas-phase uranyl, neptunyl, and plutonyl with water and methanol. *Inorganic Chemistry*, 2014. **53**(4): p. 2163–2170.

Ma, Q.S., S. Wu, and Y.C. Tang, Formation and abundance of doubly-substituted methane isotopologues ((CH_3D)-C-13) in natural gas systems. *Geochimica et Cosmochimica Acta*, 2008. **72**(22): p. 5446–5456.

Manceau, A., M. Lanson, and N. Geoffroy, Natural speciation of Ni, Zn, Ba, and As in ferromanganese coatings on quartz using X-ray fluorescence, absorption, and diffraction. *Geochimica et Cosmochimica Acta*, 2007. **71**(1): p. 95–128.

Marcus, R.A., On the theory of oxidation-reduction reactions involving electron transfer. 1. *Journal of Chemical Physics*, 1956. **24**(5): p. 966–978.

Marcus, R.A. and N. Sutin, Electron transfers in chemistry and biology. *Biochimica et Biophysica Acta*, 1985. **811**(3): p. 265–322.

Mardirossian, N. and M. Head-Gordon, Characterizing and understanding the remarkably slow basis set convergence of several Minnesota density functionals for intermolecular interaction energies. *Journal of Chemical Theory and Computation*, 2013. **9**(10): p. 4453–4461.

Martin, D.R. and D.V. Matyushov, Electrostatic fluctuations in cavities within polar liquids and thermodynamics of polar solvation. *Physical Review E*, 2008. **78**(4): p. 041206.

Matsumoto, M., S. Saito, and I. Ohmine, Molecular dynamics simulation of the ice nucleation and growth process leading to water freezing. *Nature*, 2002. **416**(6879): p. 409–413.

Maurice, P.A. and M.F. Hochella, Nanoscale particles and processes: A new dimension in soil science, in *Advances in Agronomy*, Vol **100**, D.L. Sparks, Editor. 2008, London: Academic Press. p. 123–153.

Maurice, P.A., et al., Evolution of hematite surface microtopography upon dissolution by simple organic-acids. *Clays and Clay Minerals*, 1995. **43**(1): p. 29–38.

May, V. and O. Kuhn, *Charge and Energy Transfer Dynamics in Molecular Systems*, 2nd ed. 2003, Weinheim: Wiley-VCH Verlag GmbH.

Mayer, J.M., Proton-coupled electron transfer: A reaction chemist's view. *Annual Review of Physical Chemistry*, 2004. **55**: p. 363–390.

McGuire, J.T., et al., Temporal variations in parameters reflecting terminal-electron-accepting processes in an aquifer contaminated with waste fuel and chlorinated solvents. *Chemical Geology*, 2000. **169**(3–4): p. 471–485.

McInerney, F.A. and S.L. Wing, The Paleocene-Eocene Thermal Maximum: A perturbation of carbon cycle, climate, and biosphere with implications for the future, in *Annual Review of Earth and Planetary Sciences*, Vol **39**, R. Jeanloz and K.H. Freeman, Editors. 2011, Palo Alto, CA: Annual Reviews. p. 489–516.

McLean, A.D. and G.S. Chandler, Contracted Gaussian-basis sets for molecular calculations. 1. 2nd row atoms, Z=11–18. *Journal of Chemical Physics*, 1980. **72**(10): p. 5639–5648.

Mennucci, B., E. Cances, and J. Tomasi, Evaluation of solvent effects in isotropic and anisotropic dielectrics and in ionic solutions with a unified integral equation method: Theoretical bases, computational implementation, and numerical applications. *Journal of Physical Chemistry B*, 1997. **101**(49): p. 10506–10517.

Mohammed, O.F., et al., Charge transfer assisted by collective hydrogen-bonding dynamics. *Angewandte Chemie*, 2010. **48**: p. 6251–6256.

Molina-Montes, E., et al., Water release from pyrophyllite during the dehydroxylation process explored by quantum mechanical simulations. *Journal of Physical Chemistry C*, 2013. **117**(15): p. 7526–7532.

Moller, C. and M.S. Plesset, Note on an approximation treatment for many-electron systems. *Physical Review*, 1934. **46**(7): p. 0618–0622.

Mookherjee, M. and L. Stixrude, High-pressure proton disorder in brucite. *American Mineralogist*, 2006. **91**(1): p. 127–134.

Morrow, C.P., et al., Description of Mg^{2+} release from forsterite using ab initio methods. *Journal of Physical Chemistry C*, 2010. **114**(12): p. 5417–5428.

Nanayakkara, C.E., et al., Surface photochemistry of adsorbed nitrate: The role of adsorbed water in the formation of reduced nitrogen species on alpha-Fe_2O_3 particle surfaces. *Journal of Physical Chemistry A*, 2014. **118**(1): p. 158–166.

Nangia, S. and B.J. Garrison, Ab initio study of dissolution and precipitation reactions from the edge, kink, and terrace sites of quartz as a function of pH. *Molecular Physics*, 2009a. **107**(8–12): p. 831–843.

Nangia, S. and B.J. Garrison, Advanced Monte Carlo approach to study evolution of quartz surface during the dissolution process. *Journal of the American Chemical Society*, 2009b. **131**(27): p. 9538–9546.

Nangia, S. and B.J. Garrison, Role of intrasurface hydrogen bonding on silica dissolution. *Journal of Physical Chemistry C*, 2010a. **114**(5): p. 2267–2272.

Nangia, S. and B.J. Garrison, Theoretical advances in the dissolution studies of mineral-water interfaces. *Theoretical Chemistry Accounts*, 2010b. **127**(4): p. 271–284.

Nangia, S., et al., Study of a family of 40 hydroxylated beta-cristobalite surfaces using empirical potential energy functions. *Journal of Physical Chemistry C*, 2007. **111**(13): p. 5169–5177.

Navrotsky, A., Nanoscale effects on thermodynamics and phase equilibria in oxide systems. *Chemphyschem*, 2011. **12**(12): p. 2207–2215.

Newton, M.D., Formalisms for electron-exchange kinetics in aqueous solution and the role of ab initio techniques in their implementation. *International Journal of Quantum Chemistry*. 1980. **18**: p. 363–391.

Newton, M.D., The role of solvation in electron transfer: Theoretical and computational aspects, in *Continuum Solvation Models in Chemical Physics*, B. Mennucci and R. Cammi, Editors. 2007, Chichester: John Wiley & Sons, Ltd. p. 389–413.

Newton, M.D. and N. Sutin, Electron-transfer reactions in condensed phases. *Annual Review of Physical Chemistry*, 1984. **35**: p. 437–480.

Ni, Y., et al., Fundamental studies on kinetic isotope effect (KIE) of hydrogen isotope fractionation in natural gas systems. *Geochimica et Cosmochimica Acta*, 2011. **75**(10): p. 2696–2707.

Nielsen, L.C. and D.J. DePaolo, Ca isotope fractionation in a high-alkalinity lake system: Mono Lake, California. *Geochimica et Cosmochimica Acta*, 2013. **118**: p. 276–294.

Nitzan, A., *Chemical Dynamics in Condensed Phases: Relaxation, Transfer and Reactions in Condensed Molecular Systems*. Oxford Graduate Texts. 2006, Oxford: Oxford University Press.

North, S.W., Atmospheric photochemistry: Roaming in the dark. *Nature Chemistry*, 2011. **3**(7): p. 504–505.

Panasci, A.F., et al., Rates of water exchange on the Fe-4(OH)(2)(hpdta)(2)(H2O)(4) (0) molecule and its implications for geochemistry. *Inorganic Chemistry*, 2012. **51**(12): p. 6731–6738.

Park, R.J. and S.W. Kim, Air quality modeling in East Asia: Present issues and future directions. *Asia-Pacific Journal of Atmospheric Sciences*, 2014. **50**(1): p. 105–120.

Peacock, C.L. and D.M. Sherman, Sorption of Ni by birnessite: Equilibrium controls on Ni in seawater. *Chemical Geology*, 2007. **238**(1–2): p. 94–106.

Pecher, K., S.B. Haderlein, and R.P. Schwarzenbach, Reduction of polyhalogenated methanes by surface-bound Fe(II) in aqueous suspensions of iron oxides. *Environmental Science & Technology*, 2002. **36**(8): p. 1734–1741.

Pelmenschikov, A., et al., Lattice resistance to hydrolysis of Si-O-Si bonds of silicate minerals: Ab initio calculations of a single water attack onto the (001) and (111) beta-cristobalite surfaces. *Journal of Physical Chemistry B*, 2000. **104**(24): p. 5779–5783.

Pelmenschikov, A., J. Leszczynski, and L.G.M. Pettersson, Mechanism of dissolution of neutral silica surfaces: Including effect of self-healing. *Journal of Physical Chemistry A*, 2001. **105**(41): p. 9528–9532.

Pena, J., et al., Mechanisms of nickel sorption by a bacteriogenic birnessite. *Geochimica et Cosmochimica Acta*, 2010. **74**(11): p. 3076–3089.

Perdew, J.P., et al., Chemical-bond as a test of density-gradient expansions for kinetic and exchange energies. *Physical Review B*, 1988. **37**(2): p. 838–843.

Peterson, M.L., et al., Surface passivation of magnetite by reaction with aqueous Cr(VI): XAFS and TEM results. *Environmental Science & Technology*, 1997. **31**(5): p. 1573–1576.

Piana, S. and J.D. Gale, Three-dimensional kinetic Monte Carlo simulation of crystal growth from solution. *Journal of Crystal Growth*, 2006. **294**(1): p. 46–52.

Piana, S., M. Reyhani, and J.D. Gale, Simulating micrometre-scale crystal growth from solution. *Nature*, 2005. **438**(7064): p. 70–73.

Pierre-Louis, A.M., et al., Adsorption of carbon dioxide on Al/Fe oxyhydroxide. *Journal of Colloid and Interface Science*, 2013. **400**: p. 1–10.

Pinney, N., et al., Density functional theory study of ferrihydrite and related Fe-oxyhydroxides. *Chemistry of Materials*, 2009. **21**(24): p. 5727–5742.

Qian, Z.S., et al., Density functional studies of the structural characteristics, Al-27 NMR chemical shifts and water-exchange reactions of $Al_{30}O_8(OH)_{(56)}(H_2O)_{(26)}(18+)$(Al-30) in aqueous solution. *Geochimica et Cosmochimica Acta*, 2010. **74**(4): p. 1230–1237.

Raiteri, P. and J.D. Gale, Water is the key to nonclassical nucleation of amorphous calcium carbonate. *Journal of the American Chemical Society*, 2010. **132**(49): p. 17623–17634.

Raiteri, P., et al., Towards accurate modeling of the growth and nucleation of carbonates. *Geochimica et Cosmochimica Acta*, 2010. **74**(12): p. A844–A844.

Raiteri, P., et al., Exploring the influence of organic species on pre- and post-nucleation calcium carbonate. *Faraday Discussions*, 2012. **159**: p. 61–85.

Raynaud, C., et al., Reconsidering Car–Parrinello molecular dynamics using direct propagation of molecular orbitals developed upon Gaussian type atomic orbitals. *Physical Chemistry Chemical Physics*, 2004. **6**(17): p. 4226–4232.

Reinisch, J. and A. Klamt, Prediction of free energies of hydration with COSMO-RS on the SAMPL4 data set. *Journal of Computer-Aided Molecular Design*, 2014. **28**(3): p. 169–173.

Ribeiro, A.J.M., M.J. Ramos, and P.A. Fernandes, Benchmarking of DFT functionals for the hydrolysis of phosphodiester bonds. *Journal of Chemical Theory and Computation*, 2010. **6**(8): p. 2281–2292.

Richter, F.M., et al., Isotopic fractionation of the major elements of molten basalt by chemical and thermal diffusion. *Geochimica et Cosmochimica Acta*, 2009. **73**(14): p. 4250–4263.

Roques, J., E. Veilly, and E. Simoni, Periodic density functional theory investigation of the uranyl ion sorption on three mineral surfaces: A comparative study. *International Journal of Molecular Sciences*, 2009. **10**(6): p. 2633–2661.

Rosso, K.M. and J.R. Rustad, Ab initio calculation of homogeneous outer sphere electron transfer rates: Application to M(OH2)(6)(3+/2+) redox couples. *Journal of Physical Chemistry A*, 2000. **104**(29): p. 6718–6725.

Roston, D., Z. Islam, and A. Kohen, Isotope effects as probes for enzyme catalyzed hydrogen-transfer reactions. *Molecules*, 2013. **18**(5): p. 5543–5567.

Rotzinger, F.P., Treatment of substitution and rearrangement mechanisms of transition metal complexes with quantum chemical methods. *Chemical Reviews*, 2005. **105**(6): p. 2003–2037.

Rubasinghege, G., et al., Reactions on atmospheric dust particles: Surface photochemistry and size-dependent nanoscale redox chemistry. *Journal of Physical Chemistry Letters*, 2010. **1**(11): p. 1729–1737.

Rustad, J.R. and P. Zarzycki, Calculation of site-specific carbon-isotope fractionation in pedogenic oxide minerals. *Proceedings of the National Academy of Sciences of the United States of America*, 2008. **105**(30): p. 10297–10301.

Rustad, J.R., D.A. Yuen, and F.J. Spera, Molecular-dynamics of liquid SiO_2 under high-pressure. *Physical Review A*, 1990. **42**(4): p. 2081–2089.

Rustad, J.R., D.A. Yuen, and F.J. Spera, The statistical geometry of amorphous silica at lower mantle pressures: Implications for melting slopes of silicates and anharmonicity. *Journal of Geophysical Research-Solid Earth*, 1991a. **96**(B12): p. 19665–19673.

Rustad, J.R., D.A. Yuen, and F.J. Spera, Molecular-dynamics of amorphous silica at very high-pressures (135 GPA): Thermodynamics and extraction of structures through analysis of Voronoi polyhedra. *Physical Review B*, 1991b. **44**(5): p. 2108–2121.

Rustad, J.R., D.A. Yuen, and F.J. Spera, The sensitivity of physical and spectral properties of silica glass to variations of interatomic potentials under high-pressure. *Physics of the Earth and Planetary Interiors*, 1991c. **65**(3–4): p. 210–230.

Rustad, J.R., B.P. Hay, and J.W. Halley, Molecular-dynamics simulation of iron(III) and its hydrolysis products in aqueous-solution. *Journal of Chemical Physics*, 1995. **102**(1): p. 427–431.

Rustad, J.R., A.R. Felmy, and B.P. Hay, Molecular statics calculations for iron oxide and oxyhydroxide minerals: Toward a flexible model of the reactive mineral-water interface. *Geochimica et Cosmochimica Acta*, 1996a. **60**(9): p. 1553–1562.

Rustad, J.R., A.R. Felmy, and B.P. Hay, Molecular statics calculations of proton binding to goethite surfaces: A new approach to estimation of stability constants for multisite surface complexation models. *Geochimica et Cosmochimica Acta*, 1996b. **60**(9): p. 1563–1576.

Rustad, J.R., K.M. Rosso, and A.R. Felmy, Molecular dynamics investigation of ferrous-ferric electron transfer in a hydrolyzing aqueous solution: Calculation of the pH dependence of the diabatic transfer barrier and the potential of mean force. *Journal of Chemical Physics*, 2004a. **120**(16): p. 7607–7615.

Rustad, J.R., J.S. Loring, and W.H. Casey, Oxygen-exchange pathways in aluminum polyoxocations. *Geochimica et Cosmochimica Acta*, 2004b. **68**(14): p. 3011–3017.

Saadoune, I. and N.H. de Leeuw, A computer simulation study of the accommodation and diffusion of He in uranium- and plutonium-doped zircon (ZrSiO4). *Geochimica et Cosmochimica Acta*, 2009. **73**(13): p. 3880–3893.

Saenger, C., et al., The influence of temperature and vital effects on magnesium isotope variability in Porites and Astrangia corals. *Chemical Geology*, 2013. **360**: p. 105–117.

Saukkoriipi, J.J. and K. Laasonen, Density functional studies of the hydrolysis of aluminum (chloro)hydroxide in water with CPMD and COSMO. *Journal of Physical Chemistry A*, 2008. **112**(43): p. 10873–10880.

Schauble, E.A., G.R. Rossman, and H.P. Taylor, Theoretical estimates of equilibrium Fe-isotope fractionations from vibrational spectroscopy. *Geochimica et Cosmochimica Acta*, 2001. **65**(15): p. 2487–2497.

Schroedle, S., et al., Ion association and hydration in 3:2 electrolyte solutions by dielectric spectroscopy: Aluminum sulfate. *Geochimica et Cosmochimica Acta*, 2007. **71**(22): p. 5287–5300.

Scott, M.J. and J.J. Morgan, Reactions at oxide surfaces .2. Oxidation of Se(IV) by synthetic birnessite. *Environmental Science & Technology*, 1996. **30**(6): p. 1990–1996.

Shi, W., et al., Theoretical investigation of the thermodynamic structures and kinetic water-exchange reactions of aqueous Al(III)-salicylate complexes. *Geochimica et Cosmochimica Acta*, 2013. **121**: p. 41–53.

Shimojo, F., et al., A divide-conquer-recombine algorithmic paradigm for large spatiotemporal quantum molecular dynamics simulations. *Journal of Chemical Physics*, 2014. **140**(18): p. 18A529.

Skelton, A.A., D.J. Wesolowski, and P.T. Cummings, Investigating the quartz (10(1)over-bar0)/water interface using classical and ab initio molecular dynamics. *Langmuir*, 2011. **27**(14): p. 8700–8709.

Stack, A.G., J.R. Rustad, and W.H. Casey, Modeling water exchange on an aluminum polyoxocation. *Journal of Physical Chemistry B*, 2005. **109**(50): p. 23771–23775.

Stashans, A. and Y. Flores, Modelling of neutral vacancies in forsterite mineral. *International Journal of Modern Physics B*, 2013. **27**(25): p. 50141.

Stillinger, F.H. and C.W. David, Polarization model for water and its ionic dissociation products. *Journal of Chemical Physics*, 1978. **69**(4): p. 1473–1484.

Stirling, A. and I. Papai, H_2CO_3 forms via HCO_3- in water. *Journal of Physical Chemistry B*, 2010. **114**(50): p. 16854–16859.

Stone, A.T. and J.J. Morgan, Reductive dissolution of metal oxides, in *Aquatic Surface Chemistry*, W. Stumm, Editor. 1987, New York: John Wiley & Sons, Inc. p. 221–254.

Strathmann, T.J. and A.T. Stone, Mineral surface catalysis of reactions between Fe(II) and oxime carbamate pesticides. *Geochimica et Cosmochimica Acta*, 2003. **67**(15): p. 2775–2791.

Stumm, W. and J.J. Morgan, *Aquatic Chemistry*. 3rd ed. 1996, New York: John Wiley & Sons, Inc.

Sullivan, D.J., et al., The rates of water exchange in Al(III)-salicylate and Al(III)-sulfosalicylate complexes. *Geochimica et Cosmochimica Acta*, 1999. **63**(10): p. 1471–1480.

Sung, W. and J.J. Morgan, Oxidative removal of Mn(II) from solution catalyzed by the gamma-FeOOH (lepidocrocite) surface. *Geochimica et Cosmochimica Acta*, 1981. **45**(12): p. 2377–2383.

Svelle, S., et al., Quantum chemical modeling of zeolite-catalyzed methylation reactions: Toward chemical accuracy for barriers. *Journal of the American Chemical Society*, 2009. **131**(2): p. 816–825.

Tan, X., M. Fang, and X. Wang, Sorption speciation of lanthanides/actinides on minerals by TRLFS, EXAFS and DFT studies: A review. *Molecules*, 2010. **15**(11): p. 8431–8468.

Tang, Y.C., et al., A kinetic model for thermally induced hydrogen and carbon isotope fractionation of individual n-alkanes in crude oil. *Geochimica et Cosmochimica Acta*, 2005. **69**(18): p. 4505–4520.

Tenney, C.M. and R.T. Cygan, Molecular simulation of carbon dioxide, brine, and clay mineral interactions and determination of contact angles. *Environmental Science & Technology*, 2014. **48**(3): p. 2035–2042.

Torn, M.S., et al., Mineral control of soil organic carbon storage and turnover. *Nature*, 1997. **389**(6647): p. 170–173.

Townsend, D., et al., The roaming atom: Straying from the reaction path in formaldehyde decomposition. *Science*, 2004. **306**(5699): p. 1158–1161.

Tranca, D.C., et al., Combined density functional theory and Monte Carlo analysis of monomolecular cracking of light alkanes over H-ZSM-5. *Journal of Physical Chemistry C*, 2012. **116**(44): p. 23408–23417.

Tribe, L., R. Hinrichs, and J.D. Kubicki, Adsorption of nitrate on kaolinite surfaces: A theoretical study. *Journal of Physical Chemistry B*, 2012. **116**(36): p. 11266–11273.

Truhlar, D.G., Valence bond theory for chemical dynamics. *Journal of Computational Chemistry*, 2007. **28**(1): p. 73–86.

Truhlar, D.G., et al., Incorporation of quantum effects in generalized-transition-state theory. *Journal of Physical Chemistry*, 1982. **86**(12): p. 2252–2261.

Umemoto, K., et al., A first-principles investigation of hydrous defects and IR frequencies in forsterite: The case for Si vacancies. *American Mineralogist*, 2011. **96**(10): p. 1475–1479.

Verma, A.K. and B.B. Karki, Ab initio investigations of native and protonic point defects in Mg_2SiO_4 polymorphs under high pressure. *Earth and Planetary Science Letters*, 2009a. **285**(1–2): p. 140–149.

Verma, A.K. and B.B. Karki, First-principles simulations of native point defects and ionic diffusion in high-pressure polymorphs of silica. *Physical Review B*, 2009b. **79**(21): p. 214115.

Verma, P., A. Perera, and R.J. Bartlett, Increasing the applicability of DFT I: Non-variational correlation corrections from Hartree-Fock DFT for predicting transition states. *Chemical Physics Letters*, 2012. **524**: p. 10–15.

Vladimirov, E., A. Ivanova, and N. Rosch, Effect of solvent polarization on the reorganization energy of electron transfer from molecular dynamics simulations. *Journal of Chemical Physics*, 2008. **129**(19): p. 194515.

Walsh, T.R., M. Wilson, and A.P. Sutton, Hydrolysis of the amorphous silica surface. II. Calculation of activation barriers and mechanisms. *Journal of Chemical Physics*, 2000. **113**(20): p. 9191–9201.

Wander, M.C.F., et al., Ab initio calculation of the deprotonation constants of an atomistically defined nanometer-sized, aluminium hydroxide oligomer. *Molecular Simulation*, 2013. **39**(3): p. 220–227.

Wang, Y., et al., Equilibrium H-2/H-1 fractionations in organic molecules: I. Experimental calibration of ab initio calculations. *Geochimica et Cosmochimica Acta*, 2009a. **73**(23): p. 7060–7075.

Wang, Y., et al., Equilibrium H-2/H-1 fractionations in organic molecules. II: Linear alkanes, alkenes, ketones, carboxylic acids, esters, alcohols and ethers. *Geochimica et Cosmochimica Acta*, 2009b. **73**(23): p. 7076–7086.

Wang, Y., et al., Equilibrium H-2/H-1 fractionation in organic molecules: III. Cyclic ketones and hydrocarbons. *Geochimica et Cosmochimica Acta*, 2013. **107**: p. 82–95.

Washton, N.M., S.L. Brantley, and K.T. Mueller, Probing the molecular-level control of aluminosilicate dissolution: A sensitive solid-state NMR proxy for reactive surface area. *Geochimica et Cosmochimica Acta*, 2008. **72**(24): p. 5949–5961.

Watts, H.D., Tribe L., Kubicki J.D., Arsenic adsorption onto minerals: Connecting experimental observations with density functional theory calculations. *Minerals*, 2014. **4**(2): p. 208–240.

Webb, E.A. and F.J. Longstaffe, Identifying the delta O-18 signature of precipitation in grass cellulose and phytoliths: Refining the paleoclimate model. *Geochimica et Cosmochimica Acta*, 2006. **70**(10): p. 2417–2426.

Wieclaw, D., M.D. Lewan, and M.J. Kotarba, Estimation of hydrous-pyrolysis kinetic parameters for oil generation from Baltic Cambrian and Tremadocian source rocks with Type-II kerogen. *Geological Quarterly*, 2010. **54**(2): p. 217–226.

Woodcock, L.V., C.A. Angell, and P. Cheeseman, Molecular-dynamics studies of vitreous state: Simple ionic systems and silica. *Journal of Chemical Physics*, 1976. **65**(4): p. 1565–1577.

Wu, Y.J., H.L. Tepper, and G.A. Voth, Flexible simple point-charge water model with improved liquid-state properties. *Journal of Chemical Physics*, 2006. **124**(2): p. 024503.

Xiao, Y.T. and A.C. Lasaga, Ab-initio quantum-mechanical studies of the kinetics and mechanisms of silicate dissolution: $H^+(H_3O^+)$ catalysis. *Geochimica et Cosmochimica Acta*, 1994. **58**(24): p. 5379–5400.

Xiao, Y.T. and A.C. Lasaga, Ab initio quantum mechanical studies of the kinetics and mechanisms of quartz dissolution: OH- catalysis. *Geochimica et Cosmochimica Acta*, 1996. **60**(13): p. 2283–2295.

Xu, X. and W.A. Goddard, Bonding properties of the water dimer: A comparative study of density functional theories. *Journal of Physical Chemistry A*, 2004. **108**(12): p. 2305–2313.

Yakob, J.L., et al., Lithium partitioning between olivine and diopside at upper mantle conditions: An experimental study. *Earth and Planetary Science Letters*, 2012. **329**: p. 11–21.

Yang, D.S., O.F. Mohammed, and A.H. Zewail, Environmental scanning ultrafast electron microscopy: Structural dynamics of solvation at interfaces. *Angewandte Chemie-International Edition*, 2013. **52**(10): p. 2897–2901.

Zarzycki, P., S. Kerisit, and K.M. Rosso, Computational methods for intramolecular electron transfer in a ferrous-ferric iron complex. *Journal of Colloid and Interface Science*, 2011. **361**(1): p. 293–306.

Zarzycki, P., S. Kerisit, and K.M. Rosso, Molecular dynamics study of Fe(II) adsorption, electron exchange, and mobility at goethite (alpha-FeOOH) surfaces. *Journal of Physical Chemistry C*, 2015. **119**(6): p. 3111–3123.

Zeebe, R.E., A new value for the stable oxygen isotope fractionation between dissolved sulfate ion and water. *Geochimica et Cosmochimica Acta*, 2010. **74**(3): p. 818–828.

Zewail, A.H., Femtochemistry. Past, present, and future. *Pure and Applied Chemistry*, 2000. **72**(12): p. 2219–2231.

Zhang, H.Z. and J.F. Banfield, Aggregation, coarsening, and phase transformation in ZnS nanoparticles studied by molecular dynamics simulations. *Nano Letters*, 2004. **4**(4): p. 713–718.

Zhao, Y. and D.G. Truhlar, The M06 suite of density functionals for main group thermochemistry, thermochemical kinetics, noncovalent interactions, excited states, and transition elements: two new functionals and systematic testing of four M06-class functionals and 12 other functionals. *Theoretical Chemistry Accounts*, 2008. **120**(1–3): p. 215–241.

Zhu, M., et al., In situ structural characterization of ferric iron dimers in aqueous solutions: Identification of μ-oxo species. *Inorganic Chemistry*, 2013. **52**(12): p. 6788–6797.

Zotov, N. and H. Keppler, Silica speciation in aqueous fluids at high pressures and high temperatures. *Chemical Geology*, 2002. **184**(1–2): p. 71–82.

Index

References to figures are given in *italic* type. References to tables are given in **bold** type.